American Institute of Chemical Engineers

17th Topical Conference on Ethylene Producers 2005

AA003011

Held at the 2005 AIChE Spring National Meeting

April 10-14, 2005
Atlanta, Georgia, USA

Volume 1 of 2

Printed from e-media with permission by:

Curran Associates, Inc.
57 Morehouse Lane
Red Hook, NY 12571
www.proceedings.com

ISBN: 978-1-60423-460-2

Some format issues inherent in the e-media version may also appear in this print version.

American Institute of Chemical Engineers

17th Topical Conference on Ethylene Producers
2005

Table of Contents

Volume 1

Development & Implementation of Critical Rotating Equipment Maintenance Strategies

Author: Brad Johnson

The Dow Chemical Company

Prepared for Presentation at
The 17th Ethylene Producer's Conference

Rotating Machinery Reliability

Unpublished

December 2004

ABSTRACT

Development and Implementation of Critical Rotating Equipment Maintenance Strategies

As olefin companies continue to be squeezed by higher feedstock prices and lower margins, focus continually shifts to eliminating sources of asset utilization (AU) gaps.

By utilizing available data from industry wide benchmarking studies and in-house tracking databases, 7 major contributing factors were found to impact AU. Once the AU gaps were identified, proactive equipment maintenance strategies were developed by global sub-teams. In order to maximize the value of the newly created maintenance strategies, an assessment tool was developed in order to identify gaps and assess the economic value associated with closing each one.

This paper will focus on the methodology used to identify the areas of AU loss opportunity, the global equipment strategies that were developed, and the implementation plan/tool that was used to extract the maximum value from those strategies.

BACKGROUND

With ethylene feedstock costs continuing to rise, the emphasis on being a low cost producer has never been greater. As margins tighten, focus on reducing unplanned maintenance expenses and product outages become critically important in order to maintain position in the ever competitive ethylene commodities market.

As with most global companies, Dow is continually searching for ways to cut costs and increase the operational effectiveness of its crackers. Although being a global company does afford significant economies of scale benefits, ensuring all the crackers are following the most efficient and cost effective maintenance strategies becomes a challenge.

For years, the plants had been highlighting to the fact that the critical rotating equipment within each cracker were the main sources of the asset utilization gaps. So with the emphasis on improving the company asset utilization of its cracker assets, two major questions were asked:

1) "What is causing the majority of the gaps in asset utilization"
2) "Do we have a global strategy to address these gaps".

FINDINGS

The first question required significant data mining within the AU databases and many other historical cost and event logging tools. On a macro level, the critical rotating equipment did appear to be the source of a majority of the major events within the cracker units. Figure 1 on the next page shows 3 years worth of data for all the cracker units within Dow. As can be clearly seen, the compression units are a major source of events. For the readers understanding, all events incorporate the following (with the majority coming from plant slowdown and shutdowns):

a) Near Miss for Production
b) Plant Slowdown
c) Plant Shutdown

Figure 2 breaks out just the plant shutdowns which create significant production loss and create a huge potential for extraordinary maintenance spending.

On the surface, it would appear that the critical rotating equipment must have an inherent design flaw or operating discipline gap to be categorized as "the" major contributor to the significant AU losses the business was experiencing. This first pass look at the data did show that the rotating equipment was significant, but more data mining would be required to fully understand the drivers behind the events.

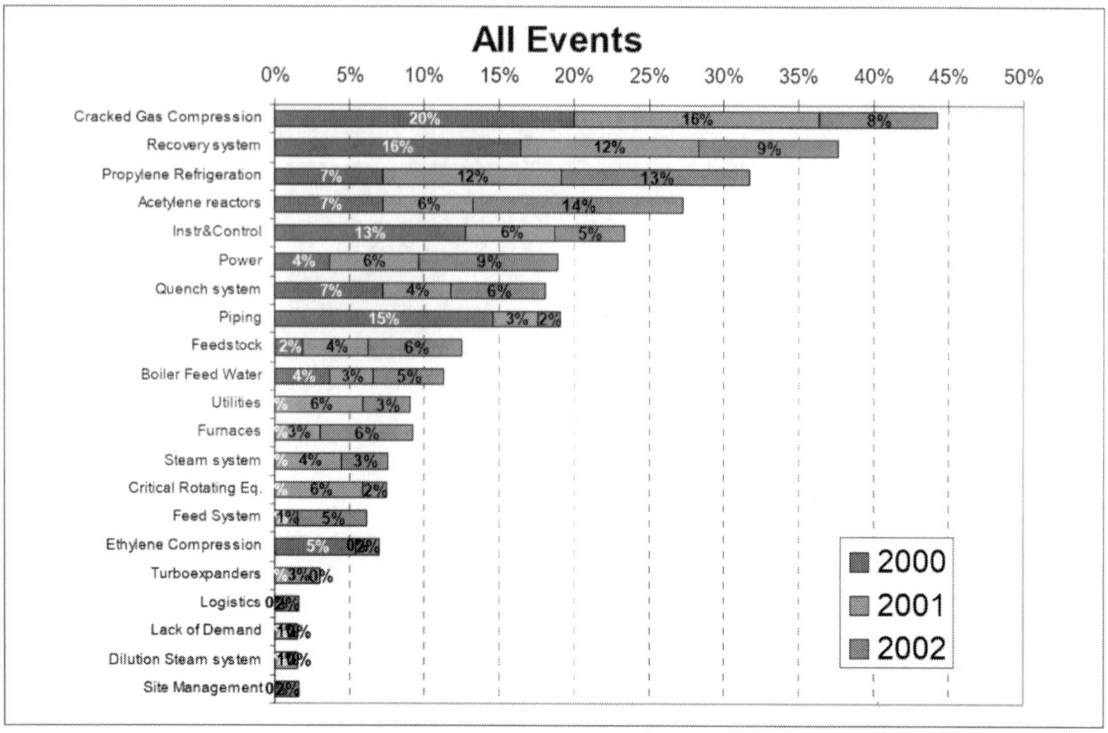

Figure 1: 2000-2002 Event Data for Dow Ethylene Crackers

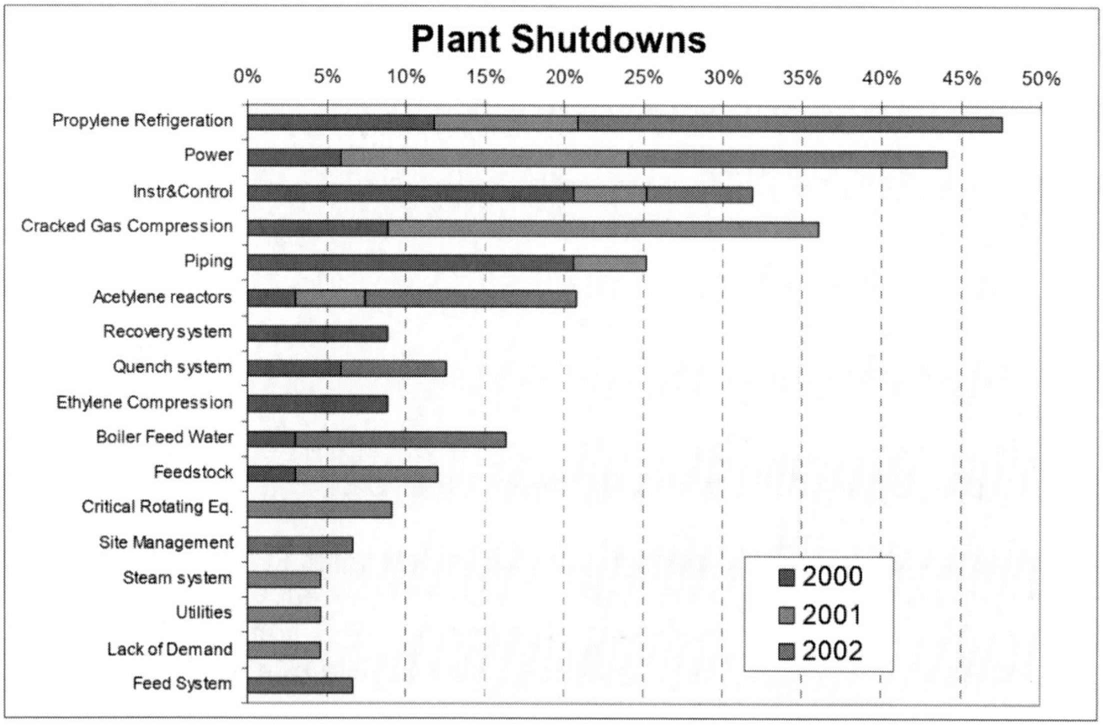

Figure 2: 2000-2002 Plant Shutdowns for Dow Ethylene Crackers

One of the biggest hurdles faced early in the discovery process was that of improper failure classification. Many instances were found where the run plant engineers would record the compressor as having failed when in reality the level transmitter on the suction drum "to" the compressor failed causing the compressor to trip off. In this case, the asset utilization loss should have been attributed to instrumentation instead of the compressor but in many times that was not the case. Figure 3 shows how most events got categorized. They were categorized by major piece of equipment which actually led to the plant shutdown instead of the piece of equipment that actually caused the rotating equipment shutdown.

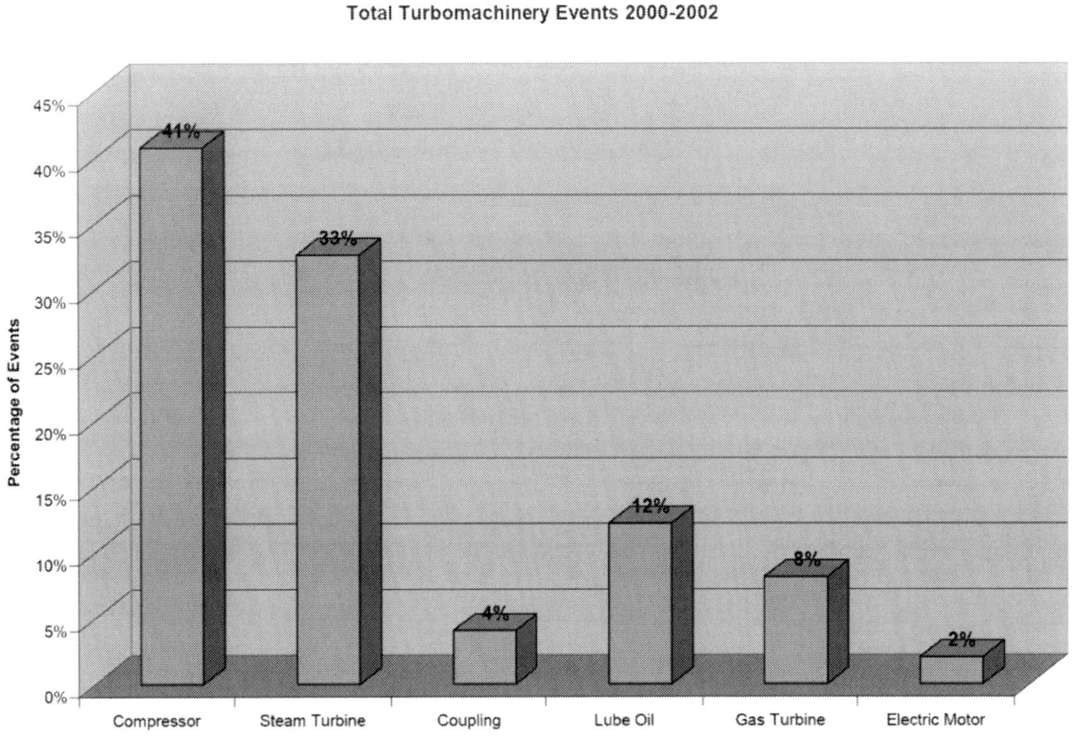

Figure 3: Total Turbomachinery Events for 2000-2002

In order to get to the heart of the failures, each failure report had to be carefully examined over the 3 year period and then categorized as either a design failure or a failure as a result of inadequate operating discipline.

Examples of design problems that led to significant plant events included:

1) Piping failures resulting from insufficient acoustical analysis after a compressor re-rate.
2) Instrumentation failures resulting from improper mounting location
3) Compressor control system power supply failure

Examples of operating discipline problems that led to significant plant events included:

1) Inadequate procedure use when installing compressor seals resulting in a leak at startup.
2) Operations induced surge causing piping stub-out to fail and shutdown the plant.
3) Improper compressor isolation allowed for reverse rotation and failed bearings.
4) Underutilized maintenance practices on existing rotating equipment instrumentation.

Figure 4 shows the breakdown of design vs. operating discipline failures for all turbomachinery events from 2000-2002.

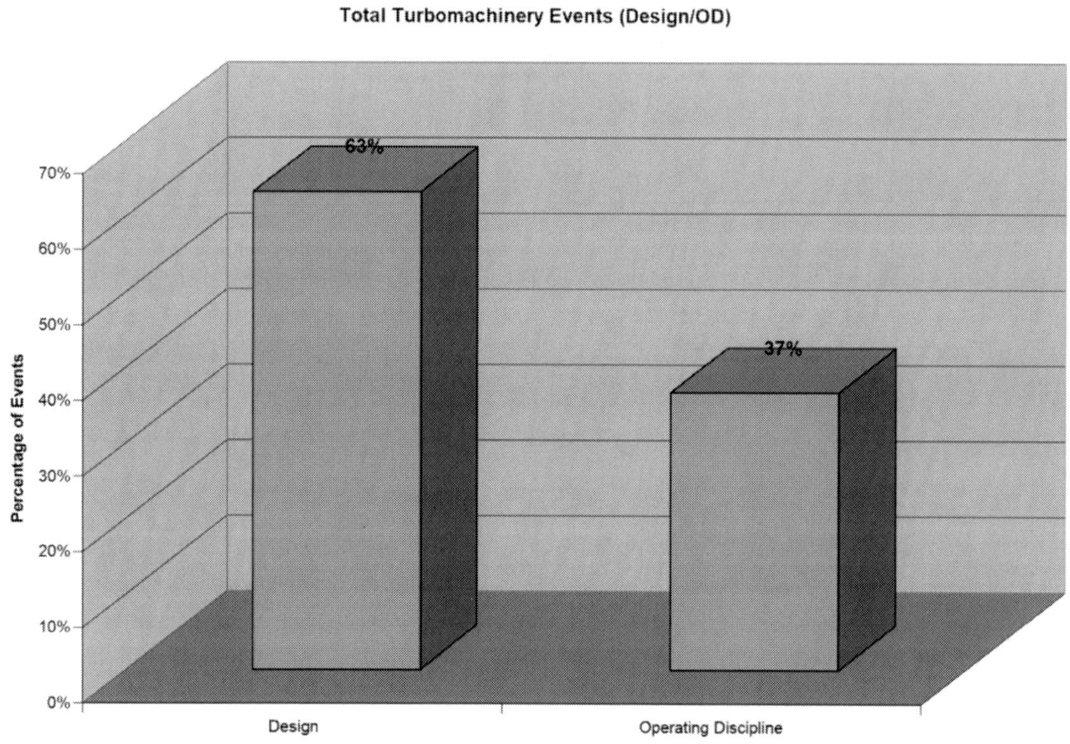

Figure 4: 2000-2002 Design vs. OD for All Turbomachinery Events

Breaking this data down even further to the component level, it started to become evident that the problems were focused more on the auxiliary systems that supported the equipment than the actual "between bearing" rotating equipment. Figure 5 clearly shows that instrumentation resulted in a significant number of turbomachinery events followed by piping and then the rotors.

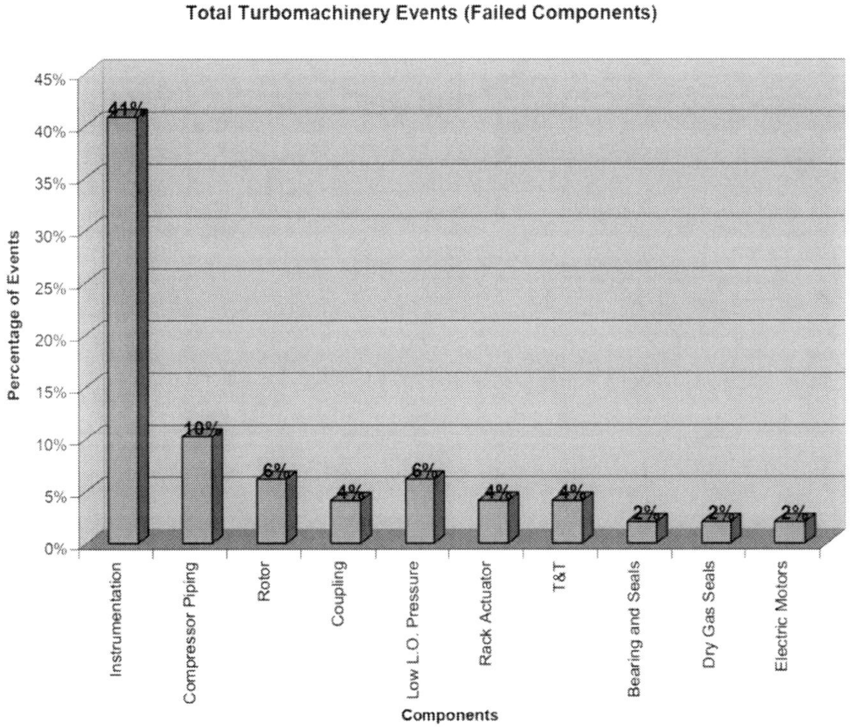

Figure 5: Component Level Turbomachinery Events 2000-2002

PLAN FORWARD

After carefully analyzing the data, it became apparent that more than just the turbines and compressors needed to be considered. In fact, compressors and turbines themselves took a back seat in priority to the auxiliary systems such as instrumentation, piping, lube oil systems, and turbine governors. One quick conclusion could have been to blame all the problems on poor design and installation practices which would have led to a significant effort to modify our upfront design criteria used when bidding a specific piece of rotating equipment. However, this would have missed the mark since the problems being found were PPM, operating discipline, procedure use, and process fouling related etc. All of these factors could not alone be addressed by simply making the procurement specifications more robust. After looking at the problems, a robust maintenance strategy was needed that outlined the best practices when it came to rotor storage, PPM, spares, procedure use, etc. for each of the key trouble areas around the rotating equipment. Not only would these maintenance strategies be robust, they would also serve as a standard criteria for all the crackers globally.

A turbomachinery task force was chartered with the sole responsibility of developing maintenance strategies, completing assessment audits, and following up on action items that were identified during the audits. The areas that were specifically targeted are shown in Figure 6 on the next page.

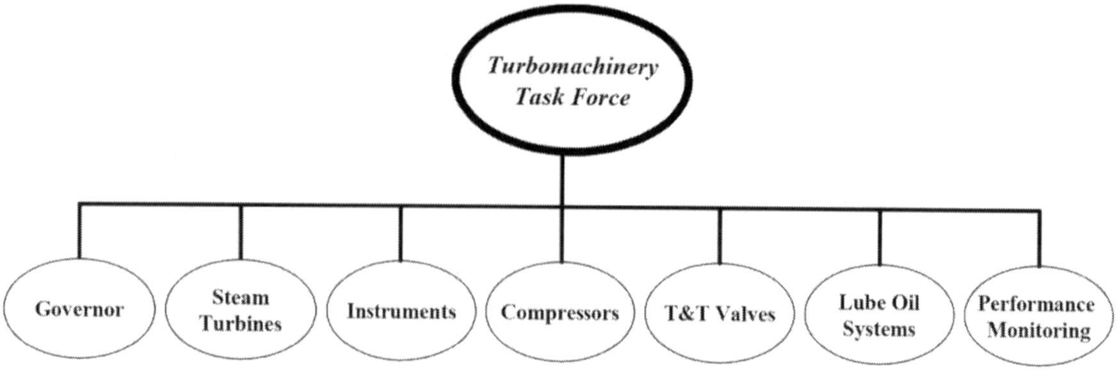

Figure 6: Turbomachinery Task Force

Each team was made up of subject matter experts within the corporation that had extensive expertise in those areas. Once a team was charted, there was still a sense of discovery that had to happen. The teams had to once again go through the data and find out where the majority of the problems in their specific area were originating from. Information came from Failure Modes Effects Analysis (FMEA), asset capability databases, customer surveys, as well as "gut feel" and experience.

COMPRESSOR TEAM

The author was the leader of the compressor team in charge of developing a global compressor maintenance strategy, creating an assessment tool, and initiating action items to close any gaps between the existing operating practices and the maintenance strategy.

Like other teams, the first place to start in the development of the strategy was to dive back into the data specifically targeting compressor failures. Figure 7 on the next page shows the breakdown by component that caused the compressor events.

Instrumentation again was at the top of the list. Faulty vibration probes, unreliable suction level transmitters, faulty bearing RTD's, etc. were major contributors to the instrumentation problems. Since the compressor team was made up of individuals whose background was primarily mechanical in nature as compared to instrumentation, the team shared their data on instrumentation with that subteam and asked them to develop the strategies to address those problems.

Compressor piping was another major contributor to large asset utilization losses which would most likely never have been categorized had this comprehensive approach to look at all the data not been taken. Piping failures had resulted due to acoustical vibration attenuation after a re-rate startup, small bore nozzles welded directly to the cases with no provisions to periodically check the integrity of the welds, and no preventative strategies to check for excessive vibrations of the piping after major overhauls (supports are sometimes moved during turnarounds and never replaced).

Rotors were also captured as a major problem but the root causes were not what the team expected. The expectation was that data would show the rotors themselves had cracks or bows, or the impellers may have cracked causing the machines to shutdown. However after looking at the data more closely, process fouling on cracked gas compressors created high vibration and consequently this was attributed to the rotors instead of to the process conditions. The team took this data and created a section in the maintenance strategy specifically focused on compressor performance monitoring that addressed on-line washing strategies for cracked gas centrifugal compressors.

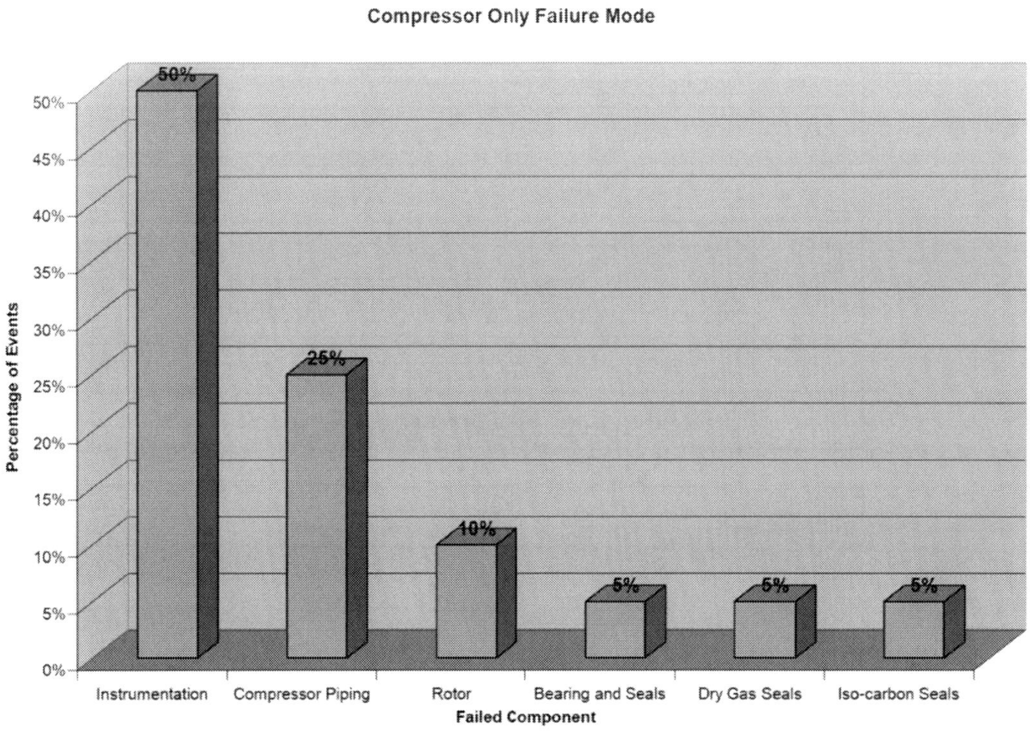

Figure 7: Compressor Only Component Failures (2000-2002)

COMPRESSOR MAINTENANCE STRATEGY

This methodology of deep drilling into the data and finding the root causes of the component failures as well as utilizing the experience base on the team, led the team to develop a comprehensive maintenance strategy for centrifugal compressors of which the outline is shown in Figure 8.

Centrifugal Compressors Maintenance Strategies

Table of Contents

The following links take the reader directly to the maintenance strategy for each topic.

Topic	Description
1	Alignment
2	Lubrication
3	Vibration and Rotor Balancing
4	Acoustics
5	Spares/Storage
6	PPM
7	Shaft End Seals
8	Overhaul Strategy
9	Cracked Gas Fouling/ Efficiency Monitoring
10	Compressor Piping Inspection
11	Coating of Internal Compressor Components

Figure 8: Outline of Compressor Maintenance Strategy

The major subtopics were determined primarily by three methods (data driven, experience driven, global consistency). As data pointed to a significant area creating asset utilization gaps, the team developed a specific section (cracked gas fouling, compressor piping inspections, acoustics). Other sections were inserted because they were deemed critical by the SME's and no comprehensive global strategy existed (vibration, lubrication, alignment, PPM). The balance of the sections were addressed not specifically because of AU losses, but because of cost inefficiencies due to inconsistencies. Having such a large fleet of compressors that are globally located, raised the question whether or not all units are operating as cost effectively as possible. Therefore the team, with input from global SME's did develop a standard approach to spares/storage, shaft end seals, overhaul strategy, and coatings which should lead to a consolidated strategy and thus a lower overall cost position for the company.

Each section of the strategy was structured the same as shown in Figure 9 below. Since the strategy document was around 50 pages long, for it to be useful, the team felt it was critical to pull out the key maintenance strategy points and place them at the very first part of the section. Then, for ease of use, internal bookmarks were used throughout the strategy to take the reader to more thorough information about a subject if they desired.

A.10. Compressor Piping Inspections

Piping NDT Maintenance Strategy	• Need to do dye penetrate inspections of welds during each major overhaul • Need to perform a piping vibration baseline analysis after each major overhaul to insure all pipe is properly supported. Analyst should also perform a walk-around and touch all piping for vibration problems. • All screwed nipples into compressor casing need to be replaced during each overhaul.
Introduction	This section documents standard piping inspection requirements for centrifugal compressor within the Hydrocarbons business.
Elements	The elements of the piping inspection section are as follows: • Failure Modes • Piping to be Inspected • Type of Inspection • Frequency of Inspection
Failure Modes	Most piping failures occur as a result of cracked welds. These cracks are a result of mechanical vibration, poor original weld, or improper support structure.
Piping to be Inspected	The following piping segments should be inspected: 1) Any branched piping connections Tee'd off the main process lines. 2) Nozzles welded to the compressor casing or end walls. Some of the newer compressor cases utilize a flanged configuration with the case. In these cases, the first piping flange weld away from the case should be inspected. 3) Pipe supports welded to the process or auxiliary piping within a reasonable distance (On deck) from the compressor.
Type of Inspection	Every effort should be made to perform a die penetrate inspection of each weld. Where it is not feasible to use die penetrate, visual inspection is required. Visual inspection should focus on sharp weld edges or undercutting which could create a weak weld and be a potential crack initiator.
Frequency of Inspection	Prior to any major turnaround, vibration services should perform a baseline analysis of the compressor piping. This baseline analysis should point to areas where supports are inadequate or missing altogether. Address these items during a shutdown. After every major turnaround, the same vibration route should be performed. The vibration technician should compare the pre and post overhaul piping vibration reports and identify any areas where the vibration has significantly increased after the turnaround. The analyst should then be notified of those areas and an action plan implemented to reduce the vibrations to acceptable levels.

Figure 9: Piping Section of Compressor Maintenance Strategy

ASSESSMENT

Once the strategy was written, reviewed, and published, it still had zero value to the company unless it was used. Simply publishing a 50 page document and e-mailing it to all the reliability and maintenance engineers and asking them to follow the guidelines, would simply not create the value that was expected. A structured assessment tool documenting the value impact and cost was needed.

The team created an assessment tool (Figure 10) which did just that. Each section of the maintenance strategy was reviewed again and only the key activities and recommendations were added to the assessment document. Experience had shown that developing an assessment tool with 200+ questions, although comprehensive, typically lacked in focus and participation because people just got bogged down with its completion. The team's goal was to be simple, direct, and to the point about which strategies needed to be implemented. As each strategy was assessed, the assessors had to make the decision whether or not they were going to follow the recommendation (if not they had to specify a reason), assign actions to the gaps that needed to be closed, and estimate the cost/benefit of each gap closure. After the assessment was completed, the summary page (Figure 11) rolled up the total cost/benefit impact as well as a total % of gaps existing against the strategy.

VISION FOR THE FUTURE

Dow Chemical is still in the early stages of turbomachinery maintenance strategy development and assessment. To date, two (compressors/governors) of the strategies have been completed with the others following quickly to completion. The assessment tool has been completed for the compressor strategy and one assessment has been formally completed.

The vision for the future is to have a comprehensive set of maintenance strategies specifically focused on those areas around turbomachinery which are creating the largest asset utilization gaps thus having the greatest value potential to the company. Once all the strategies are developed, a comprehensive assessment tool incorporating all the strategies will be developed, and each cracker will be assessed.

This paper was not designed to communicate the specific strategies as it relates to turbomachinery, but rather to communicate to the reader the findings creating significant AU gaps and the methodology of one company to close those gaps. As companies continue to expand globally, having a comprehensive set of strategies which reduce cost, focus on the areas which have historically created significant AU losses, and improve leveragable expertise, will in our opinion be one of the major differentiators between simply reacting to the reliability problems we face and proactively developing a plan to eliminate those problems before they occur.

Global Compressor Maintenance Strategy Asessment

A.1 Alignment	A.4 Acoustic	A.7 Shaft End Seals	A.10 Compressor Piping Inspection
A.2 Lubrication	A.5 Spares and Storage	A.8 Overhaul Strategy	A.11 Compressor Coating
A.3 Vibration and Rotor Balancing	A.6 PPM	A.9 Cracked Gas Fouling/Efficiency Monitoring	
Summary			

							Impact		Financial	
A.10	Compressor Piping Inspection Strategy	Is there a deviation from the Strategy?	Plan to Close the Gap	Focal Point for Gap Closure	Date for Completion	Implement Gap Closure	Energy Reduction (BTU/lb C2H4)	Value ($M EBIT /yr)	Capital Cost ($M)	Expense Cost ($M)
A.101	Dye Penetrant weld inspections are completed on branched T connections off main process piping during each major overhaul.									
A.102	Dye Penetrant weld inspections are completed on each nozzle welded to the casing during each major overhaul.									
A.103	Dye Penetrant weld inspections are completed during each major overhaul on pipe supports welded to process or auxillary piping that is located on the deck.									
A.104	A vibration baseline analysis should be performed on the compressor piping prior to shutting a machine down for a major overhaul.									
A.105	After the machine is overhauled and brought back on-line, the same vibration route should be performed and any increase in vibration should be addressed by the critical rotating equipment focal point.									
						Totals	0	0	0	0

Figure 10: Assessment Page for Compressor Piping Strategy Section

Global Compressor Maintenance Strategy Asessment

A.1 Alignment	A.4 Acoustic	A.7 Shaft End Seals	A.10 Compressor Piping Inspection	Startup Date	
A.2 Lubrication	A.5 Spares and Storage	A.8 Overhaul Strategy	A.11 Compressor Coating	Plant Location	
A.3 Vibration and Rotor Balancing	A.6 PPM	A.9 Cracked Gas Fouling/Efficiency Monitoring		Capacity (MMlbs)	
Summary					

Link to the Maintenance Strategy Document

		Impact		Financial		Assessment	
		Energy Reduction (BTU/lb C2H4)	Value ($M EBIT /yr)	Capital Cost ($M)	Expense Cost ($M)	% of Gaps	% Gaps Willing to Close
A.1	Alignment Strategy	0	50	80	220	80%	75%
A.2	Lubrication Strategy	10	35	0	10	10%	100%
A.3	Vibration and Rotor Balance Strategy	0	100	0	50	40%	100%
A.4	Acoustic Strategy	0	0	0	0	50%	85%
A.5	Spares and Storage Strategy	0	250	20	0	30%	90%
A.6	PPM Strategy	0	30	0	25	100%	100%
A.7	Shaft End Seals Strategy	15	100	0	100	10%	100%
A.8	Overhaul Strategy	0	100	0	40	5%	100%
A.9	CG Fouling and Efficiency Monitoring Strategy	0	50	30	30	10%	100%
A.10	Compressor Piping Inspection Strategy	0	150	0	0	10%	80%
A.11	Compressor Coating Strategy	0	50	0	15	65%	85%
	Totals	25	915	130	490	50%	85%

Figure 11: Assessment Summary Section

Reliability **A**ssessment of **R**otating **E**quipment

Author: Charlie Rutan

Lyondell Chemical Company

Abstract

The paper will discuss a program, **R.A.R.E.**, that is used by a consortium of companies to determine what is the risk of operating turbomachinery e.g. steam turbines and compressors for extended lengths of time.

BIO -Charlie Rutan, a native Texan, graduated from Texas Tech University in 1973 with a BS in Mechanical Engineering. He is a practicing engineer in the field of rotating equipment, hot tapping /plugging, and special problem resolution. Charlie has two patents and has consulted on turbomachinery, hot taping and plugging problems all over the world in chemical, petrochemical, power generation and polymer facilities. He has published and/or presented articles in *Hydrocarbons Processing*, *Petroleum Technology Quarterly*, ASME, AIChE, Pumping Systems, The Vibration Institute, NPRA, Houston Business Round Table, the Texas A&M International Turbomachinery Symposium, Texas A&M International Pump Symposium, Southern Power Machinery and Gas Compression Conference and Predictive Maintenance Technology Conference. He is a member of AIChE Process Gas Users Committee, the Texas Tech Academy of Engineers, Hydraulic Institute/ANSI Pump Standards Review Committee and on the Advisory Council of the Texas A&M International Turbomachinery Symposium. His present position in the Corporate Specialty Engineering Group is as a shared service within the Lyondell companies as a Senior Engineering Advisor (Engineering Fellow).

Charlie can be reached at:

713-844-7309
pager 1-800-278-6536
charlie.rutan@equistarchem.com

Origin of Initial Program and Concept

(One tool of an overall Maintenance and Turnaround Strategy of Critical Rotating Equipment)

The Hartford Steam Boiler Inspection & Insurance Company (HSB) typically followed the OEM's recommended interval for turbine overhaul, which is typically very conservative. As more and more companies were extending the interval between overhauls, HSB decided that they needed to assess the risk of extending the overhaul intervals and what recommendations should they make if an insured <<company?>>wanted to do it. Since they had no actuarial data to show them the effect of extending the time between overhauls, they decided that a research project should be developed to investigate this new trend of extending intervals. M&M Engineering was a part of the HSB Companies and was given the responsibility for the development of a comprehensive program that could be used as the assessment tool. HSB did not have any plans to use the program to structure premiums. What they were looking at was what type of risk was taken when companies wanted to extend outages. Their initial plan was that if a company wanted to go longer than the OEM recommended practice they would require a S.T.R.A.P. analysis. If S.T.R.A.P. showed that they were low in risk they had no problems, but if S.T.R.A.P. showed them high in risk they worked with them to implement some risk reduction procedures.

Consortium of Companies

With this vision, M&M Engineering pulled together a group of recognized Industry experts in the turbomachinery field. The experts brought with them their individual companies reliability assessments, although not named reliability assessment, industry standards (API, ASME, etc.), recognized best practices, new technologies and most important their personal experience. The initial companies involved were Shell, Dow Chemical Company, CF Industries, Inc., Stone Container Corp., Hickham Industries Inc. and Revak Turbomachinery Services Inc.

Starting with a blank sheet of paper, the group spent countless hours defining, developing, categorizing and weighting the question and their responses. The caveat for this group was not completely altruistic, when completed they would take this program back to their respective companies and use as a tool that could be used to justify improvement of the turbines, reliability and extend run length which resulted in the savings of major monies. All the accumulated data and information was then put into a software program, consisting of a database along with calculation algorithms to give an output as a risk of continued operation of the turbine(s) and make recommendations to minimize or mitigate the risks that were define with a rate of return on the investment. Like home and car insurance, risk cannot be eliminated, but most can be minimized, then it becomes a decision on what level of risk the plant/company is willing to accept. With this information a long term operation, maintenance, reliability, upgrades a shutdown strategy can be developed.

Steam Turbine Risk Assessment Program S.T.R.A.P.

1. General Information Data Sheet (Plant specifics)
   a. Size or Class of the turbine (5 categories)?
   b. Age of the turbine?
   c. Manufacture of the turbine?
   d. What is the turbine driving?
   e. When the turbine was last dismantled?
   f. Etc.

2. Turbine Performance (Design & Actual)
   a. Horsepower?
   b. Speed?
   c. Inlet Flow?
   d. Temperature?
   e. Pressure?
   f. Etc.

3. Site & Utility Data
   a. Location?
   b. Steam Generation?
   c. Etc.

4. Construction Features
   a. Type of turbine?
   b. Critical Speed?
   c. Control System?
   d. Etc.

5. Spares

6. Maintenance & repairs

7. Turbine Operation

8. Monitoring & Protection

9. Upgrades Steam System

10. Past Problems/Failures

11. Consequence Data

The total number of possible questions is about 3000, but for a large steam turbine 350 to 400 questions are normally answered.

Question Weighting

The team of experts then weighted each of the 3000+ questions and their respective answers based on their experience and knowledge. As part of the development of failure probabilities, the team needed to set as baseline interval for overhaul outages.

They decided to use a six -year dismantle overhaul schedule frequency for Class 1-4 turbines and a five-year overhaul schedule frequency for Class 5 turbines. Thus, the baseline probabilities developed were based on the turbine operating for six years or five years without opening the case. Since risk is calculated by multiplying probability times the consequence, the calculated risk would be for a six-year/five-year interval. Thus, the risk calculated would be in days of lost production over a five or six year interval. Because some S.T.R.A.P. users do not divulge financial data about production revenue, converting the days in lost production to dollars cannot be performed for all turbines. It was decided to create the term **R**isk **I**ndex **N**umber **(RIN)**, which would allow all turbines to be compared to each other. The RIN is the risk of failure in days of lost production over a six-year interval for Class 1-4 turbines and over a five-year interval for Class 5 turbines.

On the basis of the input data and the likelihood-consequence information, a risk for operation of the turbine may be calculated as a function of time between the dismantle inspections. In each case, a quantified list of recommendations to mitigate the risk will also be reported based on the greatest contribution to the risk. Inspection outage plans then may be tailored to optimize the time between overhauls on the basis of acceptable level of risk. The RIN is a number generated by the program based on:

- The questions and their corresponding answers
- Industry Standards (ASME, API, etc.).
- Accepted Industry Practices.
- Latest Technology.
- Relative to other turbines in the company and/or the number of turbines of the same design in the database.

Probability

The probability of failure of a component is the risk of failure in days of lost production (RIN), divided by the consequence of that component. Since risk is the product of probability times consequence, there are questions that will significantly affect the RIN if answered with a poor option or not answered at all. An example of this is the question "Testing of Over Speed Trip" and an answer of "Never Tested" would have a significant impact on the RIN and the recommendations by increasing the probability of failure.

Program Aids

The program has several aids for the user in an effort to make the out meet the users needs while providing the best output. Some the aids are:

- Check for missing answers.
- Check for inconsistent answers.
- Etc.

Results and Comparisons

S.T.R.A.P. will compare the turbines with:

- Other turbines in the company.
- All the turbines in the database.
- Turbines in the same class.
- Turbines by the same manufacture.
- Turbines in the same industry.
- Etc.

Recommendations

S.T.R.A.P. makes recommendations to improve the reliability of the turbine(s) based on **R**eturn **O**f **I**nvestment **(ROI)**. It is then up to the engineers to decide which recommendation can be executed, in what order and during an outage and/or on line.

Impact

Based upon the consequence data of the operating unit that has been input to the program the ROI can be calculated. Some of the basic questions are:

- Unit production rate?
- Effect on the plant if a unit goes down per day?
- Impact of a trip of the turbine?
- Etc.?

Reliability Assessment of Compressors (R.A.C.)

The **R**eliability **A**ssessment of **C**ompressors (R.A.C.) program has been developed in very similar manner to the S.T.R.A.P. program. Compressors are divided into the types of service they are being applied. In addition the program requires all the constituents on a percent mol weight bases. In reality the R.A.C program is far more complicated than the S.T.R.A.P. program. There are an a number of:

- Process applications (almost infinite).
- Types of end seals design in use.
- Compressor designs.
- On-stream washing or injection.
- Etc.

Conclusion

The **Reliability Assessment of Rotating Equipment (R.A.R.E.)** is the combination of both the S.T.R.A.P. and R.A.C. programs. Depending on the maintenance strategy of the company and/or the facility, either/or S.T.R.A.P. and R.A.R.E. programs can be a significant benefit in assessing the critical machinery performance and identify the areas that could or should be addressed to improve the reliability and define the obstacles to extending the run time between minor and major overhauls with an acceptable risk.

Rotor Balancing Options

Authors: Sandye Simmons – TurboCare & Chet Stroh – TurboCare

ABSTRACT

This paper discusses the merits of high (operating) and low speed balancing of rotor assemblies of critical rotating equipment that have been in service; primarily multistage centrifugal or axial compressors, multistage steam turbines, and gas turbines. Balancing is discussed in strictly layman's terms and incorporates several case examples. Actual balancing techniques are only briefly discussed.

INTRODUCTION

A properly balanced vibration-free rotor assembly is essential to maintain unit clearances, efficiency and trouble-free operation. The benefits of a balanced rotor assembly are numerous:

- Confidence in a smooth start-up
- Less rotor deflection through critical speed
- Lower rotor vibration at operating speed
- Increased bearing and seal life
- Maintain machine efficiency
- Longer periods between scheduled turnarounds
- Due to a lower initial vibration level, other components such as valve rack components, splitline bolting, etc., do not fatigue as quickly.

To address this, rebalancing must be a routine part of maintenance. In doing so, equipment operators have two options; a low speed or high (operating) speed balance. In most cases, high-speed balancing produces a smoother start-up and a more reliable rotor as well as avoiding costly repairs and unscheduled shutdowns. In some rotor assemblies, low speed balancing is more than sufficient. It is, therefore, essential to distinguish which type of balancing to apply in a given situation.

This paper is based largely on "Options for Low Speed and Operating Speed Balancing of Rotating Equipment" by Stroh, MacKenzie and Rebstock which originally appeared on pp.253-258 of the *Proceedings of the Twenty-Fifth Turbomachinery Symposium,* Turbomachinery Laboratory, Texas A&M University, College Station, Texas, Copyright 1996. Sections are reproduced with permission of the Turbomachinery Laboratory (http://turbolab.tamu.edu).

BALANCING BASICS

To understand the relative merits of the low and high speed balancing, it is important first to have an understanding of basic balancing principles and terminology.

According to ISO, unbalance is "that condition which exists in a rotor when vibratory force or motion is imparted to its bearings as a result of centrifugal force."

Centrifugal force increases with the square of the speed:

$CF = (R) \times (WT) \times w^2/g$
CF = Centrifugal Force
R = radius for center of rotation
WT = rotor weight
w = speed
g = gravity

A rotating element having an uneven mass distribution, i.e., unbalance, will vibrate due to the excess centrifugal force exerted during rotation by the heavier side of the rotor. When at rest, the excess mass exerts no centrifugal force and, therefore, causes no vibration. Yet the actual unbalance is still present. Unbalance, therefore, is independent of rotational speed and remains the same, whether the part is at rest or is rotating (provided the part does not deform due to the unbalance). Centrifugal force, however, varies with speed. The higher the speed, the greater the centrifugal force exerted by the unbalance and the more violent the vibration. A common analogy can be found in automobile tires. Everyone has at one time experienced a condition where the car drives well at 25-30 mph; however, when at highway speeds a dramatic shimmy can be felt. This is due to the fact that centrifugal force increases proportionately to the square of the increase in speed. If the speed is doubled, the centrifugal force quadruples; if the speed is quadrupled, the centrifugal force is sixteen times greater. Low speed balancing is typically performed at 500 rpm or less; most rotor assemblies operate at speeds of 3600 rpm or greater or at a minimum 5X the speed at which low speed balancing occurs.

Centrifugal force acts upon the entire mass of a rotating component, impelling each component radially outward from the axis of rotation. If the mass of a rotating component is evenly distributed about its shaft axis, the part is "balanced" and rotates without vibration. However, if an excess of mass exists on one side of a rotor, the centrifugal force acting upon this heavy side exceeds the centrifugal force exerted by the light side and pulls the entire rotor in the direction of the heavy side.

This figure shows the side view of a rotor having an excess mass (m) on one side. Due to centrifugal force exerted by m during rotation, the entire rotor is being pulled in the direction of the arrow F.

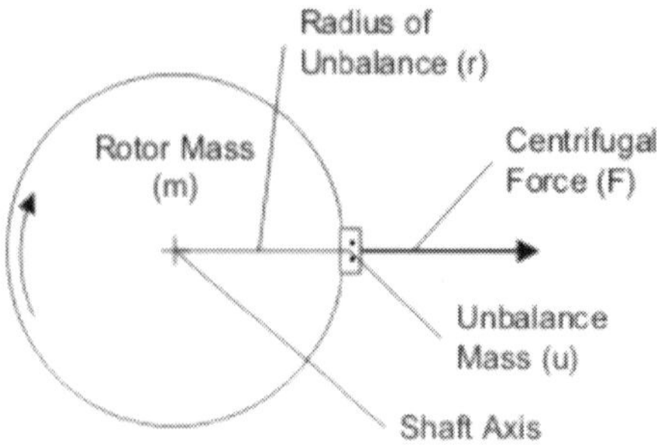

Figure 1 – Unbalance Creates Uneven Centrifugal Force and Vibration

Unbalance is usually measured by the product of the radius times weight.

UNBAL = (R) X (WT)

When making balance corrections, either a large weight can be added at a small radius or a proportionately smaller weight at a larger radius. This can be expressed in two ways: "rotor eccentricity" and "gram-inches." The term "rotor eccentricity" refers to how far the total rotor weight would have to be offset from the axis of rotation to produce the indicated unbalance amount. "Gram-inches" expresses the rotor weight and eccentricity together as a single number. To find the gram-in required, multiply the total rotor weight in grams by the given eccentricity (radius) in inches.

Another critical factor affecting balance is whether the rotor is rigid or flexible. A rigid rotor is most commonly defined as any rotor which runs at least 25 percent below its first flexible rotor critical speed where there is no significant rotor deflection. There are cases where a rotor might run above what is called a rigid rotor bearing critical speed, but all the motion takes place in the bearing clearance with no rotor deflection. This still falls under the description of a rigid rotor. A rigid rotor can typically be balanced by means of correction in any two separate planes regardless of how the unbalance is distributed throughout the rotor. For such cases, low speed balancing can satisfy all balancing requirements.

A flexible rotor is typically defined as a rotor which runs above a flexible rotor critical speed or within 25 percent below that speed, with sufficient rotor deflection to alter rotor balance behavior. Once a rotor starts to deflect, the actual axial distribution of imbalance becomes important and corrections have to be made in more than two planes, close to where the actual unbalance is located. Low speed data alone cannot supply all the necessary information on an assembled rotor to accomplish a complete balance of a flexible rotor. If the rotor can be stripped down to the bare shaft, a progressive stack and balance can be done where each piece can be balanced when it is assembled on the rotor; however, if the rotor is an integral forging of discs and shaft (as in a large number of steam turbines), this is not possible.

Rotor imbalance is the single most observed cause of vibration and must be eliminated. It results from numerous factors which occur both in normal operation, as well as during the repair cycle. Operational factors which can affect rotor unbalance response include, but are not limited to:

- Fouling (caustic buildup).
- Frequent shutdowns and startups, which can lead to loss of seal and bearing clearances.
- Erosion.
- Corrosion.
- Misalignment leading to worn areas on the rotor assembly.
- Foreign object damage.
- Distortion or dimensional changes, and shifting of parts due to rotational stresses, aerodynamic forces, and temperature changes.

During the repair cycle, blast cleaning of the rotor assembly alone can lead to imbalance. The blasting procedure does remove metal from the rotor, although in minute amounts, and imbalance can occur if the metal is removed disproportionately. Other repair procedures can affect balance, such as:

- Handworking the rotor to repair foreign object damage to components on the rotor assembly.
- Installation of new components that have not been individually balanced first.
- Improper component installation procedures, such as shrink fits, keys, taper fits.
- Rotor welding or coating.

Due to the wide variety of factors contributing to imbalance, it is impossible to eliminate them all. No matter how careful one is during operation and repair, it is still necessary and cost effective to rebalance rotors during each overhaul.

IN-CASE OR FIELD BALANCING

A rotor can be balanced successfully within its own case and bearing system, rather than in a balancing machine. If an owner chooses to do this, numerous problems can arise and these should be taken into consideration:

- Frequent stopping and starting of the driver, sometimes under load. The driver is sometimes involved with the process requiring at least a partial plant startup. Electric motors are limited to the number of starts in a certain time interval.
- Correction weight placement or removal might be difficult. Personnel may not always have proper access for making actual correction.
- The balancing process is subject the influence of other factors causing vibration such as misalignment, natural frequencies of support structures or other equipment in the area, air churning during the balancing process, feedback from the rest of the flow path when not in the operating state, difficulties around whether the unit is loaded and to what degree.
- High temperature and/or chemicals prevent access.
- A lack of room for vibration and phase measuring equipment.

- The balancing process takes longer and sometimes there is not adequate time for the job.
- Field balancing is frequently attempted to correct other mechanical deficiencies.
- It is not immediately obvious that the problem is unbalance. Several days can be lost while trying to be sure balancing is the answer.

In-case balancing, also known as field balancing, can ignore the true cause of excitation, especially on those occasions when only end plane corrections are employed because they are the only accessible balance planes in the field. In-case field balancing forces the user to adjust the rotating element to correct the combined effect of all the non-rotating members to overcome such problems as misalignment, foundation resonances, piping resonances, and any unbalance left in added-on parts at installation of the rotor into the case. If a rotor runs at undesirable vibration levels after being successfully balanced at operating speed in a high speed facility, it is not the rotor which requires scrutiny, but other factors in the system. Therefore, the balanced rotor becomes a tool for further corrective actions. Furthermore, multi-plane, in-case field balancing can be quite uneconomical because of the associated time and manpower involved.

OPERATING SPEED BALANCING

Many rotors in operation today run above the first critical speed and some run close to or even above the second critical speed. As the speed of the rotor approaches its critical, the shape of the shaft becomes sinusoidal. Any unbalance located at discs or wheels which coincides with high shaft deflection points will amplify the unbalance at those positions.

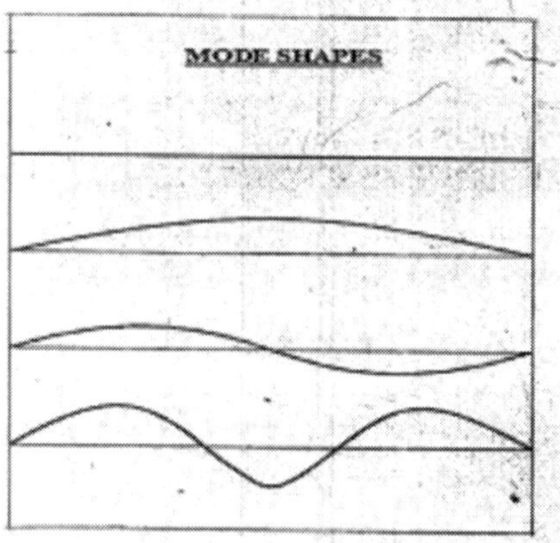

Figure 2 – Critical Speed Mode Shapes

Balancing the rotor at low speed does not take this deflection of the shaft into account. The procedures followed in balancing a rotor at its operating speed explore the sensitivity of each plane, reducing shaft deflections to minimal amplitudes where possible.

An operating speed balance is a relatively inexpensive procedure that isolates the rotor from being a source of problems in the field. From a prevention point of view, it makes sense to run the rotor at speed before installing it in its case out in the field. If an unbalance is in evidence, the cost of balancing is far less than assembly and disassembly time, and virtually insignificant when compared to lost production time. Other factors which might be considered a reason for high-speed balancing include:

- Redesign of the rotor assembly, such as an upgrade or rerate, involving newly designed components.
- Confirmation of Rotor Dynamic Analysis. Typically performed for component redesign, i.e., bearing or seal redesign.
- A large overhung weight, such as a coupling, can make a rotor more sensitive to unbalance.
- The rotor assembly may be operating on a critical speed. High speed balancing can lessen the response even at critical speed to allow satisfactory operation.
- The rotor or other similar designs have experienced field vibration problems.
- A rotor originally balanced at high speed should not be rebalanced at low speed.
- A computer simulation, rotor dynamics analysis indicates a rotor sensitive to unbalance.

A high speed balance can less the affect of unbalance at first critical speed. For rotors that run above the first critical speed, the importance of correcting the balance at the first critical speed is two-fold:

- The balance forces and the shaft deflections may be so violent that the unit will not get through the first critical speed without rubbing the shaft seals or exceeding bearing clearance and causing major damage. Even minimal contact with the seals will decrease the efficiency of the unit. This creates localized heating of the shaft, which adds to the inherent deflection at the first critical speed.
- If controlled acceleration and deceleration rates are not employed, a rotor could linger in the first critical too long, causing extensive damage to the unit.

Even though the rotor operates at a speed above the first critical, it may still have some residual mode shape from the first critical or may be beginning to form the second critical mode shape. In reality, it is usually a combination of both. Therefore, eccentricity as described above, is still a factor in a successful balance.

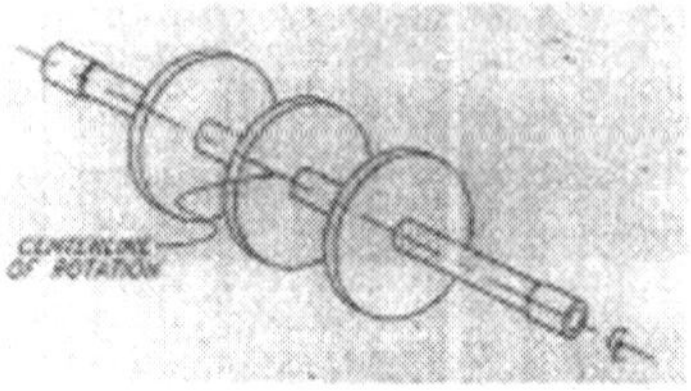

BELOW FIRST CRITICAL

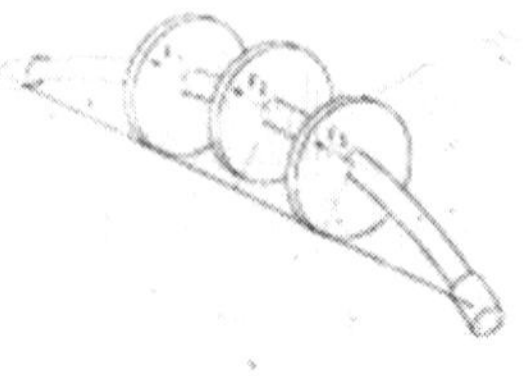

FIRST FLEXIBLE SHAFT CRITICAL

Some users believe that balancing at a higher speed than operating speed, such as at trip speed, is preferable to balancing at the actual operating speed; as if this suggests some type of margin of safety. However, this is not the case. Making corrections at trip speed means that the rotor is balanced only at trip speed and may not be within tolerance at normal operating speed.

Occasionally, the question is asked why certain integral rotors, which run above a critical speed and have never been operating speed balanced, still run with little vibration. One possible explanation is that if an integral rotor has only two major components, such as two discs, using these two discs as the correction planes will likely place the corrections where the most probable actual unbalance is located. For more complex rotor configurations, a possible explanation is that users know where their rotor critical speeds are and ramp through them very quickly, so there is no time for the system to respond to the critical. Another explanation is that some rotor bearing systems differ in their degree of sensitivity to unbalance. Even though the rotor deflects, there is still sufficient motion in the bearings to provide adequate system damping.

The fact remains that there are some flexible shaft rotors that are not balanced at operating speed and yet operate very well. The unit selections, design parameters, operating procedure and maintenance schedules are all influences for a trouble-free operating rotor. However, sometime in the future the rotor may require extensive repairs; changes in operating procedures will occur, maintenance schedules will be extended, or production demands will force operation at off-design conditions. An operating speed balance is a final verification that all variables affecting the rotor have resulted in a satisfactory system running condition.

Non-critical equipment, such as single stage steam turbines and lube oil pumps, are seldom high speed balanced. These rotor assemblies typically have short rotor lengths and operate far below first critical speed.

MODAL vs. INFLUENCE COEFFICIENT BALANCING

Both methods have the same objective: to place compensating balance weights opposite to the existing rotor unbalances. The rule of balancing is to make the corrections in the planes of unbalance. Since the unbalance distribution is usually not known and the use of a larger number of balance planes impractical, any realistic balancing procedure must obtain a satisfactory balance with a minimum number of balance planes. The two methods can be used together, resulting in the most efficient approach to flexible rotor balancing. The basic difference between these two methods is the manner in which one determines the magnitude and the orientation of unbalance.

Modal balancing is simply placing balance corrections so they only affect the mode being addressed. For instance, with a symmetrical rotor, to address only the first U mode place 10 ounces at zero degrees at the exact rotor midspan and five ounces at zero degrees on either side of the mid span halfway between the mid span and the bearing location. To address the S mode and not the first U mode, simply take away the rotor midspan correction and arrange the five ounces on either side of the mid span so they are 180 degrees out-of-phase.

In the case of influence coefficient balancing, a trial weight is applied successively at each correction plane along the rotor and the bearing reactions measured and compared to a base run. The change in bearing reactions is the "influence" caused by a known weight when applied to each of the correction planes.

For low-speed balancing, first take vibration data at each of the two bearings in terms of synchronous vibration levels (filtered for only running speed vibration) and phase angle. Then, by using a known trial weight and placing it at a known location, first in the one balance plane and then in the other, sufficient information is obtained to determine the corrections required in each of these two planes to balance the rotor. This is done by simultaneously solving the resulting vectorial equations. The process is referred to as the influence coefficient method because by using known weights at known locations, their influence can be measured and a two by two array of the resulting influence coefficients can be set up which, when multiplied with the original vibration data, yields the corrections required.

A similar process is used for operating speed balancing, when the rotor runs above a critical speed. The only difference is that now we need to obtain an array of influence coefficients at more than one speed and more than two planes of correction are now required.

PROGRESSIVE STACK AND BALANCE

An additional balancing technique, called "progressive stack and balance," is possible when working with stackable rotors. Starting with a bare shaft, parts can be mounted two at a time with a low-speed balance correction being made after each pair is installed. In this way, the balance correction is made directly on the part which has introduced an additional unbalance.

Achieving successful results depends heavily on not introducing any shaft deflections, which might or might not be relieved when the unit is put in operation, because the amount of deflection allowed is quite small. The balance stand is much more sensitive to shaft deflection than a dial indicator, so shaft runout must be given a lot of attention.

This procedure provides a reasonable level of balance, but not 100% assurance for a smooth running of the completely assembled rotor at full speed. It cannot be ruled out that some parts shrunk onto the shaft will be displaced as a result of centrifugal force. There is then a change in the balance condition as opposed to the condition at the start. If shrinkage is inadequate at certain points, some parts will loosen or internal friction will result. A loosening of parts could result in unbalances varying with time and could thus lead to vibrations of a higher amplitude.

Progressive stacking and balancing does not assure a smooth running, completely assembled rotor, through critical speeds and up to its operating speed, so they still could benefit from high speed balancing.

CASE STUDIES

Although low speed balancing can correct many problems, its effectiveness is limited. This is particularly true when rotors operate at or above critical speeds. Following are three examples where rotors that have been low speed balanced still required balancing corrections to be made at operating speed.

Seven Stage Second Body Compressor Rotor

The machine is a seven stage second body compressor. The rotor weighs about 1500 pounds, has a bearing span of 84 in, and 3.5 inch diameter journals. The rotor runs in the bunker pedestals with a first critical speed at 2200 rpm and a second critical speed at 7200 rpm. The rotor runs above the second critical at an operating speed of 8900 rpm. The rotor was stacked and low speed balanced; exceeded our target at operating speed. Three planes were used to balance the rotor, both end planes and a midspan plane. It took a total of 31 runs and over a day to balance the rotor to normal target. The next time the rotor was repaired and balanced, the rotor received a progressive stack and low speed balance. Corrections, though not as extensive, were again required to balance the rotor at operating speed.

Steam Turbine Rotor

This 15-stage, solid-forged rotor operates in the bunker pedestals at 3600 rpm, with a first critical speed at 1800 rpm. The rotor weighs about 16,000 pounds and has 7.985" diameter journals. The low speed balance was within target; however, at operating speed it exceeded ISO target specifications. The rotor needed to be ground in several locations 180 degrees opposite of corrections taken at low speed. Actual balancing data can be found in Attachment I.

Industrial Frame Gas Turbine Rotor

The machine is an industrial frame gas turbine rotor. The rotor consists of an axial compressor body between bearings and a single turbine wheel overhung on one end. The rotor weighs about 4500 pounds, has a bearing span of 78 inches, and journals are 6.0 inches in diameter. The rotor runs in the bunker pedestals with an in-phase first critical speed at 3300 rpm and an out-of-phase second critical at 6600 rpm. The compressor body of this rotor was first progressively stacked and balanced in a low speed facility. The rotor was then run in the bunker with no blades in the turbine wheel and balanced at its operating speed of 7500 rpm by making corrections only to the turbine wheel. The turbine wheel was then bladed and according to OEM specification would require no further balancing. The OEM professes that if a balanced set of blades are installed in a balanced rotor, then no further corrections to the rotor will be needed. This has not been found to be the case and usually further corrections are made to the turbine wheel. The results for over 15 identical rotors show that corrections had to made every time.

CONCLUSION

An operating speed balance can serve as a more reliable alternative to a low speed balance and more economical alternative to a field balance, especially on a repaired or redesigned rotor. In addition to demonstrating vibration levels, these tests can illuminate other potential mechanical problems such as loose fits, bent shafts and/or rotor instability. By performing an operating speed balance, the operator can eliminate vibration caused by rotor imbalance. If vibration still exists, the actual cause of that problem can then be located, rather than trying to adjust the rotor balance to compensate for other defects in the system.

BIBLIOGRAPHY

Schenck website www.schenck-usa.com

Stroh,C. and McKenzie, J. and Rebstock, J. "Options for Low Speed and Operating Speed Balancing of Rotating Equipment," *Proceedings of 25th Turbomachinery Symposium*, Turbomachinery Laboratory, Texas A&M University, College Station, Texas (1996).

Minimum Requirements for Field Testing and Monitoring of Turbomachinery

Author: D. Mark Cooper

Principal Machinery Engineer

Lyondell Chemical Company
Channelview, Texas

INTRODUCTION

There are many tools for determining the mechanical condition of turbomachinery such as vibration analysis and oil analysis, but few tools exist to determine the condition of flow paths and/or seals which can affect efficiency. Inefficiency of turbomachinery can be an indication of worn flow paths, fouling, excessive clearance or simply operation away from the design point. Evaluation of turbomachinery performance can help justify overhauls and re-rates, as well as evaluate their effectiveness. Performance analysis can also answer questions about current operation when there is a concern that the equipment is not operating as designed. Tools for the online evaluation of turbomachinery flow paths include field performance testing and performance trending.

A performance monitoring program is a combination of continuous performance trending with regularly scheduled performance testing. The concept of a performance monitoring program is fairly straightforward. Process data is used to calculate turbomachinery performance using well documented equations. While the concept is simple, the implementation of that concept is not. There are many ways that inaccurate or missing process data can affect thermodynamic calculations. In some cases, the difference between the calculated value and the real value can be 15 percent or more. While a 15 percent difference may not seem like much, on a 5,500 hp driver, the difference can be 825 hp, which is significant in terms of calculating the performance of the train.

In order to reduce errors in calculation and obtain data that can be used to evaluate the condition of turbomachinery, it is necessary to standardize the methods and instrumentation used for monitoring and testing.

Before we go into the requirements we need to define a few terms.

- Performance trending is the observation and tracking of performance data for compressors and turbines using ideal gas equations to determine fluid properties for use in traditional thermodynamic calculations. This information is corrected to design conditions for comparison to the original performance curve. The operation data used to determine fluid properties is obtained using existing instrumentation that has been calibrated within the previous six months.

Performance Testing is analysis of compressor or turbine operations under varying flow conditions using calibrated instrumentation and lab analysis of fluid properties. Fluid properties and inlet conditions are to be constant during the test. This information is corrected to design conditions for comparison against the shop performance test curve. Field performance tests are typically conducted during acceptance of a new turbomachinery train but they should also be conducted on a periodic basis to validate performance trends.

- Compressor "Section" is a row of one or more impellers (wheels) passing the same mass flow rate and gas composition.

- Compressor "Stage" is a single impeller (wheel) in a compressor. There may be one or several wheels in a compressor section.

DATA GATHERING

Shop Testing versus Field Testing

Performance monitoring of turbomachinery requires process thermodynamic data along with pressures and temperatures and flow rates. Process thermodynamic data for closed looped systems is easily obtained by correcting design data to actual pressures and temperatures. Thermodynamic data for systems, where the process composition can vary, is much harder to obtain unless there is a method to analyze the stream. Whether the system is closed loop or not, accurate inlet and outlet pressures and temperatures, along with compensated flow rates, are required in order to obtain viable results. Existing field instrumentation is typically used to gather process data. Therefore, this instrumentation needs to be evaluated for proper position, installation and calibration.

A performance acceptance test (i.e. shop test) at the manufacturer's facility will include numerous calibrated instruments located at exacting positions as outlined in either *ASME PTC 6 – Performance Test Code 6 on Steam Turbines* or *ASME PTC 10 – Performance Test Code on Compressors and Exhausters*. The data from these tests is sent to a computer program developed by the manufacturer to compensate for the differences between operation on the test stand and operation in the field. The computer program uses machine specific design information such as wheel curves and stationary component friction loss coefficients to determine the predicted performance of the equipment under design conditions.

When the equipment is installed in the field there is most often compromises made in not only instrument quality but also location and number of instruments. Compromises in number, quality and location of instrumentation are typically made on the basis of cost and maintainability. Cost is a factor in the number and quality of instruments and maintainability is a factor in the location of the instruments. For example, ASME PTC 10 requires four pressure and temperature measurement taps on the inlet and four on the outlet for a total of 16 measurement taps for a single compressor section that has a single inlet and a single outlet. There are additional taps required for flow measurement. The 16 presssure and temperature measurement taps are at specific distances from the compressor and from each other, not only axially along the pipe but radially as well. There are also requirements for measurement of both static pressure and velocity pressure. By comparison, a typical compressor installation will have a single temperature measurement device and a single static pressure measurement device on the inlet and the same number on the outlet. Also, field instrumentation is typically located to facilitate maintenance of the devices rather than to facilitate replicating a shop performance test. The difference in numbers between the instrumentation used for a turbine shop performance and a typical field installation are comparable to those for a compressor, although the number of instruments is greater. Because there are compromises in the number and placement of measurement devices, there will be differences between field performance calculations and results of the shop performance test. There are many factors that may or will affect field performance calculations, some of which are as follows:

- Number and placement of instrumentation
- Accuracy of instrumentation
- Static pressure versus total pressure measurement (Total pressure = static pressure + velocity pressure)
- Thermodynamic analysis of the process
- Design data used for comparison to actual operating conditions
- Calculation methods

We will review the minimum requirements for those factors that have the most impact on field performance analysis.

Required Operating Data

In order to calculate the performance of a compressor or turbine and do performance monitoring and testing a minimum set of data is required. For compressors the required operating data is a follows:

- Suction and discharge pressure at each section or stage
- Suction and discharge temperature at each section or stage
- Flow indication at each section or stage. If possible, flow meters should be ranged, calibrated and the bore verified prior to a performance test.
- Gas thermodynamic properties at the inlet to each section or stage (i.e. molecular weight, specific heat ratio, density, etc.) If there isn't a way to determine the gas properties in real time, then the design process gas information should be used for the performance monitoring program and gas samples shall be used for performance testing. If the gas composition is the same as specified during design and does not change during operation, then the design gas property information can be used for monitoring and testing provided they are corrected for pressure and temperature if necessary.
- If the compressor is motor driven, there should be available motor voltage, amperage and power factor. This information is desired for performance trending but is required for performance testing.

For turbines the required operating data is as follows:

- Inlet steam temperature, pressure and flow-rate

  NOTE: Due to the difficulties in measuring the flow rate of high pressure steam, where possible, the inlet steam flow rate shall be calculated as follows. For non-extraction condensing turbines the condensate mass flow rate shall be considered the inlet steam mass flow rate. For extraction/condensing turbines the sum of the extraction steam mass flow rate and the condensate mass flow rate shall be considered the inlet steam flow rate.
- Exhaust steam temperature and pressure
- Steam pressure downstream of first stage nozzle. The steam pressure downstream of the first nozzle after an admission or extraction section is also needed.
- Admission or extraction steam temperature, pressure and flow-rate if applicable
- For condensing turbines, the following is needed: condensate temperature, condenser pressure, condensate flow-rate, condenser cooling water differential temperature, condenser cooling water differential pressure and condenser cooling water flow-rate.

  There are two options if the cooling water data is not available. 1) Measure or assume (approximately 6 - 8 percent moisture) an exhaust steam quality or 2) Assume an exhaust section efficiency. The preferred method in this situation is to measure or assume an exhaust steam quality.

Accuracy of Instrumentation

Data from a pressure transmitter is preferred over data from a local gauge due to greater accuracy. When conducting a field performance test, gauges or transmitters used for data gathering should be calibrated with a minimum accuracy of +/- 0.5 percent of the span for the total span. This means that for a pressure instrument with a 500 psi range the accuracy is expected to be +/- 2.5 psi for the full range of the instrument. The accuracy of most pressure gauges varies from 0 psi to full pressure with the greatest accuracy being in the middle of the span. For this reason, it is best to use pressure gauges where the maximum pressure indication on the gauge is twice the expected operating pressure so that a 500 psi pressure gauge would be ideal for a process operating at 250 psi.

It is also best to use a gauge with a face diameter of at least 4.5 inches. A large face diameter is easier to read and it allows for an increased number of divisions between whole numbers all of which increases accuracy.

For maximum accuracy it may be necessary to calibrate a gauge in the horizontal position if it is to be installed in the horizontal position. A comparison of gauges calibrated in both the horizontal and vertical positions will reveal whether or not calibration position is a factor.

There is also a preferred device for temperature measurements. An RTD is preferred over a thermocouple because they are more accurate and more responsive, but they should be three or four wire compensated in order to maximize accuracy of the probe. If a thermocouple is to be used then it should be a type "T" because they are more accurate than a type "J". Accuracy of the probe should be +/- 1 percent of the span for the total span. Just like a pressure instrument, the accuracy of most probes is the greatest at middle of the span so a probe with a maximum indication that is twice the expected temperature should be used.

Flow measurement devices should be calibrated to an accuracy of +/- 0.05 percent of the design flow rate and full scale should be twice the design flow rate. For processes where the pressure and/or temperature at the flow device can vary from design, the measured flow rate should be compensated for the operating pressure and temperature. If possible, the bore of the flow device should be verified prior to a performance test to ensure accuracy.

PERFORMANCE MONITORING PROGRAM

As mentioned earlier, a performance monitoring program is a combination of continuous performance trending with regularly scheduled performance testing. Performance trending can be used to provide relative comparisons between similar operating points. Performance trending calculations may be performed continuously or periodically depending upon the amount of resources available. Performance Testing is used to compare current operating conditions to design or shop test information. Comparison to shop test information is preferred over a comparison to design information. Flow meters are calibrated using specific operating conditions and fluid properties. Therefore, the mass flow rate should be corrected for actual operating conditions.

Compressor Performance Monitoring

A designated person may trend compressor performance or it can be calculated and historicized by the Distributed Control System (DCS) for the operating area. Trending in the DCS is the preferred method since it requires little or no involvement by personnel. The information is readily available to everyone and the information can be easily historicized.

If the trending is executed in the DCS, calculation constants should be accessible so that the information can be updated as necessary to improve the accuracy of the calculations. Access to this information should be controlled to prevent accidental deletion or modification.

If the compressor system is a closed loop where the gas composition does not change, development of gas properties for use in performance calculations can be accomplished by correcting design gas properties to actual conditions using ideal gas laws.

If the gas composition can vary from design conditions, a gas analysis is desired for calculation of gas properties. An alternative to gas properties obtained from a gas analysis is to correct design gas properties to operating conditions using ideal gas laws just as you would for a closed loop system. If the alternative approach to obtaining gas properties is used, there is the possibility for error in the performance calculations due to differences in the calculated gas properties and actual gas properties. The preferred method for obtaining actual gas properties is to use a real time gas analyzer.

The design specific heat ratio (k), compressibility factor (Z) and molecular weight of the design gas is used in conjunction with actual pressure, temperature and corrected mass flow rate to determine polytropic efficiency, head and horsepower.

If there are operating parameters unrelated to the compressor that cause the gas properties to vary, those should be trended as well to facilitate accurate comparisons of trend points.

Compressor Performance Testing

If a real time gas analysis is not available, two free-flowing gas samples should be taken at the inlet of each section. One gas sample is to be taken at the beginning of the performance test and the other at the end of the test. If possible, the samples should be obtained using an insertion probe. The gas sample should be heated to the original process temperature before analysis is conducted to prevent errors caused by condensation.

The mass balance around all points in the equipment gas path, where the material enters or exits the system, should be validated to ensure accurate results.

The calculated k, Z and molecular weight of the sampled gas is used in conjunction with pressure, temperature and corrected mass flow rate to determine polytropic efficiency, head and horsepower.

Performance analysis should be conducted using spreadsheets or programs that are specific to centrifugal compressors and by someone trained in their use. All calculations are to be based on a polytropic process unless the equipment manufacturer has based the design on an adiabatic process.

Compressor Performance Monitoring Calculations

Once the gas properties are determined, then the next step is to calculate the performance.

These calculations are to be performed for each compressor section.

Polytropic Head:

$$H_p = ZT_1\left(\frac{1545}{MW}\right)_1\left(\frac{n}{n-1}\right)\left[\left(\frac{P_2}{P_1}\right)^{\frac{(n-1)}{n}} - 1\right]$$

Where:

$$n = \frac{\ln\left(\frac{P_2}{P_1}\right)}{\ln\left(\frac{T_1 P_2}{T_2 P_1}\right)} \quad \text{Or} \quad \frac{n}{n-1} = \frac{\ln\left(\frac{P_2}{P_1}\right)}{\ln\left(\frac{T_2}{T_1}\right)}$$

Polytropic Efficiency:

$$\eta_p = \frac{\left(\frac{n}{n-1}\right)}{\left(\frac{k}{k-1}\right)}$$

Gas Horsepower:

$$GHP_p = \frac{H_p \dot{m}}{\eta_p 33000}$$

Where:

GHP_p	= Polytropic Gas Horsepower
H_p	= Polytropic Head (ft-lbs/lbm)
k	= Isentropic exponent. Assumed to be constant for all cases and equal to design unless the actual value is determined by analysis
MW	= Molecular Weight
$\dot{m}$	= Mass flow (lb/min)
n	= Polytropic Exponent
η_p	= Polytropic Efficiency
Z	= Compressibility Factor
P	= Pressure (psia) (P_1 = Inlet Pressure, P_2 = Outlet Pressure)
T	= Temperature (deg. R) (T_1 = Inlet Temperature, T_2 = Outlet Temperature)

For constant speed machines, the performance data may be trended by calculating the difference between the expected head at the measured volumetric flow rate to the calculated head at that same volumetric flow rate, then plotting this difference versus time. Efficiency may be plotted in the same manner.

For variable speed machines, the performance data will need to be corrected for speed before comparisons can be made. The traditional method for plotting data is the use of non-dimensional curves. The subject of non-dimensional calculations for the purpose of comparing compressor operation to design or to another compressor is well documented. Some of the calculations can be complex, but for the purpose of performance trending, the calculations can be reduced to the following:

Non-dimensional Flow = Q/N
Non-dimensional Head = H/N^2

The process for trending data using non-dimensional information includes generating a non-dimensional performance curve based on the design or test performance curve. Various operating points from the design or test performance curve are used to calculate their equivalent non-dimensional point. The resulting non-dimensional curve is then used as the basis for comparison to actual operating conditions. Actual operating conditions are used to generate non-dimensional points which are compared to the non-dimensional curve by calculating the difference between the actual non-dimensional head and the expected non-dimensional head at the calculated non-dimensional flow. The head difference is then plotted versus time. Efficiency curves do not require speed correction, but there is still the need for a plot of efficiency versus non-dimensional flow. Efficiency for variable speed machines may be trended by calculating the difference between actual efficiency and expected efficiency at the measured non-dimensional flow.

Performance Monitoring of Steam Turbines

For condensing turbines, the heat balance around the condenser is the preferred method for calculating exhaust steam quality and enthalpy, but if it is not available, the design steam quality can be used to calculate the exhaust enthalpy. It must be understood that there will be some error in the horsepower and efficiency calculations when a steam quality is assumed rather than calculated.

In addition to trending horsepower and efficiency versus time there are additional parameters that should be included with each data point for comparison.

- Inlet steam valve position
- Inlet steam temperature and pressure
- Pressure after 1st stage
- Extraction/admission valve position
- Extraction/admission steam pressure and temperature
- Pressure after low pressure section nozzle
- Exhaust steam pressure, temperature and quality

All of these items need to be tracked along with horsepower and efficiency. Tracking of this information allows for the comparison of performance not only to design, but also comparison to similar operating points.

A designated person may trend turbine performance or it may be calculated and historicized by the DCS for the operating area. Trending in the DCS is the preferred method since it requires little or no involvement by personnel, the information is readily available to everyone and the information can be easily historicized.

If the trending is executed in the DCS, then calculation constants should be accessible so that the information can be updated as necessary to improve the accuracy of the calculations. Access to this information should be controlled to prevent accidental deletion or modification.

Performance Testing of Steam Turbines

In order to accurately calculate the performance of a condensing turbine, the heat balance around the condenser must be calculated in order to establish exhaust steam conditions.

For condensing turbines, inlet steam flow should be established by measuring the condensate flow rather than by using the steam flow meter. Liquid flow measurement is more accurate than steam flow measurement. If the turbine has an extraction, the extraction flow is to be added to the condensate flow.

Performance analysis should be conducted using spreadsheets or programs that are specific to steam turbines and by someone trained in their use. All calculations are to be based on an isentropic process.

Steam Turbine Performance Monitoring Calculations

These calculations are to be used for each turbine section. Inlet to extraction (or induction) is considered one section and extraction (or induction) to exhaust is considered a separate section.

Turbine Horsepower:

$$Horsepower = \frac{SteamFlow(lb/hr) * \Delta Enthalpy(Btu/lb)}{2559.15}$$

Turbine Efficiency:

$$Efficiency = \frac{\Delta Enthalpy(Btu/lb)}{\Delta Enthalpy(Btu/lb)\ @\ \text{constant entropy}}$$

Where:

$SteamFlow\ (lb/hr)$	= Inlet Steam Flow
$\Delta Enthalpy(Btu/lb)$	= Enthalpy Difference between the Inlet and Outlet Steam
$\Delta Enthalpy\ (Btu/lb)$ @ constant entropy	= Enthalpy Difference between the Inlet and Outlet Steam with Outlet Steam Enthalpy calculated at the same Entropy as the Inlet Steam

Do's and Don'ts – Pitfalls and Gotcha's

There are many things that can affect turbomachinery performance and interpretation of performance data can be misinterpreted. Here are few of the areas where the writer has experienced trouble with the process of turbomachinery performance evaluation.

It has already been mentioned in this paper, but accuracy of process data cannot be stressed enough. The adage "Garbage In – Garbage Out" can be applied to performance calculations. If the process data is not correct, then the calculated results will be incorrect. The degree of error in the process data will be magnified by the calculations due to the various multiplication and division steps. Accuracy of process data is the most important aspect of performance monitoring or testing. This applies to all types of turbomachinery.

For turbomachinery systems with variable speed drivers and a process gas that can vary in composition a rigorous and detailed performance model built using process modeling software would seem like a worthwhile effort for an accurate performance analysis. Unfortunately, the vast number of process points required to build this type of model can introduce error into the calculations that is greater than the expected increase in accuracy. The reason is that for each process variable used in a model there is a degree of uncertainty, or error, which is introduced. As more process variables are introduced, the combined degree of uncertainty increases. The use of detailed performance models can be very beneficial, but the amount of uncertainty introduced by the use of multiple process variables should be taken into consideration and the results of the model compared with the simplified methods mentioned in this paper. A comparison of the performance trends between the detailed model and the simplified calculations can help determine which method works best for that particular application.

Troubleshooting turbomachinery performance involves not only the comparison of actual performance to design but also an understanding of how differences between design operating conditions and actual operating conditions can affect actual performance. If flow rates, inlet/outlet pressures or temperatures vary from design, then the possibility exists that the capabilities and efficiency of the machinery will vary from design. Care must be taken to either operate the machinery at design conditions for the purpose of calculating performance or compensate for the effects of off-design operation.

SUMMARY

Performance trending or testing of turbomachinery installed in the field is a valuable tool for evaluating the condition of equipment, but the results can be worthless if the data used in calculations is not accurate. The first step in creating or maintaining a performance monitoring program is ensuring that the data used in calculations is accurate. The data source must be checked for accuracy each time a performance check is done to ensure that the comparisons to design are meaningful.

Even with accurate data, there may still be variation between the calculated performance and design due to differences in the type, location and number of instruments used to gather data in the field compared to what is used for shop tests.

If a consistent approach is used in the gathering of data and calculation of performance, then the benefit of performance trending and testing can be realized, which is the ability to track either an increase or decrease in performance over time.

Case Studies of Low Speed Operation on Dry Gas Seals and Bearings in Ethylene Compressors

Authors: Joe Delrahim – John Crane Inc. & Chris Stewart – Turbocare

Abstract

This paper will discuss the design features of dry gas seals and bearings that are affected during turning gear, slow roll and reverse rotation of centrifugal compressors. Data and formulas will be presented that can be used to determine the minimum desirable speed for slow roll as well as the breakaway torque for journal bearings. Case studies, field data and in-house testing data will be presented related to these various modes of operation.

Compressor

Centrifugal compressors are critical to the operation of Ethylene Plants. These compressors are un-spared and their reliability directly affects the number of years of uninterrupted operation of the Ethylene plant. The operation of the bearings and seals of these compressors directly affects the reliability of these compressors.

The compressor trains in Ethylene Plants have increased in size and many require to be slow rolled or put on turning gear when the equipment is shutdown to prevent thermal bows and the potential for rubs on restart. Slow roll and turning gear operation can cause unique problems for the design of dry gas seals and bearings of these large compressors. Another operational problem related to shutdown of these compressor trains is reverse rotation of the equipment due to leaking check valves in the discharge lines. This reverse rotation also presents unique problems for the design of dry gas seals and bearings.

Dry Gas Seals

This paper covers the effectof slow roll and effect of reverse rotation during shut down period On gas seals, as well as outlines and recommends Guidelines for pressure/speed on gas seals for Ethylene compressors required to also operate in slow roll mode during the operation.

Type 28XP

Why the gas seal is a preferred method to seal ethylene plant compressors?

- Although high purity of ethylene and propylene is an essential requirement, oil contamination from wet seals has been a major problem encountered by plant operators for decades. In response to these frequent problems, nearly 30 years ago, John Crane introduced a spiral groove non-contacting dry running gas seal to replace the wet seals that were commonly used on refrigerant compressors. According to field experience, no other sealing development has led to the operating and maintenance cost savings, not to mention the reliability improvements, than those associated with the non-contacting dry running gas seal for refrigerant compressors. In addition to no wear or process contamination, dry gas seals offer lower power consumption and improved operational safety. Today, dry running gas seals have become the preferred method of sealing, replacing wet seals as the standard product.

Slow Roll

Steam or gas turbines are used as a driver for refrigerant compressors. To allow even warm up and cool down, the unit is slow rolled to prevent casing or rotor warping due to temperature imbalances. A question commonly asked is the affect of slow roll speed on the reliability and long-term life for gas seals used for refrigerant applications. In addition, the unit may also experience a one-quarter turn ratcheting speed. The ratcheting feature is normally used to cover for power failure. A typical slow roll condition may occur at speed of 5 to 300 RPM for a period of 12 to 50 hours. In short, the larger the unit, the longer the slow roll period.

Reverse Rotation

The compressor trains may experience, for a short period of time (normally less than a minute), reverse rotation of the equipment due to leaking check valves in the discharge lines. Several field studies have revealed that the majority of refrigerant compressors equipped with the uni-directional spiral groove face seal design have experienced short periods of reverse rotation. In every reported case these compressors restarted without any noticeable change to the seal performance.

Uni-Directional vs Bi- Directional (BD) Seals

The uni-directional spiral grove gas seals are known to provide more efficient pumping action across the seal faces than bi-directional (BD) seals during operation. More efficient pumping action results in higher gas film stiffness across the seal faces. Gas film stiffness is critical to separate the seal faces from each other which is considered to be a crucial parameter to avoid face contact at all operating conditions.

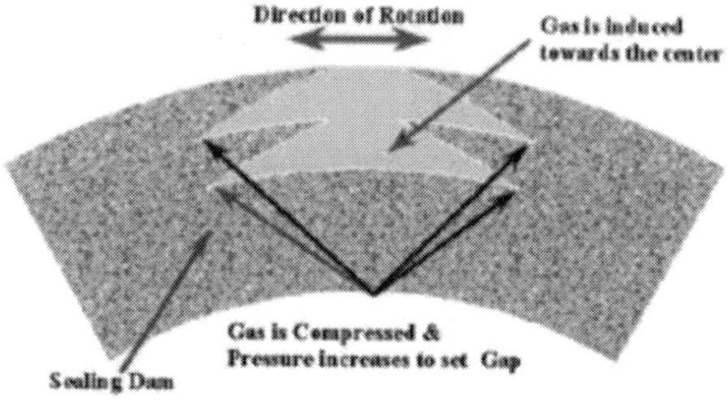

Bi-Directional

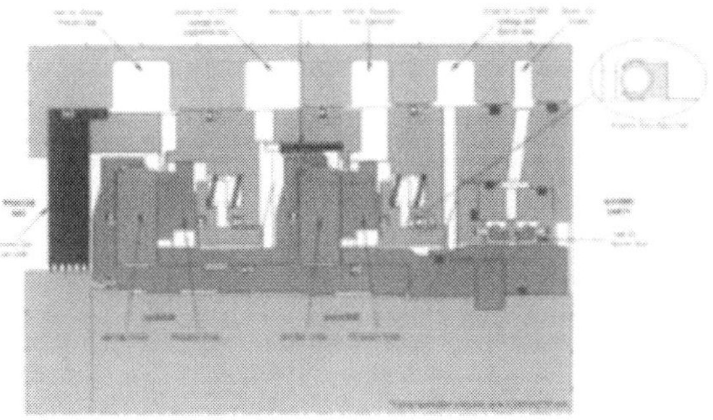

Uni-directional

Principle of Operation

The operating principle of the spiral-grooved gas seal is that of hydrodynamic and hydrostatic force balance. The rotating face of the seal (mating ring), which has a logarithmic spiral-groove pattern, is the key to generating an efficient hydrodynamic force. As the shaft rotates, the mating ring rotates with it while the stationary seal face (primary ring) remains stationary. As this occurs, gas is pumped inward to the tip of the spiral groove. The sealing dam acts as a restriction to the gas outflow.

Seal Arrangement

A tandem seal arrangement is the most commonly used on refrigerant processes. In this arrangement, two seal modules are stacked up in series orientation. The first seal (inboard seal) handles full pressure while the second seal (outboard seal) operates as a back-up seal with zero pressure differentials. The leakage from the inboard seal can be monitored and safely vented to the flare. The outboard seal prevents any leakage from reaching the atmosphere. Typically the secondary vent can be taken to a safe location/atmosphere.

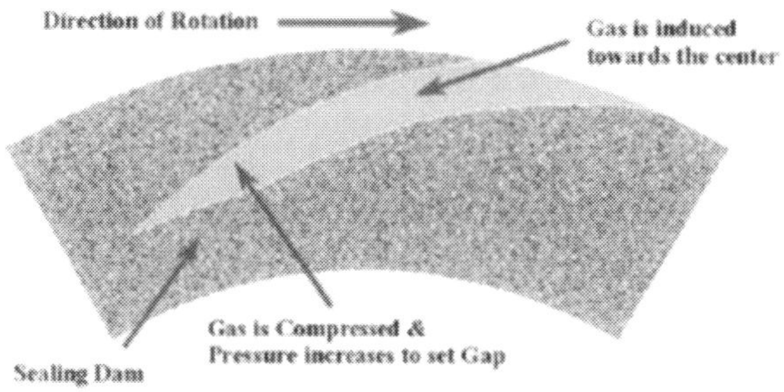

Tandem Seal Arrangement

Slow roll test at John Crane facility to establish and recommend a guideline for slow roll limit on non-contacting dry running gas seals.

The proof is in the numbers. A series of seal tests at low speeds were performed using John Crane T28XP tandem arrangement seals. The test seal used John Crane standard spiral grove profiles and carbon face material. Several tests comprised a series run at various speeds including ratcheting condition. The tests performed include:

Phase #1

Up to 50 RPM including 43 hours ratcheting. The test cycle is shown below.

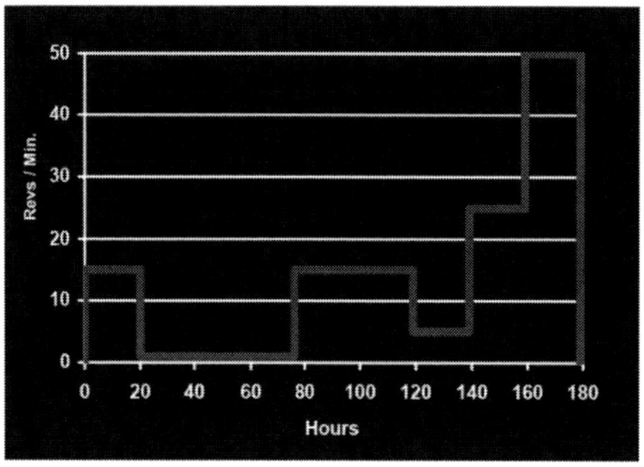

Test cycle

Phase #2

80 hours of testing under various conditions:

1. 20 hours at 15 RPM
2. 20 hours at 5 RPM
3. 20 hours at 25 RPM
4. 20 hours at 50 RPM

Phase #3

25 RPM 800 hours (33 days) to assess long term slow roll capability for John Crane spiral groove design faces.

After each stage of the test, the dynamic performance of the seal was assessed. The seal also was stripped at various stages to examine the face condition and assess the face wear or degradation. Having completed nearly 1,200,000 revolutions, which equates to approximately 6.3 years of operation, the test was considered successful and demonstrated that:

- The spiral groove face pattern has excellent resistance to slow roll.
- Ratcheting for more than 20,000 quarter turns did not adversely affect seal performance.
- Carbon vs tungsten carbide and carbon vs silicon carbide faces have excellent slow roll speed capabilities.
- Hard faces such as tungsten or silicon carbides rubbing against each other result in poor slow roll capability, galling and damage to the seal faces is expected.

Spiral Grove Face

Hard faces

Guideline and limitation on slow roll

Based on John Crane's extensive in-house testing and reports received from the field during the course of the last 30 years, the following guidelines were created for standard gas seals used on compressors experiencing slow roll conditions. Note that minor adjustments can be made to improve lift-off at various speeds.

Optimum Performance

- Material – Tungsten vs. Carbon
- Spiral groove design for maximum lift

Speed range: pressure less than 5 bars

- < 0.25m/s - No adverse effect
- 0.26 m/s to 2.3 m/s- (3.5m/s for BD) -Avoid running without pressure
- < 2.3 m/s- (3.5 m/s for BD) -No adverse effect
- **Ratcheting has no adverse effect on seal**

Pressure effects

- 5 to 10 bar pressure is required and can be used for lift-off at all speeds.

Special attention should be made to the outboard (OB) seal for proper lift-off. To ensure a safe operation, make sure that the outboard seal also is slightly pressurized. The outboard seal pressure has to be below gas seal buffer pressure.

Limitations on slow roll

- Particulate or excessive oil contamination will accelerate seal face wear and may damage the faces during slow roll.
- Processes, such as dry nitrogen or air with a dew point lower than minus 20 F is known to accelerate wear on a specific carbon graphite grade due to lack of sufficient condensable vapor across the seal faces. Therefore, adjustment to the slow roll speed may be necessary to achieve face separation during the slow roll period.

Case study #1 - Slow Roll at a major Ethylene plant

Spiral groove non-contacting dry running gas seals using carbon vs tungsten carbide faces were supplied to a major Ethylene plant in 1992. As part of a special specification, the plant required a slow roll test to be included in the seals' test procedures. A special test fixture was designed to carry the slow roll-testing requirement. The 11.187-inch seal ran at 12 rpm for a period of 48 hours at a pressure of 0 to 50 psi in John Crane test facility. At zero pressure, some carbon material transfer was observed, though the material was easily cleaned off. The subsequent testing at normal speed showed that the material transfer had no detrimental affect on seal performance. No material transfer was observed at 50 psi and 12 rpm, mainly a result of hydrostatic lift due to the higher pressure. Today, more than a decade later, these seals are still operating successfully and meeting performance standards.

Inboard

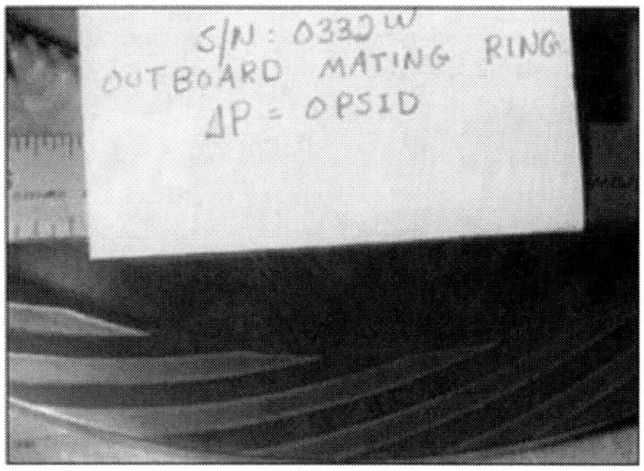

Outboard

Case study #2 - Slow roll experience in an Ammonia Synthesis plant

These seals were installed in an ammonia synthesis gas compression train in 1997. Before start-up, the compressor would operate in the slow roll for four hours and following the shutdown, it returned to slow roll for 24 hours at casing pressure of 260 psi. Prior to shipment, the seals were subject to slow roll test of 8.5 rpm for a period of 24 hours at the John Crane test facility. The pressure at the inboard seal was 75 psi and the outboard seal vent connection was closed. The seals completed the test without incident and were found to be in excellent condition. Further, since start-up in 1997, the seals have accumulated several thousand hours with numerous starts and stops during this period.

Ammonia Syngas

Ammonia Syngas

SLOW ROLL AND FLUID FILM BEARINGS

Hydrodynamic fluid film bearings are used extensively in the high speed, critical rotating equipment used in Ethylene service. Under normal operating conditions, the bearings are designed to operate without metal-to-metal contact between the rotating shaft and the stationary bearings. However, bearing wear will inevitably take place when the unit starts and stops (in the absence of a hydrostatic lift system) since the bearing's load capacity drops off as speed decreases. If the starting or stopping sequence of the unit includes a slow roll for thermal reasons, bearing wear can be greatly accelerated if the system is not designed properly.

BEARING LUBRICATION AT LOW SPEEDS

A quick review of the fundamentals of hydrodynamic lubrication will help explain the mechanisms behind wear that takes place at lower speeds. This will also offer insights into ways to compensate or eliminate the wear that can occur during slow roll.

Any surface, regardless of how smooth it is really a series of peaks and valleys. The smoother the surface, the less variance you have between the high spots (asperities) and the troughs. When a rotor is at rest there is metal-to-metal contact between the journal and the bearing surfaces along the asperities of each surface, even when the lubrication system is operating. Since there is no relative motion between the surfaces the oil will not develop any lifting pressure. The entire contact load is supported across the surface asperities. This contact forms a cold weld that will shear once there is relative motion between the surfaces. This is the source of friction. It is termed ***boundary lubrication*** and the material removal is referred to as adhesion wear. This is the wear that takes place in a door hinge and it is also takes place in your bearings during startup or shutdown. The surfaces will continue to erode as long as the separation distance is less than the RMS surface finish of the mating surfaces. The parameter that defines the relationship between the separation distance between two surfaces and their surface finishes is:

$$\lambda = h/(f_1^2 + f_2^2)^{1/2}$$

where:

λ = film parameter
h = film thickness or surface separation distance (in)
f_1 = RMS surface finish of surface (1) (in)
f_2 = RMS surface finish of surface (2) (in)

The relative motion that is causing the wear will also begin to develop hydrodynamic pressures in the lubricant that initiates separation of the surfaces. As the sliding velocity increases, the separation increases. During this period known as ***mixed film lubrication***, there will continue to be some metal-to-metal contact but the load is now carried by a combination of hydrodynamic pressure and contact pressure at the asperities. As the number of asperity contacts diminish, the source for friction is reduced. As a fluid film bearing transitions from boundary lubrication to mixed film lubrication, the coefficient of friction reduces dramatically. At some point, the pressure within the film reaches a point where it is sufficient for full separation of the two surfaces. This is termed ***full film lubrication*** and the entire load is now carried hydrodynamically. The relationship between the film parameter and the three regimes of lubrication are show in Figure (1).

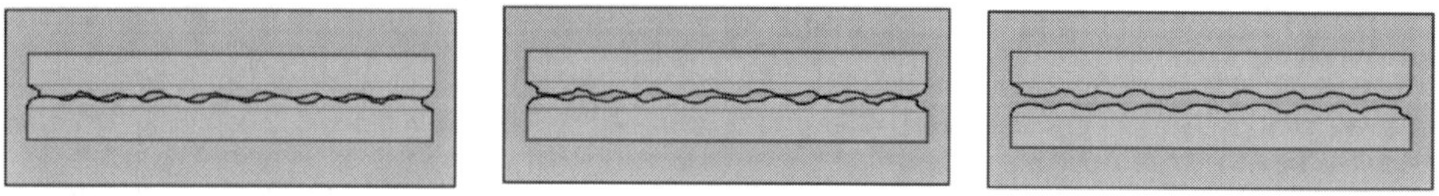

Boundary Lubrication $\lambda \leq 1$

Mixed Film $1 \leq \lambda \leq 5$

Full Film $5 \leq \lambda$

[Figure (1) Surface separation versus film parameter]

SLOW ROLL EVALUATION

A Stribeck curve is a plot of friction versus a dimensionless value called the Sommerfeld number and clearly shows the three regimes of lubrication. Gardner's [1] experiments in the early 1970's with bearings operating at low Sommerfeld numbers can be used to develop a modified Stribeck curve known as the ZN/P curve shown in Figure (2). This curve can be used to determine the lubrication regime for most industrial bearings operating at low speeds associated with slow roll. While there are many factors that influence the development of the lubrication mode, this method is useful as a first check on your system.

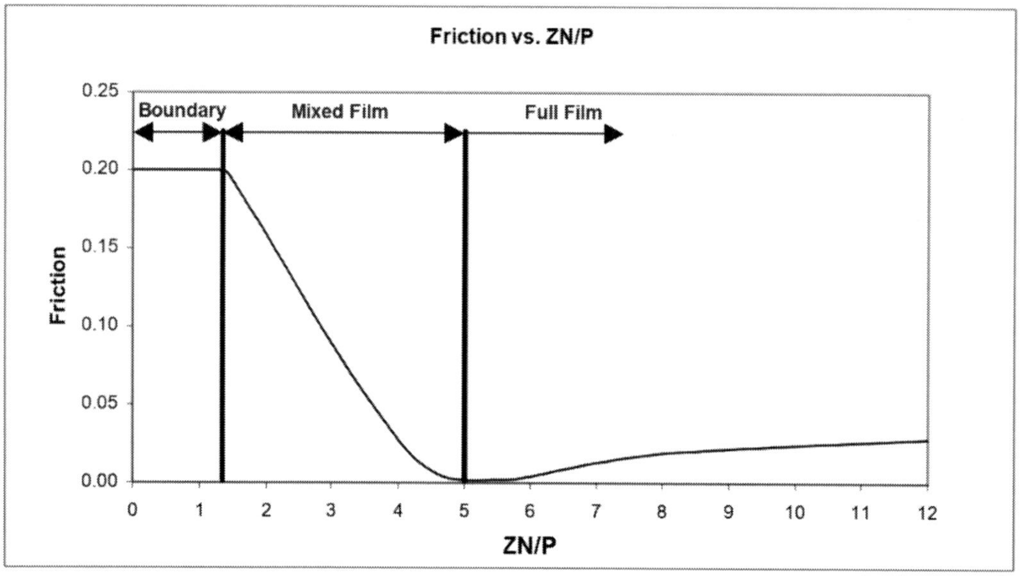

[Figure 2: ZN/P versus Friction Curve}

HOW TO USE THE CHART

The abscissa of the curve is the ZN/P parameter and the vertical axis is friction. The ZN/P number is really a hybrid Sommerfeld number that has been modified to ignore clearance effects and to incorporate units that are a little easier to use. As shown in the Figure (2), the friction is quite high for ZN/P values from 0 to roughly one. This is the boundary lubrication regime. From 1 to 5 the friction drops off quickly as the system moves from boundary lubrication into mixed film, reaching a minimum a little beyond ZN/P = 5. For values above 5 the system is in full film lubrication and once again the friction begins to increase, but at a very manageable rate. This increase in friction is due to the shearing of the lubricant in the film.

The parameters are defined as:

Z = Dynamic Viscosity of Lubricant at Supply Temperature (cp)

N = Rotational Speed (rpm)

P = Specific Load on the Bearing (psi)

P = Journal Load (lbs)/(Bearing Length (in)*Shaft Diameter (in))

This curve can be used in several ways. For an existing system, evaluate the ZN/P parameter to determine if your slow roll speed is in a range sufficient to ensure no metal-to-metal contact. One note of caution: even though the calculation indicates that there is separation of the two surfaces, if there is debris or other particulates in the oil that are equal to or greater than the separation distance, abrasive wear can take place as these particles travel though the separation zone.

To determine if your system slow roll speed is sufficient, solve for the ZN/P parameter:

Z = 20 cp [dynamic viscosity of light turbine oil (ISO VG 32) @ 120 F]

W = 2,000 lbs Shaft OD = 5" Bearing L = 2.5" Slow Roll Speed N = 50 rpm

Calculate specific pressure P = (2000)/(5*2.5) = 160 psi

Solve for ZN/P = (20)*(50)/(160) = 6.25 (full film)

The parameter can also be used to determine an acceptable slow roll speed. Set the ZN/P parameter to a value greater than 5 and solve for slow roll speed, N. Because of the various other influences mentioned previously, use a factor of safety in the calculation to account for factors such as contamination.

Set ZN/P = 10

Z = 28 cp (ISO VG 46 @ 120 F)

W = 10,000 lbs Shaft OD = 10" Bearing L = 5"

Solve for P = 10,000/(10*5) = 200 psi

Solve for N = (10)*(200)/28 = 71 rpm

FINAL NOTES

In some high speed rotating equipment, the bearings are designed for unidirectional rotation. Features such as pressure dams in sleeve bearings or offset pivots in tilt pad journal bearings are designed to operate with a specific direction of rotation. However, in a slow roll condition these bearings are normally not damaged if the shaft rotation should be reversed. These features are designed to influence the stability and dynamic characteristics of the unit at high speeds. Provided there is sufficient rotation to separate the shaft from the bearing during slow roll, unidirectional bearings should not be adversely affected.

1. Neal, M.J., The Tribology Handbook, Butterworth Heinemann, 1997, E7 – E8.

2. Gardner, W.W., "Journal Bearing Operation at Low Sommerfeld Numbers", ASLE Transactions, Vol. 19, No. 3, 1976, pp 187-194.

3. Peterson, Reino A., "Taking the Pressure Out of Sleeve Bearing Design", Power Transmission Design, 1996, pp. 27-32.

4. Gardner, W.W., Hydrodynamic Oil Film Bearings, Fundamentals, Limits and Applications, Waukesha Bearings Corporation.

5. Philipossian, Ara and Olsen, Scott, "Fundamental Tribological and Removal Rate Studies of Inter-Layer Dielectric Chemical Mechanical Planarization", Jpn. J. Appl. Phys., Vol. 32, No. 10, 2003, pp. 6371-6379.

Turboexpander-Compressor with Active Magnetic Bearing Technology In Ethylene Process Plants

Kazim Akhtar (ABB Global)
Chet Farabaugh (S2M)
Reza Agahi (GE Energy Oil & Gas)

INTRODUCTION

The cryogenic turboexpander was introduced into the ethylene process design in the 1970's. The expander is generally utilized in ethylene plants to expand hydrogen offgas from the demethanizer feed separator in the LP demethanizer system or to expand methane offgas from the HP demethanizer overhead. Since the offgas is high on hydrogen content and hence has low molecular weight, expansion is normally achieved in multiple stages, mostly in two stages and in some process applications up to four stages.

The expander loading choice varies depending on the ethylene process and plant requirements. Three loading methods have been utilized in ethylene plants: recompressor, electric generator, and oil brake.

In expander-recompressor configuration, the expanded offgas, after exiting the last exchanger, is recompressed to the required fuel gas pressure. In expander-gear-generator configuration, the expander drives an electric generator, normally an induction generator, through a reducing gear - box. In expander-oil brake configuration, each expander stage is loaded with an oil brake at the other end of the shaft. An oil brake is an inefficient pump that produces more turbulence and less pressure head. Figure 1 shows PFD for an ethylene plant process with expander-compressor.

In spite of expander's great contribution to cold box performance with deep cryogenic temperature production, some ethylene process licensors, engineering companies and end users were reluctant to incorporate turboexpander applications due to risk of oil contamination.

Turboexpander suppliers have and continue to put considerable efforts into improving sealing design, lubrication systems and monitoring systems to address the concerns of oil mist migration into the cryogenic process. The nature of ethylene process and how the expander is integrated in it, which is out their control, offsets the expander supplier improvements. This is for most part due to multiple stage expansion. Upon any expander shutdown, process related or expander related, piping and vessels between the first and last stage expanders approach an initial settle- out pressure in a shorter time than the expander lubrication and sealing systems could react and adjust to the lower settle out pressure level. As a result, oil mist migrates into the cryogenic process and causes a major headache due to freezing in gas – gas exchangers.

Table 1 depicts the expander process condition for an ethylene plant in the UK with an expander-integral gear-induction generator. Table 2 shows the expander process for an ethylene plant in Qatar with an expander-oil brake. Table 3 presents the expander process conditions for an ethylene plant in Malaysia with an expander-compressor.

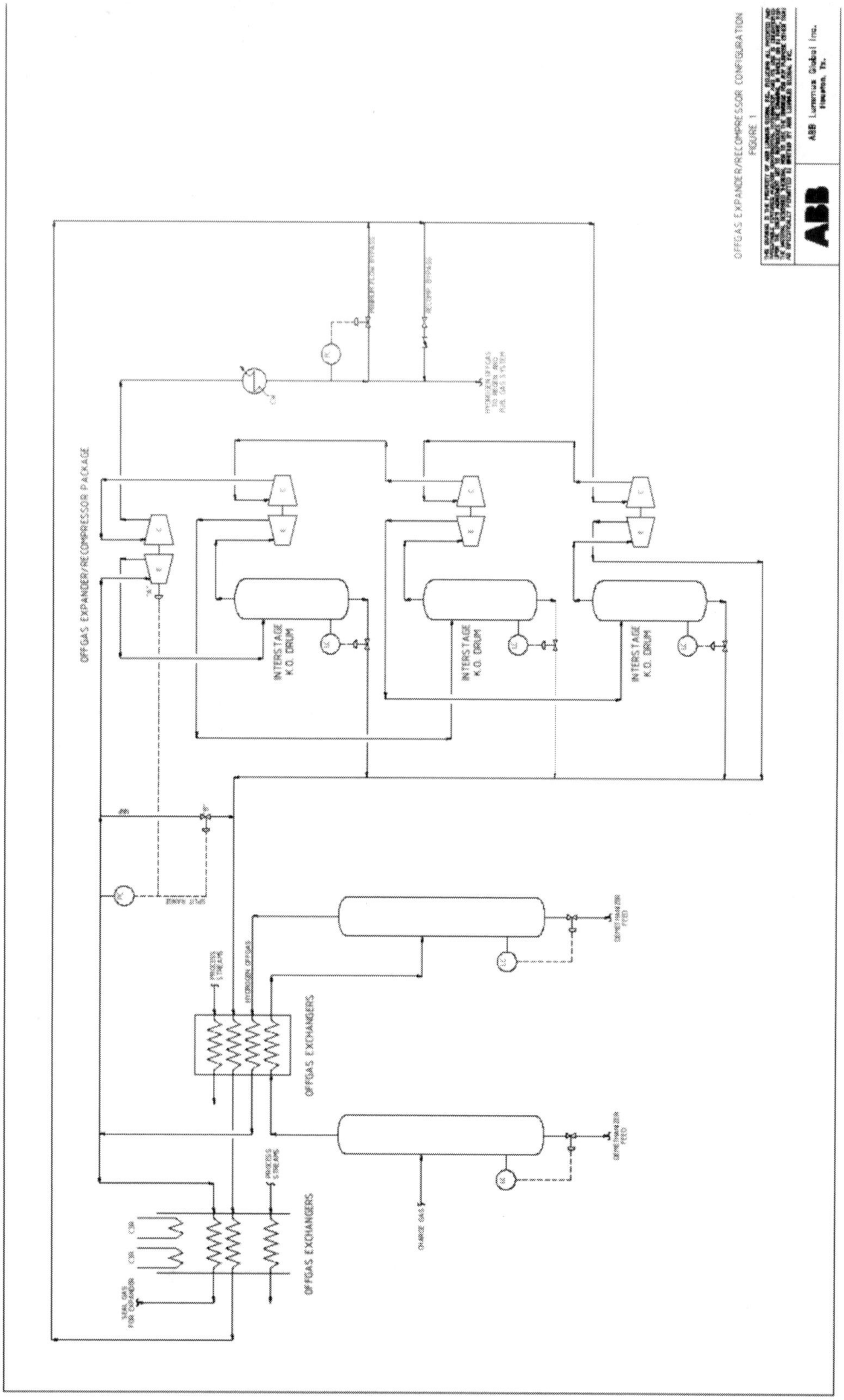

Figure 1. PFD of an Ethylene Plant with Expander-Compressor

Table 1

	Expander 1	Expander 2	Expander 3	Expander 4
Molec. Wt.	3.73	3.73	3.73	3.73
Press. Inlet, kPa A	3,000	2,100	1,500	1000
Temp. Inlet C	-105	-118	-130	-142
Press. Out kPa A	2,100	1,500	1,000	675
Temp. Out C	-118	-130	-142	-153
Speed rpm	24,000	24,000	24,000	24,000
Power KW	270	175	190	180

Table 2

	Expander 1	Expander 2	Expander 3
Molec. Wt.	3.56	3.56	3.56
Press. Inlet, Bar A	33	23	15
Temp. Inlet C	-96	-111	-128
Press. Out Bar A	23	15	9
Temp. Out C	-111	-128	-145
Speed rpm	42,000	42,000	42,000
Power KW	320	335	355

Table 3

	Exp. 1	Comp. 4	Exp. 2	Comp. 3	Exp. 3	Comp. 2	Exp. 4	Comp. 1
Molec. Wt.	2.88	2.88	2.88	2.88	2.88	2.88	2.88	2.88
Press. Inlet, Bar A	35	6.0	22	5.0	13.5	4.5	8	4.0
Temp. Inlet C	-105	60	-120	50	-135	37	-150	25
Press. Out Bar A	22	6.0	13.5	5.5	8	5.0	4.5	4.5
Temp. Out C	--120	70	--135	60	--150	50	--166	37
Speed Rpm	44,000	44,000	44,000	44,000	44,000	44,000	44,000	44,000
Power KW	160	130	160	130	160	130	160	130

To address oil mist migration concerns, turboexpander vendors initially embarked on the design of turboexpanders with dry gas seals. Later on, to eliminate lubricating oil all together, active magnetic bearing was adapted for expander–compressors in ethylene plants.

TURBOEXPANDER WITH DRY GAS SEAL

An ethylene plant owner requested an expander vendor in late 1980's to re-design multiple stages of their expanders with dry gas seal. That project, is believed to be the first expander to be designed with dry gas seal. The expander vendor and the dry gas seal vendor were both facing a major challenge. The former had the challenge of squeezing dry gas seal in a tight space and maintain an acceptable rotor dynamics behavior. The latter had the challenge of designing for cryogenic temperature and mechanical integrity at high speed.

The outcome of that project was a successful retrofit and a beginning for expander with dry gas seal in ethylene plants. Figure 2 shows cross section of an expander –integral gear with dry gas seal. Figure 3 depicts cross section of an expander-oil brake with dry gas seal.

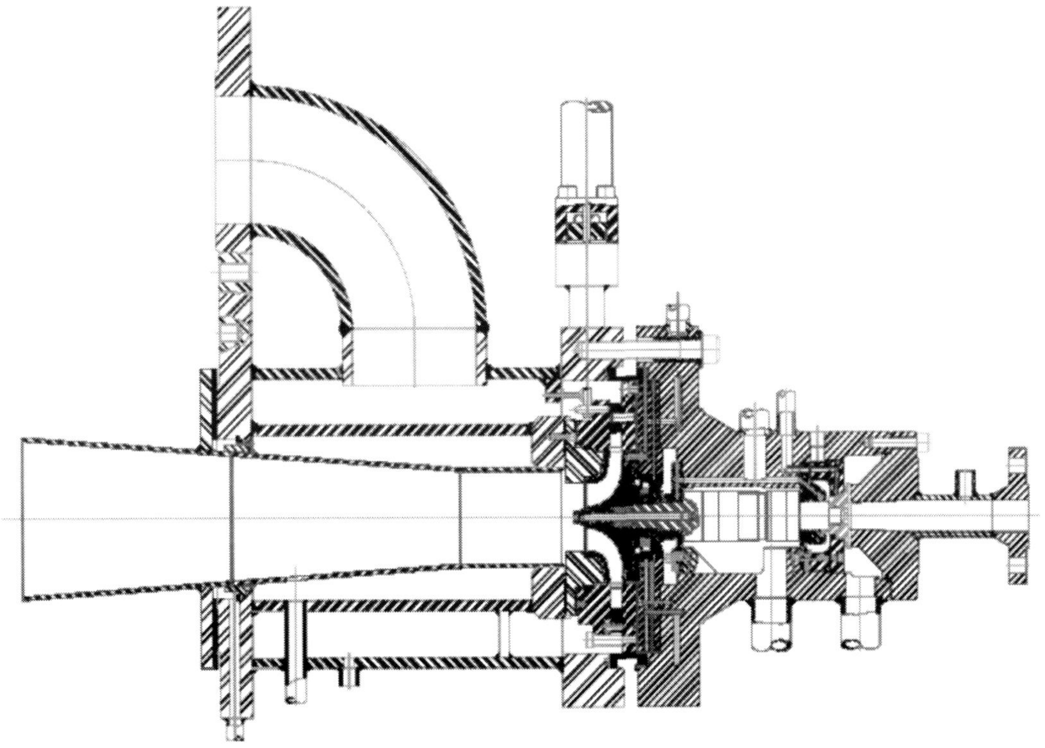

Figure 2. Cross Section of Expander-Oil Brake with Dry Gas Seal

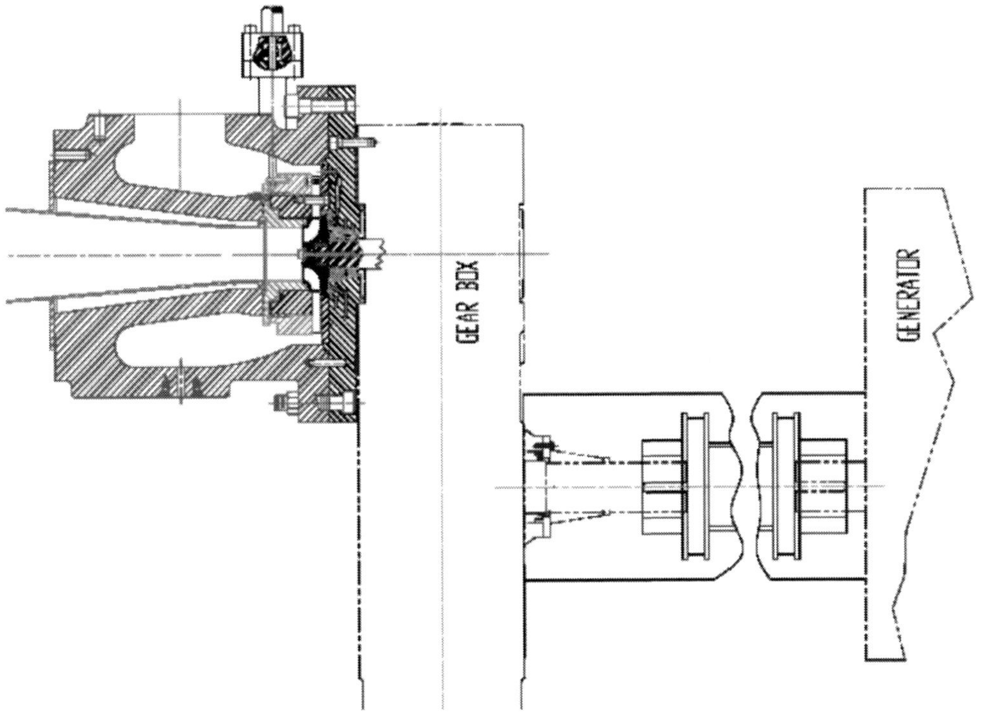

Figure 3. Cross Section of Expander-Integral Gear with Dry Gas Seal

TURBOEXPANDER WITH ACTIVE MAGNETIC BEARINGS

Magnetic bearing technology that has been around since 1980's in industrial and oil and gas applications was not widely utilized in ethylene plants until the turn of the century. The main reason, in the authors' opinion, was the fear of unknown and new technology. No plant wants to be the first ethylene plant to install turboexpander-compressor with active magnetic bearings.

In the late 1990's, expander-compressors with active magnetic bearings and digital controller technology were in operation in several plants. The operators were pleased with performance and reliability of the expanders. As a result, expander-compressors with active magnetic bearings became the only option for ethylene plants. Since the turn of century, more than six ethylene projects, grass root or expansion, with expander-compressor have adopted expander-compressors with active magnetic bearings.

The gas dynamic design of a turboexpander-compressor with active magnetic bearing is the same as the one with oil bearings. One exception is more power to boost is available, because there are fewer friction losses and hence compressor discharge pressure is slightly increased. The expander and compressor casings are also the same with the exception that their interface with bearing housing should be modified. The bearing housing for AMB is larger in diameter. The bearing housing should accommodate the shaft with laminations, AMB cartridges, power and signal cables and connectors and feed – through seals. Figure 4 shows cross section of a turboexpander – compressor with AMB.
Table 4 shows process conditions for an expander-compressor with AMB.

ACTIVE MAGNETIC BEARINGS AND ITS CONTROL SYSTEM

In addition to the mechanical hardware described above, magnetic bearings utilize an electronic control system that provides closed loop position control to keep the shaft levitated. The complete magnetic bearing system is shown in Figure 5. The bearings are made up of electromagnets that work by force of attraction to levitate the shaft in an air gap. Variable inductive position sensors monitor the position of the shaft and provide feedback to the control system. The position signal is compared to a reference (i.e. center position). If the shaft is not centered, an error signal is processed in a Proportional Integral Derivative (PID) control. The PID control provides a correction signal to the power amplifiers, which in turn provide correction currents to the electromagnets. The electromagnetic actuators provide the restoring force to re-center the shaft.

Table 4

	Exp. 1	Comp. 2	Exp. 2	Comp. 1
Molec. Wt.	4.57	26.11	4.27	26.11
Press. Inlet, Bar A	25	7	14	3.4
Temp. Inlet C	-116	9	-130	-40
Press. Out Bar A	14	13	7.5	7
Temp. Out C	--130	54	--148	9
Speed rpm	29,500	29,500	29,500	29,500
Power KW	940	930	940	930

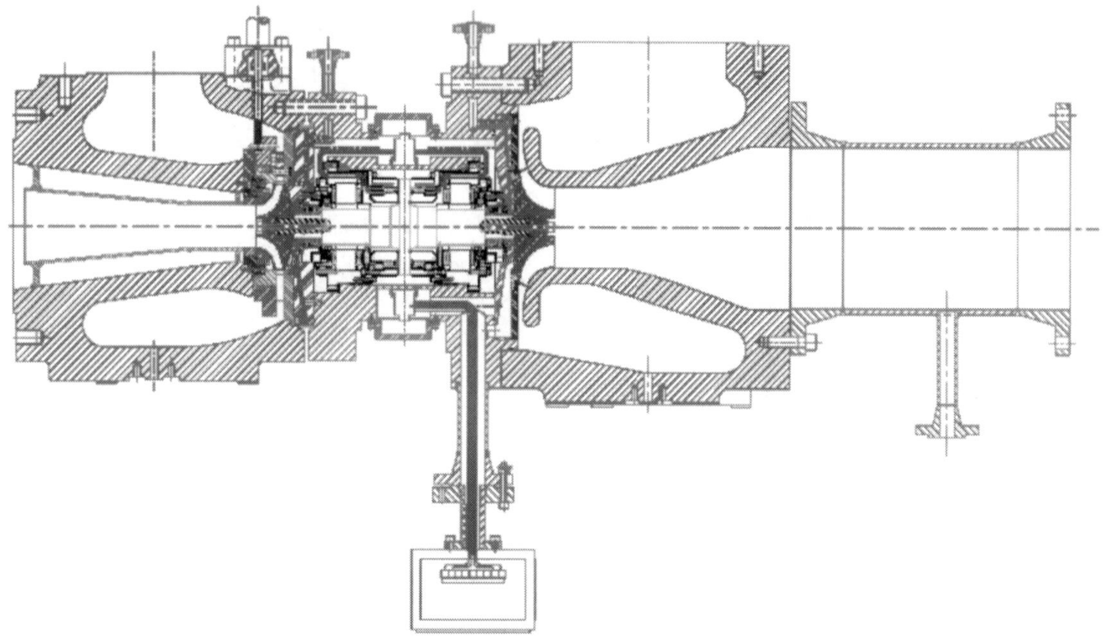

Figure 4. Cross Section of Expander-Compressor with AMB

Early magnetic bearing controls were analog systems with multiple printed circuit boards. Since 1997, digital control systems have been used where the entire PID control for 5 axes (4 radial axes and 1 axial) is done on one microprocessor. The major advantage of moving from analog to digital control is the dramatic reduction of components, which results in a higher mean time between failure (MTBF). Today's digital control systems have an actual MTBF of more than 40,000 hours or approximately 5 years continuous operation.

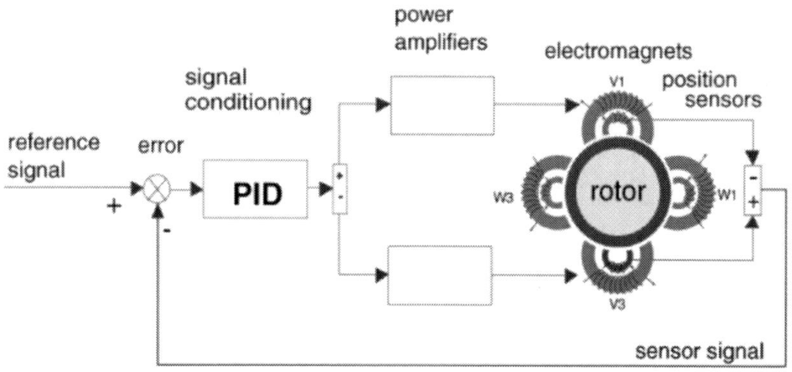

Figure 5. Magnetic Bearing System

The PID control establishes the bearing characteristics by modifying the error signal with phase and gain. Phase is synonymous with damping and gain is synonymous with stiffness. The bearing characteristics are set to provide a stable rotordynamic response over the turboexpander operating speed range, meeting acceptable vibration criteria specified by API 617. Typically, a rotordynamic analysis is completed to predetermine the bearing characteristics. If the rotor model is accurate, no further adjustment or "tuning" is required. With the digital control system, any further adjustment can be made using a laptop computer interface.

Figure 6 shows an E300/30 (300 VDC, 30 amps) digital control system. These control systems are designed to provide remote operation of the turboexpander. Typically, the controller is placed in a non-hazardous location. The cable length for the bearing power and signal wires can be up to 300 meters between the controller and the machine. The user can operate the control system either locally (at the cabinet) or remotely through a MODBUS serial link. An added advantage of the digital control system is the machine operating information that is available for monitoring, trending, diagnostics and troubleshooting. This operational data includes vibration, bearing loads, shaft unbalance, and bearing temperatures. The control system also includes a battery backup (UPS) in order to maintain levitation in the event of an AC power failure.

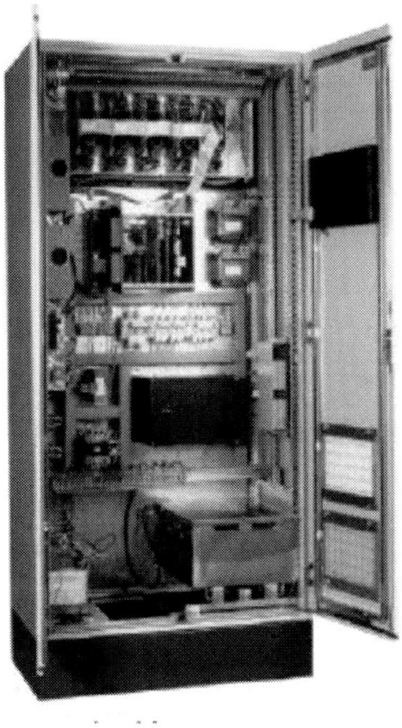

Figure 6. E300/30 Magnetic Bearing Control System

An advantage of the magnetic bearing system is the ability to run inside the process gas environment. This eliminates the need for dry gas seals that add cost and can adversely affect rotordynamics by increasing rotor length. The bearing hardware is made up of laminations on the stator and rotor with coils wrapped around the stator poles to form the electromagnets. This hardware is very robust (similar to electric motors) and can be supplied with a liquid- proof technology that allows the bearing to recover from being submersed in liquid.

Another important feature of the magnetic bearing is that auxiliary bearings provide safe coastdown of the machine in the event of a failure of the control system or an overload of the bearings. Figure 7 shows a cross- section of a radial magnetic bearing including the auxiliary bearings. The auxiliary bearings are angular contact ball bearings that support both radial and axial loads during coastdown.

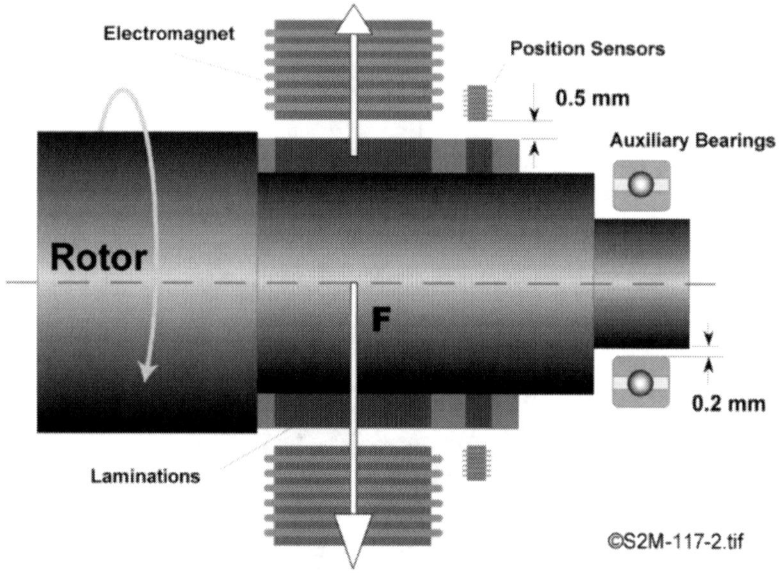

Figure 7. Radial Magnetic Bearing and Auxiliary Bearing

Note that the auxiliary bearing clearance is set at about half of the magnetic bearing clearance. Under normal operation the shaft is levitated and the auxiliary bearing does not rotate, but in the event of a failure the auxiliary bearing provides a safe coastdown of the machine, preventing contact between the magnetic bearing rotor and stator. The auxiliary bearings can provide 3 to 5 full speed, full load coastdowns of the turboexpander before being replaced. This is sufficient for the life of most machines considering a normal MTBF of 40,000 hours for the control system.

CONCLUSION

Ethylene plants owners and EPCs were the main forces behind development of the turboexpander with dry gas seal and active magnetic bearings, which has become the choice for expander-generator and expander-oil brake applications.

On the other hand, turboexpanders have become an ideal turbomachinery application for magnetic bearings due to some specific advantages offered by the technology. This is especially true for turboexpander-compressors in ethylene plants where almost every new unit installed over the last 5 years has been with AMB. Today, there are more than 300 large turbomachines in operation with magnetic bearings and about one half of these machines are turboexpander-compressors.

Mercury Contamination in Ethylene Plants: An Overview

Feedstock & Product Contaminant Issues Subcommittee

Prepared for Presentation at the 2005 Spring National Meeting

Atlanta, GA, April 10-14, 2005

UNPUBLISHED

April 11, 2005

Mercury exists in nature as a trace component in all types of fossil fuels, including coal, petroleum and natural gas. Since ethylene plant feedstocks fall in this category, mercury represents a potential contaminant. The presence of mercury in ethylene plants increased significantly with the onset of gas field condensate cracking in the early 1980s, in particular condensate from fields in Algeria. One of its manifestations was a significant reduction in cycle lengths of the C3 hydrogenation reactors. A paper by Dr. Raghu Menon, presented in this session (1), describes the impact of mercury on an ethylene plant's catalyst systems.

A second impact of mercury in ethylene plants is attack on brazed aluminum core exchangers. Liquid metal embrittlement, involving diffusion of liquid mercury into grain boundaries on an atomic level, can lead to equipment failure accompanied by uncontrolled release of hydrocarbons. A liquid metal embrittlement failure of an aluminum alloy pipe in the cryogenic section of Cain Chemical's Chocolate Bayou ethylene plant in May 1987 was the topic of a paper at the first Ethylene Producers' Conference (2). A paper by Ronald Bell, presented in this session (3), describes mercury-related equipment failures and the underlying mechanisms, and how such failures might be avoided.

In addition to the potential for damage to heat transfer equipment and catalyst systems, mercury presents an environmental challenge to plants processing feeds containing mercury. While mercury entering the ethylene plant may be in elemental, inorganic (salts, such as mercuric chloride) or organic forms, it is converted predominantly into the elemental form in the cracking furnaces and can accumulate in equipment, particularly in the cryogenic portion of a plant. During plant shutdowns, where equipment is opened for inspection, the potential exists for exposure to liquid mercury, as well as, to levels of mercury vapor in confined spaces in excess of threshold limit values. There is also the potential for accumulation of mercury-containing solids in feedstock storage tanks, and for mercuric salts to enter into water streams in desalters treating plant feeds, or into water that may phase out in storage tanks. A paper by Mark Brayden, James Graham and Mike Potter, presented in this session (4), addresses the challenges of detecting and measuring mercury at low concentrations. Another paper by Steve Coleman, George Nowowiejski and Marilyn Blaschke, also presented in this session (5), describes the various technologies available to ethylene producers for protecting their plants and the environment from mercury.

The feedstock subcommittee hopes that this session will increase understanding within the ethylene community of the hazards mercury can pose to ethylene plants, provide some insight regarding the detection and measurement of mercury and suggest some removal technologies for producers whose feedslate includes mercury-contaminated hydrocarbons.

There are several references which the subcommittee recommends to the reader interested in a more in-depth understanding of mercury. The EPA R&D report titled "Mercury in Petroleum and Natural Gas: Estimation of Emissions from Production, Processing and Combustion" (6) is an excellent "tutorial" on mercury and its impact on the environment and includes much information on naturally occurring mercury species and their disposition in refining and gas processing operations. Encyclopedias of chemical technology, such as Kirk-Othmer or Ullmann's, provide a general overview of mercury. And finally, the subcommittee commends to your reading the various reports presented in the Proceedings of the Ethylene Producers' Conference (7), including results of feedstock contaminant surveys by the subcommittee in 1992 and again in 2004 which include mercury as a contaminant of concern to ethylene producers.

References:

1. Menon, R. K., *Mercury and its Impact on Ethylene Purification Catalysts*, Proceedings of the 17th Ethylene Producers' Conference, Atlanta, GA., April 11, 2005.
2. English, J. J., *A Mercury Induced Aluminum Alloy Piping Failure in an Ethylene Manufacturing Plant*, Proceedings of the First Ethylene Producers' Conference, Houston, TX., April 4, 1989
3. Bell, R. N., *Understanding and Preventing Failure of Aluminum Equipment in the Presence of Liquid Mercury*, Proceedings of the 17th Ethylene Producers' Conference, Atlanta, GA., April 11, 2005.
4. Brayden, M., Graham, J. and Potter, M., *Literature Survey of Analytical Methods for Mercury in Matrices of Interest to Ethylene Plants*, Proceedings of the 17th Ethylene Producers' Conference, Atlanta, GA., April 11, 2005.
5. Coleman, S., Nowowiejski, G. and Blaschke, M., *Mercury Removal – An Overview*, Proceedings of the 17th Ethylene Producers' Conference, Atlanta, GA., April 11, 2005.
6. Wilhelm, S. M., *Mercury in Petroleum and Natural Gas: Estimation of Emissions from Production, Processing and Combustion*, Prepared for the EPA Office of Air Quality Planning and Standards by National Risk Management Research Laboratory, Research Triangle Park, NC., September, 2001.
7. Proceedings of the Ethylene Producers' Conferences, http://epc.omnibooksonline.com/

Paper 35b

Literature Survey of Analytical Methods For Mercury in Matrices of Interest to Ethylene Plants

Mark Brayden

Dow Chemical

James Graham

ExxonMobil

Mike Potter

Shell Chemical LP

Prepared for Presentation at the 2005 Spring National Meeting

Atlanta, GA, April 10-14, 2005

Feedstock Contaminants in Ethylene Plants - Mercury

April 11, 2005

Abstract:

The charter of the American Institute of Chemical Engineers' (AIChE) Ethylene Producers Committee (EPC) calls for the communication of non-confidential information concerning the manufacture and distribution of ethylene. In support of this objective, the EPC's Feedstock and Product Issues Subcommittee commissioned a report that was presented at the 1996 Ethylene Producers' Conference that dealt with the analytical measurement of trace mercury contaminants in ethylene plant feed and product streams. Mercury contamination continues to pose a variety of problems to the ethylene producer, including threats to safety, health and the environment. Hopefully, the dissemination of these test methods will prove valuable in alleviating such problems and, in doing so, contribute to the common good and well-being of the ethylene industry. The purpose of this report is to include new and up-to-date information on mercury measurement methods, especially those that can be used by its members.

Disclaimer:

The purpose of this report is to provide the Ethylene Producers Committee of the AIChE the most recent information on mercury analyses to enhance the safety, environmental quality and industrial hygiene in the ethylene industry. This report is a literature survey of mercury methods, which is meant to enhance and further the 1996 paper by this sub-committee. This report is not comprehensive, nor does it purport to offer the best possible mercury analytical methods. It is only meant as a starting point for its members that are interested in conducting mercury analyses. Therefore, neither the authors, the AIChE or its directors, officers, employees or members, committees or subcommittees, their employer's officers or directors warrant or represent expressly or implied, the correctness or accuracy of the content of information presented in this document, nor can they or will they accept any liability or responsibility whatsoever for the consequences of its use or misuse by anyone.

Several of the papers cited in this report deal with the handling of mercury compounds, (e.g., dimethylmercury). Before any work with any mercury compound, the industrial hygiene and safe handling of these compounds must be thoroughly investigated, and only highly trained individuals should attempt working with these materials.

Scope:

The mission of the authors of this paper was to collect information on analytical techniques used for mercury analysis, with the primary intent to document analytical improvements since the subcommittee's last report in 1996 (1). New information from the EPC's members is limited; therefore, it was decided a literature survey would be more beneficial. While this is the primary source of information for this paper, the authors also received information from those that perform mercury analyses, as well as information from instrument suppliers, mainly from their sales literature.

While there are many review articles on mercury, the authors concentrated on its measurement in streams that are of primary interest to ethylene production plants.

The sensitivities of the detection techniques have not significantly changed since the last report, but there have been significant improvements in detector design, sample preparation, sample collection, and speciation, especially in organic samples like naphtha and condensate. These improvements have led to significant improvements in lowering the limit of detection for total mercury and individual mercury compounds, as well as identifying the actual mercury compounds in products like naphtha and gas condensate.

Background:

The primary concern in ethylene plants has been the potential impact of mercury on aluminum heat exchangers and hydrogenation catalysts, as well as industrial hygiene issues when performing maintenance on vessels and lines that have been exposed to mercury-containing streams. Secondarily, issues with disposal of mercury-containing sludges, spent adsorbent materials, and wastewater have also required the need of accurate mercury analyses. More recently, there has been concern about mercury emissions to the environment, specifically via the burning of mercury-containing fuels. While the primary interest is coal-fueled plant emissions, there is also mercury has also been reported in natural gas from various sources.

Complicating the issue is the need for lower limits of detection. Measurement of total mercury in a water stream has been accomplished to very low levels (sub-ppt); however, the measurement in organic matrices requires more sophisticated techniques to reach lower limits of detection. Also, in order to identify proper removal or handling methods of streams containing mercury, the quantifications of the potential mercury compounds, like elemental mercury, inorganic mercury, non-polar dialkyl or diaryl mercury compounds, and polar mono-alkyl mercury compounds, are also important.

The cost of the analytical instrumentation is another factor to consider. Depending on the need, the cost can be relatively inexpensive to very expensive. For surveying equipment for industrial hygiene needs, relatively inexpensive instrumentation like the Jerome analyzer is still appropriate. For instance, the TLV for mercury is 0.025 mg/Nm3, and the instrument is suitable down to 0.001 mg/Nm3. For a more demanding analysis like speciation of mercury compounds in condensate, a more complex, and more expensive, analytical system is needed. For instance, to reach sub-ppb levels may require a system composed of a gas chromatograph (GC), a post column treatment system, and a highly sensitive detector like a cold vapor atomic fluorescence spectrometer (CV-AFS). In addition to the higher cost, this also requires a highly skilled analyst. In some cases, a company may already have a sophisticated analyzer, such as an Inductively Coupled Plasma – Atomic Emission Spectrometer (ICP-AES), available. In this case, depending on the limit of detection needed, the instrumentation may be used as is, or if a lower limit of detection or speciation is needed, modification may be required. Modifications may include an improved nebulizer for an ICP or interfacing a GC, or it may simply require pre-concentration of the mercury via wet chemical extraction methods.

Several papers were reviewed for this report and may be divided into areas of interest, namely:

1) Aqueous
2) Solid
3) Organic
4) Vapor

Furthermore, with each analysis type, the analysis may be sub-divided further as a function of the type of analysis needed, namely:

1) Total mercury analysis
2) Speciation of mercury compounds

Sampling:

One of the most important issues to obtain accurate results is sampling. This is dependent on the sample matrix and can be successfully accomplished with planning.

For aqueous samples, the EPA recommends collecting the samples in bottles made of fluoropolymer. In addition, preservation of the sample with an oxidizing agent or acid can fix the sample until it is analyzed (26).

For organic liquid samples, such as naphthas and gas condensates, one study showed no loss of measured mercury after 28 days for samples stored in glass or tin. However, to prevent potential amalgamation of elemental mercury, the authors recommended storage in glass containers with Teflon-lined caps (26). In another study (15), storage experiments showed that elemental and inorganic mercury were highly unstable in most shipping containers, and the author concluded that speciation be conducted at the point of origin of the sample, or at worst delivered overnight to the laboratory in a liquid-full glass container for next-day analysis.

For vapor samples the primary concern is loss of analyte on sample lines and sample containers. To minimize this samples are routinely collected on traps (2), or scrubbed into solutions (9). Additionally, in-line sampling is recommended (9,20). Furthermore, to prevent errors, a recommendation is to minimize the length of sample line to the collector (2), or to apply an on-line analyzer (20). Furthermore, one author states that the adsorption of mercury onto various surfaces is an equilibrium process; to reach a steady-state for glass may take several hours, but for metal surfaces the equilibration may take several days (2). There are also silicon-dioxide treatments of stainless steel that are commercially available that aid in achieving representative samples. The adsorption of analyte onto the sampling vessel walls is further minimized by flowing a sufficient volume through the adsorbent trap, and most representative sampling can be accomplished by sampling at process pressure (6).

Mercury in Aqueous Samples:

The source of the sample can be from several locations, and often the mercury analysis is needed for waste disposal. These samples may be part of a sludge or water layer from a storage tank, or wastewater from the production unit. The analysis may also be needed for a mercury balance or identification of proper treatment methods.

The detection limit needed can be quite low for environmental samples. For instance, the current U. S. EPA guidelines have set water-quality criteria for mercury at 0.77 µg/l for freshwater and 0.94 µg/l for saltwater (20). In some states, though, the older U.S. EPA guidelines are still being used, which set the water-quality criteria at 12 ppt for freshwater and 25 ppt for saltwater. Many of the techniques like ICP-AES, Atomic Adsorption Spectrometry (AAS), and ICP-Mass Spectrometry (ICP-MS) can not reach these levels directly (1,26). To reach these levels special preparation techniques are used. For instance, U.S. EPA Method 1631, Revision E, can measure Hg in the range of 0.5-100 ppt by proper sample handling, sample preparation, and pre-concentration using a gold-surface trap. With modification, and no interferences, the method can be adapted to go to a minimum detection limit of 0.05 ppt. Detection is accomplished with a CVAFS (25). This is significantly improved performance over that previously reported (1). Similarly, EPA Method 245.7 can also reach low detection limits (~ 1 ppt). It also uses CVAFS but uses a different technique to isolate the Hg for analysis (26). Perhaps the simplest CVAFS analysis is the European method EN 13506, since it can reach the same detection limits as EPA Method 1631 but does not require two analytical steps (26).

In summary, low detection limits in water can be achieved using tested methods used with CVAFS detectors. CVAFS instrumentation is available at a moderate cost, have gained widespread acceptance, and thus is the detector of choice, especially for environmental samples. For non-environmental samples, total mercury analysis can be accomplished via other detectors, such as AAS or ICP-MS. The limit of detection of these techniques, then, becomes the most important parameter.

Mercury in Liquid Organic Samples:

In an ethylene plant, the samples to be analyzed are from two sources: 1) feedstocks, and 2) products.

Mercury in liquid feedstocks can be in various forms, including:

Elemental mercury (Hg^0)
Inorganic mercury (Hg^{+2})
Dialkyl or diaryl mercury (R_2Hg)
Monoalkyl mercury (RHg^+)

Depending on the level of quantification needed, the analysis can become quite complicated. For instance, if the goal is to speciate the mercury compounds in a naphtha sample and to quantitate these to low limits of detection, the analysis becomes complicated and requires a hyphenated technique, such as ICP-MS coupled to a GC, which is not typically found in an industrial laboratory.

Compared to the analysis of water samples, the direct introduction of an organic may offer processing problems with the detection instrument. For instance, organic added to a plasma may increase the background, interfere with the analyte emission, or may even quench the plasma (3). The net result can be a higher limit of detection compared to water analysis, or may require special handling or instrument interface.

Mercury in feedstocks can be in several forms, but is primarily inorganic and elemental, and only trace amounts of ionic organic mercury compounds have been measured (5,8). In natural gas, the dominant species is elemental mercury, and occasionally a trace amount of dimethlmercury, while in gas condensate, the dominant species are elemental and inorganic compounds with trace levels of organic mercury compounds (18). Concerns about mercury in the ethylene plant products deal primarily with the elemental form (27).

Mercury in Solid Samples:

Various techniques have been used for solid sampling, and depending on the needed limit of detection, there are several suitable alternatives. In general, solid samples are analyzed for disposal, and a very low limit of detection is not needed. Techniques such as X-ray fluorescence and neutron activation can be conducted on the solid samples. Other techniques, such as AAS, require dissolving the sample first. The limitation and applicability of these techniques were discussed in our previous report (1).

Mercury in Vapor Samples:

In general, the best methods are those that measure mercury directly via on-line analysis, or capture via trap or scrub the mercury compounds into an extraction solvent for off-line analysis in the laboratory. Furthermore, the success is enhanced by using in-line sampling devices with inert surfaces, or at the least well-passivated surfaces. Also, if metal tubing is used, the length of tubing should be minimized, as discussed earlier.

In general, most vapor samples in an ethylene plant contain predominantly elemental mercury, so speciation is not a major concern. For stack gas analysis, speciation may be desirable. This is accomplished by scrubbing into solutions suitable for elemental and inorganic mercury, the major components of flue gases.

Literature Survey Summary:

A review of literature since 1996 was conducted, and while this is not a comprehensive review, it should be of value for those that are investigating mercury measurement methods, especially in an ethylene plant support laboratory. In the summaries that follow, often the limit of detection, or minimum detection limit, is described for a method. Often, though, no mention of the recovery or accuracy is given. In general, if not mentioned, the recovery will be close to 100 %, usually within 10 %, that is, in the range of 90-110 % recovery. Some papers did not conduct recovery studies. For the details consult the literature cited. The literature is reviewed in approximate chronological order to give a sense of the evolution of the mercury measurement methods.

McNamara (2) discusses the use of a Jerome gold foil analyzer for the measurement of total mercury in natural gas. A low limit of detection can be obtained (< 0.01 μg/m^3) because the analyte is removed from the natural gas by passing the sample stream over a gold-sand trap. Another reason for off-line trapping and a reason not to directly analyze with the Jerome analyzer are the interference due to hydrogen sulfide, which gives a false positive on the analyzer but is not an issue when a trap is used since it is converted to a non-interfering compound, sulfur dioxide.

Several papers by Snell, Frech and colleagues (3,7,13,14,25) deal with the speciation of mercury compounds, especially in natural gas condensate, and chronicle attempts to lower the limits of detection of individual compounds like dimethylmercury by using multiple, interfaced techniques, such as GC-ICP-MS. In the first paper reviewed (3), Snell, Frech, and Thomassen quantify the mercury compounds in a condensate sample, probably from Norway, using GC separation followed by trapping onto an Au-Pt trap, and subsequent thermal desorption and analysis by microwave induced plasma – atomic emission spectrometry (MIP-AES). The post-GC treatment allowed the authors to eliminate the background interference of the organic onto the MIP-AES. Direct analysis of the GC column effluent yields a higher limit of detection due to two factors: 1) carbon compounds give spectral emissions in the same region as mercury and 2) carbon compounds overload the plasma discharge, which results in reduced excitation capability. With the improved technique, dimethylmercury could be analyzed to a detection limit of 0.24 μg/l, and derivatized ionic species, monomethylmercury and inorganic mercury, could be detected down to 0.56 μg/l. This technique was compared to solid-phase microextraction (SPME) followed by dilution, and then direct analysis on the GC-MIP-AES for dimethylmercury and for derivatized samples for ionic mercury compounds. These yielded higher limits of detection: 20, 3.1, and 2.3 μg/l for dimethylmercury, monomethylmercury, and inorganic mercury, respectively.

Olsen et al. (4) discuss the use of ICP-MS for the analysis of total mercury. The analysts increased sensitivity with the application of a new type of nebulizer. While the team reported increased sensitivity, the use of the new nebulizer led to loss of highly volatile mercury compounds, like dimethylmercury. This loss of volatile mercury was not observed with the standard nebulizer. The LOD for Hg in naphtha samples was 0.13 ng/g for the standard nebulizer, and 0.02 ng/g for the ultrasonic nebulizer. The paper also discusses memory effects on the nebulizers and matrix effects, which impact the ultrasonic nebulizer to a greater extent than the standard nebulizer.

Zettlitzer et al. (5) speciated mercury compounds in North German condensate and formation brines. Various pre-treatment, pre-concentration techniques were employed before analysis, most often by CVAAS. Total mercury, for instance, was analyzed by aqua regia digestion, followed by reduction with stannous chloride and analysis by CVAAS, with a limit of detection of 0.1 mg/l. Other components required more elaborate pre-treatment. Only the non-polar compounds, like dimethylmercury, were not analyzed by CVAAS, but by GC-MS. For gas condensate the detection limits were 0.1 mg/l for total mercury, 0.2 μg/l for methylmercury, 0.4 μg/l for ethylmercury, 0.5 μg/l for phenylmercury, 0.1 mg/l for elemental mercury and inorganic mercury, and 1 mg/l for the dimethylmercury, diethylmercury, and diphenylmercury. For brine, low LOD's were demonstrated for the ionic organic compounds, namely, 0.001 µg/l for methylmercury, ethylmercury, and phenylmercury.

Müssig and Rothman (6) paid particular attention to sampling, trapping the analyte by directing natural gas samples through a silica gel tube treated with iodine for total mercury analysis. After sampling, the sample tube is extracted in the laboratory with an aqueous ammonium iodide/iodine solution, which is analyzed by flameless atomic adsorption spectrometry (AAS). This method is purported to be more accurate and repeatable for high mercury concentration (0.1 µg/m^3 to several mg/m^3), especially at higher process pressure of up to 45 MPa. The amalgamation technique is reported to be limited to lower mercury concentration (1 ng/m^3 to 3 µg/m^3) at lower pressure. The authors also discuss the analysis of mercury in solid samples, such as steel or inorganic soil (sand), using acid dissolution followed by AAS. For organic solids a more elaborate Wickbold furnace digestion followed by extraction is employed, again followed by AAS.

In (7), Frech, Snell, and Sturgeon further explored the direct introduction of samples into a plasma. The stated aim of this project was to compare the performance of GC-furnace atomization plasma emission spectroscopy (FAPES) to the performance achieved in their previous paper (3). The desired goal was the direct analysis of a GC effluent with increased sensitivity, based upon this improved design detector over the MIP-AES used in the previous study. Again, derivatization was used for ionic mercury compounds. In this study, the limits of detection for the ionic mercury compounds were comparable for monomethylmercury but a factor of 3 worse for inorganic mercury. However, for dimethylmercury, only the FAPES system could perform the measurement, as the direct injection of the sample onto the MIP led to significant interference from the organic matrix.

Tao et al. (8) used GC-ICP-MS for the analysis of mercury compounds in natural gas condensate. By pre-treating the GC column with HBr, derivitization was not necessary to achieve sharply eluted peaks. While the limits of detection for various compounds was acceptable, allowing quantification at the sub-ppb level, the recoveries for elemental mercury, mercuric chloride, and dimethylmercury deviated significantly from 100 %.

McManus et al. (9) reported on the use of the Ontario-Hydro method for the sampling and analysis of natural gas flue gas. This method employs an in-line sampling device to extract sample from the flue, with subsequent scrubbing with two different solutions to differentiate between oxidized and elemental mercury. Detection is by AAS. The aim of this work is primarily for coal-fired processes, but it can also be used for ethylene plant systems.

Shiowatana and his colleagues developed methods for total mercury in natural gas liquid and natural gas condensate (10). This group sought a method that would achieve a low limit of detection, while using available equipment. In order to overcome the limitation of electrothermal AAS (ETAAS), pre-concentration of the mercury compounds was accomplished by adsorption onto activated carbon. This improved the LOD to 2.0 ng/ml, with good repeatability, with the potential for lower LOD by increasing the concentration factor. For the Thailand liquid samples surveyed in this study, the achieved LOD was more than adequate to determine the total mercury content.

Shafawi et al. (11) report on the adaptation of the standard gas technique of gold-surface trap with subsequent analyses via CVAFS for total mercury in gas condensate. Seven different non-polar and ionic organic mercury compounds were tested in recovery studies with acceptable recoveries, ranging from 88 – 97 % in gas condensate. The technique vaporizes the sample at 400 °C, with collection on the trap at 200 °C. An acceptable limit of detection was observed for all compounds tested, with the worst being for ethylmercury chloride, which was 2.5 ng / ml.

Wilhelm and Bloom (12) is a useful review of mercury analysis up to the time the authors published the paper in 2000. It reviewed many of the same papers as this review.

In (13), Snell, Björn, and Frech investigate the use of electrothermal vaporization (ETV) to introduce sample into an ICP-MS for the determination of total mercury content of an organic liquid, such as condensate. The goal of this project is to eliminate solvent effects that cause problems with the plasma. This is a similar goal as that for direct injection micro-nebulizers, such as the direct injection nebulizer (DIN) and the direct injection high efficiency nebulizer (DIHEN), which minimize the amount of organic introduced into the plasma. While the limit of detection for mercury compounds (0.19 μg/l) in condensate is comparable to other techniques, the recoveries for volatile mercury compounds (elemental mercury and dimethylmercury) were poor.

In (14), Snell, Stewart, Sturgeon, and Frech explored new techniques for use in the analysis of mercury in condensates using GC-ICP-MS and GC coupled to a furnace atomization plasma ionization mass spectrometer (FAPIMS). The GC-ICP-MS had similar limits of detection of previous methods reported in the literature to speciate mercury compounds; this was achieved with only minor modification, the addition of an oxygen flow to the plasma, to improve operation. FAPIMS, while a specialized technique, offered the ability to identify new compounds. Furthermore, the paper details methods for preparing isotopically-enriched mercury standards from elemental mercury.

Bloom (15) used wet chemical extractions to differentiate elemental mercury, dissolved and particulate total mercury, inorganic mercury, and methylmercury, with detection via CVAFS. A detection limit of approximately 0.1 ng/g for each type of mercury was reported. This is another example of a low cost analysis, using available analytical equipment, with good results.

The objective of Ceccarelli and Picón (16) was to develop a fast, reliable method for total mercury in organic liquids like natural gas condensate and naphtha. Both AAS and AFS were explored. Limits of detection for the two methods were 32 ppb and 3.3 ppb, with linearity over about 3 orders of magnitude, for both AAS and AFS. This is another example of using on-hand instrumentation.

Boylan et al. (17) developed a field instrument for soil analysis using thermal decomposition, amalgamation, and AAS detection. This provides for on-site analysis with no pre-treatment. This procedure allows for sub-ppm determination of Hg in soil.

In two papers by Wilhelm in 2001 (18, 19), he reviewed Hg distribution in various streams, such as crude oil and condensate, to try to determine the impact of burning these fuels on the environment.

Stiltner (21) briefly describes the use of a field CVAFS analyzer, with a limit of detection of 0.01 µg/Nm^3. The paper mentions the use of a gold surface trap for collecting mercury from various hydrocarbon streams, and the need for good sampling via an in-line sampling device.

Bouyssiere et al. (22) speciated mercury compounds in gas condensates via GC-ICP-MS. The analysts used a diluted sample for elemental mercury and dialkylmercury compounds and a derivatized sample for ionic species. Detection limits of selected compounds were 0.98, 1.32 and 0.11 µg/l for elemental mercury, dimethylmercury, and dibutylmercury, respectively. Dibutylmercury is the derivative formed by Grignard reagent with Hg^{+2}.

Pavlish et al. (23) discuss Hg analysis methods for coal.

For those interested in highly specialized techniques, Huggins et al. (24) discuss X-ray absorption fine structure (XAFS) and X-ray absorption near-edge structure (XANES) analyses of adsorbed mercury compounds on activated carbon.

In (25), Björn and Frech describe the introduction of organics into an ICP-MS using a DIHEN. Good recoveries and low limits of detection were demonstrated for all mercury species, including elemental mercury, mercuric chloride, monomethylmercury chloride, and dimethylmercury. Compared to other techniques the detection limit is similar or better than other techniques previously discussed (13), including DIN and ETV. However, this technique cannot be routinely used until the problem with deposition on the nebulizer tip is completely solved. The analysts partially solved the problem through optimization of the system, but they were only able to achieve 50 sample analyses before cleaning the tip became necessary.

Stockwell (26) discussed the application of analytical methods for mercury for water samples, primarily for wastewater analysis. Older methods depend on AAS, but more recent methods typically depend up CVAFS, which gives a lower LOD. These validated methods are generally applicable to sub-ppt levels in water.

Finally, Coleman et al. (28) give further description of the in-line sampling technique used by UOP. Using an in-line sampler, the analyte is collected onto a gold-surface trap for analysis by a field CVAFS. Field measurements for total mercury in gas streams can be measured to a lower limit of 10 ng/Nm^3.

Conclusions

Since the last report in 1996, the primary improvements in mercury analyses have been in two areas:

1 Improvement in the LOD for organic samples, primarily by improved delivery of the sample to the detector (e.g., ICP micro-nebulizers) and improved sample preparation that leads to improved sensitivity, either through pre-concentration or elimination of interfering compounds.

2. The development of methods for the speciation of mercury compounds in organic liquid samples like naphtha and gas condensates.

The costs of many systems have been greatly reduced, with relatively inexpensive equipment offering low detection limits, like CVAFS. Multiple techniques like GC-ICP-MS are more common, with the development of commercial interfaces. Even with typically lower performance instruments, like AAS, the needed detection limits can be reached for many applications via pre-concentration techniques.

Bibliography

1. Brayden, Mark, Seb Corbino, Jerry Hale, and Dwight Hines, *Proceedings of the 8th Annual Ethylene Producers Conference*, 1996, 336-384.

2. McNamara, James D., and Norman J. Wagner, *Gas. Sep. Purif.* 1996, **10**, 137-140.

3. Snell, James P., Wolfgang Frech and Yngvar Thomassen, *Analyst* 1996, **121**, 1055-1060.

4. Olsen, S. D., S. Westerlund, and R. G. Visser, *Analyst* 1997, **122**, 1229-1234.

5. Zettlitzer, M., H. F. Scholer, R. Eiden, and R. Falter, *Proceedings – Society of Petroleum Engineers Symposium on Oilfield Chemistry* 1997, 509-516.

6. Müssig, S. and B. Rothman, *Proceedings – Society of Petroleum Engineers Symposium on Oilfield Chemistry* 1997, 559-569.

7. Frech, Wolfgang, James P. Snell, and Ralph E. Sturgeon, *J. Anal. At. Spectrom.* 1998, **13**, 1347-1353.

8. Tao, Hiroaki, Tadahiko Murakami, Mamoru Tominaga and Akira Miyazaki, *J. Anal. At. Spectrom.* 1998, **13**, 1085-1093.

9. McManus, Terence J., Robert O. Agbede, and Robinson P. Khosah, *Proceedings of the Air & Waste Management Association's Annual Meeting & Exhibition* 1998, Paper number 98-WP79A.07, 7 pp.

10. Shiowatana, J., A. Siripinyanond, W. Walyawat, and S. Nilmanee, *Atomic Spectoscopy* 1999, **20**, 224-229.

11. Shafawi, Azman, Les Ebdon, Mike Foulkes, Peter Stockwell and Warren Corns, *Analyst* 1999, **124**, 185-189.

12. Wilhelm, S. Mark, and Nicolas Bloom, *Fuel Processing Technology* 2000, **63**, 1-27.

13. Snell, James P., Erik Björn, and Wolfgang Frech, J. *Anal. At. Spectrom.* 2000, **15**, 397-402.

14. Snell, James P., Ian I. Stewart, Ralph E. Sturgeon, and Wolfgang Frech, *J. Anal. At. Spectrom.* 2000, **15**, 1540-1545.

15. Bloom, N. S., *Fresenius J. Anal. Chem.*, 2000, **366**, 438-443.

16. Ceccarelli, Claudio, and Ana Racquel Picón, *Petroleum Science and Technology*, 2000, 18, 1055-1075.

17. Boylan, Helen M., Robert C. Richter, H. M. Kingston, and Angela Ricotta, *Water, Air, and Soil Pollution* 2001, **127**, 255-270.

18. Wilhelm, S. Mark, *Proceedings of the Engineering Technology Conference* 2001, Paper number ETCE2001-17086, 187-196.

19. Wilhelm, S. Mark, Environmental Science and Technology 2001, **35**, 4704-4710.

20. EPA-822-R-02-047, National Recommended Water Quality Criteria: 2002, November 2002.

21. Stiltner, Julie, *Annual Convention Proceedings - Gas Processors Association* 2002, 81st meeting, 90-99.

22. Buoyssiere, Brice, Frank Baco, Laurent Savary, and Ryszard Lobiñski, *J. Chrom.* 2002, **976**, 431-439.

23. John H. Pavlish, Everett A. Sondreal, Michael D. Mann, Edwin S. Olson, Kevin C. Galbreath, Dennis L. Laudal, and Steven A. Benson, Fuel Processing Technology 2003, **82**, 89-165.

24. Frank E. Huggins, Nora Yap, Gerald P. Huffman, and Constance L. Senior, Fuel Processing Technology 2003, **82**, 167-196.

25. Björn, Erik, and Wolfgang Frech, *Anal. Bioanal. Chem.* 2003, **376**, 274-278.

26. Stockwell, Peter, *Today's Chemist*, November 2003, 27-28.

27. Baumgartner, Arthur J., Marilyn W. Blaschke, Steven T. Coleman, Robert Kohler, and Timm E. Paxson, *Proceedings of the 17th Annual Ethylene Producers Conference*, 2004, Paper 55b, 65 pp.

28. Coleman, Steve, Jason Ries, Don Campbell, and Keith R. Clark, *Proceedings of the 17th Annual Ethylene Producers Conference*, 2004, Paper 55c, 9 pp.

35c

Understanding and Preventing Failure of Aluminum Equipment in the Presence of Liquid Mercury

Ronald N. Bell

Sr. Pr Weld Engineer

Chart Heat Exchangers

La Crosse, Wisconsin

Prepared for Presentation at the Spring Meeting of the American Institute of Chemical Engineers

Atlanta, Georgia

April 10 – 14, 2005

UNPUBLISHED

February 28, 2005

Understanding and Preventing Failure of Aluminum Equipment in the Presence of Liquid Mercury

Ronald N. Bell
Sr. Pr Weld Engineer
Chart Heat Exchangers
La Crosse, Wisc

ABSTRACT

Liquid Hg in contact with aluminum is a concern for Ethylene gas processing plants that utilize Brazed Aluminum Heat Exchangers and other aluminum equipment. The concern is the potential for failure during operation and risk analysis for continuing operation once Hg is detected.

The mechanism for failure for aluminum components, primarily aluminum welds, is reviewed as well as an explanation of key terminology. This mechanism identifies the weld metal as the primary site of failure initiation as found during the analysis of field failures. Understanding how and why failure occurs is key to knowing where and how to prevent failure. This paper will present a simple test to determine if weld metal (or sensitized base metal) will be susceptible to Hg LME failure. This test can then also be used to develop weld methodology that can be used to prevent aluminum components from susceptibility to Hg attack.

Understanding the primary cause of failure can be used to develop robust weld processes that are resistant to Hg attack and to provide understanding in how to assess risk analysis for those plants which have or may have Hg present in Brazed Aluminum Heat Exchangers or other aluminum equipment.

OVERVIEW

The object of this paper is to assist in your understanding of mercury (Hg) related equipment failures, the principle modes of failure, how they occur, and what parts of the heat exchanger may be susceptible. This paper will also present recent research at Chart HXD in addressing Hg related failures in brazed aluminum equipment.

Introduction

Brazed Aluminum has been used in the Ethylene gas processing industry and in NGL and other natural gas installations for many years where Brazed aluminum components are necessary for the processing of the feed gas or other secondary gas streams. It is a well known fact that elemental Mercury, present in liquid form, can have severe detrimental affects on Brazed Aluminum equipment, primarily the headers which cover and direct flow from individual streams, the nozzles to these headers and associated piping.

ALPEMA manufacturers of brazed aluminum equipment have developed various methods and techniques to construct those components susceptible to mercury corrosion in such a manner as to be resistant to failure from mercury. Chart HXD has a long history in the development of manufacturing and welding techniques known collectively as "Mercury Tolerant Design". Some of the more common techniques involve the use of designing the exchangers to be fully drainable where possible, protective internal liners and specialized heat treatment operations.

Chart HXD has participated in the evaluation of a number of mercury related field failures. This field evaluation has significantly increased our technical knowledge as related to Hg failures. Chart HXD is considered to be the expert and world leaders in Mercury Tolerant construction.

Terminology

In order to understand how and why aluminum is susceptible to failure from mercury and to then develop the methods and technology to make aluminum resistant to Hg attack, it is necessary to understand some of the metallurgical terms used in defining mercury and how mercury attacks aluminum, specifically the 5xxx alloys.

Amalgamation: Process by which Hg forms a liquid solution with aluminum. Simple amalgamation results in surface pitting and etching in the absence of H_2O vapor.

Amalgam Corrosion: Combined action of Hg and H2O on aluminum. This process is self-propagating as long as Hg is in contact with the aluminum (breach of the oxide layer) and H2O is available. It results in more severe pitting than simple amalgamation. The familiar white feathers found on the surface of aluminum, which has residual amounts of mercury, are characteristic of amalgamation corrosion.

Grain: Periodic and orderly arrangement of atoms in a specific geometric pattern. The grain structure of aluminum is FCC (Face Centered Cubic)

Grain boundaries: The disruption of the periodic and orderly arrangement of the grains where the grains meet, forming a continuous network throughout the metal. The nominal thickness of the grain boundaries in aluminum alloys is 4.0 A

Epitaxial grain growth: Grain growth that initiates from the substrate at the fusion boundary and proceeds towards a weld (or primary energy source) centerline.

IGLME: Inter Granular LME: Modification of LME which incorporates the concept of Hg attack on Mg precipitates at the grain boundaries of > 3.5% Al-Mg alloys. Results in rapid grain boundary dissolution and resultant rapid failure. It does not require presence of H2O for failure, but presence of water could significantly increase the rate and extent of the failures. Weld metal is the most susceptible to Mercury Embrittlement because weld deposits are inherently sensitized to the more severe degree.

LME: Liquid Metal Embrittlement: Brittle failure at the grain boundaries of normally ductile aluminum alloys when liquid mercury comes in contact with susceptible grain boundaries which are sub sequentially stressed in tension. This involves the diffusion of liquid Hg along the grain boundaries and subsequent rapid brittle failure. Cracks usually initiate and propagate via the grain boundaries.

Oxide layer: The natural formation of Aluminum Oxide (Al2O3) or hydrated Aluminum Oxide (Al2O3 · xH20). This oxide has a nominal thickness of .01 µm and is comprised of an oxide layer and barrier layer that has a nominal thickness of 5-20 nm3. This layer provides a natural barrier to attack by mercury.

Phase: A physically homogeneous and distinct portion of an alloy.

Polycrystalline: A solid composed of many crystals intimately bonded together. All of the alloys as used in heat exchanger construction are polycrystalline.

Sensitization:The metallurgical precipitation of constituents or phases at grain boundaries that greatly affects the properties of the alloy(s) under certain conditions.

PMZ: Partially Melted Zone; localized melting along the grain boundary and, to a lesser extent, in the interior of the grain.

Mechanisms of Mercury Attack

The two primary modes of failure as observed in field operations are associated with 1) IGLME and 2) LME. Both of these methods result in Hg Embrittlement either through the attack of sensitized components or weld metal (IGLME) or initiated in susceptible grain structure (LME) with either mode resulting in brittle or catastrophic failures. Figure 1 illustrates a failure that represents both aspects of failure.

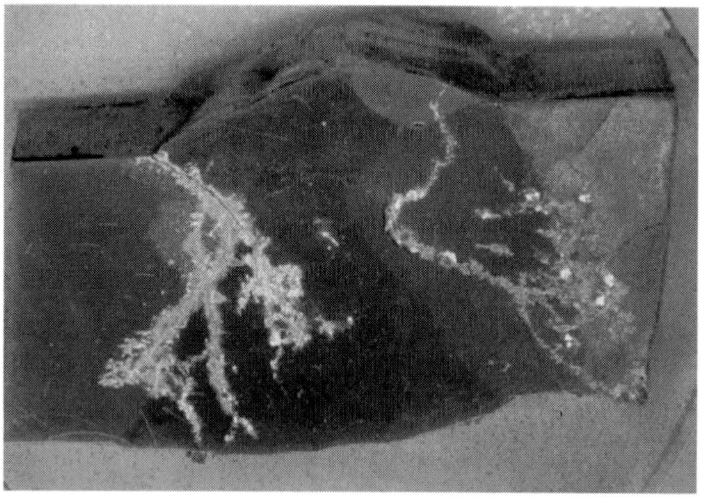

Figure 1 Hg Embrittlement

There is still some uncertainty about the exact propagation mechanism that is taking place at the crack tip at the atomic level, but the metallurgical characteristics for Hg Embrittlement or Hg failure are clearly understood.

IGLME Failure

This is a modification of LME that incorporates the concept of Hg attack on Mg precipitates at the grain boundaries. This mode of failure is primarily related to high Mg ($\geq$ 4.5% Mg) alloys. The following four conditions must be present for IGLME failure to occur:

1) The weld joint or base material must be in a stressed condition. Stresses are present as the result of process pressures, pipe loads, residual weld stresses and other factors inherent in the design and operation of cryogenic equipment.

2) A source of free liquid Hg must be present. Hg can collect over long periods of time in process equipment concentrating in low-lying areas and revert to a liquid state during warm up periods. At operation temperatures colder than -39C, Hg is present as solid matter and as a solid is inert to aluminum. It is only when Hg is warmer than –39C that it can present a potential for Hg Embrittlement.

3) The base material or weld metal must be sensitized. Sensitization in the material and especially weld metal is induced during fabrication or by field process conditions.

4) The natural formation of the oxide layer over the base aluminum must be breached or broken. The Al_2O_3 oxide layer is inert to Hg and is a natural barrier to the exposure of the aluminum substrate.

- **Grain Boundary Sensitization**

Sensitization is a key factor to the understanding of IGLME. From the metallurgical viewpoint, all of the materials used in brazed aluminum heat exchangers are polycrystalline; i.e., they are made up of many, many individual grains of aluminum. The points at which the grains meet are known as grain boundaries. In aluminum alloys with a Mg content >1%, which includes the 5xxx alloys, a separate phase i.e. a volume of material with a different chemical composition and crystal structure can precipitate out and form at the grain boundaries of the material.

The phase that forms is composed of two parts Mg for every three parts of Al and is known as the **β** -phase (Mg_2Al_3). If the alloy contains more that ~ 3.5% Mg then there is enough Mg that the phase that forms at the grain boundary can form in a continuous network rather than in discreet volumes.

If a continuous network of this secondary **β** -phase form along the grain boundaries, then the material is "sensitized". Sensitized material will, under specific conditions of stress and the presence of Hg, fail along the sensitized grain boundaries. Failure is dependant on formation

→

of the Al– Mg β -phase precipitates at the grain boundaries with Hg reacting with the Al-Mg precipitate per the reaction: $4Hg+Mg_2 Al_3 \quad 2MgHg_2 + 3 Al$.[1]

The Mg_2Al_3 phase is anodic to the grains and is subject to attack by liquid mercury. This results in grain boundary dissolution and resultant rapid failure. This failure mechanism does not require presence of H_2O for failure although moisture in the form of frost, hydrates, etc can accelerate the failure process with failure occurring in a matter of minutes in material as thick as 38mm once the failure initiates. Weld metal is most susceptible to IGLME attack due to the natural severe sensitization that occurs in multi-pass welds in high Mg alloys.

- **IGLME sensitization test**

The failure mechanism as stated above is well understood and has been confirmed by the analysis of a number of field failures. This analysis has shown that the weld metal is most prone to failure and that failure prevention should be focused on how to mitigate or prevent the sensitized weld metal from exposure to mercury. One of the initiatives undertaken at Chart HXD was to see if there was a relatively easy and simple test to determine if a weld was susceptible to IGLME.

The outcome of these tests resulted in a simple and reliable method to determine if a weld was susceptible to IGLME. The procedure basically involved placing the weld in a stressed condition and heating the stressed weld to a temperature to sensitize the weld. The combination of stress and temperature accelerates the formation of the **β** -phase precipitates. This procedure can determine if (a) the precipitates form a continuous network along the grain boundaries i.e.; susceptible to IGLME or (b) only discreet isolated precipitates i.e.; not susceptible to IGLME.

The samples were prepared by welding two plates were welded together (any thickness) by use of either the GMAW or GTAW weld process recording the KJ/in heat input. The filler metal selected is based on either mechanical properties or metallurgical properties as required by the base metal. The plates were then sectioned for examination of the weld in the as-welded condition with additional sections cut for either side bends or root and face bends as per ASME IX or EN 288-4.

The specimens were bent to an 180^0 angle by use of a wrap-around jig and heated in a 350^0 oven to sensitize the weld metal. After cooling, the weld was sectioned, polished and etched. The resulting microstructure clearly shows that the grains are outlined with the **β** -phase precipitates [Figure 2] in high Mg alloys and only isolated precipitates [Figure 3] in low Mg alloys.

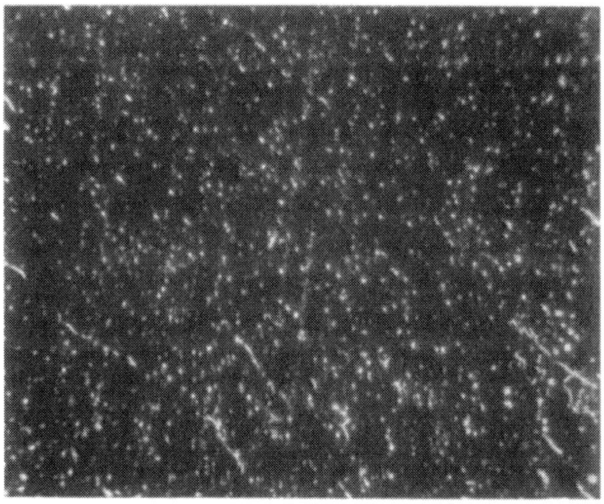

Figure 2 100X
ER5554 weld metal, strained and heated

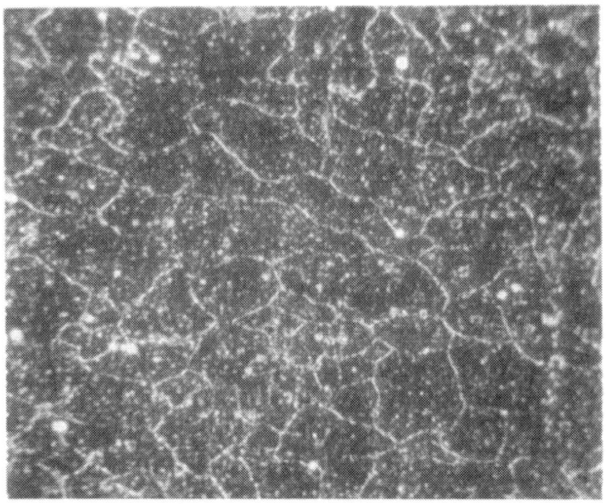

Figure 3 100X
ER5183 weld metal, strained and heated

The sensitized sections were then solution heat treated to 750^0 for 4 hrs. , air quenched and re-polished. Examination of these sections confirmed the network of the **β** -phase precipitates were broken with the Mg going back into solution in the grains.

This analysis can evaluate selection of the weld process, weld parameters, heat input and filler metal as critical elements in determining if a weld is susceptible to IGLME. The analysis can also evaluate effectiveness of solution heat treatment as an effective means to mitigate IGLME.

- **Methods of construction to mitigate IGLME**

In general, use of mercury guard beds should be considered as mandatory if Hg is suspected to be present in the feed gas or other product streams during operation of the aluminum BAHX. The best method to mitigate Hg Embrittlement is to keep the mercury out of and away from the aluminum components.

In operation, however, it is not always possible to offer 100% assurance that some or all parts of the equipment components will not come into contact with liquid Hg at some time during the expected life or service of the equipment. Respective components of the BAHX can be designed, fabricated and welded to mitigate Hg Embrittlement:

- 5083 alloy welded with ER5183. For this alloy, the high Mg content (nominal 4,5% Mg) will result is weldments susceptible to IGLME failure. As such, internal construction methods are employed to prevent the liquid Hg from coming in contact with the 5083 material or susceptible ER5183 weld metal. These construction methods include:
    - Use of non Mg alloys such as 3003 for internal liners, seal welding of liners

- Use of non Mg filler metals such as ER4043 or ER4047 where applicable and properly qualified to the applicable design code for weld deposits that could come in contact with mercury.

- Solution heat treatment of the final welded assembly if size of assembly permits controlled heating. The solution heat treatment breaks up the network of Mg_2 Al_3 precipitates and in effect de-sensitizes the 5083 base materials and the ER5183 weld metal.

- Use of low Mg filler metal in conjunction with a high Mg filler metal. The low Mg filler metal provides an inner weld deposit, which is on the side of the weld that is in contact with the process gas or fluid. This low Mg weld deposit provides a chemistry that is not susceptible to IGLME. The outer high Mg weld deposit provides the desired chemistry to achieve the required mechanical properties.

- Minimal use of permanent backing. If permanent backing is required (i.e. non accessible pipe runs) the backing is seal welded on the upstream side or in the flow direction to minimize entrapment areas and welded with proprietary weld processes such that the root pass is deposited with a filler metal which is not susceptible to IGLME.

- Sloping of assemblies to promote drainage. This is becoming less of a factor as plant operations become more sophisticated. Recent data over the past few years indicate that mercury is more likely to be dispersed as fine droplets as opposed to pools of liquid Hg. The high surface tension of the droplets causes the Hg droplets to adhere to the internal surface in any position and pooling of Hg is unlikely to occur. As such, sloping and drain ability features become less significant.

• 5454 alloy welded with ER5554. For this alloy, the low Mg content (nominal 2,5% Mg) will result is weldments which are not susceptible to IGLME failure. The lower Mg content does result in lower allowable design stress, which can then limit the application of the 5454 alloy. Internal construction methods to protect the ER5554 weld from contact with mercury such as use of 3003 liners and solution heat treatment of the final welded assembly are minimal or not required. Some common sense fabrication methods for this alloy do include:

  - Seal welding of liners to prevent entrapment areas
  - Minimizing KJ/ in heat input by appropriate selection of weld processes
  - Minimal use of permanent packing. The weld deposits, by definition, are not susceptible to IGLME

• 6061 alloy welded with ER4043. This alloy as well as the ER4043 filler metal is resistant to sensitization and is not susceptible to IGLME because of the low Mg content. The use of this alloy is limited because of low allowable design stress and higher propensity for weld metal cracking if not welded properly. This alloy does have applications for small diameter assemblies with low pressure and thin gauge material. The common sense fabrication methods for this alloy are the same as for 5454 alloy.

Various publications mention the benefits of an enhanced oxide layer as a means to mitigate Hg failure. There is substantial corrosion information in various publications (Hg and otherwise) that indicates that just because a particular oxide is thicker, it is not necessarily better. Differential expansion-contraction between oxide and metal make oxide thickness a very tenuous barrier to rely on for resistance to IGLME. Figure 4 illustrates the porous nature of the oxide layer and the actual barrier layer contiguous with the aluminum substrate. [2]

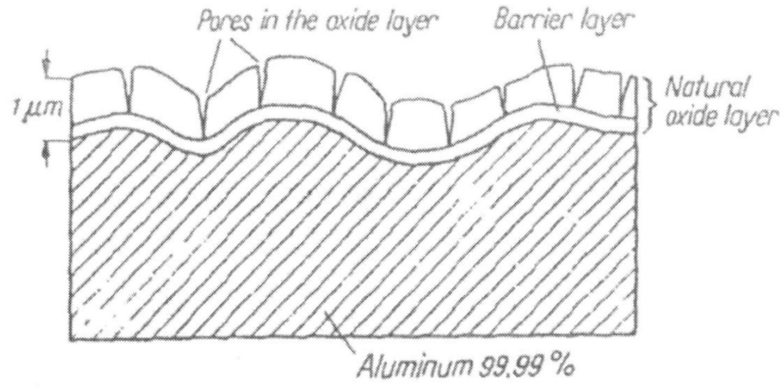

Figure 4: Illustration of the Al_2O_3 oxide layer and barrier layer

Studies of how to predict resistance to IGLME are much more important than an Hg mitigation method based on enhancing the oxide thickness. Chart HXD does not promote enhanced oxide formation as a mitigation feature for IGLME except as follows:

- Anodizing 6061 flange alloys. The anodizing process forms a very thick layer of tenacious aluminum oxide that does not crack after being placed in operation. This anodized layer protects the flange gasket surface and internal surface of the flange from pitting due to Hg amalgamation.

LME Failure

Failure due to LME is the more rare mode of failure in aluminum although LME may have been a secondary mode of failure in some previous failures where IGLME failure has been the primary mode of failure. Observations of failure analysis indicate that the frequency of this mode of failure is significantly less than the IGLME mode of failure. This failure process involves the liquid diffusion of Hg atoms along grain boundaries; similar to IGLME except that formation of the **β** -phase precipitates from high Mg alloys is not a factor in failure initiation.

The LME mode of failure is not alloy specific. A grain boundary site is required for initial crack propagation to begin. Research at Chart HXD has observed such a grain boundary site to be an anomalous and susceptible grain structure at the surface of the base metal, adjacent to the toe of a weld. Micro fissures or surface cracks propagating along the anomalous grain structure exposed to the surface coupled sufficient energy at the tip of the micro fissure and a source of finely dispersed mercury provide the necessary driving energy for the Hg to cause dissolution of the grain boundaries. In the absence of any of the above, Hg failure would not occur.

It is also important to note that LME failure will most likely be initiated in the base metal adjacent to the toe of the weld as opposed to IGLME which is more likely to originate and propagate thru the weld metal.

- **Anomalous Grain Structure**

Research at Chart HXD has shown that failure initiation site(s) for LME failure are present as micro fissures in a coarse epitaxial grain structure at the surface of the base metal adjacent to the toe of the weld. Chart HXD has not fully identified the mechanism for this anomalous grain structure. In some aspects the coarse epitaxial structure appears to be similar to or related to the grain structural identified in the PMZ (Partially Melted Zone) in aluminum alloys.[3] The structure also has aspects of secondary recrystallization as related to anomalous grain growth.[4]

Polishing and etching with Barkers etch and examination in polarized light reveals an area with a very large grain structure nucleating from the normal substrate grain structure. This anomalous grain structure provides the site for the formation of micro fissures at the surface of the base metal.

Embedded is this large columnar grain structure are micro and proto cracks propagating along the grain boundaries of this very large epitaxial grain structure. Examples of this grain structure are identified in Figures 5a and 5b.

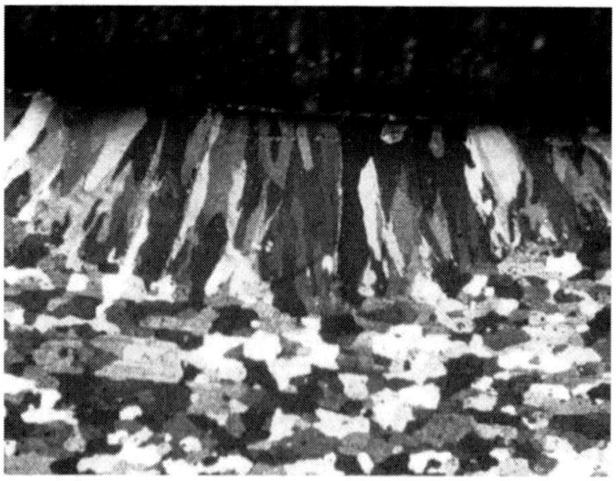

Figure 5a **Figure 5b**

Anomalous grain structure nucleating from fine grain substrate (Barkers Etch, 100X)

Laboratory studies at Chart HXD have identified certain conditions necessary for the formation of the anomalous coarse epitaxial grain structure. The primary conditions appear to be driven by the combination of heat and stress.

The micrographs all indicate normal fine grain structure in the substrate below the columnar grains. At undefined locations, the heat from the weld arc combined with the residual stress from the welding operations provide the necessary driving force to cause anomalous grain growth or secondary recrystallization at the surface of the base metal adjacent to the toe of the weld.

The surface appearance of these areas is similar to the appearance of a hot short crack at the surface. Figures 6a and 6b illustrate this hot short crack appearance at the toe of the weld.

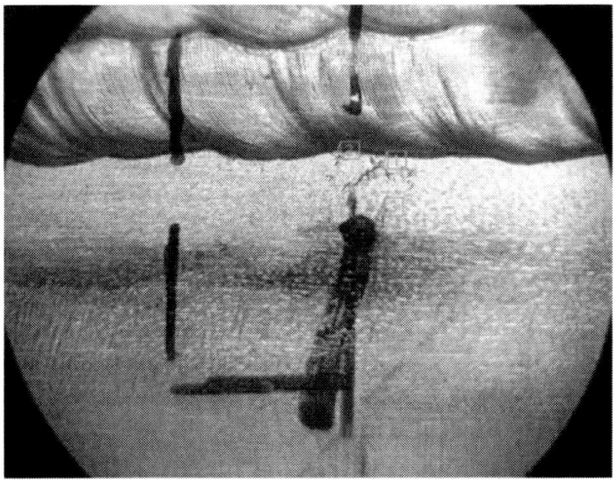

Figure 6a

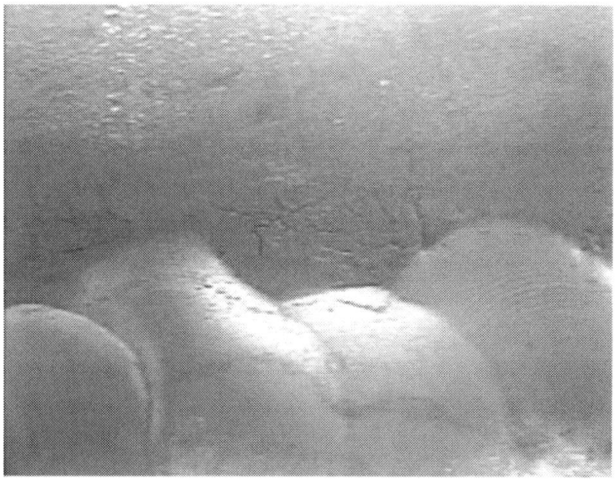

Figure 6b

Anomalous grain structure as exhibited by appearance of hot short cracks at surface of base metal

Proto cracks and micro fissures can then develop along the grain boundaries of these large columnar grains extending to the surface of the base metal. Residual weld stress from welds placed in a restrained condition result in very high stress intensification factors at the tip of the micro fissures. Figures 7 and 8 illustrate examples of these micro fissures.

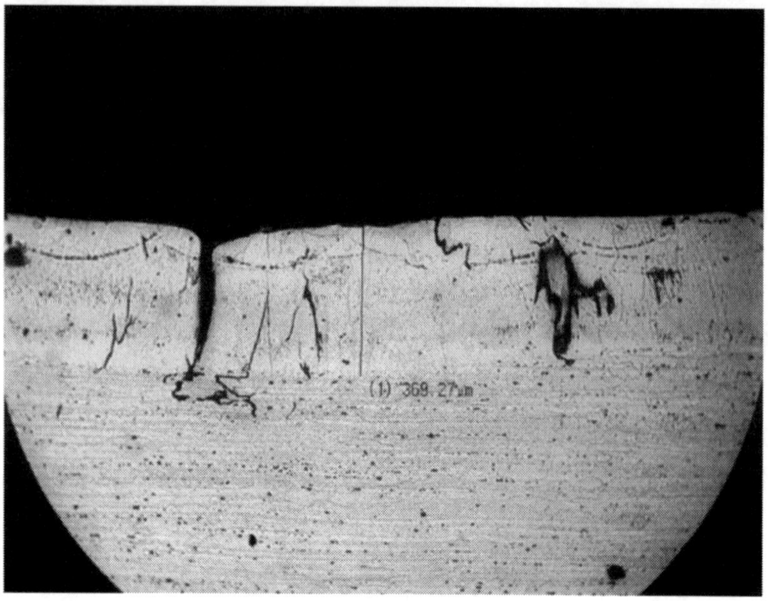

Figure 7 Micro fissure at surface 100X

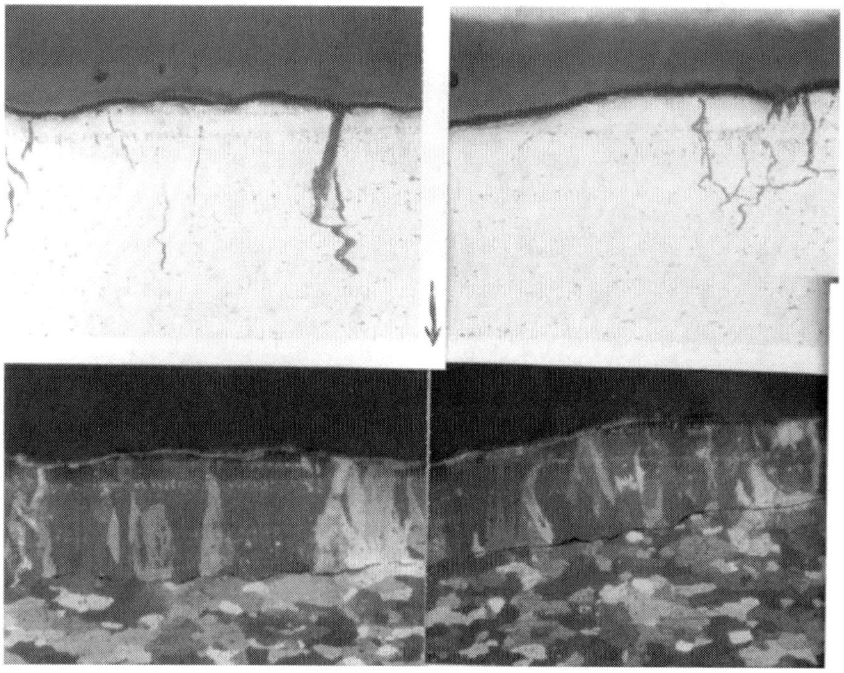

Figure 8 Micro Fissures (Top) located at GB of anomalous grain structure (Bottom)

The oxide barrier formed at the tip of the micro fissures has to be breached. This breaching of the oxide layer can be cracked or opened by the application of additional stress. This additional stress can be in the form of pipe loading, thermal stress or other mechanical stress.

The fine grain structure in the base metal provides the nucleation sites for this anomalous grain growth. Once began, the grain formation or grain growth is epitaxial to the fine grain substrate. The epitaxial growth of the large anomalous grains forms a very coarse columnar structure that is very weak and it is this weak structure which provides the initiation sites for micro fissures and proto cracks at the columnar grain boundaries.

The boundary between the epitaxial secondary recrystallization and the fine grain substrate is very well defined. In no cases were any grain boundary anomalies or interstitial grain defects noted in the fine grain substrate. The depth (or height) of the anomalolous grains was always limited to a nominal depth of approximately 1mm or less.

For LME failure to occur, the Hg has to penetrate the openings in the surface micro fissures. The surface micro fissures, coupled with the driving energy from residual weld stresses or other stress, are sufficient to allow the Hg to wet the grain boundaries in the fine grain substrate. The LME failure can propagate very fast with a crack velocity which can have approach 100 cm / sec. The dissolution of the grain boundaries ceases when the stress intensification factor falls below the limiting threshold of approx 5 ksi $[in]^{1/2}$. [5]

- **Methods of construction to mitigate LME**

There has to be an initiation side exposing a susceptible grain boundary in order to precipitate LME failure. In the absence of a site for failure to initiate, LME is indeed a rare occurrence. Similar to IGLME, use of mercury guard beds should be considered as mandatory if Hg is suspected to be present in the feed gas or other product streams during operation.

The potential failure sites have to be addressed during fabrication and welding by implementation of suitable manufacturing processes. Some of these manufacturing considerations would include:

- Limitation and control of heat input and interpass temperature.
- Consideration of weld sequence and repairs to minimize residual weld stress
- Solution heat treatment where applicable to reduce residual weld stress
- Suitable NDE inspection techniques. RT inspection will normally not detect surface micro fissures.

Other causes of Hg failure

There are several other failure mechanisms related to mercury that are not common in the BAHX industry.

1) Failure due to amalgam corrosion: This is a mode of failure due to the combined action of Hg and H_2O on aluminum. The process is self-propagating as long as Hg is in contact with the aluminum (breach of the oxide layer) and H_2O is available. This results in more severe pitting than simple amalgamation. This is not a normal failure mode since moisture or water is seldom in contact with liquid mercury.

2) Failure due to amalgamation: This is a process by which Hg forms a liquid solution with aluminum and results in surface pitting and etching in the absence of H_2O vapor. Because true amalgamation failure requires significant quantities to dissolve a very small amount of aluminum, this mode of failure is usually discounted and has not been observed in field failures.

Summary:

The science of how to use aluminum in a corrosive environment such as mercury has been and continues to evolve. The principles of Hg embrittlement and understanding the metallurgy of aluminum and interaction of Hg with aluminum are critical. It is this understanding which provides the basis for sound engineering selection of alloys and weld filler metals, methods of construction to keep Hg from coming into contact with a sensitized aluminum alloy, heat treatments to place base material and weldments back into a non sensitized condition, selection of weld processes and a host of other details which result in a brazed and welded aluminum equipment which can very resistant to Hg Embrittlement.

References

1. David R. Nelson "Mercury Attack of Brazed Aluminum Heat Exchangers in Cryogenic Gas Service" Annual GPA Convention / 73; New Orleans, Louisiana, Mar 1994.

2. George E Totten, D. Scott MacKenzie "Handbook of Aluminum, Vol 1 Physical Metallurgy and Processes" Chapter 2 "Properties of Pure Aluminum" by Alexey Sverdlin; pg 60.

3. Sindou Kou, "Welding Metallurgy" Chapter 12 "Formation of the Partially Melted Zone"; Chapter 13 "Difficulties associated with the Partially Melted Zone"

4. George E Totten, D. Scott MacKenzie "Handbook of Aluminum, Vol 1 Physical Metallurgy and Processes" Chapter 4 "Recrystallization and Grain Growth" by Wiemin Mao.

5. S. Mark Whielheim "Methods to Combat Liquid Metal Embrittlement in Cryogenic Heat Exchangers" Annual GPA Convention / 73; New Orleans, Louisiana, Mar 1994.

Mercury and its Impact on Ethylene Purification Catalysts

Author: Dr. Raghu K. Menon
Süd-Chemie Inc.

AICHE – Ethylene Producers Conference,
Atlanta, GA.

April 10 – 13, 2005

ABSTRACT: Mercury and its Impact on Ethylene Purification Catalysts

Mercury has come to be recognized as a significant contaminant in condensate, naphtha and even gas cracker feed stocks. Due to the high volatility of mercury and Hg compounds, Hg migrates to areas in the plant that are typically considered poison-free. Performance of precious metal (palladium) catalysts used in selective hydrogenation of acetylenes and dienes are affected by mercury contamination. This paper will present the impact of mercury contamination on selective hydrogenation catalysts and suggest some means of protecting the precious metal catalysts by mercury sequestration and removal.

Introduction: Mercury has come to be recognized as an important contaminant in steam cracker processing for the production of olefins. Increasingly, ethylene producers are utilizing higher hydrocarbon feedstock, such as naphtha and condensate in place of ethane gas, in order to balance their risk of exposure to the volatilities of the raw materials markets. While such a move offers economic advantages in feedstock sourcing, it presents new challenges to the safe and protective operation of downstream processes for olefin purification. Due to the high volatility of elemental mercury and organic mercury compounds, they can migrate into areas in the plant that are typically considered poison-free. Performance of precious metal catalysts used in olefin purification is thereby affected by mercury contamination. This paper will discuss the impact of mercury contamination on selective hydrogenation catalysts and suggest some means of protecting the precious metal catalysts by mercury sequestration and removal.

Source of Mercury Contamination: Mercury is ubiquitous in various hydrocarbon and petroleum sources. Mercury compounds tend to concentrate in the higher hydrocarbon fractions such as natural gas condensates and gas oil. Increased use of such materials as cracking feedstock for olefins production exposes the steam-cracker process to increased mercury contamination.

Gas Crackers: Due to the high volatility of elemental mercury, it can be present in a variety of hydrocarbon sources; no hydrocarbon source is truly exempt. Since ethane is recovered through hydrocarbon processing that minimizes the presence of mercury in the ethane fraction, traditional use of ethane as the cracking feedstock in gas crackers tended to minimize the impact of mercury and mercury compounds. Changes to the mix of cracking feedstock from pure ethane to Ethane/Propane in gas crackers, and recycle of vent streams increases the risk of exposure to heavy metal contaminants such as arsenic and mercury.

Liquid Crackers: Operators using liquid cracking feedstock have already recognized the concern of contaminants in the feedstock. Straight-run naphtha tends to be the cleanest, and natural gas condensate tends to be the most contaminated among liquid feedstock. Certain sources of condensate feedstock are known to be high in mercury contamination, and ethylene producers have come to anticipate the impact of using such sources upon their down-stream processing. Increased utilization of such feedstock in recent times has heightened the risk of mercury contamination among ethylene producers using condensate feedstock. However, no feedstock is truly exempt, since mercury contamination is cumulative in how it manifests its effects on safety, environment and operational efficiency.

Alternate Sources of Olefins:

Olefins plants often co-process cracked gas with olefin streams that are made available from other sources such as refinery propylene. These streams bring in their own sets of contaminant profiles that are distinct from what is traditionally expected in cracked gas. Recent developments in olefin production technologies envision the use of heavier fractions such as gas oil and olefin recovery from the off-gas of FCC refinery and coker. When such feeds rich in heavy metal contaminants are co-processed with cracked gas, they exacerbate the vulnerability of olefin purification catalysts to mercury contamination. Sourcing of olefin-rich streams from refineries can add to the complexities of olefin purification in the steam cracker process.

Ethylene Purification Processes: Petrochemical plants engaged in olefin production can be broadly categorized as Front End (FE) or Back End (BE), based on sequence of down-stream processing of cracked gas. Front End (FE) process purifies cracking feedstock by selective hydrogenation to remove acetylenes and some diolefins before separation of fractions, while Back End (BE) process carries out the separation of cracked gas in the "cold box" into H2, C2, C3, C4.... fractions, and each of these fractions are then individually purified by selective hydrogenation to remove acetylenes and some diolefins. FE process operates under relatively high H2 concentration, while BE process operation requires the controlled addition of small quantities of hydrogen. The "hot section" where the steam cracking takes place is common to both types of processes, and the various chemical and thermal transformations of mercury contaminant are also similar. However, the vulnerability of selective hydrogenation catalysts to mercury contaminant in down-stream processing of cracked gas is different for FE and BE processes, due to the difference in sequence of distillation and selective hydrogenation.

Selective Hydrogenation Catalysts in Steam-Cracker Process:

1. Sulfided Nickel Catalyst for FE processing:

Early development of ethylene purification catalysts in selective hydrogenation of olefins relied upon the relatively high selectivity exhibited by sulfided nickel catalysts towards acetylene hydrogenation. Sulfided nickel catalysts are operated at higher temperature (~200 C) under continuous sulfiding, and the wet raw gas contains a relatively high level of carbon monoxide. Attenuation of hydrogenation reactivity achieved under these operating conditions made it possible for sulfided nickel catalysts to obtain 99+ % conversion of acetylene while minimizing the extent of ethylene hydrogenation. The relatively high concentration of nickel (~ 1%) and higher operating temperature result in the capture and sequestration of a significant fraction of mercury contaminants.

2. Palladium Catalyst for FE processing:

Further upstream purification and drying of the wet raw gas made it possible for the use of more selective palladium catalysts in this service. In this type of process, unmodified Pd or silver-modified Pd catalysts are used to purify dry raw gas that has been fractionated to remove heavier hydrocarbons. Since the process feed is hydrogen-rich, a lower temperature operation and smaller concentration of active Pd (~ 0.02%) are necessary to achieve high selectivity. The low Pd concentration makes this FE catalyst highly vulnerable to the poisoning effects of mercury contaminants in dry raw gas. In FE processes that employ de-propanizer or de-ethanizer overhead fractions, this vulnerability is somewhat mitigated, but not eliminated.

3. Palladium Catalysts for BE processing:

Back End Olefin processing takes place down-stream of the "cold box", where the various fractions are separated and H2 is removed from the raw gas. Mercury compounds are mostly condensed and sequestered in the "cold box". Therefore, palladium catalysts used in BE converters to remove acetylene are relatively protected from mercury contamination, as long as the process feed is strictly sourced from the "cold box". Co-processing of other streams that may not have had sufficient pre-treatment to remove Hg contaminants can raise the vulnerability of the Pd-catalyst (e.g.: MA/PD converters).

Reactions of Mercury Contaminants:

1. Elemental Mercury

Elemental mercury contained in the cracking feedstock is relatively un-reactive. A fraction of the elemental mercury is expected to pass through the "hot section" of the steam cracker unscathed. Some of the Hg that exits the hot section may be entrained in the acid gas scrubber and the molecular sieve dryers. Since this absorbed elemental Hg is not sequestered, it could be released to the vent systems during dryer regeneration step. Further downstream, a fraction of elemental Hg may exit the dryer can contaminate the Pd-based FE catalyst down-stream.

In a BE process, the fraction of Hg that exits the dryer will enter the "cold-box", and pool in the low points. While this protects the BE converter catalyst from contamination, it poses a serious safety concern for the operation of the cold box. Condensed Hg can weaken the aluminum metallurgy of the cold-box by formation of amalgam.

2. Mercury Halides

Mercury halides (and pseudo-halides) are volatile compounds that may be present in small amounts in the cracking feedstock. They may also be created in the cracking furnace due to thermal reactions of elemental mercury with other halide sources. Recycling of hydrocarbon vent gases from halide-rich down-stream processes (e.g. VCM production) can introduce organic halides into the cracking furnace. Elemental mercury may then undergo atom-transfer reactions with these to generate organic mercury halides. Such compounds are relatively inert towards caustic, and can exit the charge gas dryers without capture. Organic mercury halides will react with nickel and palladium catalyst to result in permanent poisoning of the active sites.

3. Organic mercury:

Organic mercury refers to a class of compounds with the general formulae R2Hg, R-Hg-R' and R-Hg-H, where R=alkyl groups. These compounds may be generated from thermal reactions of alkyl radicals with mercury; conditions favorable to the formation of such compounds are present in the cracking furnaces. Low-sulfur operation of the cracking furnace will exacerbate the formation of such compounds, since sulfur tends to suppress atom transfer reactions that lead to their formation. Organic mercury compounds are inert to most pre-treatment steps, and will result in poisoning of the FE catalyst.

Handling of Mercury Contaminants:

Mercury contaminants pose a unique challenge to ethylene producers. Mercury is a cumulative poison to most catalysts employed in down-stream processing of olefins. Unless mercury is removed, it can condense in the "cold box" and create a safety hazard. Mercury poses serious HS&E risk during maintenance turn-around operations in an olefins plant. Due to it high volatility, it is prudent to assume that no part of the Olefins plant is truly safe from mercury contamination. Several methods have been devised to address mercury contamination.

1. Absorption of mercury:

Absorption of mercury is a widely used method to remove mercury from process feeds. Typically, molecular sieves or other zeolite materials are used to "capture" mercury from process feeds. While this method is effective in removing elemental mercury, it is not clear whether alkyl mercury compounds are removed with equal efficiency. Operation of such absorbers is typically cyclical - an absorber cycles through absorption and de-sorptive regeneration steps. If the absorbed mercury is released upon regeneration, it will enter the vent systems. Therefore, absorption of mercury only moves the concern from the process feed to the vent gases, rather than elimination of the contaminant.

2. Sequestration of mercury:

Mercury can permanently react with metallic species to form non-volatile compounds. If this were not true (with Pd or Ni), mercury poisoning of these catalysts would not be such a concern. Therefore, chemical sequestration of mercury from process feeds is a viable option of eliminating mercury contamination. Since elemental mercury and alkyl mercury compounds are relatively inert to chemical reactions, they pose a challenge to chemical sequestration. Sequestration of mercury becomes more efficient at higher temperature operation, due to thermal decomposition of alkyl mercury compounds. Also, transition metals that can access a variety of oxidation states (e.g. Mn) and those that can easily capture mercury in the form of metal solutions (amalgam) tend to enhance the efficiency of capture via. atom transfer reactions.

Süd-Chemie has leveraged some of these chemistries to develop mercury removal catalysts that are capable of sequestration of mercury. Generally speaking, these materials involve transition metals such as copper and silver, and oxidation state shuttles such as iron and manganese. We believe that these materials are effective in sequestering mercury in its elemental form, as well as organic mercury compounds.

Mercury Disposal:

Absorption and de-sorption of mercury contaminant does not offer a viable method of removing and disposal of mercury. If absorption technique is employed, a down-stream secondary catalyst capable of mercury sequestration will become necessary, in order to truly contain the mercury contaminant. Sequestered mercury may be considered to be sufficiently non-volatile as to pose a safety hazard in handling. Disposal of spent catalyst that has mercury sequestered within it will have to follow local HS&E regulations.

Plant Operating Philosophy:

There are approaches to the handling of mercury contaminants in ethylene production that are based on how to solve this new contaminant problem that has crept into the operation of an established technology. However, contaminated cracking feedstock is a reality that is here to stay. Olefins plants have to accept the reality of varying amounts of heavy metal and other contaminants will be present in their feedstock. Chemically, the most efficient way to remove these contaminants is at a stage when the hydrocarbon matrix is chemically inert (before steam cracking), as opposed to a reactive matrix that contains olefins, acetylenes and hydrogen. Therefore, the operating philosophy for sequestration and removal of heavy metal contaminants must locate such processes directly where the feedstock is first introduced into the plant. This approach will eliminate side-reactions that lead to conversion of elemental mercury to other species in the cracking furnace. It will also expand the chemistries that can be employed to trap mercury species from an essentially inert hydrocarbon matrix.

Conclusion:

Elemental mercury and Organic mercury compounds have come to be recognized as a significant contaminant in Olefins plant feed stocks. Precious metal (palladium) catalysts used in selective hydrogenation of acetylenes and dienes are permanently poisoned by mercury contamination. Absorption removal of mercury only shifts the mercury contamination issue to the vent section; sequestration and removal of mercury presents a sustainable method. A viable plant operating philosophy will focus upon sequestration and removal of these contaminants from the feedstock directly as it enters the plant, before any further thermo-chemical transformations have a chance to spread the problem.

Paper 35e

Mercury Removal – An Overview

Steve Coleman

Equistar Chemicals, a Lyondell Company

George Nowowiejski

Equistar Chemical, a Lyondell Company

Marilyn Blaschke

Baker Petrolite

Prepared for Presentation at the 2005 Spring National Meeting

Atlanta, GA, April 10-14, 2005

Feedstock Contaminants in Ethylene Plants - Mercury

April 11, 2005

Abstract

An overview of commercially available mercury removal technologies will be presented. This overview will serve as an introduction to a poster session where various mercury removal suppliers will be available to discuss their technology.

Introduction

Elemental mercury and mercury compounds are trace contaminants in all types of fossil fuels. Ethylene plants which crack natural gases have been aware of mercury contamination for some time. With the movement to more heavy feedstocks, even plants which crack liquids exclusively are beginning to see more problems with mercury contamination. Although the dominant mercury species encountered in the ethylene plant is elemental mercury, other mercury compounds are possible depending on where the plant feedstock originates. For example, mercury exists almost exclusively as elemental mercury in natural gas, whereas crude oil and gas condensates can contain several forms of mercury. Table 1[1] shows various chemical forms of mercury in naturally occurring hydrocarbons.

Table 1 – Approximate Natural Abundance of Mercury Compounds in Hydrocarbons

Compound	Coal	Natural Gas	Gas Condensate	Crude Oil
Hg^0	T	D	D	D
$(CH_3)_2Hg$	?	T	T, (S?)	T, (S?)
$HgCl_2$	S?	N	S	S
HgS	D	N	Suspended	Suspended
HgO	T?	N	N	N
CH_3HgCl	?	N	T?	T?

Abundance: D (dominant) – greater than 50% of total
S (some) – 10 – 50%
T (trace) – less than 1%
N (none) – rarely detected
? indicates that data not conclusive

Operating Problems due to Mercury Contamination

Mercury presents several challenges to the ethylene plant operator.

- Potentially damaging to aluminum heat exchangers – When elemental mercury comes in contact with aluminum metal it makes a liquid metal solution with the base metal. This is known as amalgam corrosion. Liquid metal embrittlement (LME) occurs when liquid mercury diffuses along grain boundaries in the base metal. LME can lead to rapid propagation of cracks.[2] Corrosion of aluminum alloys in steam cracker cold boxes has been reported by various companies.[3,4]

- Catalyst poison for selective hydrogenation systems – Mercury is well known as a poison to palladium hydrogenation catalysts. These catalysts, commonly used for the selective hydrogenation of C_2-C_4 acetylenes and pyrolysis gasoline, are areas of concern when mercury contaminated feedstocks are cracked.[5]
- Presents a health hazard when it accumulates / concentrates in equipment, particularly during plant turnarounds where equipment is opened for inspection.

To avoid these problems, there are several options which have been developed for removing mercury from the hydrocarbon streams. One approach is to treat the feedstock before it enters the ethylene plant. This method has that the distinct advantage of keeping mercury from entering the plant, so all downstream concerns are mitigated. The main disadvantage is it can involve treating high volumes of material. Secondly, the mercury in the feed may be in several forms, some of which may be more easily removed (i.e. elemental) than others (i.e. organic). Finally, the feedstocks may contain other contaminants which can compete with mercury in the removal system. If competing contaminant levels are high, treatment of feedstock may not be a viable option.

For NGL crackers, most any of the technologies can be made to remove elemental mercury from the incoming feed. For plants cracking liquid feedstocks, the treatment is more challenging. Metal sulfides on alumina or carbon supports have been used in these services and are effective if the mercury is in its elemental form. IFP has developed a process for removing mercury in any form (organic and/or ionic as well as elemental) via a two step process involving hydrogenation of the feed to convert the mercury to an elemental form followed by a second step where the mercury is adsorbed onto a bed containing metallic sulfides on a support medium[9].

Another option is to treat specific streams within the ethylene plant. The treatment method can be tailored to the portion of the plant being treated. For example, to protect brazed aluminum core exchangers from mercury attack, the gas stream entering this portion of the plant can be treated. The sulfur-impregnated carbon bed technology has been used to treat the cracked gas downstream of the dryers to protect aluminum exchangers in the cold end of the ethylene plant.[4] Alternatively, silver-containing molecular sieve has been added to cracked gas dryers to keep mercury from contaminating the cryogenic separation equipment within the ethylene plant.[7]

Other locations where mercury removal internal to the ethylene plants has been successfully employed are in front of catalytic hydrogenation beds. Several catalyst suppliers have guard bed technologies specifically for protecting catalyst systems. These include Sud-Chemie (silver on alumina), CRI KataLeuna, and Axens (former IFP). An advantage with some of the guard materials is that they can be installed in the reactor, eliminating the need for a separate pretreatment vessel.

Mercury Removal Processes

Table 2[6] shows the various methods for removing mercury from hydrocarbons.

Table 2 – Mercury Removal Systems for Hydrocarbons

Reactant	Substrate	Complex Form	Application
Sulfur	Carbon	HgS	Gas
Metal Sulfide	Al_2O_3; Carbon	HgS	Gas, Condensate
Iodide	Carbon	HgI2	Condensate
Hydrogen, Metal Sulfide	Al_2O_3	HgS	Condensate
Ag	Zeolite	Ag/Hg amalgam	Gas, Condensate
Metal Oxide	Sulfided metal oxide	HgS	Gas, Condensate

Some of the more commonly used methods are discussed below:

Process: Reaction of Mercury with Sulfur
Process involves passing the feedstock over a bed of adsorbent material, consisting of an inert support on which a reactive compound has been bonded. The reactive species chemically reacts with the mercury to form a stable compound which is retained on the bed. Reactive compounds include sulfur or metal sulfides, which react with mercury to form mercuric sulfide. Supports include activated carbon and aluminum oxide.

Commercial Offerings:

- ***Sulfur-Impregnated Activated Carbon***
    - Patents: U.S. 4,500,327 (Nishino, 1985), U.S. 4,708,853 (Matviya, 1987)
    - Suppliers of Technology: Calgon (HCR®), NUCON (MERSORB®)
    - Process Considerations: Works on gas streams which are free of moisture and higher (C_5+) molecular weight species[8]. Liquid hydrocarbon will over time wash the sulfur off of the carbon support, rendering the the bed ineffective. Removes metallic mercury (Hg^0). Resultant mercuric sulfide is retained on the bed.

- ***Metal Sulfide on Activated Carbon***
    - Patents:
    - Suppliers of Technology: NUCON (MERSORB®)
    - Process Considerations: Will remove metallic mercury (Hg^0) from liquid streams. Resultant mercuric sulfide is retained on the bed.

- ***Metal Sulfide on Alumina***
    - Patents: U.S. 4,094,777 (Sugier, 1978)
    - Suppliers of Technology: Johnson Matthey(PURASPEC)
    - Process Considerations: Can treat liquid and gas streams. Moisture does not adversely affect bed performance. Mercuric sulfide is formed by

reacting with sulfur donated by the metallic sulfide. The resultant mercuric sulfide is retained in the bed.

Process: Hydrogenolysis of Organometallic Mercury followed by Adsorption
Process involves first passing liquid hydrocarbon feedstock over a nickel on alumina bed along with hydrogen. Organometallic Hg is converted into metallic mercury. The effluent then passes over a bed of adsorbent material consisting of metallic sulfide bonded to an inert support. The resultant mercuric sulfide is retained on the second bed.

Commercial Offerings:

- Patents: U.S. 4,911,825 (Roussell, Boitiaux and Cosyns, 1990)
- Suppliers of Technology: IFP / Axens (RAM™)
- Process Considerations: Process is designed specifically to remove mercury in either metallic or organometallic form from liquid hydrocarbon feedstocks. The 2-step process also claims to remove arsenic, another problematic ethylene plant contaminant.

Process: Reaction of Mercury via Amalgamation with Metal on a Molecular Sieve Support.
Process involves the reaction of mercury with another metal, such as silver, bonded to a support, such as a molecular sieve. The mercury is released into the regeneration gas when the molecular sieve bed is regenerated. This technology has been widely used in natural gas processing, but has been installed in at least 15 ethylene plant dryers worldwide.

Commercial Offerings:

- Patents: U.S. 4,874,525 (Markovs, 1988)
- Suppliers of Technology: UOP (HgSIV™)
- Process Considerations: Can treat gas and liquid streams. The mercury is removed coincident with removal of moisture and other adsorbed contaminants in the molecular sieve dryer, and has no effect on the sieve's capacity to adsorb moisture or other contaminants. A second step may be required to remove the mercury from the dryer regeneration gas.

Process: Reaction of Mercury with Iodide-Impregnated Activated Carbon
Process involves passing mercury-containing hydrocarbon liquid over a carbon bed impregnated with iodide. The mercury must be oxidized to react with the iodide.

Commercial Offerings:

- Patents: U.S. 5,336,835 (McNamara, 1994)
- Supplier of Technology: Calgon (HGR®-LH)
- Process Considerations: Designed for liquid phase treatment. Streams should be free of separated water prior to treatment.

Disclaimer:

The purpose of this report is to provide the Ethylene Producers Committee of the AIChE an overview of mercury removal methods. This report is not comprehensive, nor does it claim to offer the best possible mercury removal methods and/or suppliers of that technology.

Bibliography

1. Mercury in Petroleum and Natural Gas: Estimation of Emissions from Production, Processing and Combustion. Prepared for the EPA Office of Air Quality Planning and Standards. Prepared by nation Risk Management Research Laboratory; Sept 2001; page 28.
2. Rios, Julio, Coyle, David A., Durr, Charles A., Frankie, Brian M., *Removal Of Trace Mercury Contaminants From Gas And Liquid Streams In The LNG And Gas Processing Industry*; Proceedings, 77th Annual Convention – Gas Processors Association, 1998.
3. Frenken, P. and Hubbers, T., AIChE Spring Nation Meeting, Houston, paper 27a, April 1991.
4. English, J. J., *A Mercury Induced Aluminum Alloy Piping Failure in an Ethylene Manufacturing Plant*, Proceedings of the First Annual Ethylene Producers Conference, Houston, paper 74c, April 4, 1989.
5. McPhaul, D. R. and Reid, J. A., *Ethylene Plant Feedstock Contaminants Treatment*, Proceedings of the 7th Annual Ethylene Producers Conference, Paper 20e, Houston, March 1995.
6. Mercury in Petroleum and Natural Gas: Estimation of Emissions from Production, Processing and Combustion. Prepared for the EPA Office of Air Quality Planning and Standards. Prepared by nation Risk Management Research Laboratory; Sept 2001; page 42.
7. Coleman, Steve, Jason Ries, Don Campbell, and Keith R. Clark, *Mercury Removal from Cracked Gas and Liquid Streams*, Proceedings of the 17th Annual Ethylene Producers Conference, Paper 55c, New Orleans, April 2004.
8. Sugier, A. and F. La Villa, 1978, *Process for Removing Mercury from a Gas or Liquid by Absorption on a Copper Sulfide-Containing Solid Mass*, U.S. Patent 4,094,777.
9. Sarrazin, P., Cameron, C., Barthel, Y. and M. E. Morrison, IFP, *Arsenic and Mercury Removal from Hydrocarbons Streams,* Proceedings of the 4th Ethylene Producers' Conference, March 31, 1992.

Operating Company Experiences with Mercury

"Opportunity Feedstock" Strikes Again

Anonymous Contributor

Experience with the Decontamination and Change-out of a Carbon Bed Treater

Jason Cooper and Victor Trivitt
Equistar Chemicals, a Lyondell Company

Mercury Removal from a Raw C4 Stream

Brian Dillingham
Johnson Matthey

Prepared for Presentation at the 2005 Spring National Meeting
Atlanta, GA, April 10-14, 2005
UNPUBLISHED

Abstract: Several brief descriptions of incidents where mercury has been encountered in operating facilities and how the situation was handled are described.

"Opportunity Feedstock" Strikes Again

Anonymous Contributor

International Refining and Petrochemicals Company

Prepared for Presentation at the 2005 Spring National Meeting

Atlanta, GA, April 10-14, 2005

UNPUBLISHED

April 11, 2005

In the mid-1990s our company was scouting the world's condensate markets for inexpensive ethylene plant feedstocks and came across a very attractively priced South East Asian condensate. Its predicted yields of ethylene and propylene were not particularly good, however the very low price more than made up for that from a profitability point of view. As a consequence, we shortly found ourselves in possession of a tanker load of it, courtesy of our feedstock traders.

A mercury absorption bed downstream of the driers had been previously installed to protect our aluminum cold box exchangers so we felt that from a safety standpoint, some mercury contamination would be acceptable. Unfortunately, we should have more closely considered the distribution of mercury in a hydrocarbon distillation system as it distributes mainly to the C_3 and C_4 fraction, and our backend methyl acetylene / propadiene (MAP) converter was poisoned much more rapidly than we could regenerate converter beds. That created an off spec condition which was difficult to recover from.

Take home lessons from our experience are:

- Ethylene Plants cracking condensates or C_3 and C_4 NGLs should have feedstock specifications covering mercury, keep them current, test all feedstocks and allow violation of those specifications only after thorough review.

- Understand the paths that mercury is most likely to follow and how this distribution might impact sensitive areas in your plant.

  The refining experts in our company report that while some mercury distributes to propane in refineries, it primarily distributes to the iC_4 fraction which is not surprising considering previous reports of mercury being found in BD plant feed systems. Again, before considering high mercury feedstocks, its full distribution needs to be examined. Many systems may be affected in addition to well known cold box areas of accumulation.

- Incorporate the advice of your engineering contractor

  A review of the above mentioned mercury poisoning incident showed that the engineering contractor who designed and installed the mercury absorption bed specifically mentioned that the MAP converter would still be subject to poisoning by mercury since a C_3/C_4 stream bypassed the dryer area where the mercury absorption bed was installed. Obviously, the importance of this comment was not fully appreciated and incorporated into our feedstock specifications.

- Palladium based MAP converters can be regenerated from mercury poisoning; however, the rate of poisoning needs to be considered.

  The advice given by DSM in a previous EPC paper on the regenerability of Pd catalysts from mercury poisoning is good. However, the rate of poisoning needs to be taken into account also, as it may not be feasible to regenerate poisoned MAP converters quickly enough.

Experience with the Decontamination and Change-out of a Carbon Bed Treater

Jason Cooper, Process Engineer

and

Victor Trivitt, Chemist

Equistar Chemicals, a Lyondell Company

Prepared for Presentation at the 2005 Spring National Meeting

Atlanta, GA, April 10-14, 2005

UNPUBLISHED

April 11, 2005

Background

Lyondell Chemical Company's Chocolate Bayou olefins plant began cracking significant volumes of Algerian condensate, a feedstock with trace mercury contamination (10 – 50 ppb) in 1985. In 1987, during a planned shutdown the weld between an eight-inch pipe and the first brazed-aluminum core exchanger failed due to metal embrittlement; the cause being its contact with liquid mercury. The plant responded to this incident by installing a bed of sulfur-impregnated, activated carbon upstream of the feed to the cold box for mercury removal. At the time of the incident, sulfur-impregnated, activated carbon was considered 'best available' technology for inorganic mercury removal in several industry applications, but Chocolate Bayou was one of the first facilities to use carbon technology on the cracked gas dryer effluent gas. The new carbon bed, since being placed on-line in early 1989, performed well and without incident until the pressure drop across the bed increased significantly following a unit upset in late 2001. Due to the economic impact and concerns around channeling or other detrimental effects the pressure drop might have on bed performance, the decision was made to bypass and change out the carbon bed.

Changing out of the carbon bed required three general steps. First, the spent carbon had to be decontaminated, such that the containing vessel could be safely opened. Second, the spent carbon had to be removed and disposed of according to environmental requirements, and then the vessel could be recharged with fresh carbon. Third, the new carbon had to be placed on-line. Each of these steps had specific engineering challenges to be overcome for the plant to maintain a safe and environmentally responsible operation.

Decontamination

After a scheduled olefins turnaround in the fourth quarter of 2002, planning began for the decontamination; the object being complete hydrocarbon desorption from the spent carbon so the change out and maintenance could be done in a safe atmosphere and so the spent carbon could be disposed of as a non-hazardous waste. The carbon bed at Chocolate Bayou, located immediately downstream of the cracked gas dryers, was designed for a very long life (35 to 40 years) and, as a result, the vessel was not installed with any connections or piping for decontamination. The bed was also designed for the same gas velocity as in the dryers, which resulted in a short and squat bed design with a length to diameter ratio of less than one. In addition to these design issues, properties of both the adsorbent and the components in the cracked gas would add to the difficulty of decontamination. Activated carbon has a very broad pore size that allows the adsorption of many different types of compounds by varying amounts. The hydrocarbon components passing through the carbon bed range from hydrogen through C_5 hydrocarbons, and include saturates, olefins and diolefins. Of these the diolefins, iso-olefins and any aromatics that may be present will adsorb very strongly due to van der Waals forces. These would be the last to desorb completely during decontamination.

Since Equistar Chemicals, a Lyondell Company, did not have any previous experience decontaminating a bed of this type and the plant had several options for the gas stream to be used to decontaminate the bed. Equistar's corporate process development group performed several test desorptions during the planning stage to ascertain the minimum temperature required for desorption. These tests provided data pertaining to hydrocarbon desorption and the possible sublimation of the impregnated sulfur at elevated temperatures. The tests involved exposing a small, pilot-sized carbon bed to the cracked gas for a 24-hour period and then heating the spent bed in a gas chromatograph oven with a nitrogen sweep.

During the heating, a thermal conductivity detector provided real-time analysis of the decontamination progression, and bag samples were gathered at regular intervals and analyzed to provide a hydrocarbon fingerprint of the components being desorbed. As a result of the tests wherein (whereas) the chromatograph oven was maintained at 212°F and 300°F, total desorption of the hydrocarbons was achieved. As expected, desorption followed a normal exponential decay with butadiene and isobutylene being the last components to completely desorb. Analysis of the carbon before and after testing showed no appreciable loss of sulfur from the carbon surface, even at 300°F.

Based on this data and vessel metallurgy, the decision was made to conduct the decontamination at 200°F using plant nitrogen. Expectations were high that this would complete the decontamination in a reasonable time, while mitigating the loss of impregnated sulfur from the activated carbon. It was also planned that after the carbon was removed from the vessel, the actual bed life would be estimated by the amount of unreacted sulfur remaining on the carbon surface.

The 200°F plant nitrogen was introduced into the carbon bed in late March 2003 to start the decontamination. This flow was routed into the bottom of the bed and exited through a top relief valve bypass to the olefins main flare. All connections for the nitrogen flow to and from the bed were two-inches, which would thus limit the flow through the bed. These small connections combined with the bed dimensions ($L/D < 1$) meant that flow through the bed would probably not be well distributed. However, the premise was that getting the entire bed up to temperature was most important for hydrocarbon desorption and sufficient perturbation would be created by the bulk nitrogen flow to sweep away hydrocarbon vapors. Batch sampling of the effluent gas would be relied upon as a measure of decontamination progress.

While the initial warm-up was slow to prevent polymerization in the pores, bed temperatures rose to 190°F within four days of introducing the nitrogen and, at this point, sampling of the effluent gas began. The actual desorption curve was very similar in both exponential decay and hydrocarbon components to the curves generated during the pilot testing, but desorption time for the bed was much longer due to the flow restrictions. Bed temperatures had to be held at 190°F for a little over 15 days before sample analyses indicated sufficient decontamination. At this point, the bed was cooled and moist air was introduced and vented to atmosphere to prepare the bed for opening. The purpose of the moist air was to mitigate three safety concerns for vessel entry: excessive dust, static electricity and oxygen content of the atmosphere.

Very shortly after introducing the moist air, temperatures became elevated as if a mild exotherm was moving through the bed. A strong sulfur odor was also noticed at the bed vent. The suspected source of the exotherm and sulfur odor was the hydrolysis of carbonyl sulfide that may have been adsorbed on the carbon; the hydrolysis reaction produces both hydrogen sulfide and heat. Carbonyl sulfide (COS) production from the reaction of carbon dioxide (CO_2) and hydrogen sulfide (H_2S) in the alkaline atmosphere of molecular sieve is well documented in natural gas processing and, due to its polar properties, COS would adsorb onto carbon very strongly. At Chocolate Bayou during the 13 years the carbon bed was in operation immediately downstream of the dryers, it is very likely that COS was produced from CO_2 and H_2S that slipped past the caustic tower either continuously or during unit upsets.

Since it was suspected that the moisture was reacting away the COS, the purge with moist air needed to be continued, but the offensive odors could have created complaint problems. Due to these concerns, the bed was routed back to the flare for purging with ambient nitrogen. This was repeated (moist air and then purge) twice and on the third time the exotherm and odors subsided and breathable oxygen levels were detected at the vent. Upon removing a vessel manway, hydrocarbon vapors were immediately detected along with both the sulfur odor and a mild exotherm, all indicating incomplete decontamination. Insufficient flow and maldistribution of the nitrogen flow through the bed was concluded as the likely cause.

The next decontamination medium used was low pressure steam, which was routed in the same manner as the nitrogen but provided two additional benefits. First, it was much warmer, at 365°F, and second it would provide the water necessary to hydrolyze any COS. The steaming continued for almost 30 days. Outlet condensate samples initially showed a hydrocarbon content, but decreased quickly to zero in a few days. However, the sulfur odor in the condensate and from bleeds around the system persisted.

As a last effort to wash out the sulfur, the vessel was twice filled with condensate and drained to waste water. After the second fill and drain, no sulfur odors or hydrocarbons were detected in any of the water samples. The condensate washes took two days and manways were immediately removed afterwards with no odor or hazardous atmosphere issues. Including all of the maintenance activities between the decontamination phases, the total effort took almost four months.

Removal and Replacement of the Carbon Bed

The removal of the spent carbon and recharge with new adsorbent was executed meeting all safety and environmental goals. No hazardous atmosphere work was required, no mercury exposure issues were encountered and the spent carbon was analyzed and disposed of as a non-hazardous waste due the successful decontamination efforts. Upon removing the spent carbon, the cause of the excessive pressure drop was discovered to be the blinding of two spun-glass filter blankets in the bottom of the carbon bed with carbon dust. The source of this carbon dust would have been from pulverizing some of the carbon granules during unit upsets when the pressure at the discharge of the process gas compressor is suddenly decreased. These carbon fines could migrate down through the bed and get caught on these filter screens. A particular upset in late 2001 created the majority of the fines or moved existing fines down to the screens. Due to this behavior, a decision was made during the change out not to reinstall these filter screens. The risk of making this design change was getting carbon fines caught in the core exchanger passes downstream, but this risk was deemed minimal since the unit has filter screens or strainers in front of each exchanger and in various other places throughout the cold end. This design change in the carbon bed internals would prove to have some environmental and operational consequences in the first month after the fresh carbon was brought on-line.

Commissioning of the Fresh Carbon Bed

Placing the fresh carbon bed in service without any safety or operational issues was also an activity that, like the decontamination, had some very definitive engineering challenges. First, the exotherms generated as different hydrocarbons were adsorbed by the carbon had to be controlled both for the long-term operability of the fresh carbon and the integrity of the containing vessel. The vendor recommended that the sulfur-impregnated, activated carbon be kept under 180°F because temperatures above this would risk sublimation of the sulfur from the carbon surface, which would directly impact bed life. Second, the fresh activated carbon contained both water (about 3 percent by weight) and was also saturated with oxygen. Since neither of these is desirable in the cold separations end of an olefins plant for both safety and operability reasons (oxygen in the carbon could cause polymerization as the hydrocarbons adsorb) both had to be completely removed from the carbon while adhering to the temperature limitation of 180°F. Operations personnel at Chocolate Bayou chose to meet these challenges in 2003 using a slightly updated version of the start-up procedure vis-à-vis that was used when the original bed was placed in service in 1989.

Removing the water and oxygen from the carbon without excessive heat, required pulling vacuum on the vessel containing the carbon while sweeping oxygen or water vapor to atmosphere. The dew point of the exiting nitrogen was measured to judge the dryness of the new carbon. In 1989, the vacuum was pulled using steam jets, but in 2003, the services of a pipeline contractor with a vacuum pump were used.

The final step of the start-up with the introduction of hydrocarbons to the bed had to be very slow and deliberate due to the very large heats of adsorption that would be generated. Given the broad pore size distribution of the carbon, each hydrocarbon would adsorb to a certain extent with compounds of low molecular weight adsorbing first and with adsorption increasing with increased pressure. The exotherms generated by the introduction of a stream such as olefins cracked gas can be described as waves moving through the bed with the methane wave passing first, then ethylene and so on. In anticipation of this behavior, unit tail gas, comprised of mostly methane and hydrogen, was introduced first at a very small flow and to the main flare to minimize pressure. By carefully raising the flow and pressure, methane was fully adsorbed onto the carbon up to the process gas compressor suction pressure of 10 psig. The process gas was introduced in the same manner as the methane, except all flow was recycled back to the compressor suction to minimize flaring. Larger exotherms were noticed with the process gas than with the methane due to its olefinic content. Bed temperatures did reach 200°F for almost an hour and the pressure on the bed had to be reduced back down to 10 psig. With the passing of this initial large exotherm, the pressure was slowly, but easily increased to the full 520 psig discharge pressure. Full flow was established shortly afterwards, ending the start-up efforts that lasted three days.

Overall the decontamination, change-out and start-up of the Chocolate Bayou sulfur-impregnated, activated carbon bed was successful despite the challenges discussed above. Additionally, several changes were made to the original plans. All safety and environmental goals for the activities were achieved with no injuries, no hydrocarbon or mercury exposure and no environmental consequences other than the disposal of the spent carbon as a non-hazardous waste. Only the bed temperature upper limit of 180°F was violated during the start-up, but this violation was well within the 300°F limit established in pilot testing (to mitigate sulfur loss).

The new carbon bed appears to be performing well, as no mercury break-through has been observed. In light of these successes, however, there were some lessons learned and several items that will be changed in the event that maintenance on the carbon bed must be done in the future. First, the decontamination plan should not be as much trial and error. The combination of hot nitrogen, steam and a condensate flush worked very well and planning for these three events in sequence should reduce decontamination time and the overall time that the carbon bed is out of service. Second, the spun glass filter screens in the bottom of the carbon bed should be replaced. In the first month after the carbon bed was placed in service, filters downstream of the bed plugged several times and required back flushing to the main flare. These environmental events could have been avoided had these screens, which were removed to prevent another high-pressure drop from occurring, been in place. Reinstalling the screens during the next change out risks having to repeat the recharge more frequently, but the environmental costs and potential operational consequences of the plugged strainers outweighs the cost and inconvenience of a bed recharge as infrequently as every 10 or 15 years.

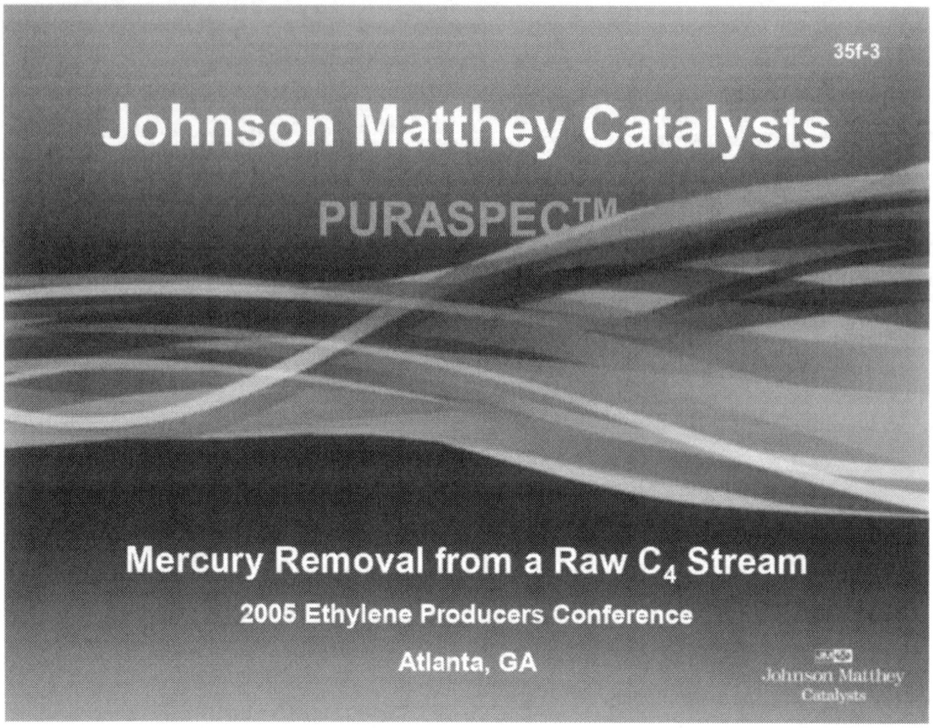
35f-3
Johnson Matthey Catalysts
PURASPEC™
Mercury Removal from a Raw C_4 Stream
2005 Ethylene Producers Conference
Atlanta, GA
Johnson Matthey
Catalysts

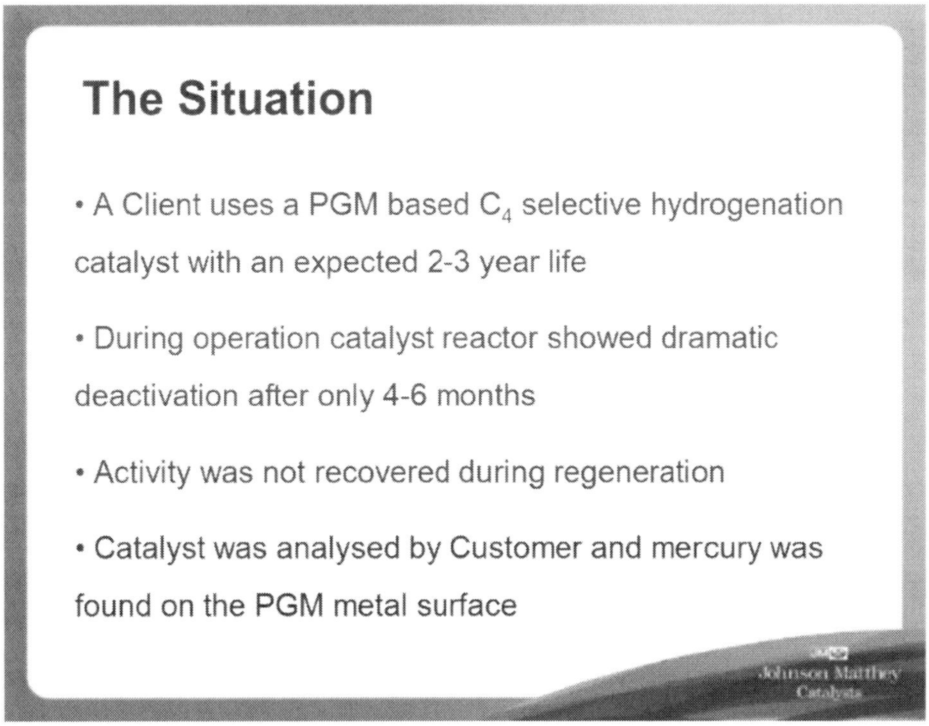

A Client is operating naphtha fed ethylene cracker. The unit recovers the C4 fraction rich in di-olefins and selectively hydrogenates it with a PGM based catalyst. The move to heavier cracker feedstocks has caused the catalyst bed to experience much shorter lifetimes. The planned life for the bed was 2-3 years. After just 4-6 months on line the bed had lost activity and required regeneration. When the client analyzed the catalyst, several hundred ppm of mercury was found on the catalyst surface causing the premature catalyst deactivation.

An investigation of commercially available mercury removal methods was conducted and Johnson Matthey's Puraspec 7156 was chosen to provide protection to the unit. Puraspec 7156 was shown to be particularly effective in this liquid duty over other technologies.

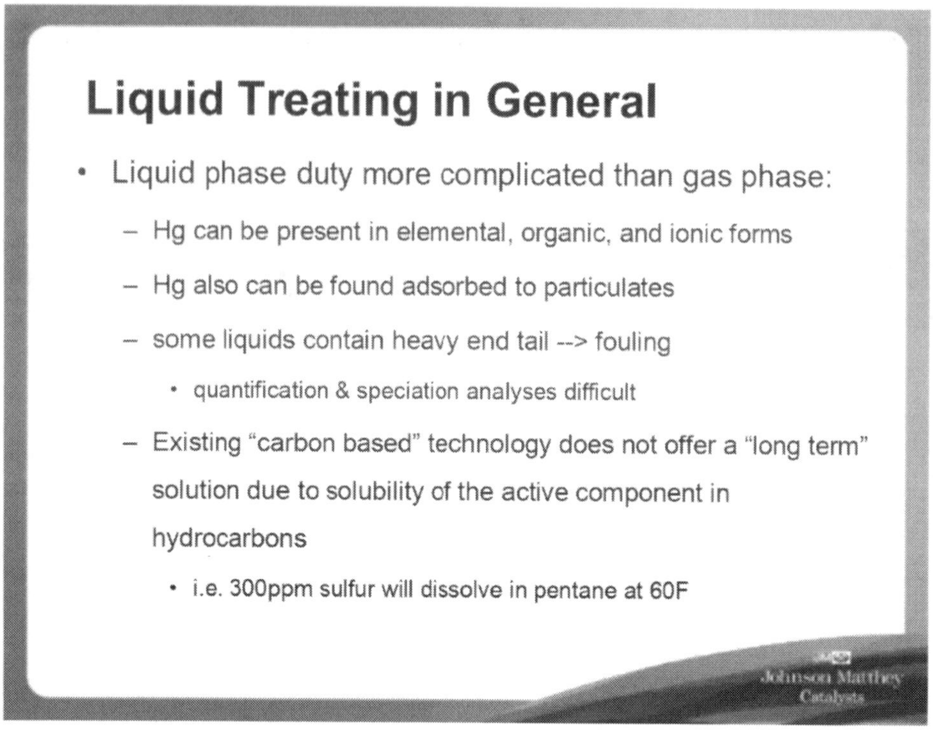

Johnson Matthey has over 10 years of experience removing elemental and organic Mercury species from natural gas, liquefied natural gas, and refinery streams. Through our PURASPEC program, Johnson Matthey has spent considerable time and effort analyzing and testing the removal efficiencies of our sorbent range on mercuric species in various condensates from around the world. Developing this database of knowledge makes us uniquely qualified to handle this challenge.

The objective of removing Mercuric species from a liquid fraction was a particular challenge for the following reasons:

- In liquid phase, as opposed to gas phase, mercury can be found in organic, ionic and elemental form.
- Mercury can also be found adsorbed to particles.

Some liquid streams contain a heavy tail that promotes fouling and can make quantification and qualification of the mercury species difficult.

Existing sulfided carbon technologies failed due to removing of the active sulfiding component in the liquid hydrocarbon. i.e. 300ppm sulfur will dissolve in liquid pentane at 60F

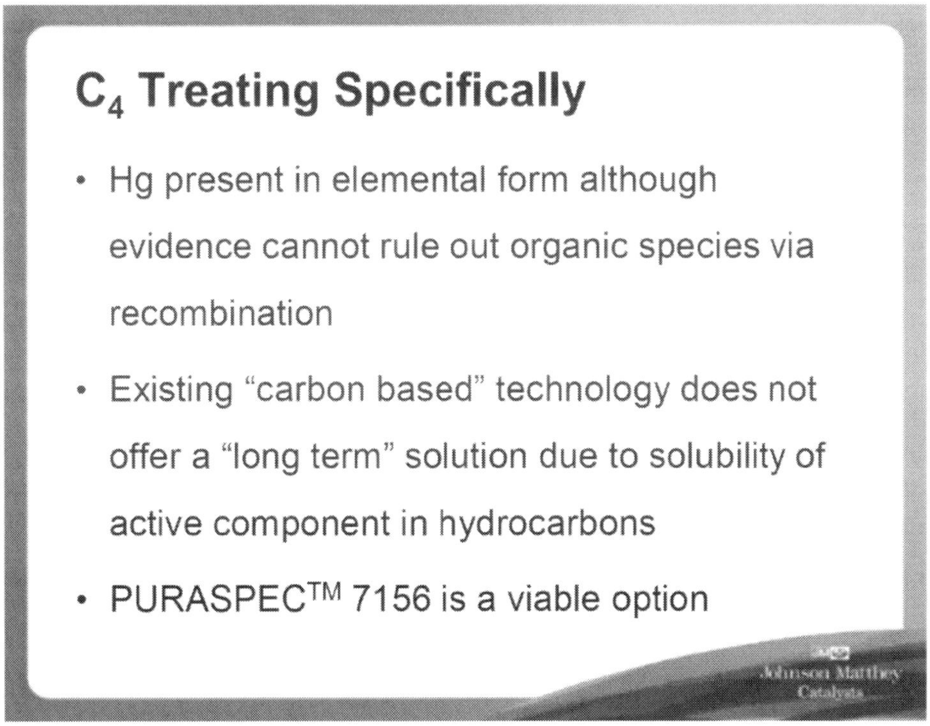

We believe that the Mercury species found in this stream are elemental. This is confirmed by the low effluent concentration and the sharp bed profile demonstrated during the pilot study. However, recombination if mercury with organic species in the stream is possible.

The Client had experience with sulfided carbon in gas phase treating applications. However, in liquid phase C4 streams, the solubility of sulfur limits the usefulness of traditional carbon based technologies. Puraspec 7156 addresses this concern by incorporating a fixed sulfided metal complex that is effective for removing mercury but not soluble in liquid hydrocarbons.

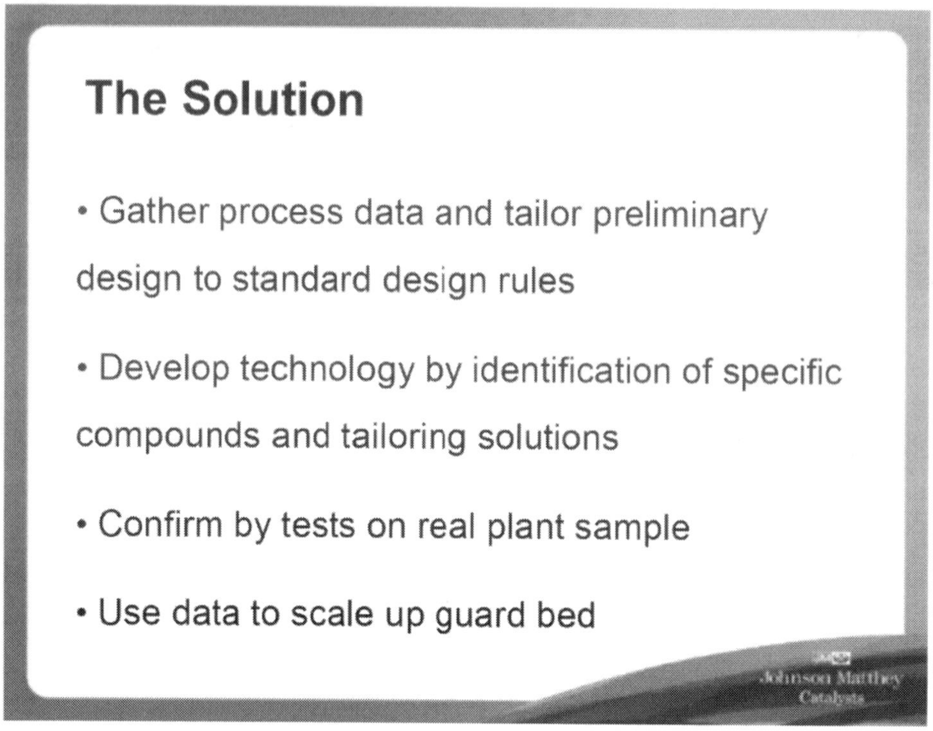

The initial prediction provided by Johnson Matthey combined the use of the current unit conditions matched with our PURASPEC 7156 sorbent to meet an LHSV of 5. This created an economic framework for the Client. Once agreed, a slipstream unit was established on the plant site to allow proof of concept in close to ideal conditions. This testing was performed over a period of 22 months in which time extensive analytical work was conducted on the inlet and outlet streams to determine overall efficiency and further justify/alter the economic model. All outlet analyses consistently showed <1ppbw mercury. "Accelerated" testing via mercury "doping" was deemed unrealistic in predicting performance.

The Solution

Olefin Cracker, C_4 Purification

50% Butadiene	Pressure: 12 barg
25% i-Butene	Temp: Ambient
17% n-Butene	LHSV: 5 h^{-1}
7% Butane	
1% Ethylacetylene	

Hg Inlet: 40-150 ppbw (70 ppbw average)

Hg Exit: Undetected

2 test done:

1. Continuous side stream 18 months 5h^{-1}, then 10h^{-1} for additional 4 months with no slip
   Granules at top contained 3.5% Hg
2. Test reactor with 15 zones, real feed and determined profile (all Hg in top 45% of bed)
3. Post mortem analysis showed low carbon content

Johnson Matthey Catalysts

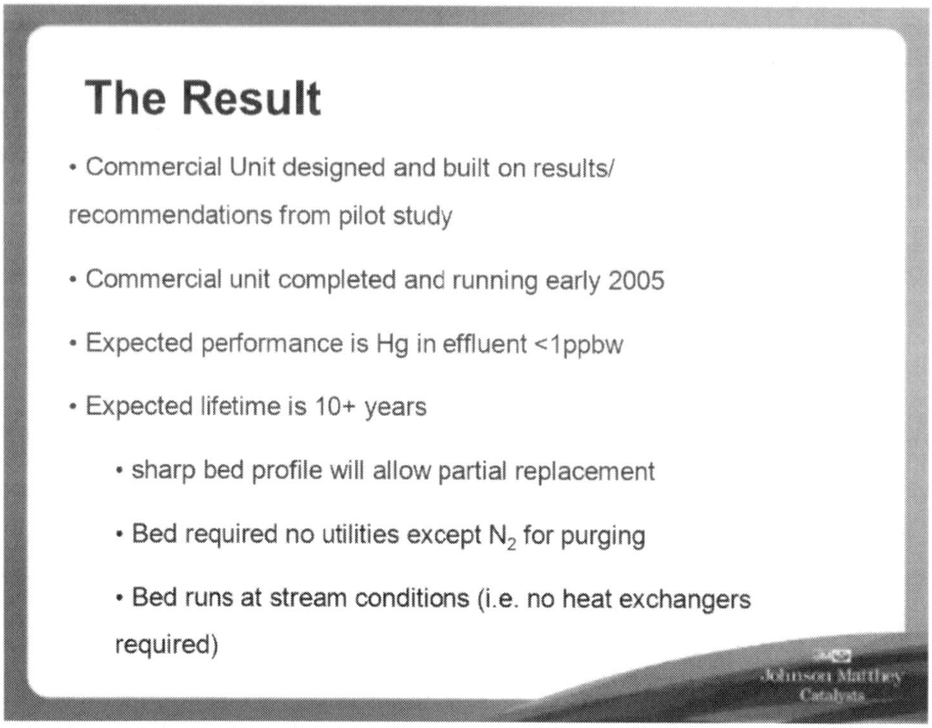

As a result of this collaboration, commercial scale units have been built based around Puraspec design recommendations and are now in operation. The expected performance of the guard bed is <1ppbw in the effluent stream and a conservative life of >10 years. The bed requires no special loading or activation procedures and requires no utilities except nitrogen gas for purging. The bed operates at stream conditions.
Confidentiality bars further discloser of unit details.

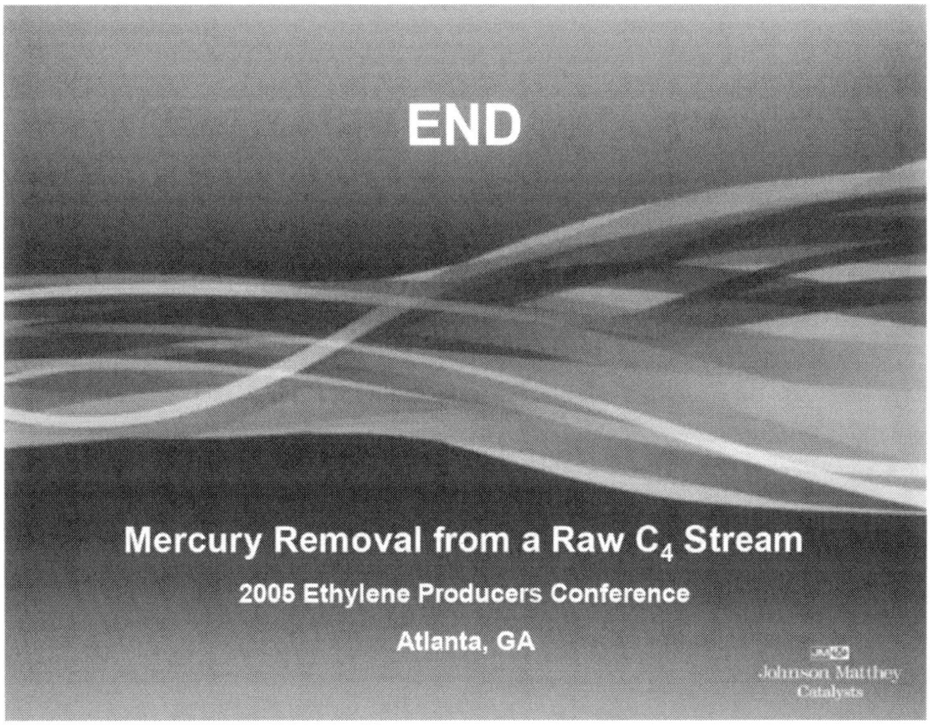
END
Mercury Removal from a Raw C_4 Stream
2005 Ethylene Producers Conference
Atlanta, GA
Johnson Matthey
Catalysts

The Energy Outlook

Stephen P. Hart
Vice President - Global Olefins Business Unit
ExxonMobil Chemical Company

Prepared for Presentation at the 2005 Spring National Meeting, Atlanta, GA April 11-14
17th Annual Ethylene Producers Conference
Session TA001 – Keynote Speech

April 11, 2005

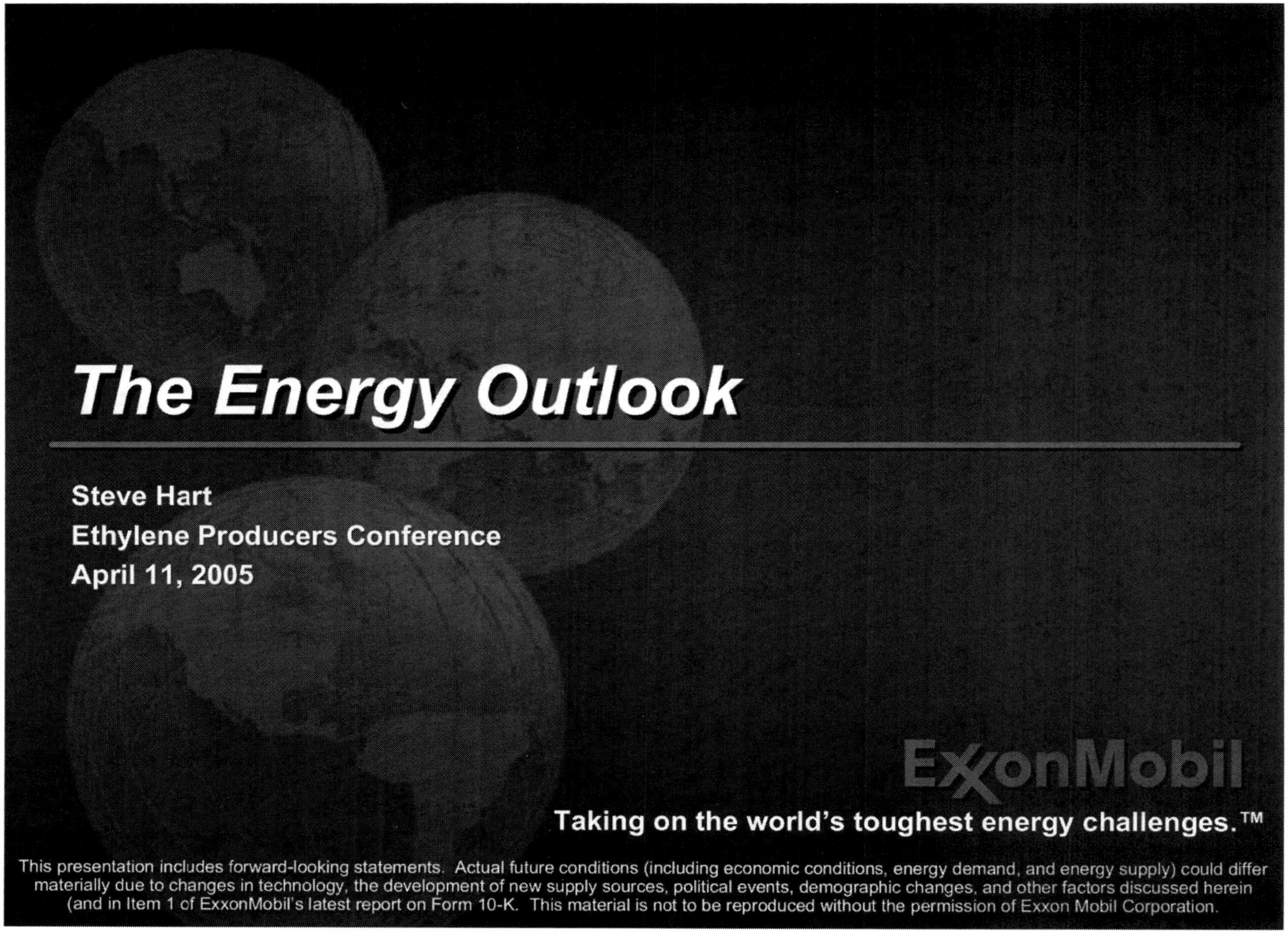
The Energy Outlook
Steve Hart
Ethylene Producers Conference
April 11, 2005
ExxonMobil
Taking on the world's toughest energy challenges.™
This presentation includes forward-looking statements. Actual future conditions (including economic conditions, energy demand, and energy supply) could differ materially due to changes in technology, the development of new supply sources, political events, demographic changes, and other factors discussed herein (and in Item 1 of ExxonMobil's latest report on Form 10-K. This material is not to be reproduced without the permission of Exxon Mobil Corporation.

Long-Term Energy Outlook

- Developed by ExxonMobil economists & energy planners.
- Results consistent with IEA and DoE studies.

Includes:

- Forecasts of energy demand through 2030.
- Expected sources of energy & challenges.
- Role of technology.

ExxonMobil

Summary

- Economic growth generates increased demand for energy.
- Oil and gas remain the primary sources of energy through the mid-century.
- Natural gas is the fastest-growing primary energy source.
- LNG imports required to meet North American gas demand growth.
- Higher gas prices will impact competitiveness of North America's chemical producers.

ExxonMobil

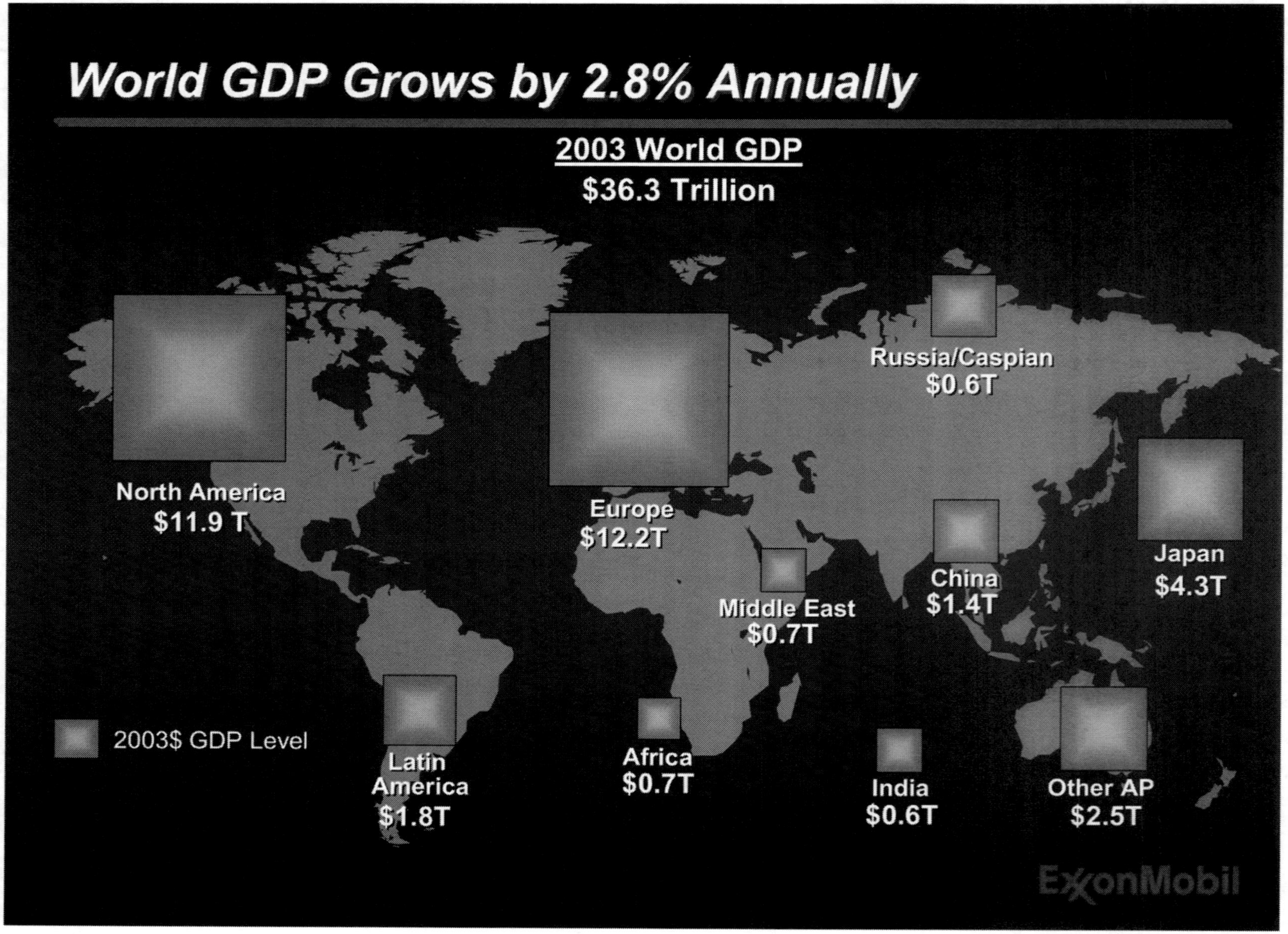
World GDP Grows by 2.8% Annually
2003 World GDP
$36.3 Trillion
North America
$11.9 T
Europe
$12.2T
Russia/Caspian
$0.6T
Middle East
$0.7T
China
$1.4T
Japan
$4.3T
Latin America
$1.8T
Africa
$0.7T
India
$0.6T
Other AP
$2.5T
2003$ GDP Level
ExxonMobil

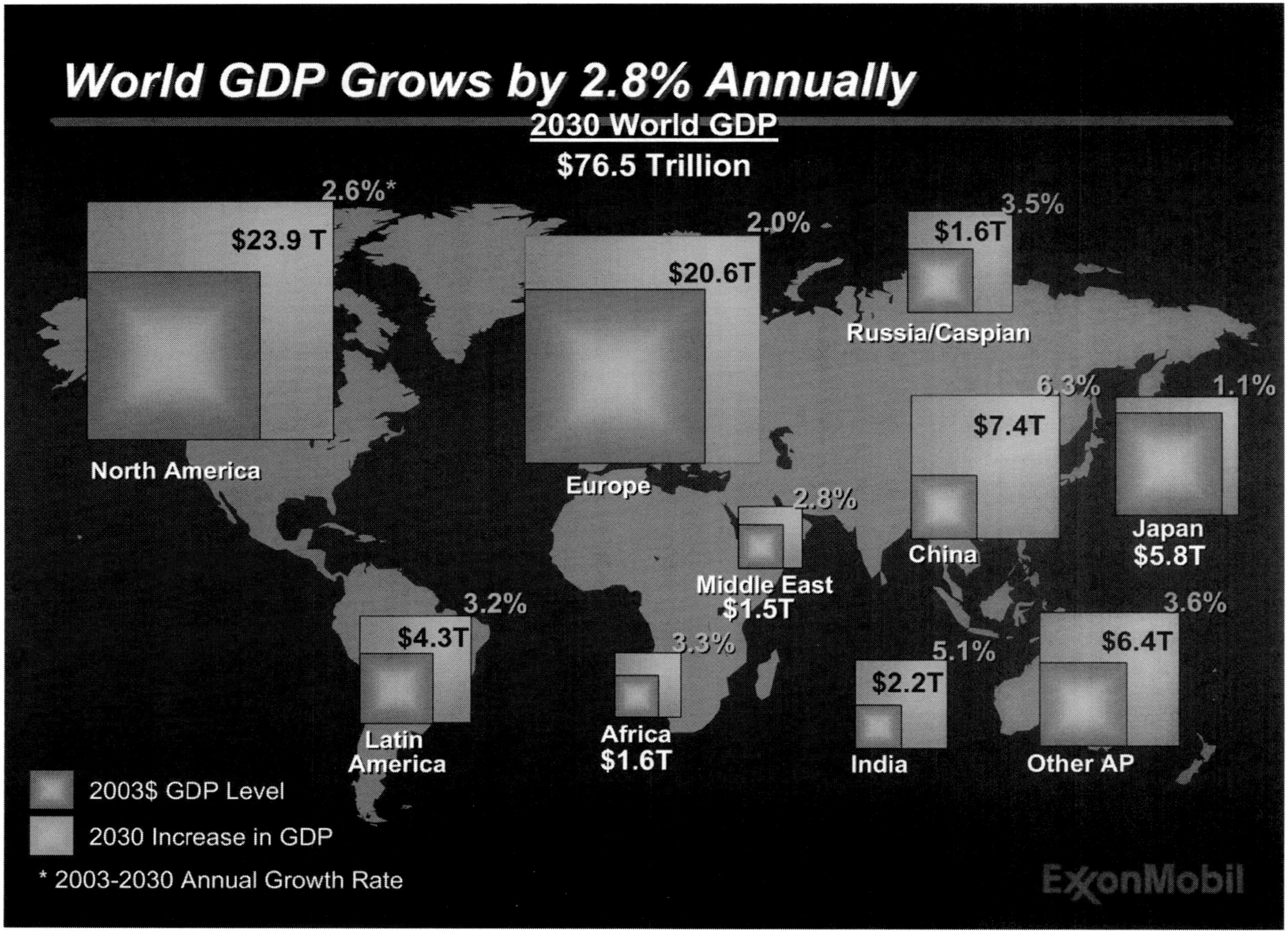
World GDP Grows by 2.8% Annually
2030 World GDP
$76.5 Trillion
2.6%*
$23.9 T
North America
2.0%
$20.6T
Europe
3.5%
$1.6T
Russia/Caspian
6.3%
$7.4T
China
1.1%
Japan
$5.8T
2.8%
Middle East
$1.5T
3.2%
$4.3T
Latin
America
3.3%
Africa
$1.6T
5.1%
$2.2T
India
3.6%
$6.4T
Other AP
2003$ GDP Level
2030 Increase in GDP
* 2003-2030 Annual Growth Rate
ExxonMobil

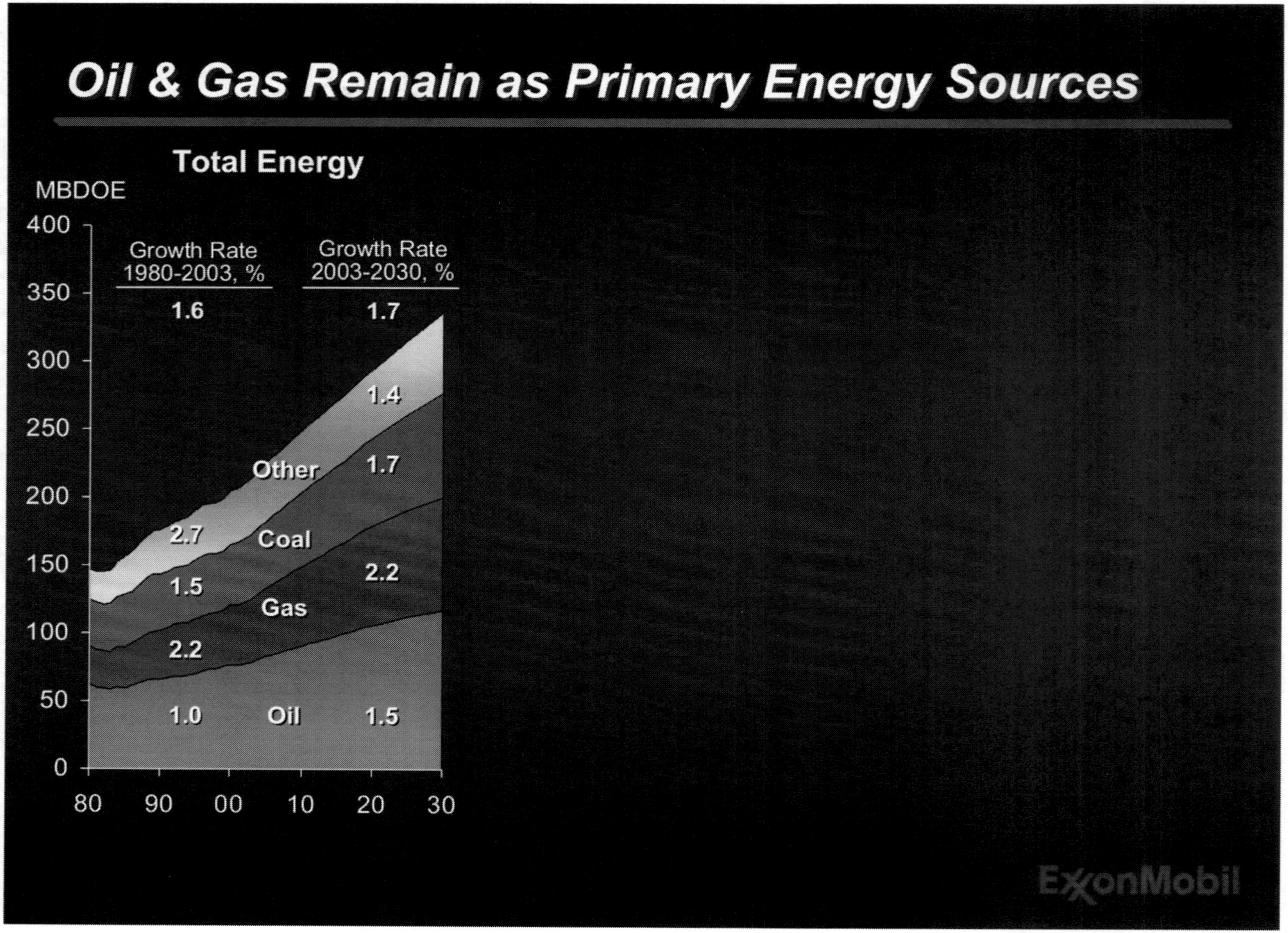
Oil & Gas Remain as Primary Energy Sources
Total Energy
MBDOE
400
350
300
250
200
150
100
50
0
80
90
00
10
20
30
Growth Rate 1980-2003, %
Growth Rate 2003-2030, %
1.6
1.7
2.7
Other
1.4
1.5
Coal
1.7
2.2
Gas
2.2
1.0
Oil
1.5
ExxonMobil

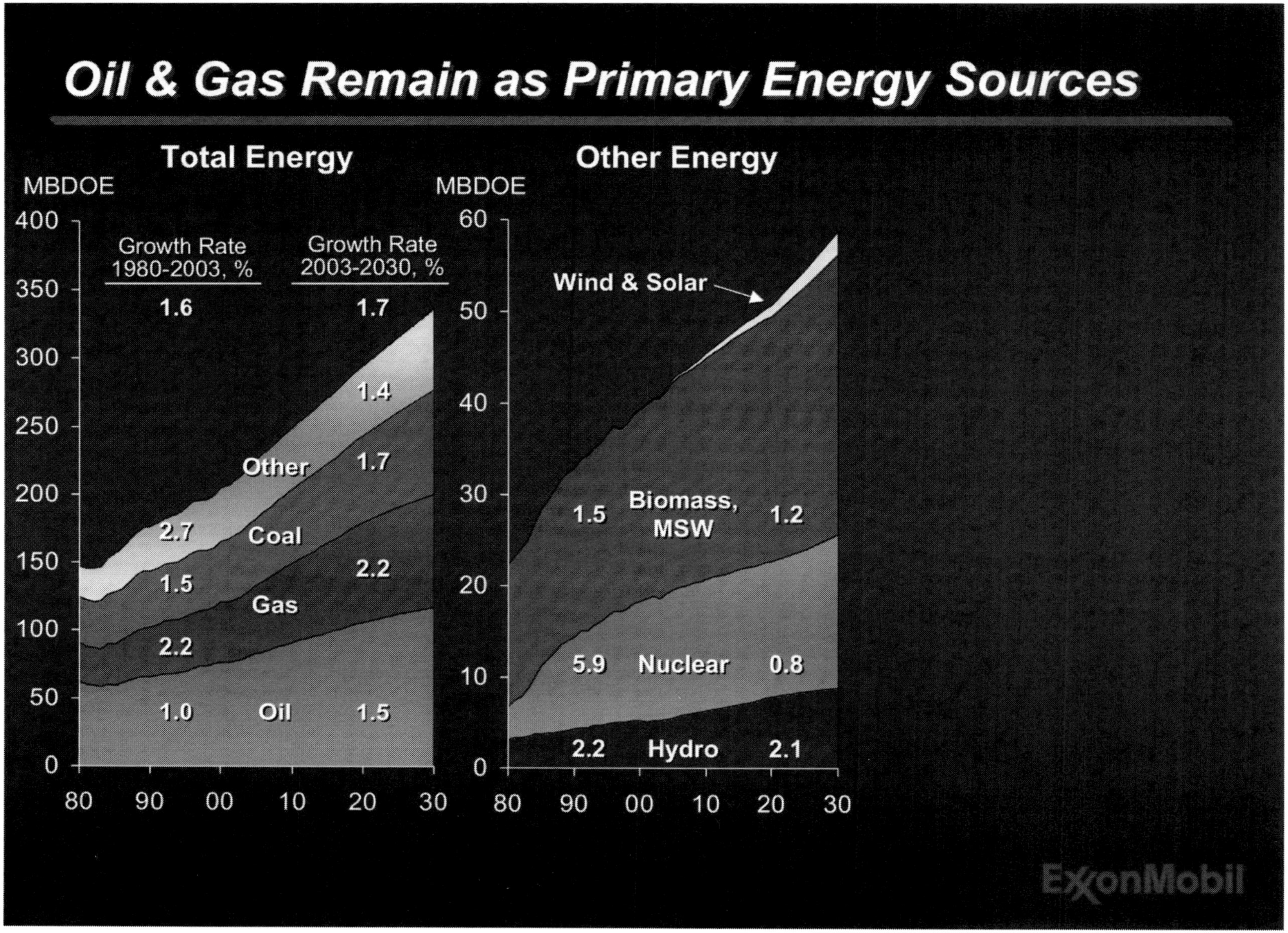

Oil & Gas Remain as Primary Energy Sources
Total Energy
MBDOE
Growth Rate 1980-2003, %
Growth Rate 2003-2030, %
1.6
1.7
2.7
Other
1.4
1.5
Coal
1.7
2.2
Gas
2.2
1.0
Oil
1.5
400
350
300
250
200
150
100
50
0
80
90
00
10
20
30
Other Energy
MBDOE
Wind & Solar
1.5
Biomass, MSW
1.2
5.9
Nuclear
0.8
2.2
Hydro
2.1
60
50
40
30
20
10
0
80
90
00
10
20
30
ExxonMobil

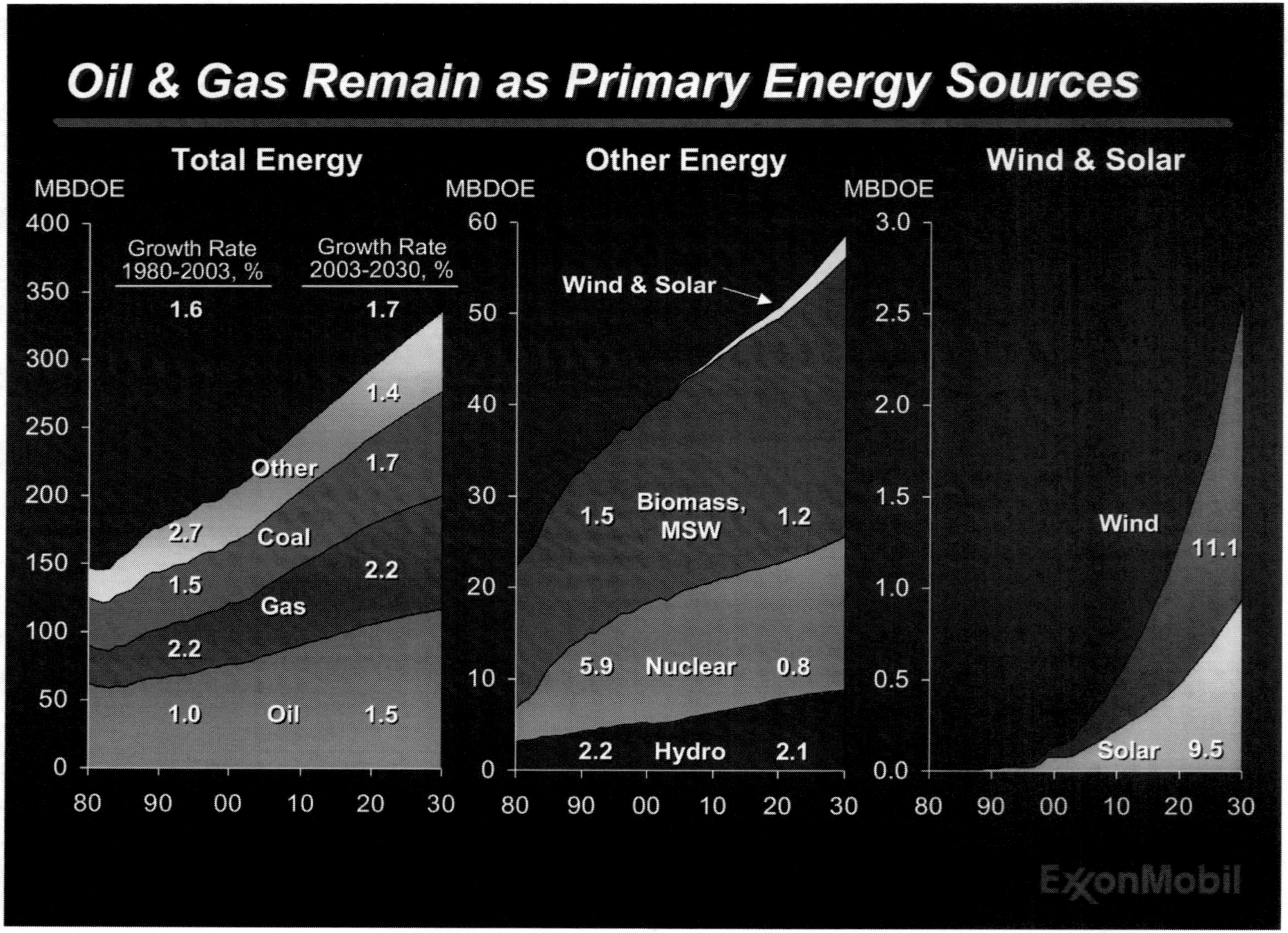
Oil & Gas Remain as Primary Energy Sources
Total Energy
MBDOE
Growth Rate 1980-2003, %
Growth Rate 2003-2030, %
1.6
1.7
2.7
Other
1.4
1.5
Coal
1.7
2.2
Gas
2.2
1.0
Oil
1.5
Other Energy
MBDOE
Wind & Solar
1.5
Biomass, MSW
1.2
5.9
Nuclear
0.8
2.2
Hydro
2.1
Wind & Solar
MBDOE
Wind
11.1
Solar
9.5
ExxonMobil

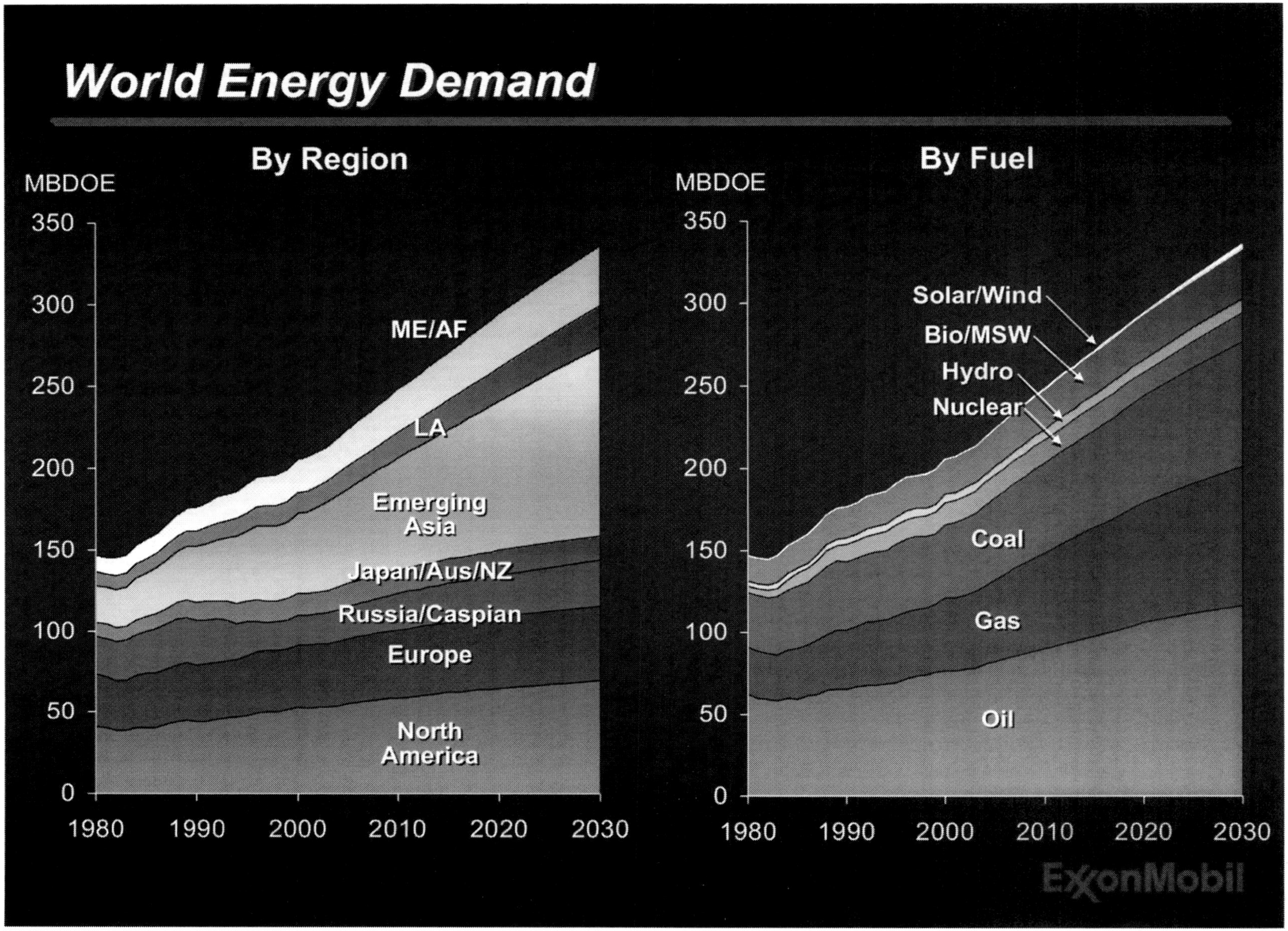
World Energy Demand
By Region
MBDOE
350
300
250
200
150
100
50
0
1980
1990
2000
2010
2020
2030
ME/AF
LA
Emerging Asia
Japan/Aus/NZ
Russia/Caspian
Europe
North America
By Fuel
MBDOE
Solar/Wind
Bio/MSW
Hydro
Nuclear
Coal
Gas
Oil
ExxonMobil

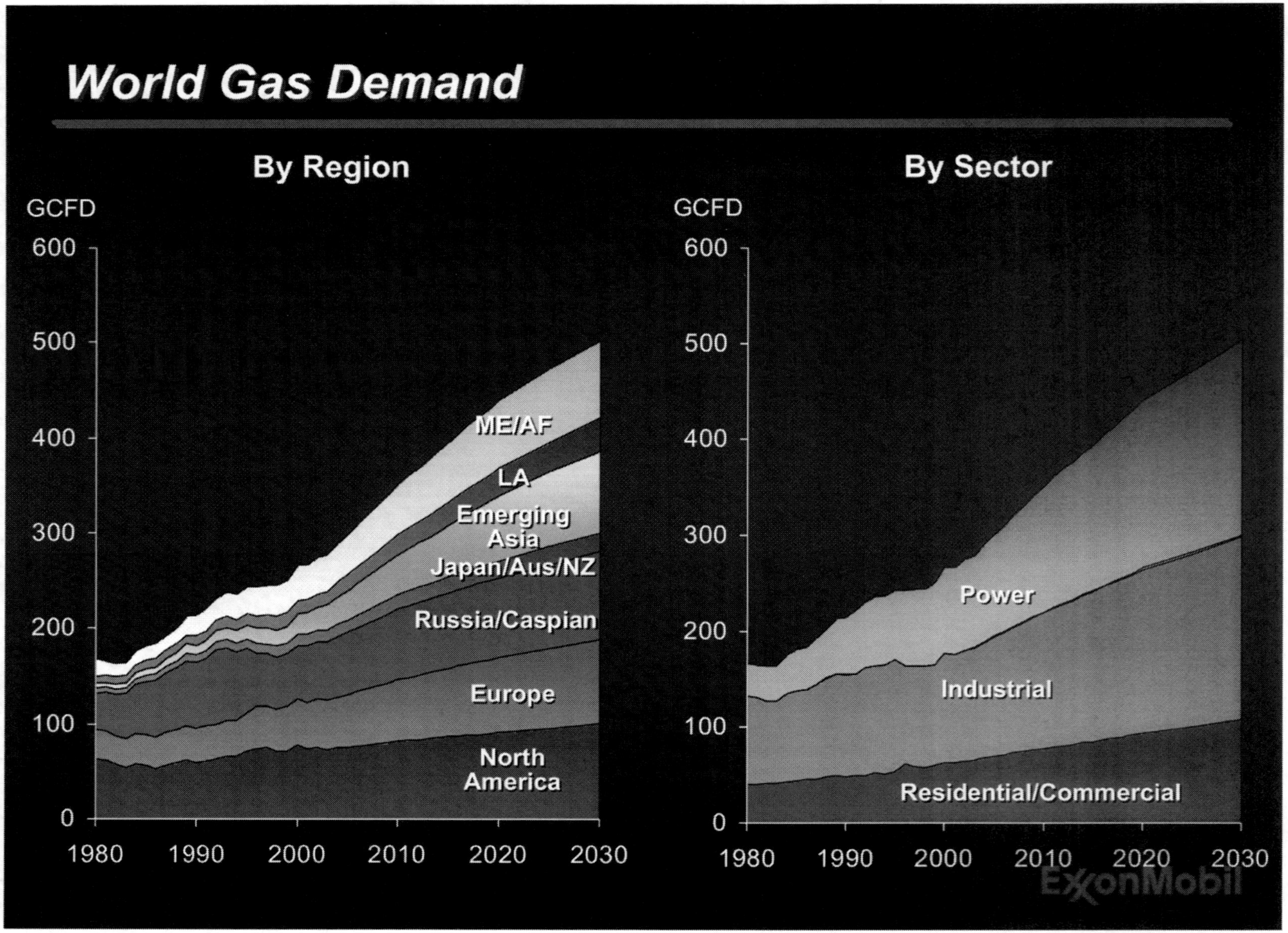
World Gas Demand
By Region
GCFD
600
500
400
300
200
100
0
1980
1990
2000
2010
2020
2030
ME/AF
LA
Emerging Asia
Japan/Aus/NZ
Russia/Caspian
Europe
North America
By Sector
GCFD
600
500
400
300
200
100
0
1980
1990
2000
2010
2020
2030
Power
Industrial
Residential/Commercial
ExxonMobil

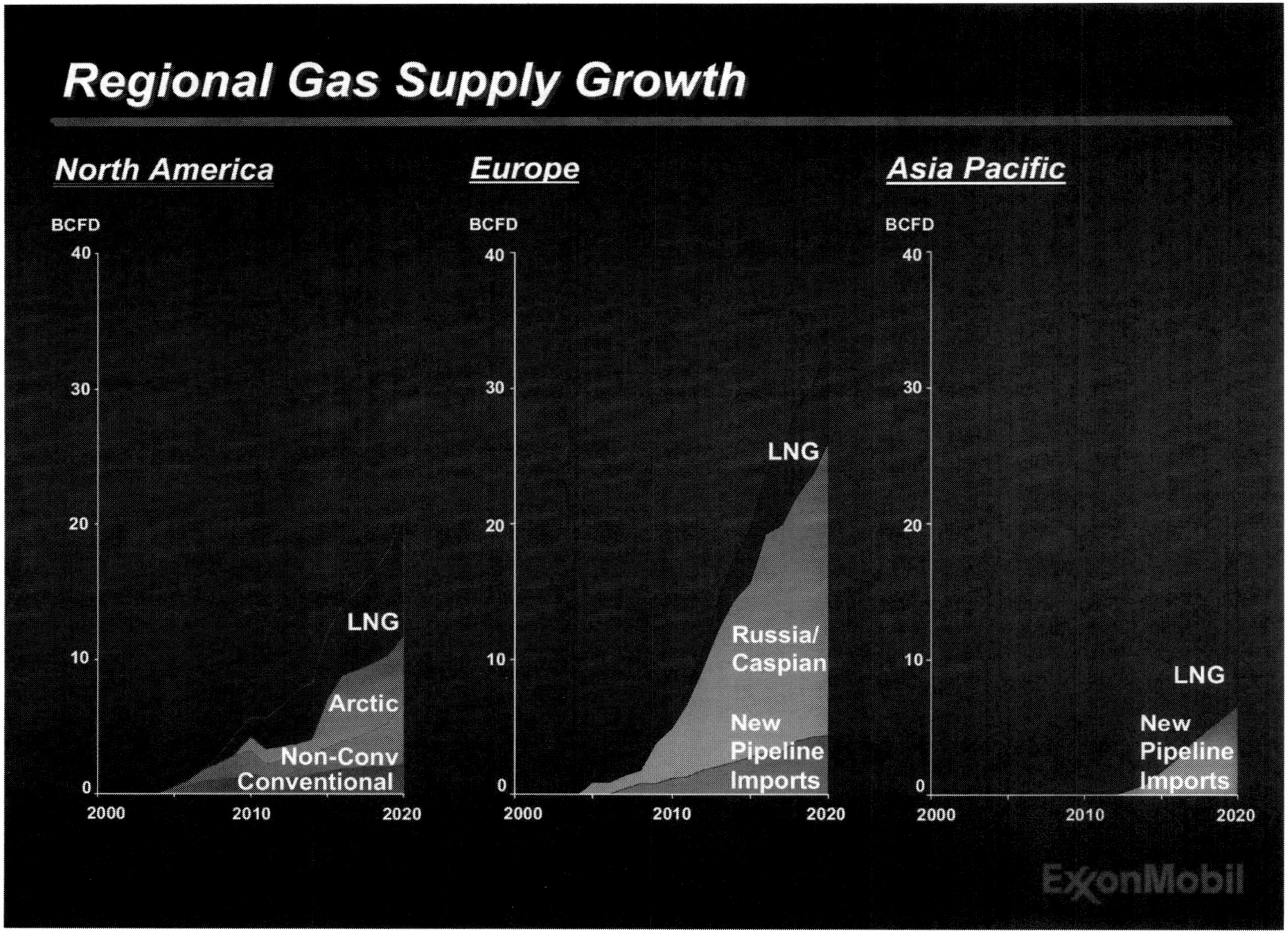
Regional Gas Supply Growth
North America
BCFD
40
30
20
10
0
2000
2010
2020
LNG
Arctic
Non-Conv
Conventional
Europe
BCFD
40
30
20
10
0
2000
2010
2020
LNG
Russia/
Caspian
New
Pipeline
Imports
Asia Pacific
BCFD
40
30
20
10
0
2000
2010
2020
LNG
New
Pipeline
Imports
ExxonMobil

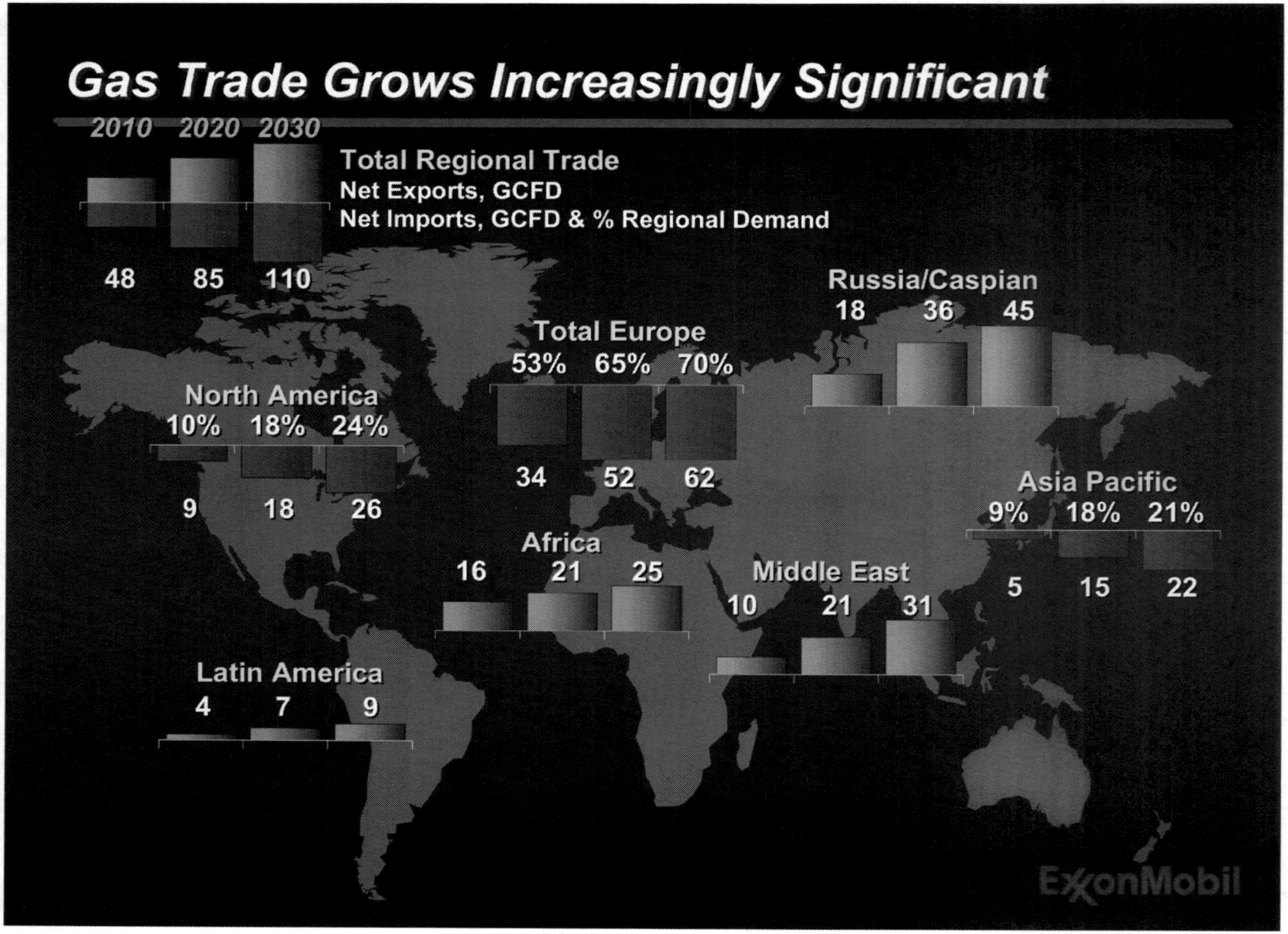
Gas Trade Grows Increasingly Significant
2010 2020 2030
Total Regional Trade
Net Exports, GCFD
Net Imports, GCFD & % Regional Demand
48 85 110
North America
10% 18% 24%
9 18 26
Latin America
4 7 9
Total Europe
53% 65% 70%
34 52 62
Africa
16 21 25
Russia/Caspian
18 36 45
Middle East
10 21 31
Asia Pacific
9% 18% 21%
5 15 22
ExxonMobil

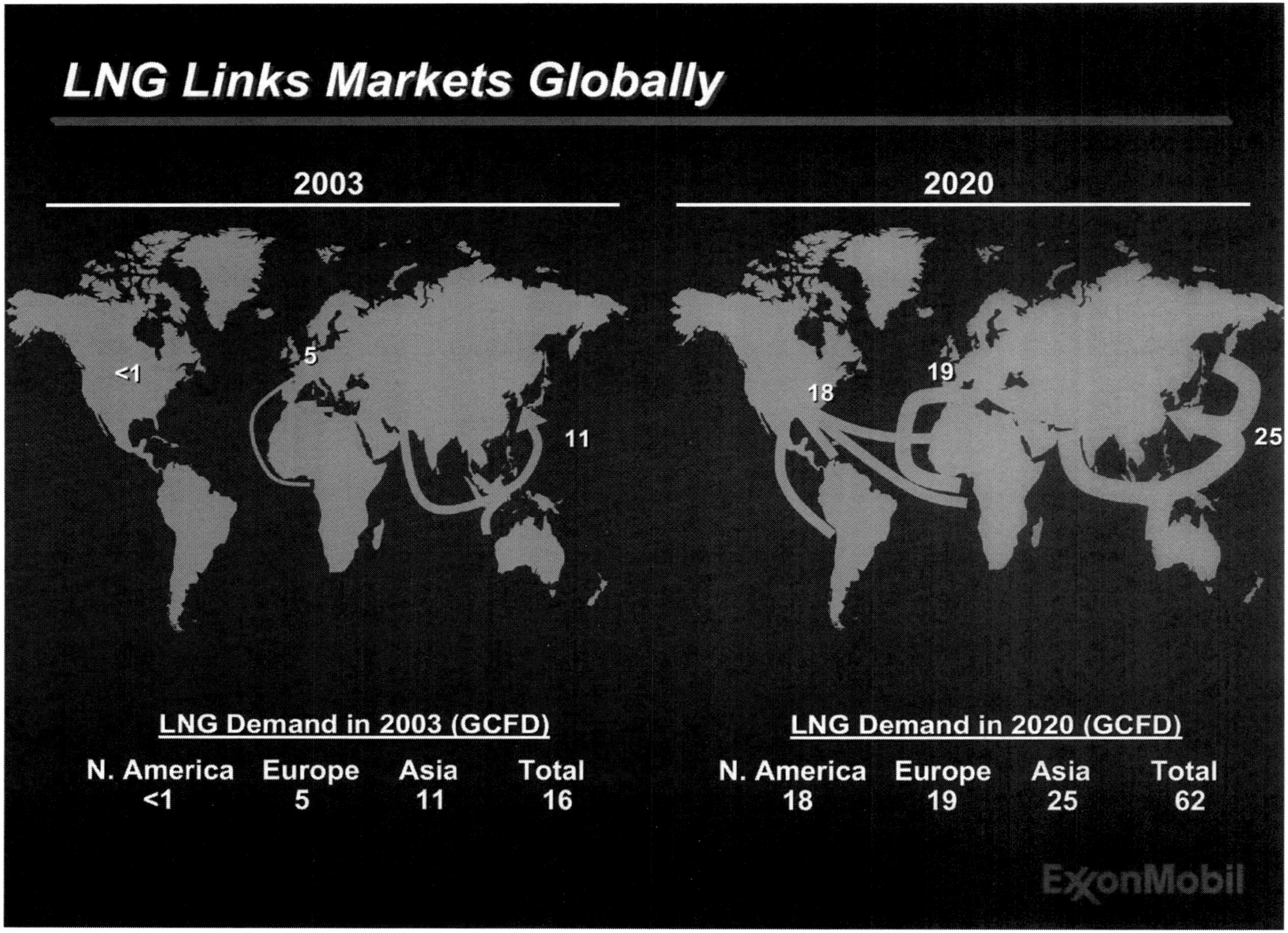
LNG Links Markets Globally
2003
2020
<1
5
11
18
19
25
LNG Demand in 2003 (GCFD)
N. America <1
Europe 5
Asia 11
Total 16
LNG Demand in 2020 (GCFD)
N. America 18
Europe 19
Asia 25
Total 62
ExxonMobil

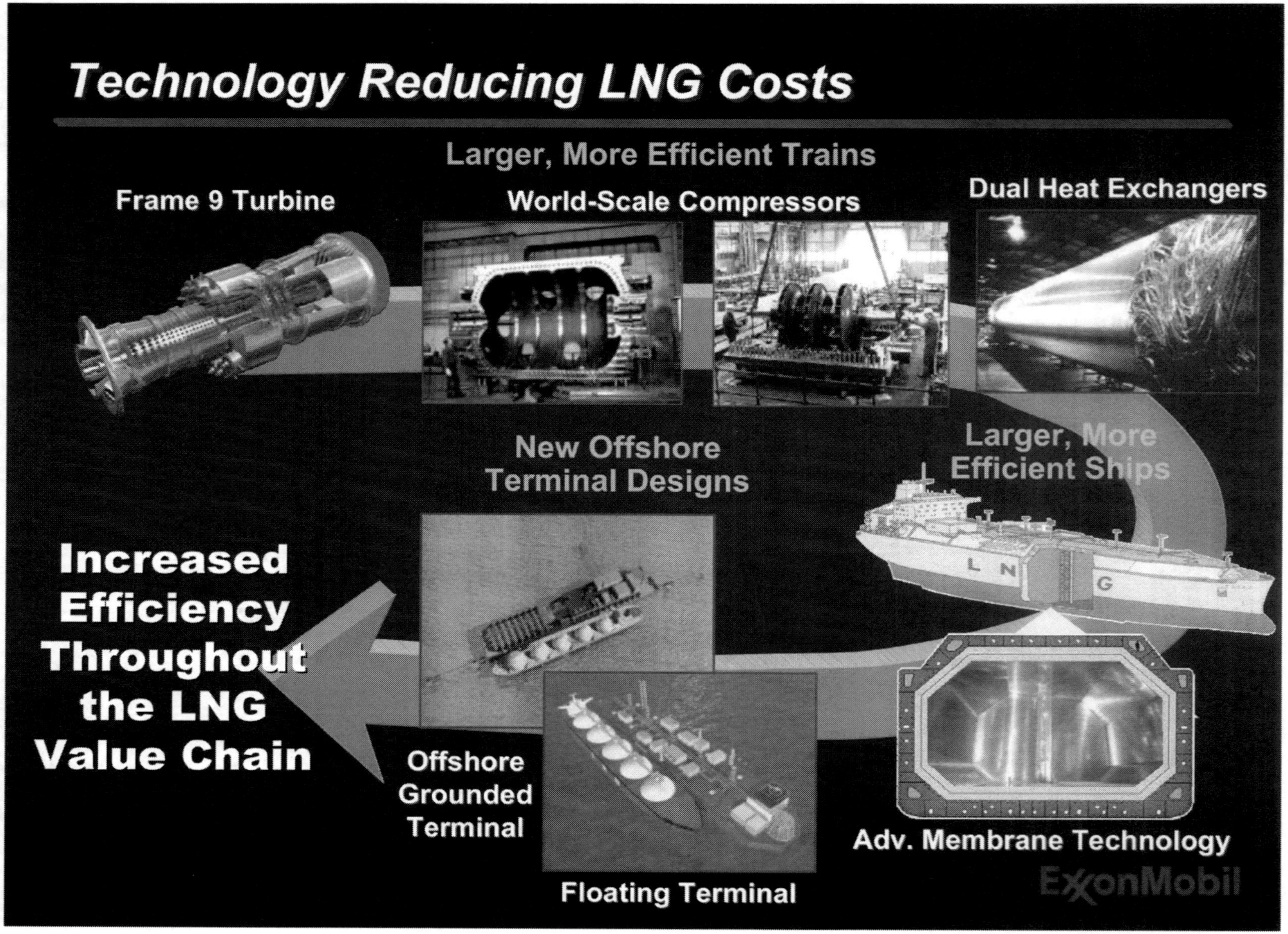
Technology Reducing LNG Costs
Larger, More Efficient Trains
Frame 9 Turbine
World-Scale Compressors
Dual Heat Exchangers
New Offshore
Terminal Designs
Larger, More
Efficient Ships
Increased
Efficiency
Throughout
the LNG
Value Chain
Offshore
Grounded
Terminal
Floating Terminal
Adv. Membrane Technology
ExxonMobil

North American LNG Imports

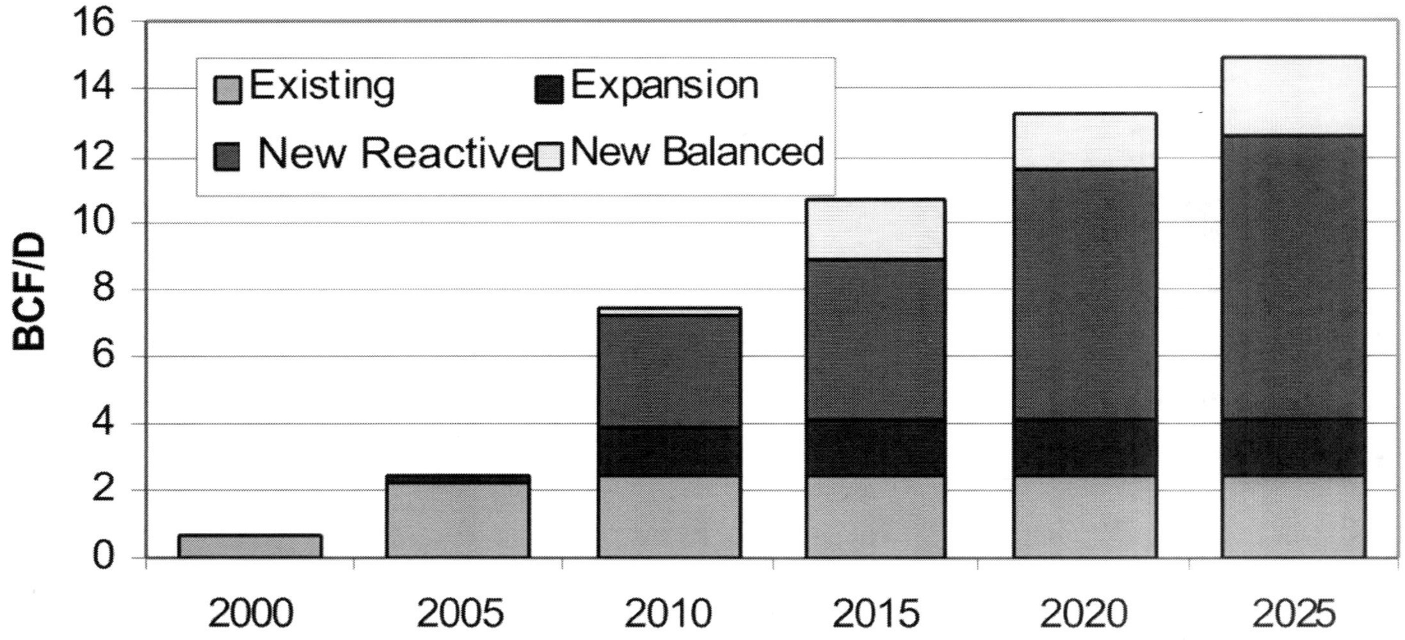

NPC Findings, p. 48

ExxonMobil

Existing & Proposed Terminals
Everett
Cove Point
Elba Island
Lake Charles
ExxonMobil

NA Gas / Crude Price Trend

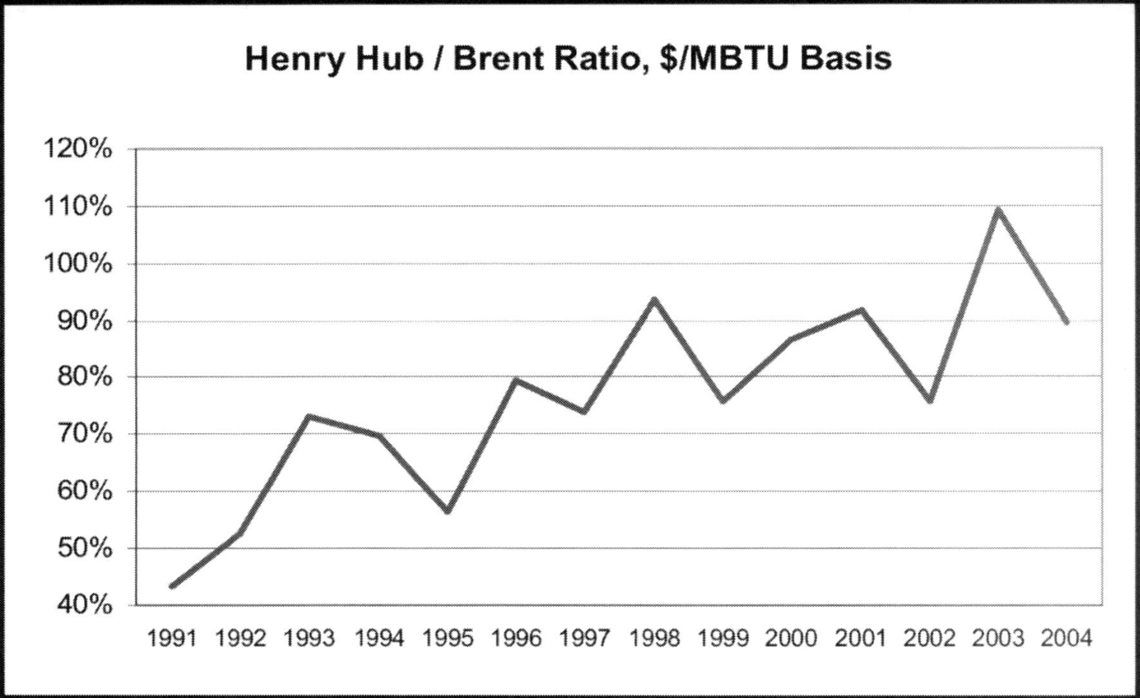

Source: Platts

ExxonMobil

NA Gas / Crude Price Trend

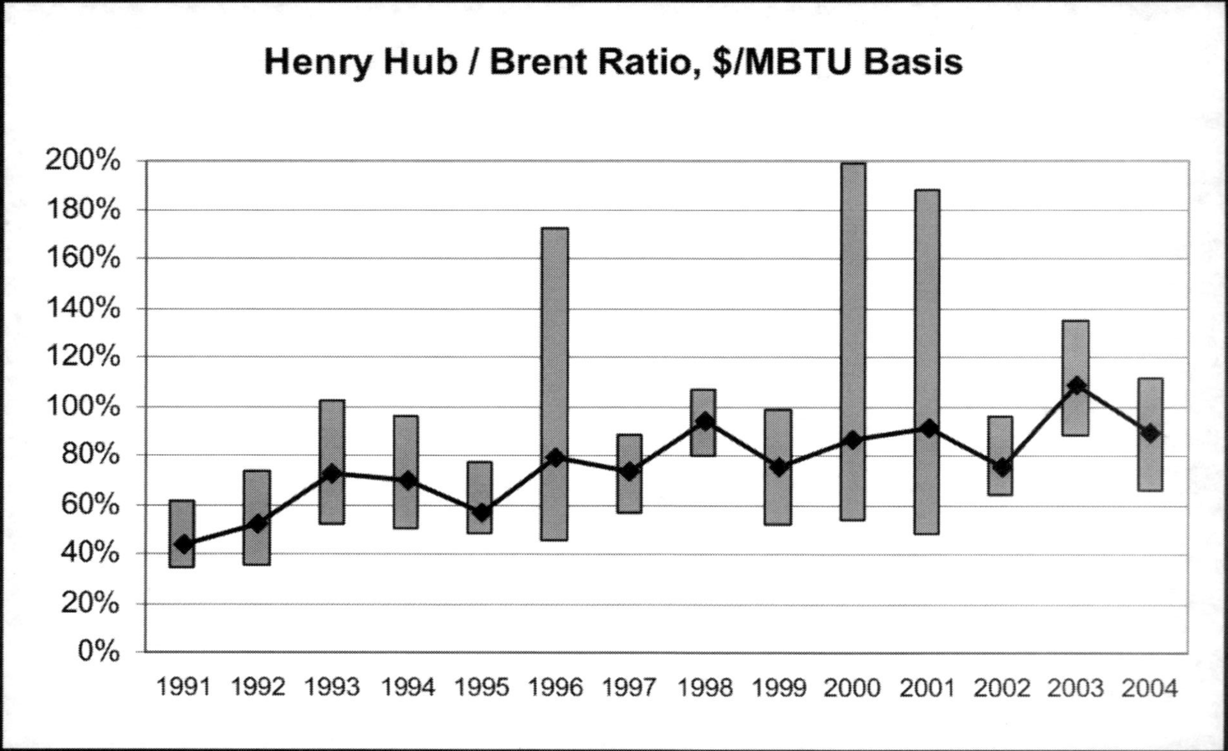

Source: Platts

Petrochemical Competitiveness

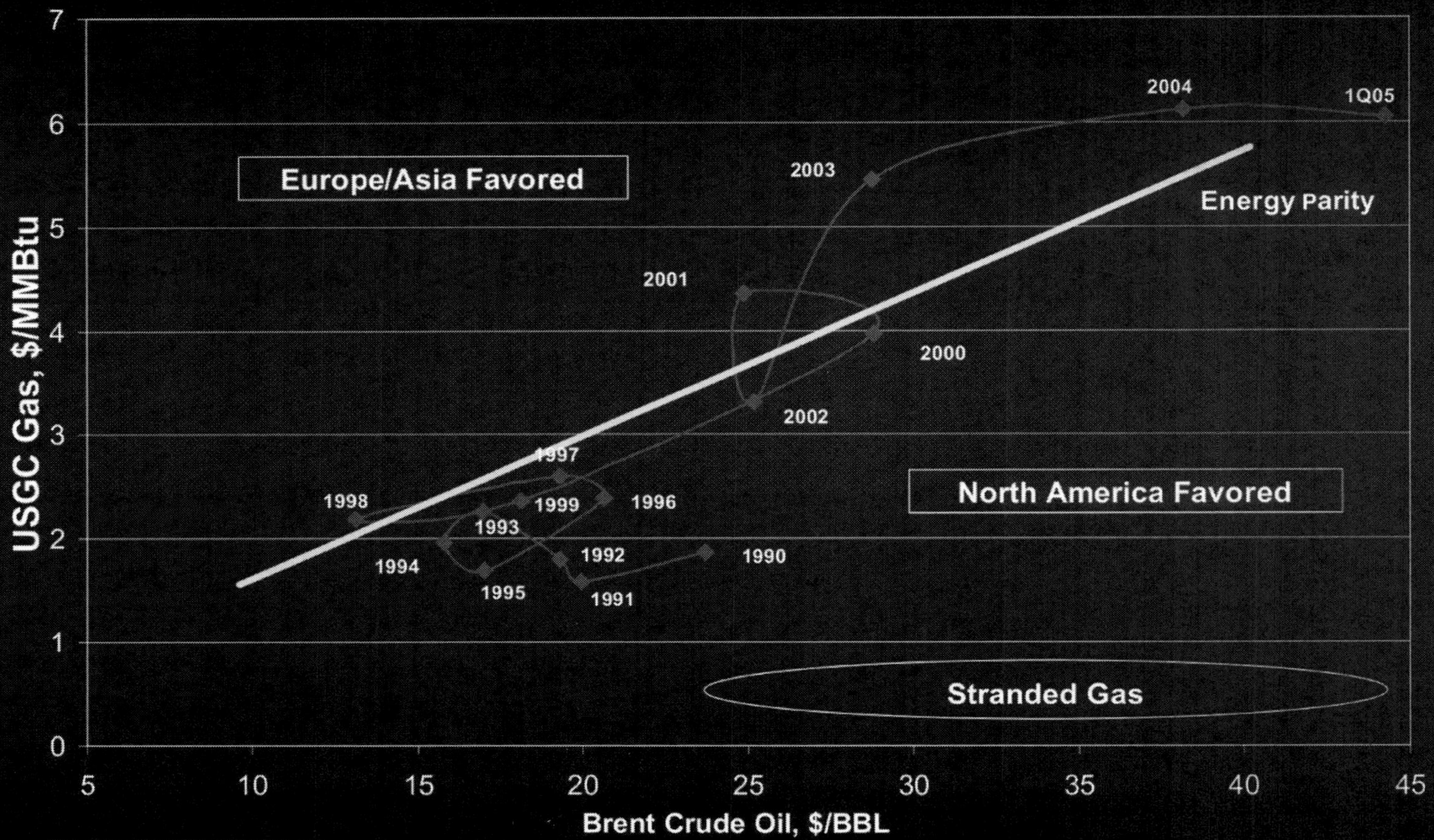

Source: CMAI

ExxonMobil

Summing Up

- Energy critical to economic growth, especially in developing markets.
- Oil and gas will remain the primary sources of energy through the mid century.
- Gas is the fastest-growing primary energy source.
- Lower LNG costs driving increasing gas globalization.
- North America gas demand growth relies on new LNG imports.
- Gas-intensive industries are most vulnerable to higher gas prices.
- North American chemical producers may become less competitive in global market.
- Domestic chemical growth may be moderated by imports.

ExxonMobil

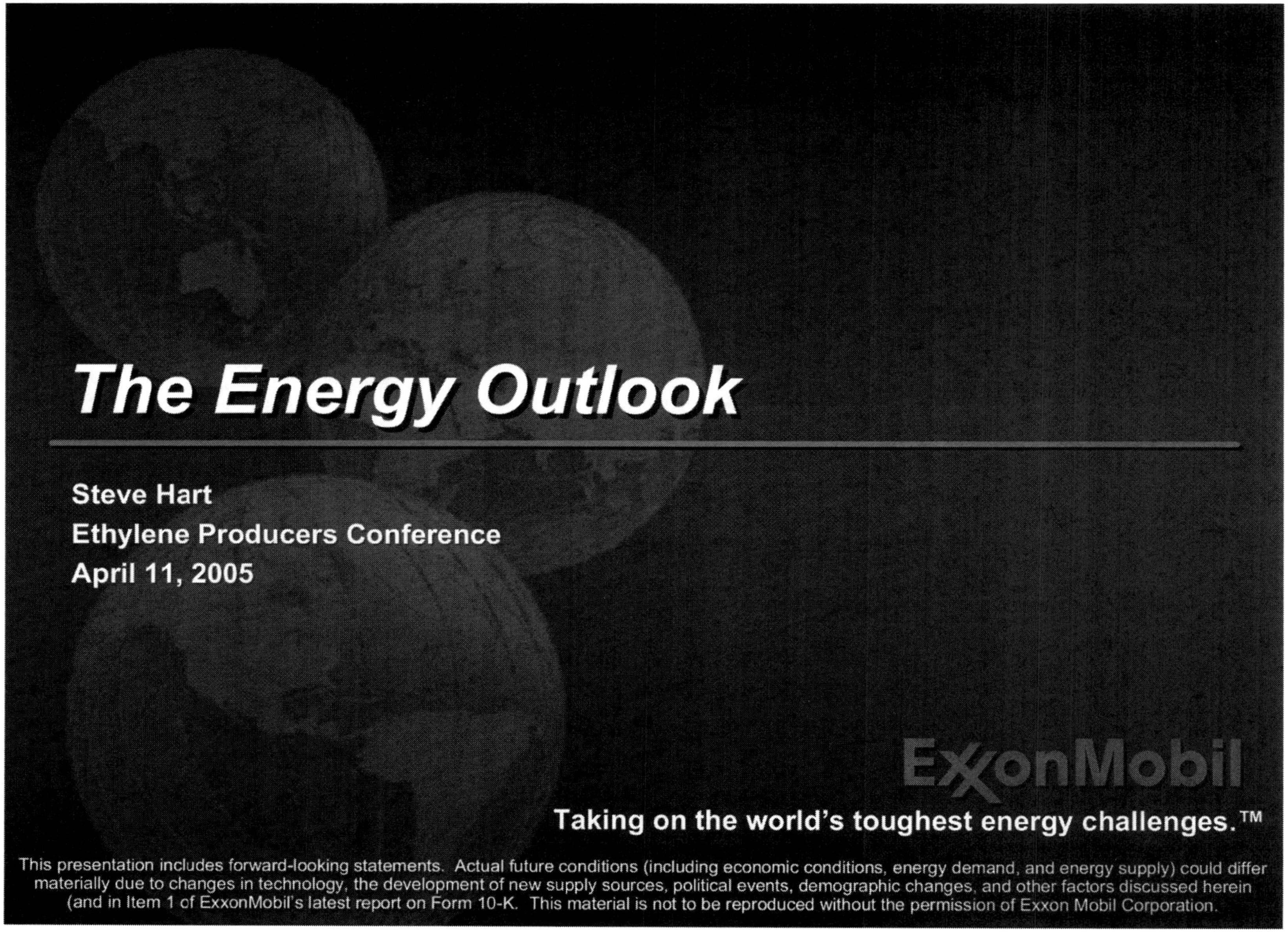
The Energy Outlook
Steve Hart
Ethylene Producers Conference
April 11, 2005
ExxonMobil
Taking on the world's toughest energy challenges.™
This presentation includes forward-looking statements. Actual future conditions (including economic conditions, energy demand, and energy supply) could differ materially due to changes in technology, the development of new supply sources, political events, demographic changes, and other factors discussed herein (and in Item 1 of ExxonMobil's latest report on Form 10-K. This material is not to be reproduced without the permission of Exxon Mobil Corporation.

Paper No. 38A

THE REACTIVITY OF ETHYLENE CONTAINING MINOR AMOUNTS OF ACETYLENE

M.E. LEVIN, N.O. GONZALES, and L.W. ZIMMERMAN
Shell Global Solutions (US) Inc.

And

R. RODRIGUEZ, T.E. PAXSON, and J.E. FISCHER
Shell Chemicals, LP

Prepared for Presentation at the 2005 Spring National Meeting, Atlanta, GA, Apr 10-14
17th Annual Ethylene Producers' Conference
Session TA005 – Ethylene Plant Safety

UNPUBLISHED

THE REACTIVITY OF ETHYLENE CONTAINING MINOR AMOUNTS OF ACETYLENE

by

M.E. LEVIN, N.O. GONZALES, and L.W. ZIMMERMAN
Shell Global Solutions (US) Inc.

And

R. RODRIGUEZ, T.E. PAXSON, and J.E. FISCHER
Shell Chemicals, LP

ABSTRACT

A project to deal with off-specification ethylene involved transfer of "off-spec" ethylene to caverns or salt domes for storage and reprocessing at a later date. This off-spec system would operate at conditions approaching 2000 psig pressure and storage temperatures in excess of 120 deg F. This system would have to be able to store a mixture of ethylene contaminated with acetylene concentrations of as much as 1.5 mole percent. Previous literature suggests that for safe operation, no system should be designed and operated with an acetylene partial pressure in excess of 8 psig. Evaluations of the potential hazards associated with storage of a mixture of ethylene and up to 2 mole percent acetylene were reviewed. Based on experimental data obtained, a literature review, and a thermodynamic evaluation, it was concluded that storage of an ethylene mixture with up to 2 mole percent acetylene does not pose any new or unexpected hazards due to the presence of acetylene.

Keywords: Acetylene, Ethylene, Decomposition, Oligomerization, APTAC, Calorimetry

THE REACTIVITY OF ETHYLENE CONTAINING MINOR AMOUNTS OF ACETYLENE

1 INTRODUCTION

An evaluation has been performed of the behavior of ethylene at elevated pressure containing small amounts of acetylene. This work addresses an inquiry from a chemical plant considering a project to transport and store pressurized "off-spec" ethylene at a remote site. The contemplated facilities would allow for "normal" processing of off-specification ethylene with the exception that the "off-spec" ethylene would be diverted away from normal pipeline customers. The ethylene would be a stream containing acetylene in concentrations up to as much as 1.5%mol. The ethylene would be compressed up to 2000 psig at -20°F, heated to 60°F or so, and then sent to caverns or salt domes where the temperature can reach 140°F.

It was recognized that the partial pressure of acetylene in the stream would be on the order of 5 psig (20 psia) – approaching the maximum typically practiced in acetylene handling (although usually applied to acetylene-rich streams). To proceed with the project, an assessment was sought of whether the presence of acetylene changes the reactivity (particularly, oligomerization and decomposition) behavior of the ethylene. As a conservative approach, mixtures containing as much as 2%mol acetylene were evaluated.

2 BACKGROUND WORK

2.1 Literature Review and Previous Experience

To address the potential impact of acetylene on ethylene reactivity, information available in company literature, open literature, and from contacts at other petrochemical companies was collected and analyzed. While some data were available from the open literature for acetylene-diluent mixtures, none of the data were directly applicable for low acetylene concentrations, less than 2 mole percent, at pressures comparable to 1800-2000 psig [1-7].

Data are available for the behavior of ethylene at elevated temperatures and pressures [8]. For the near-2000 psig pressure envisioned in this operation, a temperature of about 260-270°C is needed for ethylene oligomerization/decomposition to occur at observable levels at adiabatic conditions.

Anecdotal evidence and experience from other operating petrochemical companies suggests that no reactivity issues have been encountered with acetylene concentrations approaching 2 mole percent at conditions in the vicinity of 2000 psig and 140°F, but no data have been provided to substantiate the claims.

2.2 Thermodynamic Calculations

Given the gap between the expected processing outlet temperature (normally about 140°F = 60°C) and onset of ethylene decomposition at a pressure of about 2000 psia (260-270°C), a first-pass screen was performed to evaluate if acetylene decomposition in itself could bridge this gap. Accounting for the heat of reaction and stream heat capacity, decomposition of 2 mole percent acetylene in an ethylene stream can be readily shown to contribute 80-100°C to the stream temperature. The extent of temperature rise depends on whether the decomposition favors formation of hydrogen plus carbon

$$C_2H_2 \rightarrow 2\ C + H_2$$

or methane plus carbon,

$$C_2H_2 \rightarrow 1.5\ C + 0.5\ CH_4$$

or some combination of these reactions. For either reaction pathway, the temperature rise is insufficient to bring the stream from the normal salt dome or compressor outlet temperature to the onset temperature for ethylene oligomerization or decomposition reactions.

2.3 Kinetics Considerations

While acetylene decomposition itself might not evolve sufficient heat to raise the mixture temperature to where ethylene oligomerization and decomposition could take place, an additional concern was that reaction of acetylene at moderate temperature could result in generation of species, such as free radicals, that could promote ethylene reactions. These species could, in principle, substantially lower the temperature for ethylene oligomerization and subsequent decomposition.

In the absence of any available reactivity data applicable to the targeted concentrations and conditions, it was decided to embark on a testing program to evaluate the potential for low temperature reaction. Initially, testing of acetylene/ethylene mixtures through use of an igniter was contemplated. However, it was realized that the observed reactivity of ethylene with acetylene is a strong function of the type and power of igniter employed. In fact, with a suitable igniter, ethylene alone could be ignited [9-10].

Attention then focused on utilization of adiabatic calorimetry to examine the reactivity of acetylene/ethylene mixtures.

3 EXPERIMENTAL EVALUATION

3.1 Experimental Evaluation

Testing for this phase of the study was carried out in the Automatic Pressure Tracking Adiabatic Calorimeter (APTAC™ available from TIAX LLC described in previous studies [11-12]). The APTAC, illustrated in Figure 1, is designed to provide an adiabatic environment by maintaining the temperature outside of a test sample/container to match the sample temperature. Ethylene and mixtures of 2 mole percent acetylene in ethylene at pressures in the 1800-2000 psig range were each subjected to elevated temperatures while a search for exotherms was conducted. Searches were conducted in a heat-wait-search mode in which the sample is heated to a pre-selected temperature and, upon reaching this temperature, the instrument waits for a period of time (usually 25 minutes) for the cell and containment vessel temperatures to stabilize. After the wait period, the instrument "searches" at that temperature (for another 25 minutes) for any exothermic activity. During this time, the temperature of the containment vessel gas space is adjusted to match that of the sample thermocouple. If no heat-up activity exceeding a 0.04-0.05° C/min self-heating rate threshold is observed, it is concluded that there is no exotherm. The sample is then heated to the next temperature and the process repeated. If an exotherm is detected, the APTAC tracks the sample conditions and adjusts the temperature and pressure of the containment vessel accordingly.

Since the series of tests conducted in this study were gas-phase in a closed cell, the pressure is expected to rise as the temperature is increased in the heat-wait-search steps or in an exotherm. The reactivity of ethylene is known to be dependent on pressure as well as temperature so that for this gas-phase system, there is no unique onset temperature for reaction. Toward the end of the testing program, a back-pressure regulator was added to enable more control of the sample pressure at the expense of loss of material and thereby, some adiabaticity. A schematic for the gas handling system is provided in Figure 2.

3.2 Samples

Two sources of ethylene were employed in the tests. A cylinder with 1200 psig ethylene at a purity of 99.9%mol was obtained from Airgas Southwest and a larger ethylene cylinder (minimum of 1200 psig) at a purity of a 99.9%mol was purchased from Scott Specialty Gases.

Acetylene at a purity of 99.6%mol was purchased from Praxair. The cylinder pressure, as delivered, was about 250 psig.

3.3 Procedures

Experiments were set up by mounting a 130 cm^3 stainless steel APTAC cell to the instrument and introducing the test gases to the cell at room temperature and pre-specified pressures. For tests involving both acetylene and ethylene, the cell was charged with acetylene to the target partial pressure followed by ethylene as balance to the target total pressure.

No gas samples for compositional analysis were taken at the end of any of the tests. A summary of test characteristics can be found in Table 1.

3.4 Testing Program

Two series of experiments were conducted: (1) ethylene alone and (2) acetylene plus ethylene. The intent of the ethylene tests was to provide a benchmark for comparison with the acetylene plus ethylene tests. In addition, tests were initially set up for low pressure to serve as a basis for evaluating the safety and performance of the APTAC in this service. The target pressure range for tests was just below 2000 psia (the instrument has a maximum high pressure shutdown of 2000 psia). Initial cell pressures were usually chosen so that the total sample pressure would approach 2000 psia in the 200-300°C temperature range. A chart illustrating the compressibility factor as a function of temperature and pressure based on data in Perry's Chemical Engineering Handbook [13] is presented in Figure 3.

Figure 1: APTAC™ Calorimeter

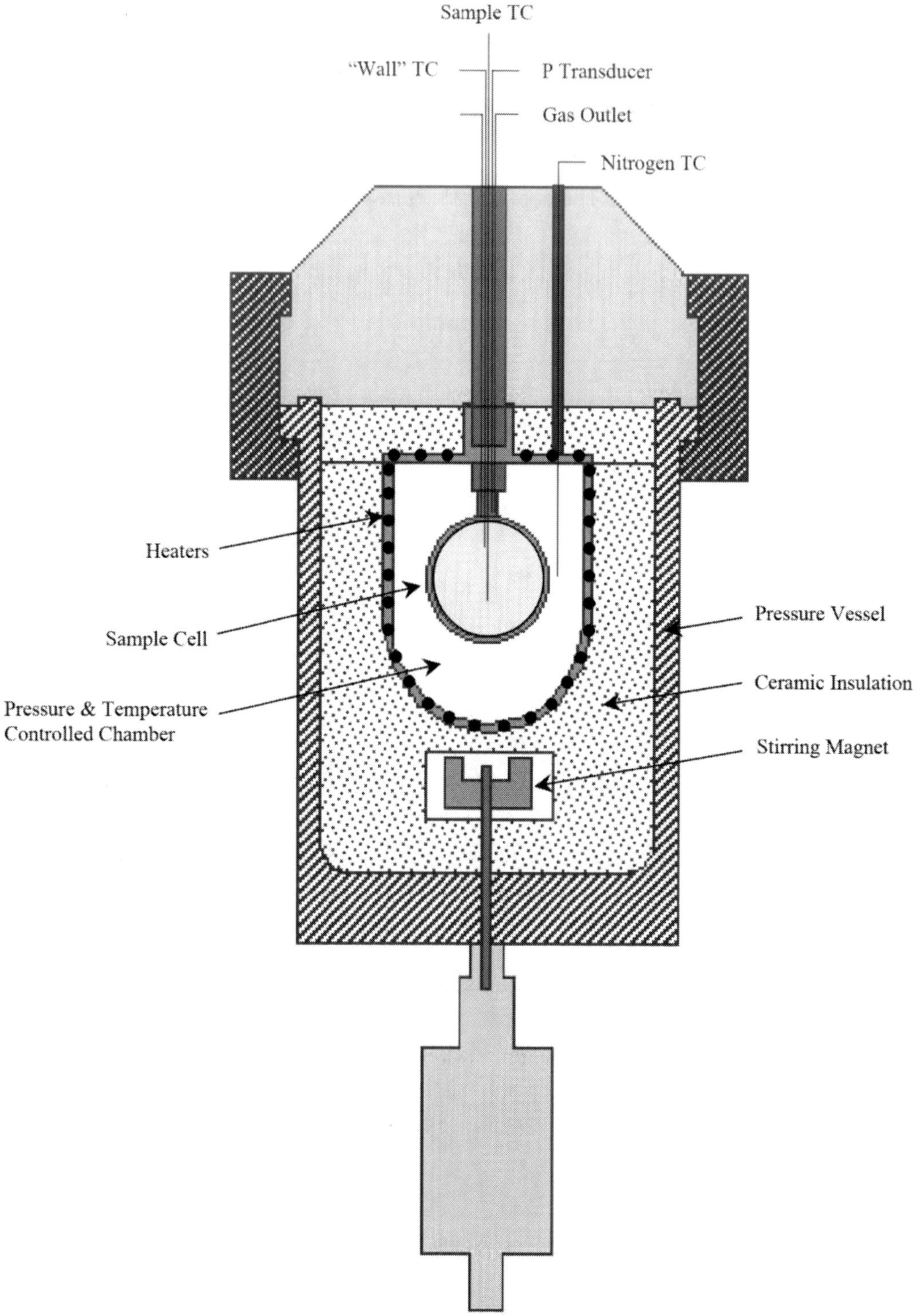

Figure 2: Schematic of Gas Handling System

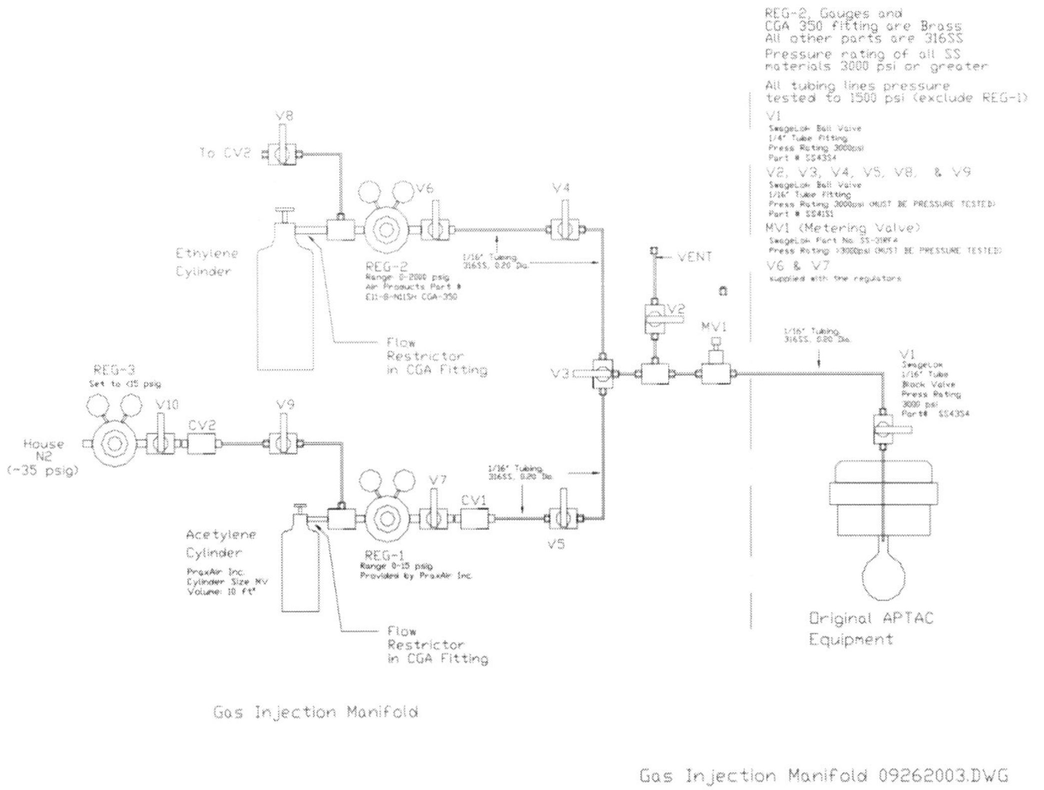

Figure 3: Ethylene Compressibility Behavior with Temperature and Pressure

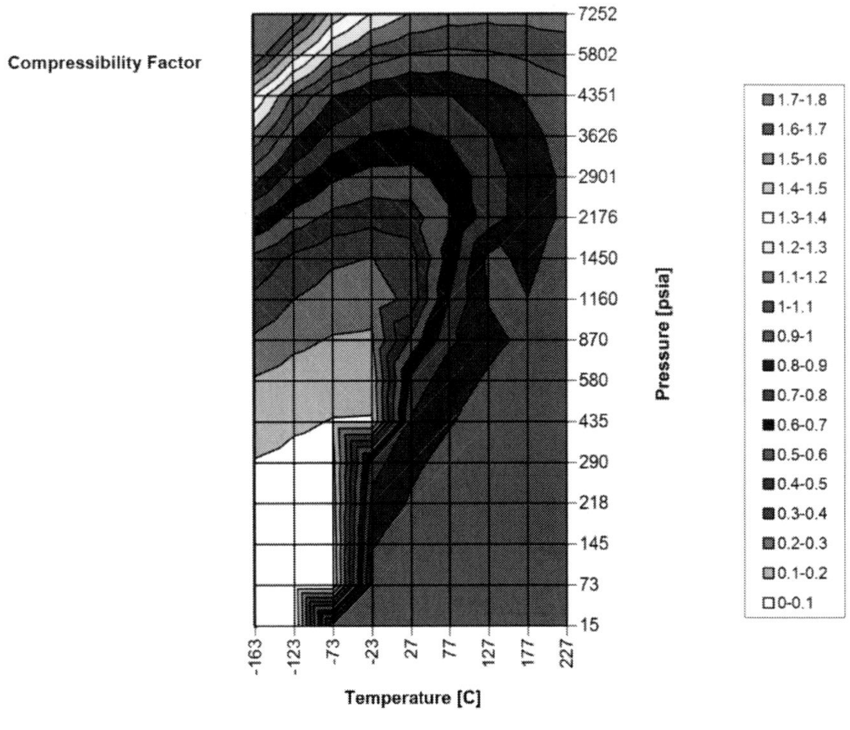

Table 1: Summary of APTAC Test Conditions and Results for Ethylene + Acetylene

Run ID	A00638	A00639a	A00643	A00644	A00645	A00646	A00647	A00649
Date	Sep. 17, 2003	Sep. 18, 2003	Sep. 22, 2003	Sep. 23, 2003	Sep. 23, 2003	Sep. 24,2003	Sep. 26, 2003	Oct. 2, 2003
LR	LR25703-131	LR25703-131	LR25703-134	LR25703-134	LR25703-135	LR25703-135	LR25703-137	LR25703-138
Ethylene Source								
Initial Acetylene Pressure [psia] (At temperature [°C])	---	---	---	---	---	15.1 (22.2)	14.7 (22)	17.7 (22.2)
Initial Ethylene Pressure [psia] (At temperature [°C])	310 (21.5)	600 (ca. 22)	788 (22.5)	738 (22)	738 (22)	720 (28)	721 (23.1)	868 (22.9)
Sample Cell Mass (Stainless Steel) [g]	51.36	51.36	51.39	51.39	51.05	51.05	50.70	50.70
Backpressure Regulator	No	No	No	No	No	No	No	Yes
Expt (Search) Start Temp [°C]	200	200	280	280	240	220	240	230
Expt Final or Max Temp [°C]	340	400	400	400	400	400	400	400
Heat-Wait-Search Increment [°C]	10	10	5	5	5	5	5	5
Expt Exotherm Limit (N_2) [°C]	440	440	440	440	440	440	440	440
Expt Temperature Shutdown [°C]	450	450	450	450	450	450	450	450
Expt Pressure Shutdown [psia]	1800	1800	2000	2000	2000	2000	2000	2000
Expt Heat Rate Shutdown [°C/min]	400	400	400	400	400	400	400	400
Expt Press Rate Shutdown [psi/min]	750	750	750	750	750	750	750	750
Exotherm Threshold [°C/min]	0.05	0.05	0.04	0.04	0.04	0.04	0.04	0.04
Number of Exotherms	0	1	0	1	1	1	1	1
Initial Observed Exotherm Temperature [°C]	---	323	---	287	308	312	274	283
Max Observed Temp [°C]	340	696	235	297	633	902	313	372
Max Observed Pressure [psia]	653	2676	2000	2000	2679	2408	2000	1957
Max Obs'd Self-Heat Rate [°C/min]	---	13,870	---	0.40	11,070	22,620	0.96	10.5
Max Obs'd Press. Rate [psi/min]	---	3878	---	1.2	2050	4178	1.7	3.8
Temp at Max Self-Heat Rate [°C]	---	675	---	223	629	902	313	368
Temp at Max Press. Rate [°C]	---	660	---	287	616	838	312	368
Expt Duration (before S/D) [min]	867	798	107	297	893	439	607	426
Expt Shutdown Cause	Max Search T	Temp S/D	Pressure S/D	Pressure S/D	Temp S/D	Exotherm Limit	Pressure S/D	Manual
Comments		Premature S/D of A00639 at 1800 psia – restart as A00639a						Premature S/D of A00649 at 2000 psia – restart as A00649a; back pressure regulator in use

[1]Self-heat and pressurization rates sufficiently high to cause possible temperature, pressure reading lags

Nitrogen pad gas

3.5 Experimental Results

The temperature-time behavior for neat ethylene and for acetylene-ethylene mixtures is displayed in Figure 4. For the first ethylene test, which was conducted at a relatively low pressure of about 660 psia, no reaction was found up through 330°C. For higher pressures of ethylene or of ethylene containing acetylene, the upward temperature deviation from the stair-step pattern of the heat-wait-search steps indicates the appearance of exothermic reaction in the 270-320°C temperature range. In some tests (A00639a A00646), temperatures rose rapidly, eventually reaching about 700°C or higher. Evidently, acetylene is not required for this behavior; reaction of ethylene alone can generate such rapid temperature rises. In other cases (A00644, A00647, A00649, A00649a), tests ended upon reaching the 2000 psia shutdown criterion before high temperatures could develop from the exothermic reaction.

The corresponding pressure profiles with time are given in Figure 5. For the most part, pressures range from 1200 to 2000 psia and vary with the sample temperature (in presumed accordance with ethylene compressibility). The pressure in test A00646 (an acetylene-ethylene mix) decreases noticeably through the series of heat-wait-search steps from 220 to 310°C. Although one could conceivably attribute this reduction in pressure to consumption of ethylene (along with acetylene), a leak in the tubing connected to the cell was found after the test. In tests A00639a and A00645, there is a slight decrease in pressure just prior to the pressure spike. This feature probably does reflect consumption of ethylene. It is also worth noting that in the tests where a significant temperature spike occurred, the pressure also rose sharply, yielding pressures ranging from 2400 to nearly 2700 psia.

The pressure vs. temperature behavior is depicted in Figure 6. Included in the graph are also results from a few tests that ended prematurely (including a couple of restarts, such as for A00639 and A00649). In principle, it should be possible to extract compressibility data from the data collected between test start-up and just prior to observed self-heating. In a few tests (A00647, A00649/A00649a), the end-of-run pressure decreased from the starting level, possibly reflecting consumption of ethylene. In contrast, the increase in pressure observed in tests A00639a and A00645 can be attributed to non-condensibles formed from the significant degree of decomposition, as evidenced by the pressure spikes in Figure 5.

Figure 4: Temperature-Time History of Acetylene Plus Ethylene
APTAC Experiments; in Nitrogen

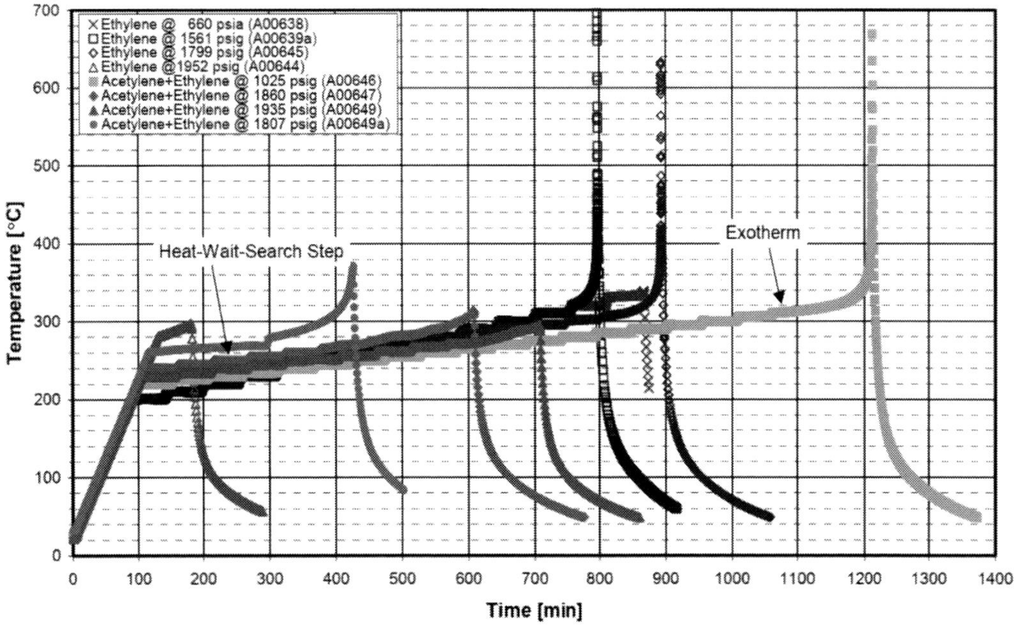

Figure 5: Pressure-Time History of Acetylene Plus Ethylene
APTAC Experiments; in Nitrogen

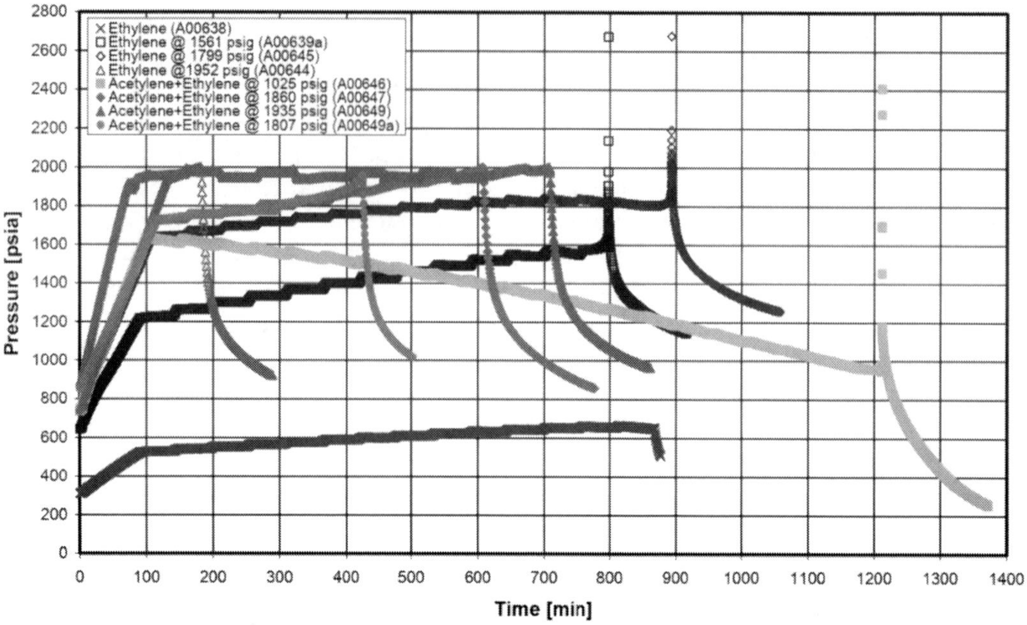

The self-heat rate plot found in Figure 7 typically reflects the kinetic behavior of the sample. This graph displays temperature rise rate vs. negative reciprocal temperature (with the corresponding temperatures in degrees Celsius shown). The direct link to an Arrhenius plot can be obtained by multiplying the self-heat rate by the sample heat capacity and dividing by the heat of reaction to yield the observed reaction rate.

Figure 6: Pressure vs. Temperature Behavior of Acetylene Plus Ethylene
APTAC Experiments; in Nitrogen

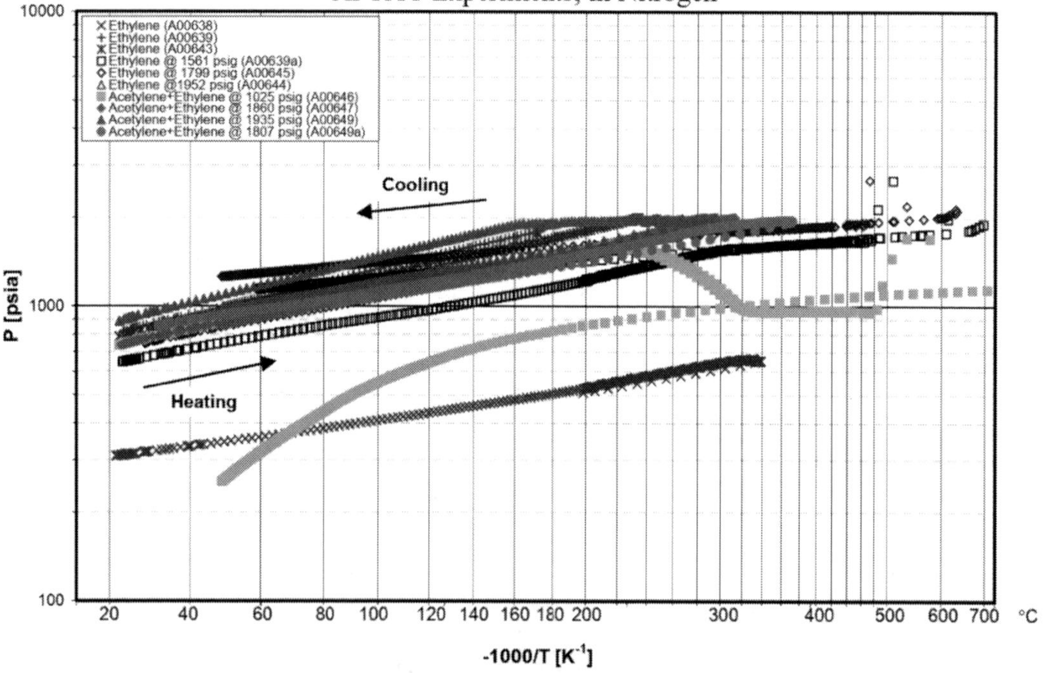

Figure 7: Self-Heat Rate vs. Temperature Behavior of Acetylene Plus Ethylene
APTAC Experiments; in Nitrogen

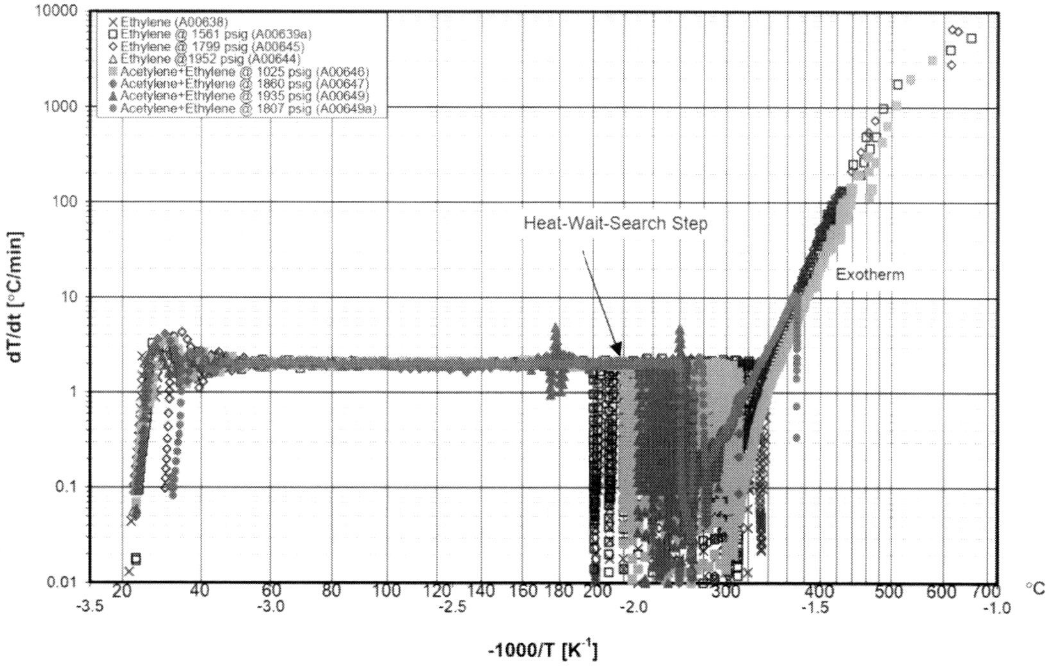

The cycling of heat rates above 200°C simply reflects the phases of the heat-wait-search steps. Beyond this cycling behavior, the slopes of the curves reflect the activation energy of the reaction. Figure 8 displays the same information as Figure 7, except that the initial heat-wait-search steps have been removed for clarity.

From the calorimetry results in Table 1 and Figure 8, ethylene is seen to exhibit exothermic behavior in the 280-300°C temperature range. In fact, reaction would likely be observed at somewhat lower temperatures since the onset of reaction was caught in these tests at a rate of at least 0.1°C/min (that is, higher than the 0.04°C/min target self-heat rate threshold). At higher pressure, reaction for ethylene alone occurs at lower temperature. When 2 mole percent acetylene is included in the mixture, the onset of reaction for 1800+ psia pressure is comparable to that for ethylene alone at about 1950 psia. For acetylene in ethylene at just above 1000 psia, the rate is slightly lower than that for 1560-1800 psia ethylene. In short, acetylene might increase the reactivity of the overall mixture, but only modestly.

Figure 8: Self-Heat Rate vs. Temperature Behavior of Acetylene Plus Ethylene APTAC Experiments; in Nitrogen; Heat-Wait-Search Steps Removed

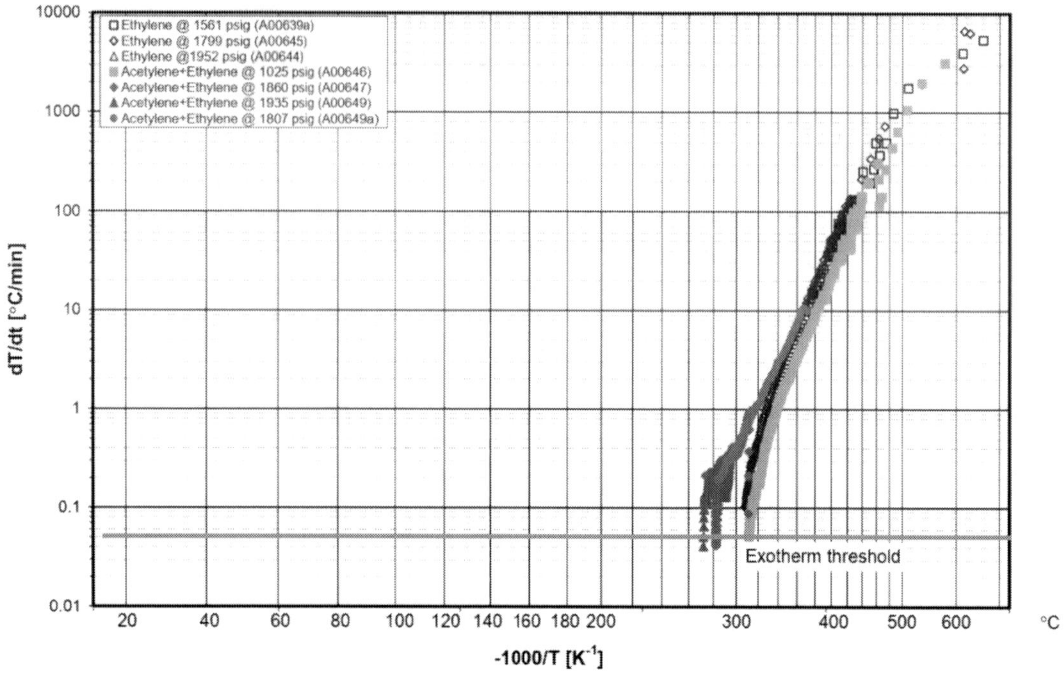

3.3 Thermal Inertia Effects

The thermal inertia factor relates the total amount of heat absorbed by the sample and container (cell) relative to that absorbed by the sample alone. In commercial-scale equipment, the thermal inertia factor or phi factor, ϕ, approaches a value of unity. In this study's experiments, ϕ, ranges between 4.2 and 4.4. These high values arise from the gaseous state of the ethylene (and acetylene). The values indicate that the sample container takes up more than three times the heat as the sample. The effect of this thermal inertia is to dampen substantially the extent of temperature rise, pressure generation, temperature rise rate, and pressurization rate. This thermal inertia will also lead to a somewhat higher onset temperature than in low thermal inertia equipment.

To adjust the current study's results properly for equipment with a lower thermal inertia, a dynamic simulation that takes into account the observed reaction kinetics coupled with material and equipment properties would be required.

4 SUMMARY

A project was being developed to deal with off-specification ethylene, without flaring the material, for a Shell ethylene plant. The "off-spec" ethylene would be transferred to and stored in caverns or salt domes for reprocessing at a later date. This off-spec system would operate at normal high pressure pipeline/salt dome conditions: pressures approaching 2000 psig and storage temperatures in excess of 120 deg F. This "off-spec" system would have to be able to store a mixture of ethylene contaminated with acetylene concentrations of as much as 1.5 mole percent. Previous literature suggested that for safe operation, no system should be designed and operated with an acetylene partial pressure in excess of 8 psig. Additionally, some data suggest that ethylene oligomerization reactions, at high pressures, can be accelerated by the presence of catalysts, including free radical sources. Evaluations of the potential hazards associated with storage of a mixture ethylene and up to 2 mole percent acetylene were reviewed. Based on experimental data obtained, a literature review, and a thermodynamic evaluation, it was concluded that storage of an ethylene mixture with up to 2 mole percent acetylene does not pose any new or unexpected hazards due to the presence of acetylene.

5 REFERENCES

1. J.W. Haun, *Petroleum Refiner*, 37 (11), November 1958, pp. 204-207.
2. A. Baumeier, D. Conrad, S. Dietlen, W. Pezold, T. Schendler, and H.-P. Schultze, *Chem.-Ing.-Tech.*, **64**(3), 1992, Nr. 3, pp. 260-261.
3. T. Schendler and H.-P. Schulze, *Chem.-Ing.-Tech.*, **62**(1), 1990, pp. 41-43.
4. D. Lietze, *Amts- und Mitteilungsblatt der Bundesanstalt für Materialprüfung (BAM)*, **16** (1), 1986, pp. 23-28.
5. W. Reppe, Chemie und Technik der Acetylen-Druck-Reaktionen, Verlag Chemi GmbH, Weinheim/Bergstraße, 1951.
6. S.A. Miller, "Acetylene: Its Properties, Manufacture, and Uses", Volume 1, Ernest Benn Limted, 1965.
7. "Kirk-Othmer Encyclopedia of Chemical Technology", 2nd Edition, John Wiley & Sons, 1963.
8. L.G. Britton, D.A. Taylor, and D.C. Wobser, *Plant/Operations Progress*, **5**(4), October 1986, pp. 238-251`
9. G.S. Scott, R.E. Kennedy, I. Spolan, and M.G. Zabetakis, "Flammability Characteristics of Ethylene", Bureau of Mines Report RI 6659 (1965).
10. L.G. Britton, *Process Safety Progress*, **15**(3), 1996.
11. M.E. Levin and A.D. Hill, "Reactivity of Unsaturated Hydrocarbons via Adiabatic Calorimetry," Proceedings of the 2000 Mary Kay O'Connor Process Safety Center Annual Symposium.
12. M.E. Levin and A.D. Hill, "The Reactivity of Butadiene with Acetylenic Hydrocarbons via Adiabatic Calorimetry," Proceedings of the 2001 Mary Kay O'Connor Process Safety Symposium (October 30-31, 2001).
13. D.W. Green, "Perry's Chemical Engineers' Handbook", 7th Edition, McGraw-Hill, 1997.

Paper No. 38b

Autorefrigeration Near-Miss Incident at Equistar Chemicals, a Lyondell Company

Michael J. Kennedy
Principal Engineer

Equistar Chemicals
Chocolate Bayou Chemical Complex
P.O. Box 2917
Alvin, TX 77512

Prepared for presentation at the 17th Ethylene Producers' Conference
2005 AIChE Spring National Meeting
Atlanta, GA
April 10th – 14th, 2005

Unpublished

Abstract

During the startup of the Equistar Chemicals Chocolate Bayou Olefins Plant following an eight day outage, an autorefrigeration near-miss incident occurred involving the unit deethanizer column. Autorefrigeration, i.e., the unintentional and uncontrolled phase change from a liquid to a vapor of a hydrocarbon that results in refrigeration, occurred when the deethanizer column was being reinventoried. The column's brittle fracture minimum allowable temperature was nearly reached. This autorefrigeration caused a failure of deethanizer reboilers resulting in extended downtime. Causes and corrective actions for this incident are presented.

Background

The columns in the back end of the Olefins Unit were in the process of being reinventoried with ethylene and propylene in preparation for startup. The unit had shut down five days earlier due a trip of the unit propylene refrigeration compressor. Mechanical damage to the propylene refrigeration compressor had occurred during this trip. The necessary compressor repairs were completed during this five day period.

During this five day period, some other unit maintenance activities, which required a unit outage were quickly planned and completed. One of these maintenance jobs was to clean the process strainers, located on the demethanizer bottom stream feeding the deethanizer. In preparation for cleaning these strainers, the deethanizer was deinventoried and depressured to approximately 30 psig. This depressuring of the deethanizer allowed this maintenance job to be safely completed. This work was completed during the first and second days of this five day unit outage.

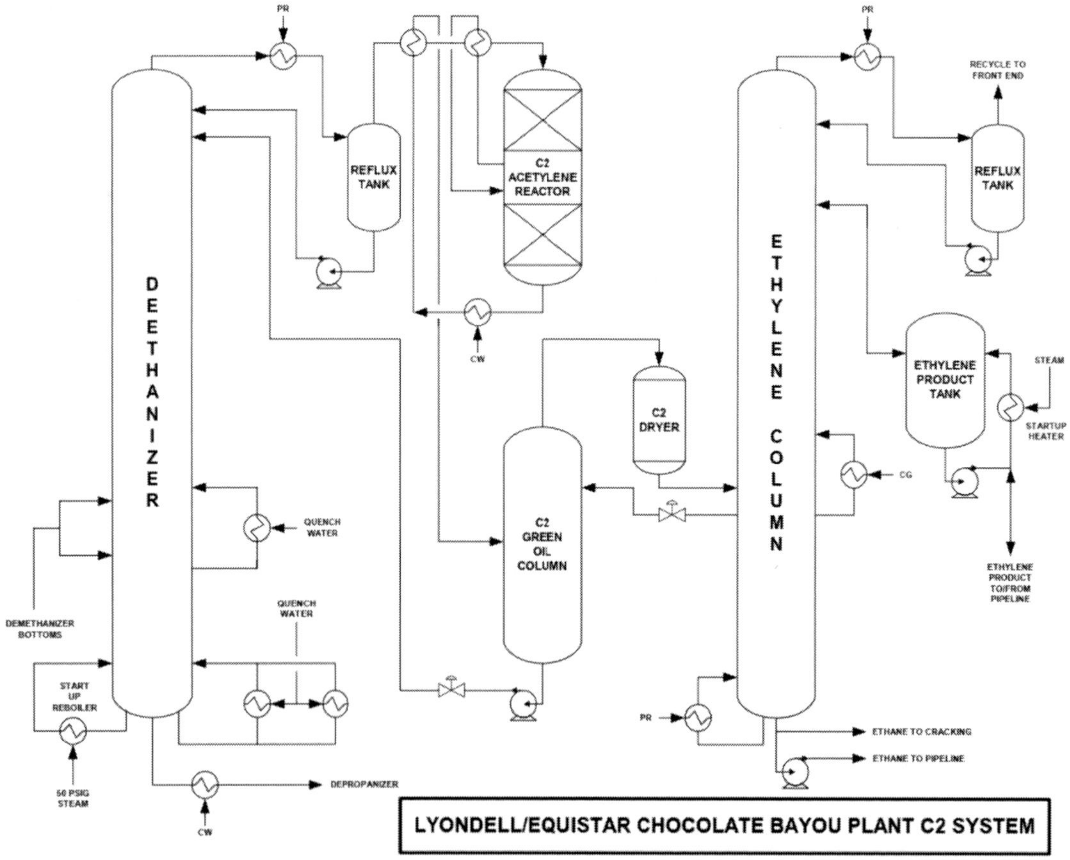

LYONDELL/EQUISTAR CHOCOLATE BAYOU PLANT C2 SYSTEM

Unit operating procedures exist for restarting the unit following a shutdown. Included in these startup procedures are the many steps, e.g., reinventorying the unit back end columns, required for a smooth startup with minimal flaring. The reinventory of the unit C2 system and restart of the deethanizer is normally accomplished by this sequence of basic steps following a unit turnaround or extended outage:

1.) Ethylene from the product pipeline is pressured back into the unit and heated using a startup heat exchanger.
2.) The ethylene column is inventoried with ethylene vapors.
3.) The ethylene column is pressured up with ethylene vapors. It then used as a pressure source for vapor filling and pressurizing the ethylene refrigeration system and other C2 system process equipment.
4.) The ethylene column is inventoried with liquid, started up, and put on total reflux.
5.) A liquid draw is taken off of the ethylene column to inventory the C2 green oil column with ethylene liquid after it is pressured up with ethylene vapors.
6.) After the deethanizer is pressured to 200 psig with ethylene vapors from the ethylene column via the C2 green oil column, the deethanizer sump is inventoried by backing propylene in through the bottoms piping. A temporary jumper piping is required to make this connection between a propylene rerun line in the unit and the deethanizer bottoms piping.
7.) The deethanizer overhead condensers are inventoried with propylene refrigerant.
8.) When a level appears in the deethanizer sump, the deethanizer startup reboiler is started up using steam. The propylene begins to vaporize, which increases the column pressure.
9.) The deethanizer is then inventoried with ethylene by pumping ethylene from the C2 green oil column sump. Column pressure must remain above 200 psig while adding ethylene.
10.) The deethanizer is pressured up to 335 psig, started up, and put on total reflux.

Incident

Following maintenance and restart of the propylene refrigeration compressor, the following occurred during this particular incident with regard to reinventorying the back end C2 area of the unit:

1.) The ethylene column was reinventoried with ethylene liquid, started up, and put on total reflux.
2.) The C2 green oil column sump was reinventoried with ethylene liquid from the ethylene column.
3.) With an indication that the deethanizer pressure was 200 psig, ethylene liquid from the C2 green oil column sump was pumped into the deethanizer.
4.) A sump level was established in deethanizer and its pressure increased to 315 psig.

Soon thereafter, when furnace feed was reintroduced into the unit and the deethanizer began accepting its feed, it quickly became apparent to operations personnel that all three of the deethanizer reboilers, i.e., two quench water reboilers and the one steam startup reboiler, were not providing the necessary amount of heat. Operations personnel were unsuccessful in starting any quench water flow through the tube side of its two normal reboilers.

Initial Investigation

Soon after it became apparent that the deethanizer was experiencing startup problems, operations personnel reported that much of the deethanizer bottoms piping had frosted up. A check of the deethanizer temperature profile via the DCS display confirmed very cold temperatures throughout the column. A process data trend analysis quickly indicated that the column temperatures had fallen sharply during the earlier reinventorying process. When ethylene liquid was initially pumped into the top section of the column, the tray 9 (numbering from the top) temperature dropped 159 degrees (71F → -88F) during a 60 minute time period. As ethylene continued to be pumped into the deethanizer, the column tray 64 temperature, located near the column sump, later dropped 107 degrees (95F → -12F) during a 39 minute time period. During this reinventorying process, there was no flow of quench water through the deethanizer main reboilers. With these cold temperatures, which remained below 32F for 24 hours, it was quickly determined that the quench water inside the reboiler tubes had frozen. The freezing water did cause mechanical damage to the two quench water reboilers, which required repair.

Operations personnel then completed an initial investigation in an attempt to determine how this apparent autorefrigeration incident could have occurred. The board man responsible for coordinating the reinventorying of the deethanizer and other operations personnel believed that the column had been pressured up to 200 psig before ethylene liquid was pumped into it. They made this assumption because the column's primary pressure indicator, whose output is located on the primary DCS display for the deethanizer showed 200 psig. However, this pressure transmitter was actually reading below scale. Since the normal operating pressure of the deethanizer is approximately 335 psig, this primary pressure had been ranged from 200 to 400 psig. The actual column pressure was then trended using a second pressure transmitter located on the deethanizer. This pressure transmitter with a much wider range is not used for control and its output is shown on a secondary deethanizer overhead DCS display. It indicated that the column pressure was actually only 22 psig when liquid ethylene was pumped into the vessel, resulting in this autorefrigeration incident. These actual temperature and pressure changes inside the deethanizer during this particular reinventorying are shown in the below trend.

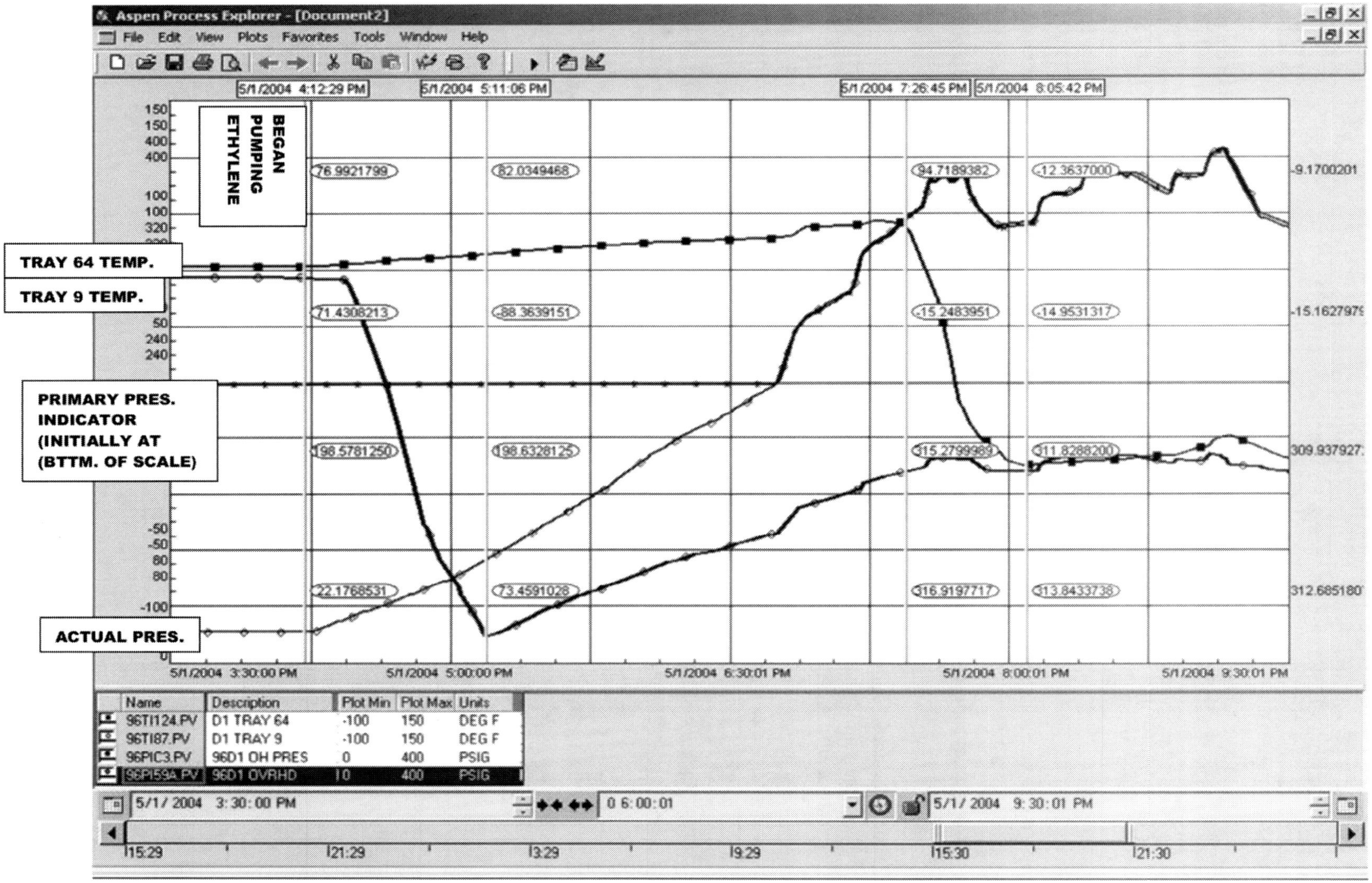
Aspen Process Explorer - [Document2]
BEGAN PUMPING ETHYLENE
TRAY 64 TEMP.
TRAY 9 TEMP.
PRIMARY PRES. INDICATOR (INITIALLY AT (BTTM. OF SCALE)
ACTUAL PRES.
5/1/2004 4:12:29 PM
5/1/2004 5:11:06 PM
5/1/2004 7:26:45 PM
5/1/2004 8:05:42 PM
5/1/2004 3:30:00 PM
5/1/2004 5:00:00 PM
5/1/2004 6:30:01 PM
5/1/2004 8:00:01 PM
5/1/2004 9:30:01 PM
Name | Description | Plot Min | Plot Max | Units
96TI124.PV | D1 TRAY 64 | -100 | 150 | DEG F
96TI87.PV | D1 TRAY 9 | -100 | 150 | DEG F
96PIC3.PV | 96D1 OH PRES | 0 | 400 | PSIG
96PI59A.PV | 96D1 OVRHD | 0 | 400 | PSIG

Site engineering personnel then assembled to determine if the deethanizer's Minimum Allowable Temperature (MAT) had been reached while it was being reinventoried, chilled, and pressured up. Analysis of the vessel using API 579 Section 3, Method 2A resulted in a Minimum Allowable Temperature of +43°F which is above the normal operating temperature. Further evaluation determined that this vessel passed all of the requirements for generation of a "Grandfather" Minimum Allowable Temperature (GMAT – Method 2C). This evaluation yielded a -2°F GMAT. This is simply at the equalization flash temperature for ethylene. Since pressurization of the deethanizer system was accomplished very slowly, i.e., limited by the vaporization rate of the liquid ethylene over a significant length of time (over 2 hours from 22psig to 150psig), the vessel wall temperature tracked the process temperature. Vessel conditions stayed just above the vessel GMAT by approximately 6°F. Had any of the deethanizer reboilers been available, repressurization of the tower would have occurred much faster with the potential to drop below the vessel GMAT.

Brittle Fracture Analysis Report for 96D1 DeEthanizer Tower

MAT is acceptable for the vapor pressure curves and the User operating pressure/temperature combinations entered.

Min. acceptable (coldest) temperature (MAT) when the pressure is at MAWP is -2°F.
Maximum permitted pressure when the temperature is at -155°F is 151 psig
Recommended minimum field hydrotest temperature = 72°F

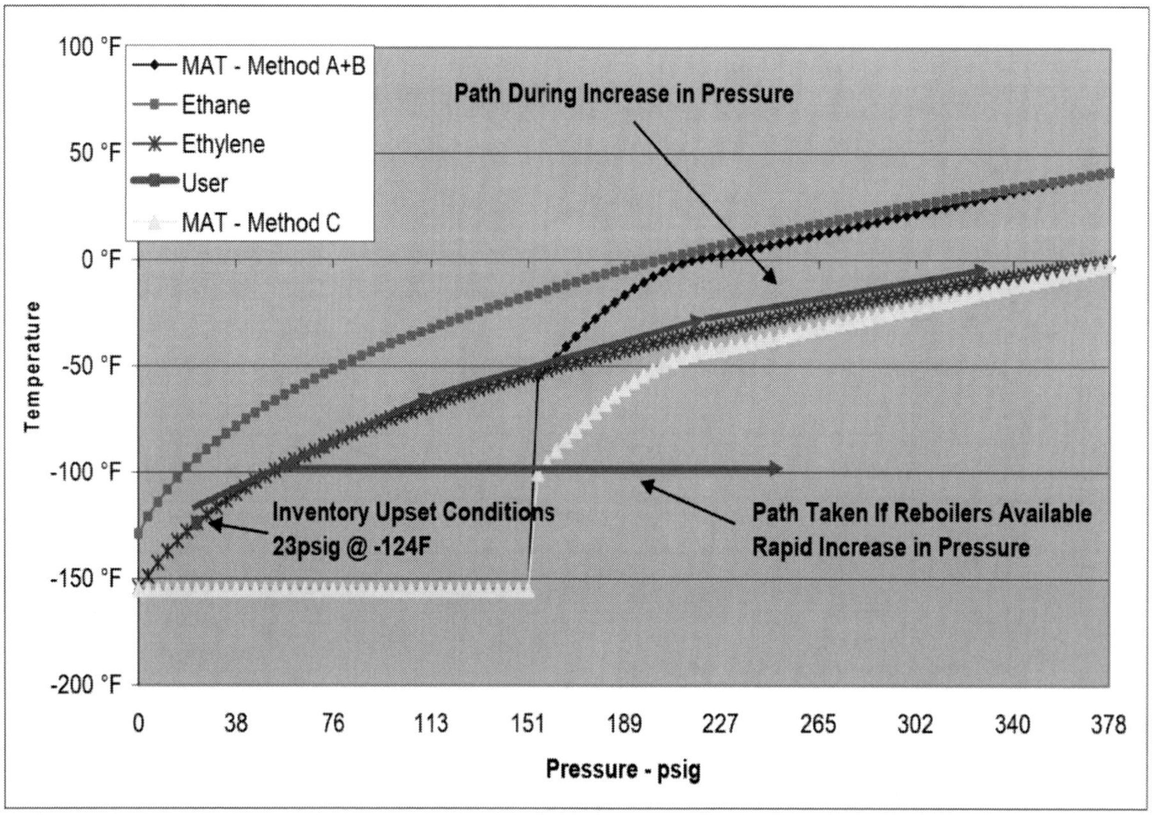

Formal Investigation

Since this autorefrigeration incident was a safety near-miss with regard to reaching the deethanizer's brittle fracture MAT and also an environmental incident due to increased flaring during the restart of the unit, it was formally investigated. This was the first known incident involving freezing the water inside the deethanizer reboilers during a reinventory following a unit shutdown. The Apollo® Root Cause Analysis process was utilized during this formal investigation in an effort to identify causes and appropriate corrective actions. From the cause & effect chart generated by the investigation team, the following causes were selected and addressed with corrective actions:

Cause:

1.) The deethanizer had been reinventoried with only ethylene, i.e., not with any propylene.

Contributing Causes:

This was done because it had become the normal practice during the unit back end reinventorying and startup following a unit trip and soon thereafter restart, i.e., outside of a startup following a major unit turnaround. This had become the practice because following a unit trip, the deethanizer would normally remain pressured up with its sump partially inventoried with propylene. Also, no permanent piping existed that could be used to quickly inventory the sump with propylene. Because simply pumping ethylene into the deethanizer had become the practice, this step was included in a unit operating procedure entitled Olefins Unit Startup After a Shutdown. This was an abridged type of procedure, which outlined the unit basic startup steps without providing much detail or safety precautions. This procedure could easily be copied and distributed to the operations personnel during a unit restart. The normal deethanizer reinventory process used following a major unit turnaround is explained in detail in a second unit operating procedure entitled Operation of Deethanizer Column. This procedure notes all of the autorefrigeration concerns.

Corrective Actions:

a.) Modify the existing procedure entitled Olefins Unit Startup After a Shutdown as needed to ***always*** reinventory the deethanizer with propylene prior to ethylene. Ensure this procedure is consistent with the Operation of Deethanizer Column procedure.

b.) Develop an "official" Olefins Unit Startup Checklist, which includes all required pre-startup and startup steps. This checklist would be consistent with the above two unit operating procedures and could be assigned to specific operations personnel to ensure all necessary steps were executed.

c.) Install the necessary permanent piping to allow the deethanizer sump to be reinventoried with propylene liquid when needed.

2.) The deethanizer contained only ethylene vapors at low pressure when reinventorying with ethylene liquid began.

Contributing Causes:

As explained above, the deethanizer had been purposely deinventoried and depressured earlier in the week to clean a process strainer. When the deethanizer was being reinventoried, the responsible operations personnel were not aware that the vessel was depressured. This depressuring had occurred days earlier and was executed by other operations personnel. The atypical process conditions inside the deethanizer had not been communicated from shift to shift via some formal process during this extended unit outage. Because the unit was in a shutdown state for an extended time period, many other process conditions more closely resembled those during a restart following a major unit turnaround rather than those resembling a quick restart following a unit trip. As explained above, the operations personnel incorrectly assumed that the displayed deethanizer pressure indication that is normally utilized was the true pressure. Operations personnel did not question the possibility of this pressure transmitter being below scale. The correct pressure reading was not located on the primary but on a secondary display for the deethanizer.

Corrective Actions:

a.) Use this actual incident to illustrate to all Olefins Unit operations personnel the potential for autorefrigeration and brittle fracture during the ***reinventorying*** of Olefins Unit back end process equipment.

b.) Provide additional low pressure and temperature alarms for the deethanizer.

c.) Provide a more visible DCS indication or flag whenever any displayed process variable value is incorrect because it is outside of its instrument's calibrated range.

d.) Include all widely ranged process condition indicators, which are present for reducing autorefrigeration risk, on the primary DCS displays.

Paper No. 38C

Pipeline Integrity of Ethylene Pipelines

Greg Nordquist, P. Eng.
NOVA Chemicals Corporation
PO Box 5006
Red Deer, Alberta, Canada, T4N 6A1

Prepared for Presentation at the 2005 Spring National Meeting, Atlanta, GA, Apr 10-14
17th Annual Ethylene Producers' Conference
Session TA005 – Ethylene Plant Safety

UNPUBLISHED

ABSTRACT

Pipeline integrity is a broad process in which continual condition monitoring, assessment, evaluation, and repairs mitigate integrity threats to pipeline fitness for service. Almost all pipeline segments traverse sensitive areas where the consequence of failure is prohibitive. NOVA Chemicals operates several ethylene pipelines ranging in size from NPS 4 to NPS 12, in the heartland of Alberta, Canada. The integrity of these pipeline segments is vital to the continued economic success of the NOVA Chemicals Joffre petrochemical site. The NPS 12 Ethylene Mainline runs from the Joffre site to Fort Saskatchewan, Alberta. Built in 1978, the line is 176 km (109.6 miles) long and is coated with a double wrap of polyethylene tape. The pipeline has never had an operational nor hydro-test failure. This paper will focus on key elements of the NOVA Chemicals Pipeline Integrity Management Plan and summarize the 2004 high-resolution in-line-inspection and follow-up investigation of the NPS 12 Ethylene Mainline.

NOVA Chemicals operates approximately 1800 km of ethylene, ethane, NGL, hydrogen, and nitrogen pipelines in Alberta. These systems are of great importance to the success of the petrochemical industry in Western Canada. The pipeline system supplies ethylene, as an essential feedstock, to various users in the Joffre, Edmonton, and Fort Saskatchewan areas. Damage to the pipeline system that could potentially result in a loss of ability to transport ethylene could have serious business implications in terms of lost production for downstream users as well as loss of reputation for NOVA Chemicals as a reliable supplier of essential feedstock. It is in the interest of NOVA Chemicals to maintain a safe and reliable pipeline system both from a Responsible Care® stand point as well as a business impact perspective. This paper will examine all key elements of the NOVA Chemicals Pipeline Integrity Management Plan.

FIGURE 1

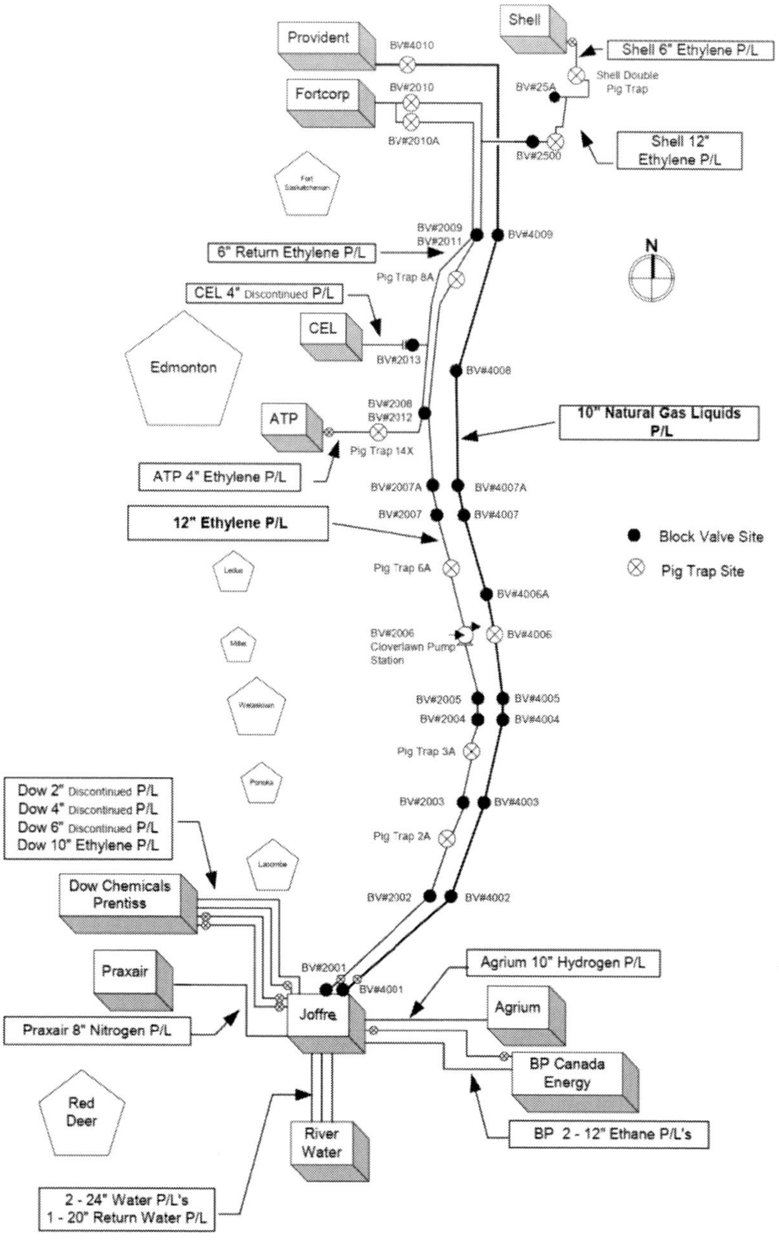

The Pipeline Integrity Management Plan (IMP) for NOVA Chemicals Corporation pipeline facilities in Western Canada is an integrated plan for all activities related to preventing pipeline integrity incidents. The IMP closely adheres to the guidance of CSA Z662, Annex N, "Guidelines for Pipeline Integrity Management Programs". As such, the plan is a core element in the management of risk for pipeline systems and serves to pull together other components such as the Pipeline Emergency Response Plan, pipeline Risk Assessments, Pipeline Rapid Repair Plan, and the Pipeline Failure and Damage Incident Investigation Manual. All of these components combine to reduce the consequences of a potential pipeline failure. In concert these plans and activities make up the overall Integrity Management Plan for the Alberta Ethylene Distribution System, Joffre Feedstock Pipeline, Alberta Ethane Gathering System, and all other pipeline facilities operated by NOVA Chemicals.

Pipeline systems operated by NOVA Chemicals Corporation in Alberta encompass varied configurations and product service fluids. This diversity requires an overall IMP to bring together the various elements in the management of risk related to pipeline system integrity. Past and current maintenance and integrity programs and practices meet or exceed all regulatory requirements. Drivers for improvement stem from ensuring the highest Responsible Care performance; perceived opportunities to increase the efficiency and effectiveness of integrity and maintenance activities; and reduce the high business risk associated with an interruption in pipeline service. It is also expected that future regulatory changes in Canada will require the development and execution of an IMP by all pipeline operators.

NOVA Chemicals IMP is structured to meet the guidance of CSA Z662, Annex N. Individual Program areas are described that relate to the management of specific threats to Pipeline Integrity or to specific Maintenance or Development needs. A general description of the IMP in terms of Principles, Structure and Planning Cycle is presented below. All elements are presented and described in detail in the Plan:

- IMP Program Records and Information Management
- Personnel Competency and Training
- Management of Change
- Failure and Damage Incident Investigation
- Hazard Identification and Control & Risk Reduction
- Risk Assessment
- Program Planning and Execution
- Inspection, Testing, Patrol, and Monitoring
- Evaluation of Inspection
- Mitigation and Repair
- Facilities Integrity

- Operating Pressures Management
- Program Review and Evaluation
- Program Audit

The IMP describes in detail; activities relating to each integrity threat area, and it presents an overall priority for integrity activities. A Responsible, Accountable, Consult and Inform (RACI) structure for the planning and implementation of the work is offered as guidance, as is long-term view for planning into the future. Objectives and targets are identified for future review in an annual "Plan, Do, Review" cycle. The pipeline Area Work Team (AWT) will develop Key Performance Indicators (KPI) to ensure adherence to the IMP. All elements and KPI's are reviewed quarterly by the AWT.

OBJECTIVE OF THE IMP

NOVA Chemicals' Pipeline IMP includes guidance on: management systems, corrosion control, maintenance and inspection, Right Of Way monitoring and surveillance, integrity inspection, monitoring, and mitigation elements. Management strategies to control integrity hazards on operating pipelines form the foundation of the IMP. Individual program elements have been specifically developed to mitigate one or more identified integrity threats presented to the pipeline facilities.

Objectives of the program are:

1. To ensure activities undertaken under the plan, such as: inspection, maintenance, and repair of pipeline facilities, physical modification, and alteration to operating parameters on pipeline facilities are done in accordance with this quality management program.

2. To contribute to the NOVA Chemicals commitment to the Responsible Care Policy and guiding principles.

3. To ensure that pipelines and associated equipment are operated within design and code limitations as required by the Pipeline Operating Limits.

4. To ensure pipeline maintenance procedures provide safe, consistent, code and regulatory compliant procedures for inspection and maintenance activities in support of integrity programs. In all cases, the procedure manuals provide a focus on maintaining the integrity of the pipeline system while maximizing safety.

Overall, the objective of the plan is to provide NOVA Chemicals Manufacturing West with a comprehensive Pipeline Integrity Management Plan for current activities with a future plan that integrates all activities related to preserving the integrity of all pipeline systems.

CORPORATE GUIDANCE

As with any Quality Assurance Program, this Management Plan has full support of NOVA Chemicals management, who ensure that adequate resources and trained personnel are provided to effectively implement the program. Management support is established through stating guiding principles, establishing high level Key Performance Indicators and statements:

- NOVA Chemicals views any operational failure of the pressure containing envelope of the pipeline system or contact damage to the pipe body by a third party as preventable.
- Unplanned or uncontrolled release of process fluid is considered a Critical Incident.
- Any unapproved excavation within the specified pipeline excavation zone is considered a Serious Incident.

The IMP is a controlled document with accountabilities and responsibilities clearly identified and established.

ESTABLISHING PRIORITY

The overall drivers of the IMP are the Pipeline Risk Assessments completed on a 5-year interval. The Risk Assessment completed in 2003 on the NPS 12 Mainline and NPS 6 Return Line identified probabilities of failure based on each failure cause established from historical failure rates for high vapour pressure pipelines in Alberta and data gathered by the Alberta Energy and Utilities Board. Each failure rate cause was evaluated and adjusted by application of factors to better reflect the present integrity level of effort and known condition of the NOVA Chemicals pipeline facilities. The chart below indicates the calculated failure probabilities resulting from this evaluation.

FIGURE 2

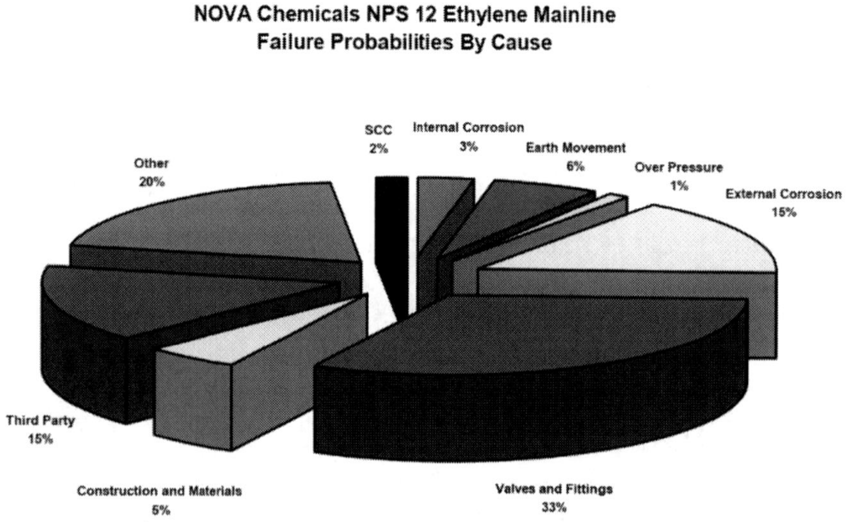

INTEGRITY MANAGEMENT PLAN KEY ELEMENTS

Integrity Management Plan Records Management

As part of the requirements of the code annex, operating companies must prepare and manage records related to pipeline design, construction, operation, and maintenance that are needed to perform the activities included in a pipeline integrity management program. Effective pipeline integrity management depends on having ready access to accurate and complete information. Past and current records must be available for key parameters such as:

a) Location of the pipeline with respect to crossings and nearby land developments;
b) Class locations;
c) Design of the pipeline, including limits to pressure, temperature, loading and other operating conditions;
d) Standards and specifications for the pipe, components, bolting, and coating materials;
e) Material test reports;
f) Joining and inspection records;
g) Coating and inspection records;
h) Terrain, soil type, backfill material, and depth of cover;
i) Pressure testing;
j) Cathodic protection system design and performance;
k) Records of methods and results from activities described in the pipeline integrity management plan;
l) Inspection records;
m) Incident reports;
n) Homeowner contacts;
o) Emergency responder contacts;
p) Code requirements; and

NOVA Chemicals records management uses legacy document tracking/storage systems in electronic and hard copy files. Electronic data records required for day to day integrity management are being transitioned to a GIS spatial format in a Pipeline Open Data Standard (PODS) database. In-Line Inspection (ILI) features and excavation inspection information, repair, or inadequate cathodic protection potentials that must be addressed are tracked using fit for purpose database applications. The information system is set up to accept new information as it is acquired in a cost effective and convenient manner. The system is readily expandable to include new types of information in the future. Compatibility with third party databases, e.g. spatial database information and weather records held by government agencies should be considered. NOVA Chemicals ongoing program reviews all database applications and their uses to ensure integration into a single solution is effective. It is the intent of the program to ensure records retention meets the requirements of code and all data is integrated into a single data storage solution.

Personnel Competency and Training

As part of NOVA Chemicals' continued commitment to the principles of Responsible Care, NOVA Chemicals has developed extensive mandatory training requirements. All company personnel, contractors, and consultants must have appropriate knowledge and skills to perform tasks associated with the execution of the pipeline integrity management plan and/or pipeline maintenance activities.

NOVA Chemicals maintains and reviews an extensive skills matrix that identifies and tracks all required training. Training requirements for all personnel are reviewed and updated on an annual basis by the Pipeline AWT; any deficiencies noted are added to the training matrix with an appropriate completion timeline. Individual employee training status is reviewed on a monthly basis. Specialized training related to pipeline integrity topics is included for personnel directly responsible for those activities as well as those individuals that express an interest in performing additional activities. It is the responsibility of the assigned Training Coordinator to ensure all training records are tracked and reflect the most up to date status for each employee. NOVA Chemicals has extensive training processes in place for all employees, it is not the intent of the IMP to influence these existing processes, only to exploit them to the fullest.

Management of Change

NOVA Chemicals uses a Management of Change process known as the Facility Change Management (FCM) process. The IMP recognizes the FCM process as a company standard and endorses its application to pipeline modifications that might affect the integrity of pipelines or NOVA Chemicals' ability to properly manage pipeline integrity. All activities or projects that meet the guidelines outlined within the process must be executed through the process.

The Facility Change Management process has been put in place to ensure:

- Changes are reviewed for safety, environmental, production and other hazards
- Changes receive the required level of review
- Follow-up is tracked
- Reviews are properly documented (audit trail)
- Records are updated

The FCM procedure is used for:

- Capital projects
- Temporary Facilities
- Repair or replacement using components, equipment, or material, which are not replacement in kind (including piping diameter, routing, joint connection changes).
- Changes to the operation of a facility where equipment is to be operated outside of its established design range(s), including temperature, pressure, chemistry, catalyst, etc.

- Safeguarding systems (alarm set points, trip systems, PSV's, firewater, deluge, etc.).
- Control application software and PLC programming.
- Instrument calibration or range change; DCS emergency alarm; display changes involving links or new equipment; minor software changes.

The FCM procedure does not apply to:

- Repair or replacement using components, equipment or material, which are replacement in kind (including maintenance "Quick Response").
- Test runs or introduction of new resins where equipment is to be operated within its established design range(s), including temperature, pressure, chemistry, catalyst, etc. (Refer to EOI procedure).

All pipeline inspections and modifications are executed through the FCM process without exception.

Failure and Damage Incident Investigation

NOVA Chemicals Corporation Manufacturing West (MW) has formalized a “Failure and Damage Incident Investigation Manual” a process containing procedures for the investigation of all pipeline facilities incidents and failures. A key element of an effective integrity program is to fully understand the causes of pipeline incidents and failures so that program changes can be made to prevent reoccurrence. All incidents that meet the guidelines outlined below must be investigated through the procedures described in the manual and must be fully documented with recommendations for mitigation of potential future incidents.

The scope of the process encompasses any and all incidents involving damage to pipeline facilities on NOVA Chemicals Manufacturing West pipeline systems. Failures include: equipment failures (other than normal wear and tear), leaks, ruptures, third party line strikes, catastrophic geotechnical failures, and geotechnical failures that affect the integrity of the pipeline pressure containing envelope. The manual is intended to be a guide only, materials testing and investigative procedures not outlined in the guide but deemed necessary to the investigation and analyses are encouraged. An incident resulting in damage includes, but is not limited to, corrosion, coating damage, cracking, overpressure beyond code tolerance, and mechanical/contact damage caused by NOVA Chemicals Corporation or third party personnel, as well as geotechnical soil movement.

The guidelines outline the tasks and activities performed by, or overseen by, the Pipeline Area Work Team to determine the cause of failure; to document the results of the investigation; to recommend and monitor implementation of appropriate preventive measures; and to enter the information into the appropriate risk assessment predictive modeling program and other databases as applicable. Methods described in the Failure and Damage Incident Investigation process for the gathering, review, and documentation of failure evidence are intended to provide traceability of action and completeness of information, as well as to meet the requirements of the Alberta Pipeline Act and Regulation.

Hazard Identification and Control

The IMP identifies hazards that have the potential to lead to a failure or a damage incident. Methods and data used for hazard identification are documented within the IMP, considering or determining the primary causes and sub-causes from any additional failure or damage incident that are relevant are also of critical importance. Where hazards that might lead to a failure or damage incident with significant consequences are identified, the IMP identifies actions to monitor for conditions that could lead to a failure. The hazards and associated control measures identified within the IMP include the following:

- Hazard of Third Party Damage, Surveillance and mitigation actions
- Hazard of External Corrosion and Manufacturing Defects, Inspection and remedial actions
- Hazard of Soil Movement, Investigation, Assessment and remedial actions
- Hazard of Stress Corrosion Cracking (SCC), Inspection, Assessment and remedial measures and actions
- Maintaining Cathodic Protection and Corrosion Control, Meet CSA and OCC-1 requirements
- Managing Operating Pressures, Meet the requirements of code and operating license
- Facility Integrity Management, Safety devices, Pressure Vessel codes, OEM maintenance requirements, operational limits

Actions are identified, standards and procedures developed to eliminate or mitigate conditions of high risk. As part of hazard identification and control, risks associated with all hazards are analyzed and quantified to ensure activities undertaken reduce the risk exposure to acceptable levels. These outputs are then applied to the pipeline risk assessment to assess the level of overall system risk reduction associated with all hazard control measures.

Risk Assessment

NOVA Chemicals Manufacturing West has completed several historical risk assessments of the Ethylene Mainline and 6" Return Pipeline. Many of the recommendations from past assessments have been implemented or considered for implementation. Population growth and development adjacent to the pipeline Right Of Way has continued and will require sustained re-evaluation of the risk assessment to ensure Manufacturing West is meeting NOVA Chemicals allowable risk exposure.

NOVA Chemicals benefits from completing risk assessments on pipeline facilities by having clearly identified areas on the pipeline systems that require mitigation efforts or even physical changes identified to reduce the overall risk profile. This program identifies physical pipeline Right Of Way sections where risks are outside of the tolerance level as set by NOVA Chemicals, Regulatory Guidelines, Code requirements, or industry accepted practice. NOVA Chemicals views risk reduction as a strategic component to the stewardship of all of the pipeline systems that they operate, and a necessary element of operational due diligence. The latest revision of the Risk Assessment was calculated using Alberta pipeline industry failure statistics gathered by the Alberta Energy and Utilities Board. The analysis has proven somewhat problematic given that significant effort on the part of NOVA Chemicals on areas that significantly influence risk levels are difficult to quantify and that may in fact reduce the risks to acceptable levels. More work and understanding of these influencing factors is needed to more accurately model the risk profile.

The Pipeline AWT is constantly monitoring the pipeline Right Of Way for development and evaluating and communicating any growth that will impact pipeline facilities, area classification, or risk exposure. These observations and communications trigger population studies and class location evaluations along the pipeline Right Of Way in areas of concern as per the requirements of CSA Z662. These evaluations will also be included in future Risk Assessments to ensure the most accurate information is used in the re-assessment. A deliverable from the Risk Assessment is to identify recommendations that will reduce the risk profile of the operating pipeline facilities. It has been an internal NOVA Chemicals policy and is identified in the IMP to re-calculate the risk assessment on a five-year interval. Risk levels are applied within the NOVA Chemicals Integrated Risk Assessment Process model for verification against established risk tolerance thresholds.

Program Planning and Execution

The IMP is an evolving tool for the Area Work Team under the accountability of the Pipeline Team Leader. It is the responsibility of the AWT to execute the plan. Implementation activities establish appropriate means to track progress and keep the plan evergreen through its annual plan, do, review cycle. Maintenance of the plan includes gathering information necessary to assess factors that trigger decisions to carry out activities in support of the plan. While a number of annual targets and objectives have been identified within the IMP for yearly review, the AWT monitors progress through a quarterly KPI review process where one of the team KPI's will focus on execution of the IMP activities. It is the responsibility of the Pipeline Contact Engineer to maintain the review history and corrective actions and recommendations from these reviews.

Routine maintenance activities have not been included in the IMP at this time. A comprehensive Maintenance Program is critical to overall pipeline system and IMP performance. Maintenance has its own planning processes and activities focused specifically on the availability of pipeline systems. These are driven by a world-class reliability effort.

Pipeline Inspection and Testing

A strategic requirement of the IMP is to provide guidance in the development of procedures for assessment and control of pipeline integrity and systems hazards. Procedural requirements such as: potential risks, safety considerations, code, regulatory, frequency, shutdown, potential findings, defect tolerances, assessment alternatives, observations, repairs, documentation, etc. are identified and described in detail. Areas developed include:

- Cathodic protection systems (OCC-1)
- Coating assessments and evaluation
- Leak detection methods and devices
- Shutdown devices and systems
- Pressure-control, pressure-limiting, and pressure-relieving systems
- Pipeline valves
- Pipeline patrol
- Inspection of exposed piping for corrosion and other types of imperfections
- Inspection of components by indirect methods (i.e. ILI, UT, Pulsed Eddie, etc.)

Continual development and improvements of existing procedures is also a critical requirement of the IMP to keep monitoring and mitigation activities current. All procedures are reviewed and revised on a prescribed schedule.

Evaluation of Inspection Results

Where results of inspection, testing, patrol, and monitoring activities indicate the presence of conditions or imperfections that could lead to a failure incident with significant consequences, it is a requirement within the IMP to perform an engineering assessment or carry out corrective actions in accordance with standards, procedures, and code requirements. The purpose of this section is to outline standards and guidelines complete with acceptable defect tolerances and defect acceptance criterion to assist in the evaluation of pipeline inspection results and to document the condition of the pipeline for continued service. Areas developed:

- In-Line-Inspection data interpretation and defect sizing
- Direct Corrosion Defect Assessment and acceptance criteria standards
- Direct SCC Assessment and acceptance criteria
- Direct Mechanical Damage Assessment and acceptance criteria standards
- Incident Investigation process
- Geotechnical Assessment and acceptance criteria
- Defect repair alternatives selection

One major component in the IMP that contributes to the continued success of these efforts is the "Failure and Damage Incident Investigation Manual". The manual details activities required to complete an engineering assessment, evaluations of failures, damage, and defects to determine cause. All assessments must include consideration of:

a) Knowledge and experience of the performance capabilities and limitations of the inspection method;
b) The types of imperfection that might correspond to the indications reported;
c) The accuracy of reported dimensions and characteristics needed for the evaluation of such imperfections;
d) The likelihood of unreported defects, such as cracking, to be associated with an imperfection indication;
e) The piping design and material properties;
f) Service conditions

This approach provides consistency in what is assessed and how it is assessed. It ensures follow-up recommendations, documentation, and communication to all stakeholders. Many of the requirements and specifications in this section of the IMP refer to industry established inspection protocols and procedures or integrity hazard assessment practices.

Audit of the IMP Program

To close the review cycle the management program requires an audit of the program itself. This section of the IMP describes the internal audit procedure that is followed in order to ensure that the intent of the program is met. The Manager of Manufacturing Infrastructure establishes a program, procedure, and team to ensure that the audit of the program is conducted in accordance with NOVA Chemicals procedures. The internal audit of the program adheres to the following:

- An audit of the program will ensure all areas are subjected to an audit process at least once per 24-month cycle.
- An audit procedure will verify that the intent of this program is followed in all areas and that the details of all procedures are adhered to in principle.
- An audit team to perform internal audits shall be defined from time to time. The team may consist of any number of members, and may include both NOVA Chemicals personnel and/or specialist contract personnel.
- The Manager, Manufacturing Infrastructure, maintains audit reports and any deficiencies noted are tracked within appropriate systems, together with the corrective action taken with review and follow-up.
- The Team Leader, Pipeline, ensures corrective action is taken to rectify all deficiencies found through the internal audit process.

Audit of the program is critical to ensuring success of the plan. As with any quality assurance program audit by an impartial, independent audit team is very important. Audit findings and recommendations must be assigned and rectified in a timely manner.

RECENT IN-LINE-INSPECTION OF THE ETHYLENE MAINLINE

The ethylene pipeline system that is operated and maintained by NOVA Chemicals transports high-pressure ethylene bi-directionally between the NOVA Chemicals manufacturing plants at Joffre and the Fort Saskatchewan storage cavern facilities. The pipeline has been inspected previously in 1990 and again in 1999, both inspections used the most up-to-date Magnetic Flux Leakage tool technology to ensure that the most accurate dataset could be gathered for analysis. The purpose of this in-line inspection was to accurately characterize existing and newly formed anomalies (such as metal loss and dents), compare corrosion depths with those reported by previous inspections, and determine whether any recent third party mechanical damage had been experienced. Secondly, NOVA Chemicals desired to acquire highly accurate pipeline geometry and geo-spatial pipeline coordinates from a geometry and inertial equipped inspection tool. NOVA Chemicals NPS 12 ethylene pipelines are fully piggable without the need for modifications to launch or receipt facilities. Repair programs have been undertaken following past inspections and no known diameter restrictions had been introduced, so passage of the cleaning and gauge tools was expected not to encounter any problems. A decision was made not to run a caliper tool for this reason; the contractor supplied dummy tool was run in all sections as well as 6% and 10% ID restriction indicator plates on the cleaning tools. The Mainline internal surfaces are kept relatively clean through non-laminar product flow and the nature and purity of the ethylene product being transported. Although some natural formation of a polymer layer on the inner surface of the pipe is experienced, it was easily removed prior to inspection by means of cleaning tool runs. The polymer removed from the pipe inner surface was collected at temporarily connected filter vessels, installed at the pig receipt stations.

Total length of the pipeline inspected was 185 km (115 miles), split into 6 sections. The project contract documents detailed all inspection activities, tool specifications, and data analysis requirements to be completed by the contractor. Pipeline flow varies greatly depending on derivative plant consumption at the Joffre end of the pipeline system. It was determined that the best pigging results would be obtained at maximum pipeline throughput. The in-line inspection schedule was set to coincide with plant shutdowns that would ensure adequate tool velocity. This was of concern to the production plants in that virtually all production was flowing through the pipeline at the time of the pigging, with the potential for a stuck pig to cause significant production curtailment. This was overcome by using a cleaning tool that allowed flow by-pass to act with a flushing effect on any polymer build up to keep any solids in suspension. This worked very well; cleaning of all pipeline segments was completed without any pigs becoming lodged in the pipeline from polymer build up. Pig slowing and stopping was encountered and expected at pipe wall thickness transitions.

Pipeline inspection with the high resolution Magnetic Flux Leakage (MFL) tool was completed following the cleaning and gauging runs. Starting at the south end and working north, all sections were inspected with pig tracking to ensure velocity and pig progress were monitored, documented, and reported. At the end of each section the MFL tool was removed from the pig trap and initial data verification was completed. All sections required a single run to collect 100% of line data. Tool performance was 100% with no data loss or degradation. The contractor was initially concerned with the cryogenic effect of high-pressure ethylene de-pressurization on the MFL tool electronics and components. As part of initial evaluation and preparation, various components of the MFL tool were tested in liquid ethylene to confirm materials compatibility. These tests were successful, with tool performance being excellent.

NOVA Chemicals has a requirement to receive the metal loss inspection data in raw format and in Morrison Scientific format. This allows for direct evaluation as well as analysis between subsequent in-line inspection runs to quantify corrosion growth rates of matched defects by statistical analysis. The data collected in the 2004 ILI runs correlated to previous inspections; both in-line inspection data and direct defect assessment has been excellent, well within the requirements of the contract. We continue to perform correlation digs and evaluate corrosion and pipe defects identified in the 2004 ILI program in our 2005 Corrosion mitigation program.

References

1. Canadian Standards Association Z 662-03 Oil& Gas Pipeline Systems

Acknowledgements

1. Tom Jack, NOVA Research & Technology Corporation, Calgary

2. Fred Henselwood, Dendritic Consulting Ltd, Calgary

3. NGTL Pipeline Integrity Plan, PIP Work Team, NOVA Gas Transmission Limited, Calgary

4. Iain Colquhoun, Bob Coote, David Diakow, David Dorling, Bob Sutherby, Michael Wong, NOVA Gas Transmission Limited, Calgary

Paper No. 38d

Waste Stream Incineration Results in Combustion in Decoke Piping

By
Randy Saunders

NOVA Chemicals Corporation

Prepared for presentation at

2005 Spring National Meeting
Atlanta, GA
April

Waste Stream Incineration Results in Combustion in Decoke Piping

By:

Randy Saunders (Nova Chemicals)

Abstract

Prevention of Process fires is critical to successful and safe operation of any Chemical Plant.

Ethylene plants generate various small waste gas streams that must be collected and recycled or incinerated to minimize VOC emissions and to ensure that these waste streams do not create hazards that could result in unintended combustion. In many cases multiple process systems and streams are connected to one common disposal system and innovative methods have been developed to ensure destruction of the various compounds. The design and operation of these systems warrant very close scrutiny as the multitude of scenarios created may result in some very unexpected problems including unplanned combustion events.

This paper describes a series of incidents resulting from the operation of such a disposal system. The incidents occurred during two separate plant outages at the Nova Chemicals Ethylene 3 plant at Joffre, Alberta, Canada. In each case the system was being operated under very low load with some of the normal components absent from the combined waste stream.

Background is provided on the disposal system process and how it influenced the events.

The incident investigations, conclusions, and mitigation are explored as well as some of the present and future challenges that may be ahead with operating this type of system as part of an integrated Ethylene plant.

Introduction

Ethylene 3 is an Ethane Cracker designed to produce 2.81 Billion lb/year Ethylene. It is operated by Nova Chemicals at the Joffre Site in Central Alberta, Canada. The plant is one of three Ethane Crackers at this site and was commissioned in August of 2000.

After three years of operation the Ethylene 3 (E3) plant was taken down for a scheduled Turnaround in May of 2003. At the end of this outage, Operations personnel were in the process of lighting off and warming up the 7 Cracking Furnaces when the decoke effluent piping connected to the number 6 Furnace (H260) was found to be glowing red hot and insulation and cladding melting off of the piping. The piping had heated sufficiently that considerable sagging and bowing had occurred. The piping was subsequently purged and cooled with nitrogen and an investigation was initiated and recommendations made to prevent re-occurrence.

Six months later a plant trip occurred due to a power failure. During the preparation for the plant re-start overheating occurred in three furnace decoke effluent headers on number 5, 6, and 7 furnaces - (H250, 260, and 270), all within a 12-hour period. Another investigation was completed and a new scenario was revealed that was not uncovered during the investigation of the first overheating event.

Both investigations identified that, indeed, combustion of significant amounts of coke was occurring within the decoke piping, causing the overheating and damage to the piping.

Background

Several decisions made in the design of the E3 Plant had an influence on the events so it is worth mentioning these and what factors influenced the decisions.

Firebox Decokes:

The Ethylene 3 Furnaces consist of a twin radiant cell design with a common convection section. They are balanced draft with air preheat. The furnaces are also designed for online decoke of the radiant coils with the decoke effluent directed back into the floor of the firebox through a distribution header and 8 nozzles per cell for a total of 16 per furnace.

Operating experience gained from the operation of the other two Ethane Crackers at the Joffre Site was used extensively in the development of the Ethylene 3 design.

Erosion of decoke piping had been a significant issue in the Ethylene 1 (E1) and Ethylene 2 (E2) plants over the years so during the design of the E3 Furnaces the piping was significantly upsized to minimize this erosion potential. An average velocity of 175 m/s in the E2 furnaces was reduced to an average of 45 m/s in the E3 furnace decoke piping.

To minimize erosion issues with refractory within the firebox and minimize disruption of the convection currents, symmetrical piping was used to distribute the decoke effluent evenly to the 2 cells and the flow was distributed through 16 nozzles evenly spaced along the floor of the two cells.

Feedstock CO2 Disposal:

The Ethane feed entering the E3 Plant contains approximately 6% CO_2 and 1 ppm to 30 ppm H_2S. The CO_2 and H_2S are removed upstream of the furnaces in an Amine Contactor. Something had to be done to dispose of this 15,000 kg/hr acid gas stream. It could not simply be vented to atmosphere due to the H_2S content and the air assisted flare would require significant amounts of natural gas to accompany this stream to ensure destruction of the H_2S.

The acid gas stream in Ethylene 2 had been successfully incinerated in 3 of the 11 cracking furnaces in that plant for several years. As well, in the Ethylene 1 plant the acid gas stream had been incinerated in the fired superheaters for 20 years, so the decision was made to route this stream to the fireboxes of the 7 E3 furnaces. Only furnaces operating in the online mode (cracking ethane) would be used for incineration. The furnace would be isolated from acid gas incineration service prior to being taken offline for decoke and returned to acid gas service after completion of the decoke.

Hydrocarbon Tank Vapour Incineration:

Many low-pressure waste vapour streams also had to be collected and, either recycled back into the process or disposed of via incineration. These included API tanks vents, compressor dry gas seal vents, analyzer purge streams, pump secondary containment seal vent streams, Waste Caustic Degasser vent, Aromatic Concentrate Vacuum Tower ejector stream, some pressure vessels in cyclic pressure operation, etc.

The options considered were:

- Vent streams to the existing E1/E2 Vapour Combustion Unit (VCU)
- Install an E3 Vapour Combustion Unit
- Vent streams to E3 Flare
- Vent the streams to the Cracking Furnace Fireboxes

E1 and E2 had installed a stand alone Vapour Combustion Unit (VCU) for the incineration of these types of streams. However it did not have the capacity to handle the additional E3 streams with an estimated flow rate of 500 kg/hr. E3 could install a VCU, but as it had been found in operating the VCU in E1 and E2 over a 10-year period that these units were not without there own operating problems. VCU trips resulted in these streams being vented to atmosphere until the unit could be repaired and placed back online. Without any kind of heat recovery, they are also not inexpensive to operate. Cyclic loads on the system sometimes resulted in incomplete destruction of some hydrocarbons due to air and fuel ratio problems. The capital cost, reliability of a single unit and ongoing operational and maintenance issues did not make this option particularly attractive.

The E3 Flare System includes twin 84-inch diameter elevated air assisted tips. One flare stack and tip is in primary service with the second stack and tip staged online, only during high flare loads. Large air assisted tips have a history of very short life if subjected to significant durations of low level flaring so a decision was made to eliminate any continuous hydrocarbon streams to the flare system. This would allow the flare to normally be under a nitrogen purge with no flame at the tip except 4 pilots. Use of the flare would then be limited to emergency relief, start-up, shutdown, upset and equipment maintenance flaring requirements. Venting these small vapor streams to such a large flare tip would result in either very little destruction of the hydrocarbons vented or significant amounts of Natural Gas would need to be added to the flare to assist in the combustion. Not only would this be very inefficient but also it was believed that it would have a severe impact on the life of the tip.

The other option was to route these streams along with the Acid Gas into the fireboxes of the 7 cracking furnaces. This allowed redundancy due to the multiple furnaces, allowed heat recovery, although minor, from the hydrocarbons being incinerated and eliminated the requirement for another piece of fired equipment, and much of the capital and operating costs that went with it.

This still left the issue of dealing with disposal of these vapour streams during a plant outage when all of the furnaces are shut down and, during an emergency scenario where all of the furnaces have tripped due to a complete power failure.

For planned outages a connection was provided on the discharge of the vapour collection blowers for installation of a temporary carbon bed for adsorption of the hydrocarbon vapours or a temporary rental VCU for incineration. For the unlikely and very infrequent complete power failure scenario, a common pressure relief valve was provided in a safe location with combustible gas detection at the vent location. On a complete furnace trip scenario the vapours would be vented at this location until furnaces could be made available for incineration. It was believed that in almost all emergency scenarios a furnace could be made available for incineration within a few hours of a trip.

So the final design for the **Vapour Collection System (VCS)** includes one common low pressure header collecting the various streams from the entire plant, a liquid knock out drum, two vapour collection blowers (100% spare) with variable frequency drives controlling the collection header pressure to slightly above atmospheric. The blowers discharge into the high-pressure header operating at about 5kpa. The Acid Gas from the Amine Stripper and the vapour from the Waste Caustic Degasser also discharge into the high-pressure header. This header is then connected to each of the furnace decoke headers through individual trip valves and flame arresters. (Figure 1).

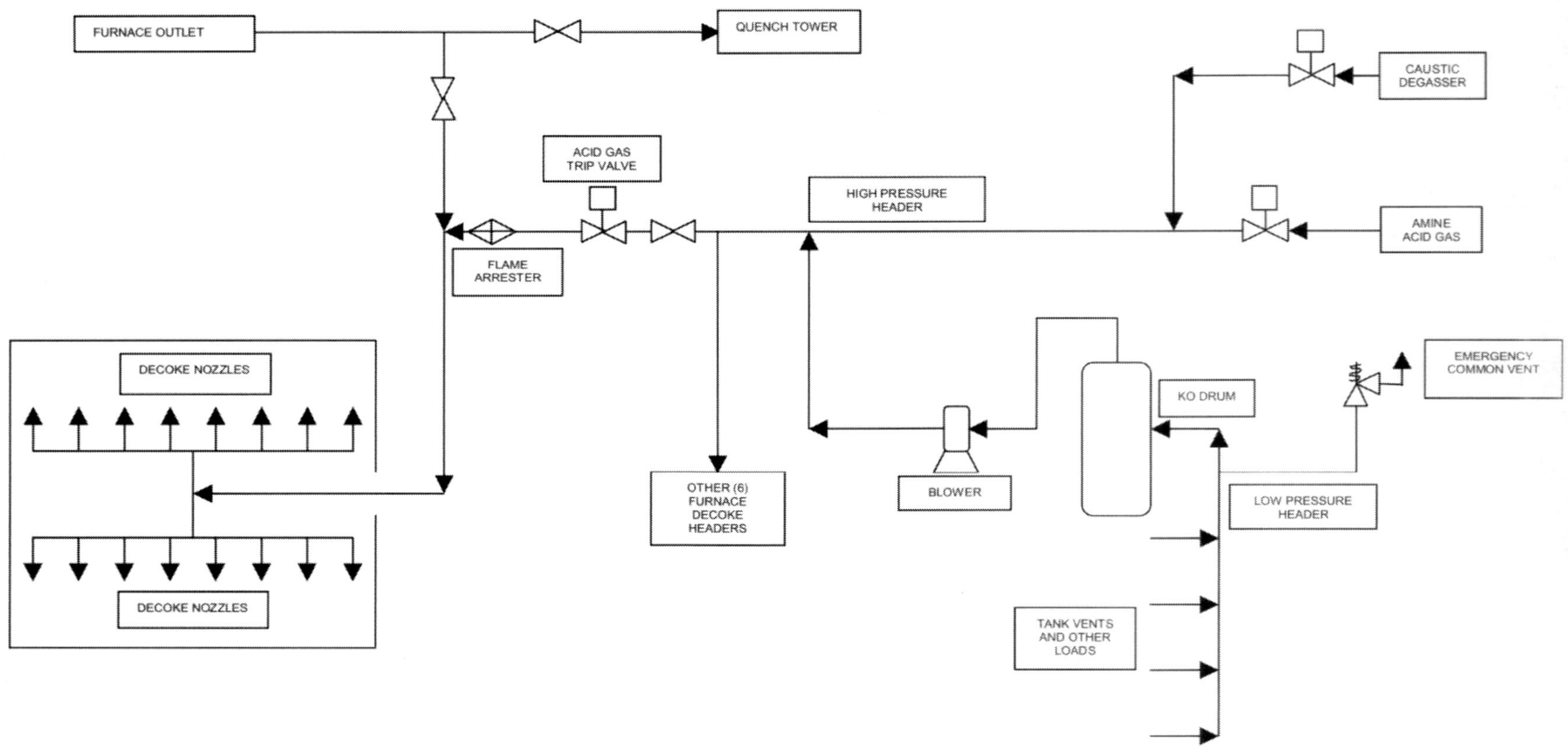

FIGURE 1
VAPOUR COLLECTION
SYSTEM SCHEMATIC

Description of the Incidents

First incident on the Number 6 Furnace (H260)

On May 28th, 2003 ethane feed was removed from E3 Plant in preparation for the planned Turnaround. Due to a limited Decoke Air supply and a requirement to maintain steam flow to the Quench Water Tower to aid in the preparation of that system H260 was not decoked immediately, but was kept in a Hot Standby condition with Dilution Steam flowing through the coils and into the Quench Water Tower. It was also maintained in a condition to receive hydrocarbon vapours from the Vapour Collection System (VCS) for incineration. The furnace was held in this state for two days. Then the VCS was isolated from the furnace and the furnace was swung into decoke. The decoke was completed in about 36 hours and the furnace was cooled down and isolated for the Turnaround. There was no significant work done on this furnace during the outage.

During this particular plant outage the VCS was temporarily piped to the flare since there was no complete flare outage planned.

On the afternoon of June 25th, 2003 the FD and ID Fans were started and the pilots were lit on H260 in preparation for the plant re-start.

Due to the plant re-start schedule the furnace was held in a "pilots only" condition for an additional 30 hours. On the morning of June 27th the Area Operator noticed that the whole main east tee section of the header was glowing hot with some cladding and insulation melting and falling to grade. At the time, the furnace firebox temperature was less than 200°C. The furnace pilots were immediately shut in and a nitrogen purge was initiated on the decoke header to snuff out the combustion and cool the piping.

The investigation followed with the furnace remaining shut down until the cause of the overheating was understood and the integrity of the piping could be determined. The damage to the piping was assessed and it was determined that the furnace could be restarted and the damaged section of piping would be replaced prior to the next decoke.

Upon internal inspection of the decoke piping, significant amounts of coke material was found in the decoke headers. In some areas the 16-inch diameter piping was almost half full. There also seemed to be other debris mixed in with the coke. A second furnace (H270) was also inspected with a remote camera to view the material that had not been exposed to fire and obtain some samples of the (pre-fire) material for analysis.

Initial Investigation

An investigation group was pulled together to collect data and investigate theories on how the combustion and overheating had been initiated.

The coke that was in the piping had probably accumulated over several decokes since the start-up of the plant in 2000.

The piping is normally in Acid Gas service with about 83% CO_2, 1% Ethane, 6%H_2O, and 200ppm H_2S. The acid gas and decoke piping is constructed of carbon steel. Due to the iron available from the piping and dusting of the furnace coils during decoke, it was suggested that the coke material could contain varying amounts of iron sulphides.

The standard practice when furnaces were shutdown was to leave the low point drains (3 inch valves) on the decoke piping open to ensure no water accumulated in the headers. The iron sulphides exposed to the atmosphere over the plant outage of 3 weeks could slowly start to oxidize and build up heat in the coke bed.

When the furnace fans were started an additional draft would be created drawing air through the drain valve over the heated coke bed. The additional oxygen available could accelerate the oxidation and result in the coke igniting.

Another variation of this scenario included possible adsorption of Ethane and/or heavier hydrocarbons during the period prior to the furnace being shutdown. When the plant is in the final stages of shutting down, the Ethane content in the Acid Gas increases significantly until Amine Circulation is shutdown completely. The furnace was also held in a Hot Standby condition for 2 days after the plant feed had been removed with only VCS vapour flowing through the decoke header. The adsorption of these materials on the coke bed surface could have lowered the flash point sufficiently to enable the Iron Sulphides to ignite the hydrocarbons and coke.

The one other scenario considered possible was that some hot pieces of carbon, possibly from one of the lit pilots, fell into one of the decoke nozzles and ignited the coke bed.

Samples of coke from 3 furnaces (H210, H260, and H270) were collected and analyzed at Nova Chemicals Research and Technology Center in Calgary Alberta. The findings from this analysis indicated that the coke did contain some levels of pyrophoric iron sulphides such as Mackinawite (FeS), Greigite (Fe_3S_4), and Pyrite (FeS_2). Bulk analysis of the various samples using Proton induced X-ray emission (PIXI) was completed. This analysis indicated that the total concentration of iron and sulphur were significant in the samples. Iron ranged from 2.73 wt % to 23.94 wt % while sulphur ranged from 0.38wt% to 9.96wt%. The major component in the samples, of course, was Carbon that ranged from

76wt% to 96wt%. There were also low levels of Chromium, Magnesium, and Nickel present.

Samples of coke were also taken from the coke traps of H210, upstream of the decoke piping to compare composition with the decoke piping samples. These samples indicated undetectable levels of iron or sulphur and were virtually 100% Carbon.

Testing was also done to measure the levels of Ethane adsorption by the coke that could occur during the conditions of the plant shutdown, to determine if this could have been a factor in initiating the combustion. The level of adsorption was found to be insignificant at the operating conditions in the piping.

Investigation of the hot particles from a pilot dropping into a decoke nozzle and igniting the coke was ruled out due to the recessed location of the pilots and the distance from the decoke nozzle where the combustion in the piping actually occurred.

The conclusion from analysis and testing was that the formation of the iron sulphides was a result of the cyclic operation of the decoke piping from decoke to acid gas service. During acid gas service the carbon steel is subjected to a water-saturated CO_2 stream with 300ppm H_2S at approximately 40°C. Iron fines present in the piping from furnace coils and the piping surface and scales oxidize. These oxides further convert to sulphides due to the contact with the H_2S. The coke material lying in the pipe remains intact due to the relatively low velocities in the piping

.

During the decoke cycle, the decoke piping is subjected to steam, air and CO_2 at 200°C. Coke removed from the upstream furnace tubes accumulates in the piping along with metal dusts due to the spalling action, erosion, and thermal shock. Some of the accumulated sulphides oxidize and are carried out to the furnace firebox with the steam/air stream. The heat generated by the oxidation is removed by the relatively large steam/air volume flowing over the coke bed. This cycle is repeated many times until the levels of iron and sulphur impregnated coke accumulates to the levels found in the sample furnaces.

The combustion in the piping, then, was very likely caused by the sulphides being exposed to oxygen for several weeks during the outage. The reaction of the sulphides with the oxygen slowly built up heat in the localized area. When the furnace fans were started up, the open drain on the decoke piping allowed additional oxygen to be drawn over the heated area, accelerating the reaction to the point of ignition of the coke. Once the coke was ignited, and with the continuous supply of air, the heat continued to build resulting in the damage to the piping.

Recommendations:

Once the most likely cause was identified, action plans needed to be put in place to ensure the incident did not re-occur. At least one of the components (oxygen, coke, or the sulphides) had to be removed from the equation.

The source of the sulphur is the acid gas stream. At least in the short term, removing the acid gas stream or the H_2S from the acid gas was not an option. The furnace fireboxes were the only means of disposing of the acid gas stream available.

The coke, unfortunately, was going to continue to accumulate in the piping. The coke traps upstream only collect a portion of the coke flowing out of the coils and the low velocity designed into the decoke piping, and the multiple changes in direction provide natural locations for the solids to drop out and collect. The coke could be removed periodically to prevent the accumulation but the piping was not designed with cleaning in mind. This would require cutting caps off the ends of the piping and installing flanges with blinds that could be removed.

The oxygen could be controlled to a certain degree. During acid gas service no oxygen is present. During decoke the oxygen is present at levels as high as 5.5% in the steam but the considerable steam flow over the coke bed should remove the heat generated by any oxidation of the sulphides, preventing the ignition of the coke.

Based on this, the high-risk period remained during a furnace outage when no acid gas or steam flow is present in the decoke piping. To address this, Operating and Maintenance procedures were modified to ensure that the decoke header drains were always closed after shutting down the steam flow through the furnace coils. The decoke headers would then be watered down to ensure the coke remained wet as long as the furnace was shutdown. Limited available oxygen and maintaining a wet environment, it was felt, would prevent any further incidents of combustion and overheating in this area.

Second incident on the Number 5, 6, and 7 Furnaces (H250,H260, and H270,)

The new operating and maintenance practices were adopted and there were no further indications combustion occurring in the decoke piping until the plant tripped due to a total power interruption in the early morning of January 7, 2004. In this event all of the 7 E3 cracking furnaces tripped due to loss of FD and ID Fans. On January 8, 2004, during preparations for the restart of the plant, overheating of the decoke piping occurred in H250, H260 and H270.

As stated in the background, in the event of a trip of all furnaces the vapour collection system vapours are vented to atmosphere, as the furnace fireboxes are the only means of incineration.

H250 had just completed a decoke cycle three hours prior to the trip. The plant trip occurred at 03:14. By 04:17 H250 was relit and placed in a hot standby condition (Primary burners online, firebox temperature above 600°C, dilution steam flow through the radiant coils with the effluent lined up to the Quench Tower). The vapour collection system vapours were lined up to H250 at that time. There was no acid gas flow as the ethane feed had been pulled from the plant.

It is worth noting that H210, although not one of the furnaces involved in the incident, had been in the online mode since January 1 and was relit and placed in a stand-by mode by 04:16. The VCS vapours were lined up to this furnace by 04:46. It was Operations personnel's intent to ensure that two furnaces were available for vapour incineration in case one of the incineration furnaces accidentally tripped. A second furnace would ensure that the vapours would not need to be vented to atmosphere.

H260 was in decoke mode at the time of the plant trip. By 04:45 this furnace was relit and back in decoke mode.

H270 had been in the online mode (since December 21st) at the time of the trip and was relit and placed in decoke mode by 05:18.

The other 3 furnaces were all placed in decoke mode.

On January 8 H260 had been swung into stand-by mode and opened to receive VCS vapour at 04:45 that morning, meaning three furnaces were lined up to receive the vapour for incineration (H210, H250, and H260). At 12:00 H270 decoke was complete and it was swung into a stand-by mode and lined up to incinerate VCS vapour. H250 was then isolated from the VCS vapour and swung into decoke mode. At this time it was noted that the insulation and cladding had melted on the west end of the A Cell decoke piping. Obviously, combustion had been occurring within the decoke piping in that area, overheating the piping. The temperature of the piping was checked and it appeared that the combustion was no longer taking place with the furnace in the decoke mode.

At 15:00, while the overheating of H250 was being investigated, a section of hot decoke piping was found on H260. H260 was immediately isolated from the VCS vapours and a nitrogen purge was placed on the piping, venting the nitrogen into the firebox. The piping quickly started to cool, indicating that the nitrogen purge was effective in preventing the combustion from occurring.

There were no indications that combustion might be occurring in any of the other furnace decoke piping; including H210 which was open to the VCS vapour. Monitoring continued.

It was determined at this time that the VCS vapour incineration was the main factor in these events so all furnaces except H270 ware isolated from the VCS vapour. It was believed that multiple furnaces open to the VCS system and the low flows involved, could be contributing to the problem. To minimize any risk of coke in the piping igniting or reigniting, nitrogen purges were initiated on all of the isolated furnaces.

This left only H270 in VCS incineration service and it was believed that this furnace was still at risk, so plans were in place to initiate a nitrogen purge on this decoke piping as well. This would ensure some minimum velocity in the piping and would also dilute the VCS hydrocarbon stream and prevent firebox excess air entering the decoke piping. Unfortunately, ignition of the coke had already occurred before the purge could be established.

At 21:00 localized hot coil outlet temperatures were noted on H270. When the Furnace Operator investigated, 6 to 8 foot flames were found exiting from the west decoke nozzles of the B Cell. A nitrogen purge was initiated immediately and the flames disappeared within about 5 minutes. Minor damage to insulation and cladding was noted on the west end of the B Cell piping.

Inspection of the piping on H250, H260, and H270 revealed only minor damage (sag and oval deformation). Hardness readings taken of the pipe wall in the damaged areas indicated the material integrity was still intact. There was some binding of the decoke nozzles in the floor penetration sleeve in a couple of areas which were addressed at a later date with some minor piping rework.

Investigation findings

Some things we knew about these events:

- Each of these Furnaces were lined up to the VCS when the combustion occurred in that particular Furnace.
- There was no flow in the acid gas line at the time other than VCS vapour.
- The plant, in the 3 years since commissioning, had gone through several short plant outages and operated in this mode using the furnaces for VCS incineration without incident, including as recently as November of 2003.

What Had Changed?

Upon investigation of the VCS operation, it was identified that there had been a change in the operation of that system that had occurred in August of 2003.

During normal plant operation there is enough flow into the VCS from the various streams to maintain one blower in operation continuously, averaging about 350 kg/hr flow rate. The variable frequency drive is modulated to maintain a constant pressure on the header. Occasionally, during periods of high venting from a part of the process, the flow will exceed the capacity of one blower and as the second blower comes on-line the flow will briefly exceed 500 kg/hr. Analysis of this stream shortly after plant commissioning indicated that it typically consisted of 80 -90 wt% nitrogen with the remaining 10 wt% to 20wt% being various hydrocarbons, mostly C_3 and heavier.

There are several small nitrogen purges flowing into the system, both continuous and intermittent. All of the API Tanks connected to the system were intended to be set-up so the individual nitrogen pad regulators would not open if the header was at design conditions. They would only open if the normal pressure could not be maintained above the regulator set pressure. However, in August 2003 it was discovered that two of these tank regulators were set to a pressure higher than the operating pressure and were operating continuously in the full open position, false loading the VCS header with a considerable amount of nitrogen. This flow was amounting to approximately 100 kg/hr nitrogen so as a cost saving measure these regulators had been isolated until new regulators could be purchased and installed. The existing regulator springs would not allow a pressure setting as low as was required.

On a plant trip, depending on what activities are taking place and the number of systems shut down, the load on the VCS system can drop off to some degree.

During this particular trip in January, the load on the VCS had been reduced enough that the single blower was tripping periodically on low suction pressure. When the blower shuts down the VCS low pressure header pressure starts to build until it reaches 0.8 kPa, approximately 0.5 kPa above the normal operating pressure. At this point the Blower restarts and a sudden rush of VCS vapour is sent to the Furnace Firebox as the blower ramps up to full speed to return the header back to setpoint.

Between the time of the plant trip and the day of the decoke header incidents fifteen of these Blower trips and re-starts occurred. Twelve of them occurred on January 8, 2004 between 02:00 and 21:00, the period when all the overheating occurred.

Some theories were developed from this information:

During this period of low flow without the acid gas stream providing velocity and backpressure on the system there may be potential for cross communication between fireboxes through the Acid Gas/VCS header. All the furnace fireboxes are connected through this header. Furnaces can be operating at different draft levels due firebox temperature or control set point. It is also possible that different areas of the same firebox are at different drafts due to air flow distribution or uneven temperatures. This could create a differential between firebox decoke nozzles on the same Furnace due to slight variations in draft pressure, resulting in a cross flow of firebox flue gases, entering the decoke nozzle at one point on the firebox floor, flowing through the decoke header and back into the firebox through another nozzle. Note that Operators observed flames exiting from only the western most decoke nozzles in the case of the H270.

Through further firebox testing since these events, it is also known that the furnaces can be operating at quite elevated oxygen levels in the floor area of the furnace, especially in a stand-by condition, even though bridge wall oxygen is in the 2% range.

Conclusion reached in the investigation:

During the plant outage the hydrocarbon rich VCS vapour became the only source of flow into the Acid Gas Header and consequently, the furnace decoke headers lined up to receive Acid Gas. The flow was very low with sudden loss of flow on several occasions due to the VCS Blower shutting down (logic shuts the blower down on low suction pressure). This loss of flow, in each case, was followed by a sudden increase in flow a few minutes later as the VCS Blower re-started and ramped up speed to bring the VCS header pressure back to set point. It is believed that as the blower shut down, the combustion of the vapour was allowed to be drawn back into the decoke header. When the combustible vapour flow to the decoke nozzles is very low and there are multiple openings to the firebox the air can be drawn in on one or more of the nozzles mixing with the hydrocarbon vapour prior to exiting into the firebox, allowing the combustion to continue within the decoke piping. Contact of this flame with the coke material accumulated in the decoke piping eventually resulted in the coke igniting and sustaining combustion in the header. Once a hot spot was created in the coke bed, combustion was maintained by drawing combustion air from the firebox through the network of decoke nozzles. The sustained combustion and elevated temperature enabled the flame to generate additional draft and draw additional air in from the floor area of the furnace, further accelerating the combustion.

One piece of information that did not support this theory was the history of previous plant trips. The plant had gone through several short outages similar to this in the initial 3 years of operation and no similar incidents had occurred, one as recently as November 2003. This was after the excess nitrogen had been removed from the VCS that August. After reviewing the history, the one difference that was identified was that during all of the other outages where the VCS vapour was the only stream entering the decoke piping, the VCS blower always ran continuously. There always appeared to be enough load on the system to maintain the single blower in operation. As there are so many streams entering the VCS and no individual flow measurement it is possible that in the November event some other source of nitrogen was open into the system that maintained the additional flow and possible dilution.

Measures to prevent Re-occurrence:

Based on the events in January 2004 with H250, H260, H270 and the occurrence in June 2003 with H260 it was evident that there were multiple scenarios causing the incidents and that the highest risk period was during initial warm-up of a Furnace (prior to introduction of steam to the coils), and when the plant is down and furnaces are in Hot Stand-by being used for VCS incineration. In other words, any time there is not significant steam or acid gas flow through the decoke piping.

To address all the scenarios the following actions were taken:

For VCS Incineration Risk:

During Plant Feed-out / Feed-in
Procedural Changes:

During the final stages of removing the feed from the plant or during initial feed-in the Acid Gas becomes ethane rich. The Acid Gas, at this point, is now routed to flare and isolated from the furnaces to prevent the ethane rich acid gas from entering the decoke piping and the furnace fireboxes.

It is ensured that the Acid Gas Analyzer is in service during feed-in and CO_2 content is greater than 90% prior to lining up Acid Gas to a furnace for incineration.

Additional furnaces are only placed into Acid Gas incineration service as required to maintain backpressure on the high pressure VCS header between 5 kPa and 20 kPa. This prevents any possibility of cross communication between furnace fireboxes through the decoke/acid gas piping system.

During a Plant Outage (no feed in plant and Hot Stand-by Furnace used for VCS Incineration)

Procedural Changes:

Only one Furnace is lined up to receive VCS vapour. All other furnaces to have the Acid Gas/VCS lines isolated with nitrogen purges on the decoke piping into the firebox.

Additional nitrogen purges are added to the VCS header to dilute the VCS vapour and ensure continuous blower operation. This also provides additional nitrogen purge on the decoke piping of the furnace in VCS incineration service.

For Furnace Shutdown and Start-up Risk

Furnace Shutdown

Procedural Changes:

As the furnace is being cooled down, the dilution steam in the coils is directed back to the firebox. Nitrogen purges are placed on the coils prior to removing all of the dilution steam. This ensures that a continuous sweep of the decoke piping is maintained until all burners and pilots are out. The nitrogen purge is maintained until turnover of the furnace to Maintenance.

Decoke piping drains are left closed to minimize possibility of air circulation into the decoke piping as the furnace is brought down.

Maintenance Personnel wet down the decoke piping as soon as possible after opening the firebox to prevent any pyrophoric material from initiating a combustion of the coke.

Furnace Start-up

Procedural Changes:

A nitrogen purge is initiated through coils and decoke header back into firebox prior to ignition of any pilots. The nitrogen purge is maintained until Dilution Steam flow is initiated into the coils.

Future Changes

Although all of the mitigations to date have been procedural, modification of the decoke piping is still under consideration to enable periodic removal of the coke from the piping, thus removing one of the fuel sources.

Lessons Learned

Incident Investigation - The first incident which occurred on H260 in June 2004 revealed accumulations of combustible material (coke) in the decoke piping. There was a very concentrated effort put into understanding this particular event and it is believed that the pyrophoric nature of the material did, in fact, initiate that event. The experiences from the subsequent investigation reinforce the need to consider all possible causes for an incident. It is also important to evaluate what other operating modes might lead to a similar event.

Design or Operating Changes in the Plant - If design or operating changes are made in the plant the impacts to the waste streams need to be carefully considered. In this scenario, the discovery of the 100kg/hr of excess nitrogen venting to the VCS was considered an economic issue. The fact that the hydrocarbon content of the stream would increase significantly when the nitrogen was removed was not considered to be creating additional risk. The system had operated reliably for 3 years, dropping the load on the system by 20% was not considered to be an issue.

Incompatible Service Cycles – equipment in services such as acid gas present the risk of sulphide formation. Cycling this equipment periodically into an oxidizing environment can present a risk of fire if significant accumulations of fouling material are present.

Fix One Problem / Create a New One – As is sometimes the case when trying to address old ongoing operating problems, changing the design of the piping for the E3 Furnaces created new ones. Accumulation of coke in the decoke piping in the E1 and E2 plants had never been a problem due to the high velocity, but piping erosion was. So, when the design team decided to reduce that velocity significantly, accumulation of coke in the piping was not considered. The focus was on minimizing piping and fitting erosion during decoke.

Conclusion

Thorough incident investigations and follow-up are critical to ensuring re-occurrence is prevented. When new information is gained about the process or equipment during the investigation, the new information needs to be fed back into the plant / process risk analysis to ensure the design and operating practices are adequate. New scenarios may need to be evaluated that were not considered in the previous risk assessments.

Operating fired process heaters and using them for waste gas incineration as well can be a very effective and economical method of waste stream incineration. This practice can create risks, however, the must be fully investigated and understood with adequate controls put in place to ensure safe operation.

If waste gas incineration is to be incorporated into a fired process heater, whether a Cracking Furnace or some other type of process heater, careful consideration needs to be given to the mechanical design and the multiple process scenarios that may occur through the operating cycles of the plant.

These should include but not be limited to:

- If there are multiple waste streams of different types of composition (combustible / non combustible) (high BTU value / low BTU value) (intermittent / continuous) (large flow / small flow) careful review and risk assessment of the streams and possible scenarios need to be completed. Are there scenarios where one or more streams are not present? Does the composition of the streams change significantly?

- Consideration must be given to both how the streams are delivered to the heaters for incineration and in which heaters to incinerate the streams. It may not be appropriate for multiple streams to be incinerated in the same firebox. Some heaters may be more suited to some streams than others due to the stream properties or the design and operating parameters of the heater.

- Each stream may require a different delivery method. Some may require a simple nozzle while others may warrant an actual burner and treated as fuel with all the controls and safeguards utilized for fuel systems.

- Consider fouling potential of the streams and combinations of the streams.

- Flame arrestors to provide protection of the piping and process systems feeding into the heater firebox should be provided as close to the entry into the firebox as possible to prevent any backflash or fires occurring in the piping.

- Careful consideration needs to be given to all the operating modes of the fired equipment. Can the heater safely incinerate the stream(s) in all these modes? What instrumented systems are required to protect the heater and equipment in certain scenarios? Does the heater operate below auto ignition temperature of the fuel or any of the streams being incinerated?

Paper No. 38e

Don't Learn the Hard Way

David S. Flett
Huntsman Petrochemicals UK Ltd.

Prepared for Presentation at the 2005 Spring National Meeting
Atlanta, GA
April 11-14

17th Annual Ethylene Producers Conference
Session TA005 – Ethylene Plant Safety

EEPC EUROPEAN ETHYLENE PRODUCERS COMMITTEE

Don't Learn the Hard Way

We are trying not to!!

EEPC EUROPEAN ETHYLENE PRODUCERS COMMITTEE

Background

- Formed in 1996
- Membership open to
  - All European Ethylene Producers
  - All Technology Contractors with an ethylene design capability in Europe

Aims / Objectives

Exchange non-confidential information relating EHS aspects of manufacturing and handling ethylene

- Promote continuous improvement in performance of our Industry
- Create expanding network of professional engineers sharing best practice and experience

Membership

- All 20 operating companies represented
- All 5 ethylene technology contractors represented
- Active membership of enthusiastic experts

EEPC EUROPEAN ETHYLENE PRODUCERS COMMITTEE

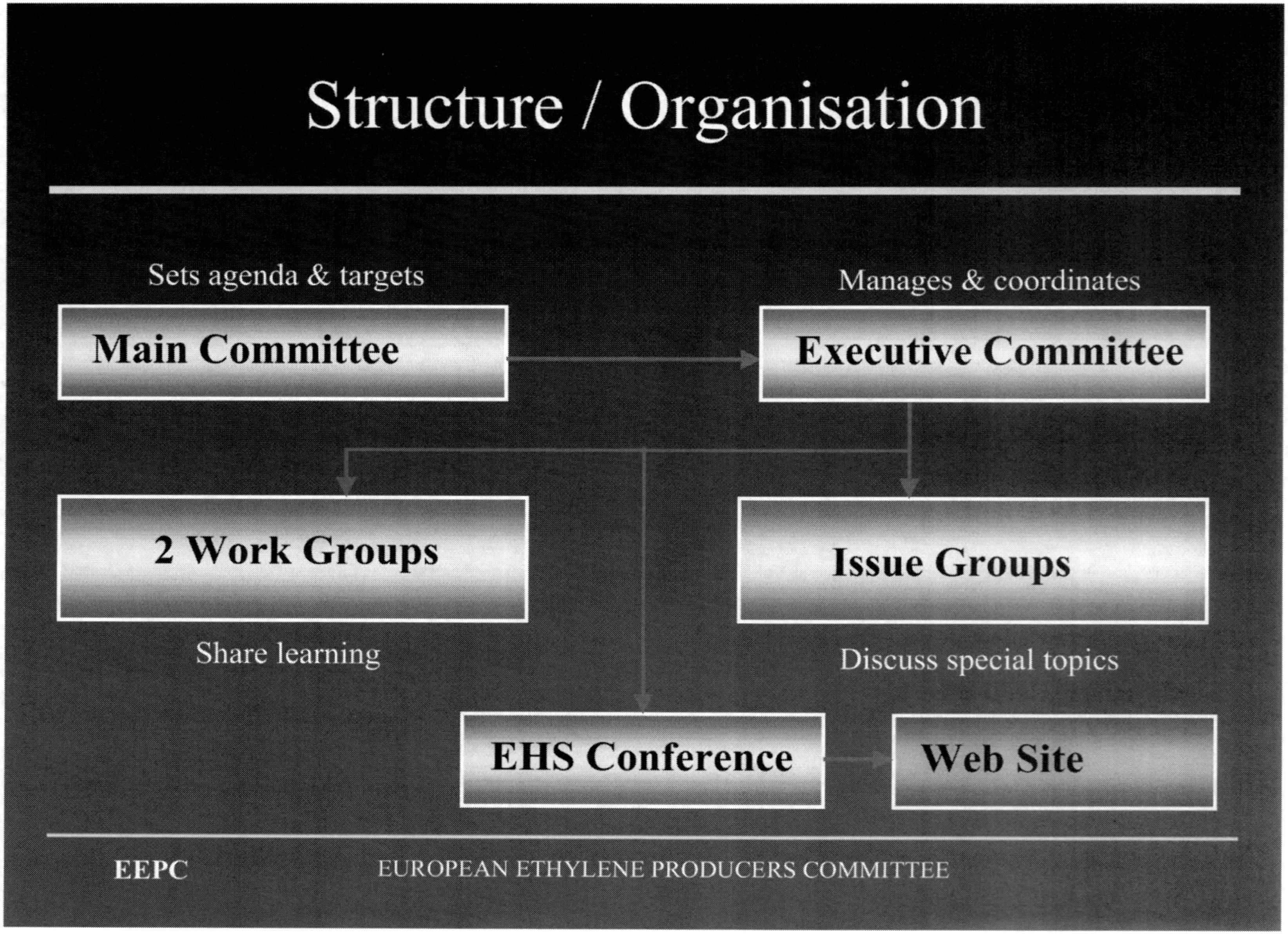
Structure / Organisation
Sets agenda & targets
Main Committee
Manages & coordinates
Executive Committee
2 Work Groups
Share learning
Issue Groups
Discuss special topics
EHS Conference
Web Site
EEPC
EUROPEAN ETHYLENE PRODUCERS COMMITTEE

Approach to Sharing

- Critical to learn from each other
- Not only major incidents
- Share all incidents however trivial
- Reporting of near misses encouraged
- Differs from EPC approach

EEPC   EUROPEAN ETHYLENE PRODUCERS COMMITTEE

How is it done?

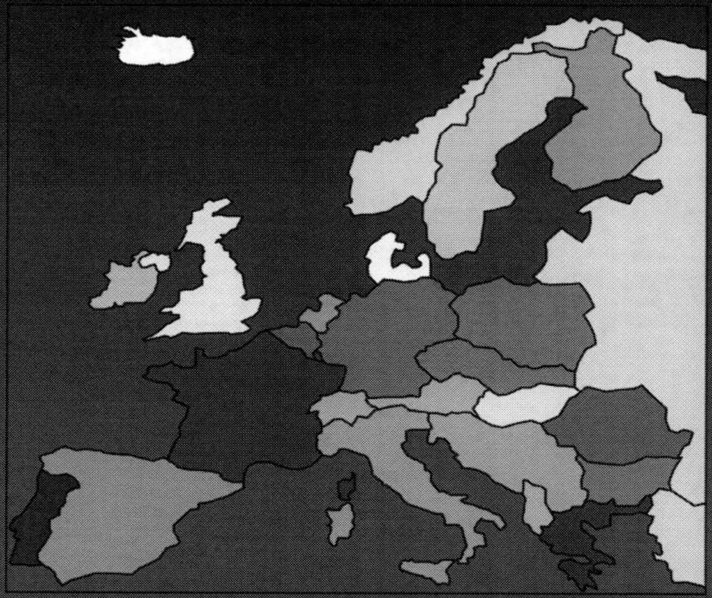

- Working Groups share incident learning
- Data discussed and reported in standard format
- Used to avoid similar incidents in other plants
- Information available for internal use by member companies

EEPC EUROPEAN ETHYLENE PRODUCERS COMMITTEE

Standard Reporting Format

- Brief description
- Circumstances leading to Incident
- Assessed Cause
- Actions taken to avoid recurrence
- Include sketch if helpful

Do not make recommendations

EEPC EUROPEAN ETHYLENE PRODUCERS COMMITTEE

Database of Incidents

- Extensive database generated since 1996
- Used by member companies
  - to raise awareness / promote discussion
  - test robustness of own management control processes
- Being used by EEPC
  - to identify emergent themes
  - target discussion topics for Working / Issue Groups

EEPC EUROPEAN ETHYLENE PRODUCERS COMMITTEE

Sort by Cause

- Design
- Operation
- Maintenance
- Corrosion / Erosion

EEPC EUROPEAN ETHYLENE PRODUCERS COMMITTEE

Sort by Equipment

- Furnaces / Boilers
- Control Systems
- Flare Systems
- Piping
- Rotating Equipment
- Static Equipment
- Utility Systems

Single (principal) causes equally spread among Design, Operation & Maintenance

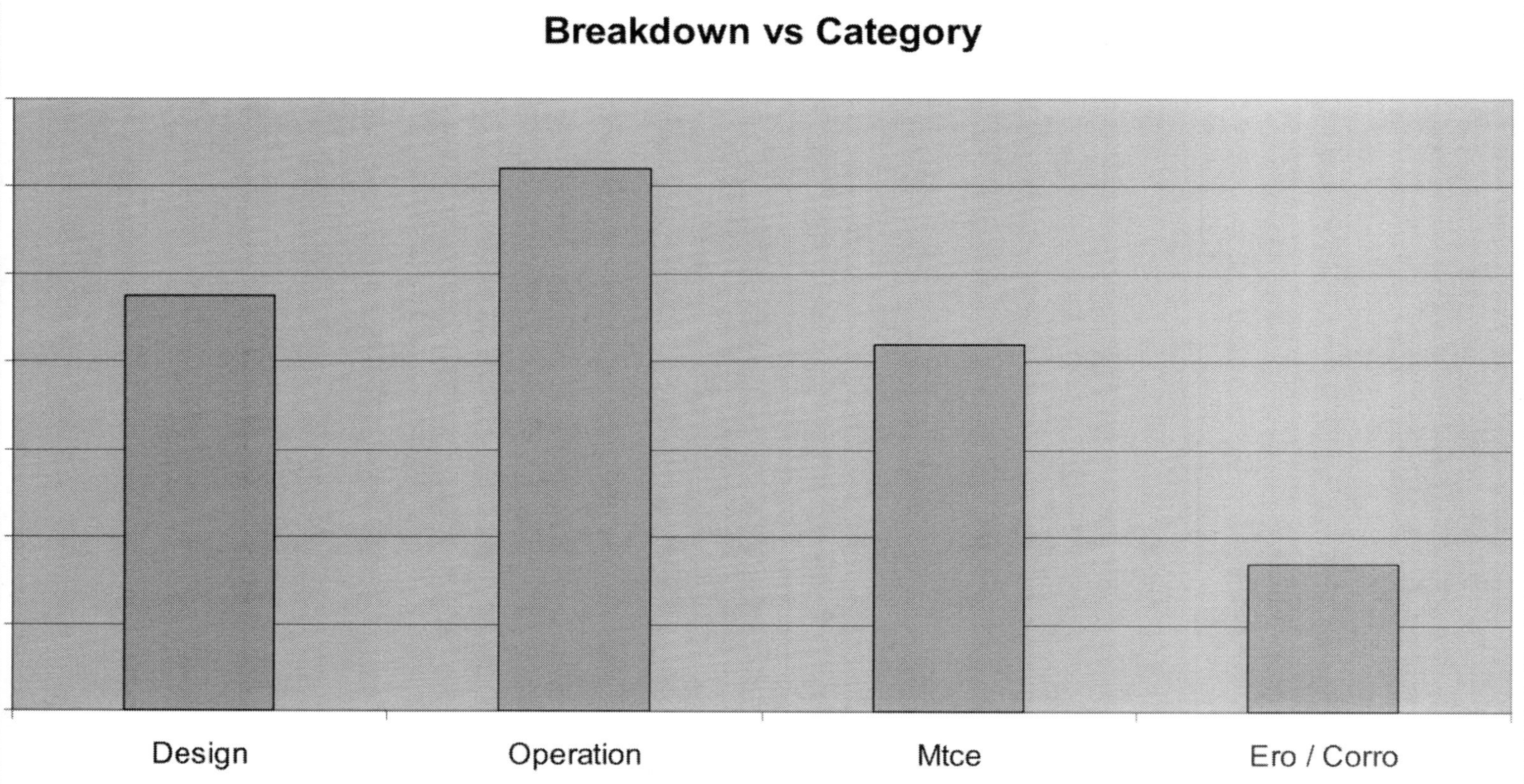

EEPC EUROPEAN ETHYLENE PRODUCERS COMMITTEE

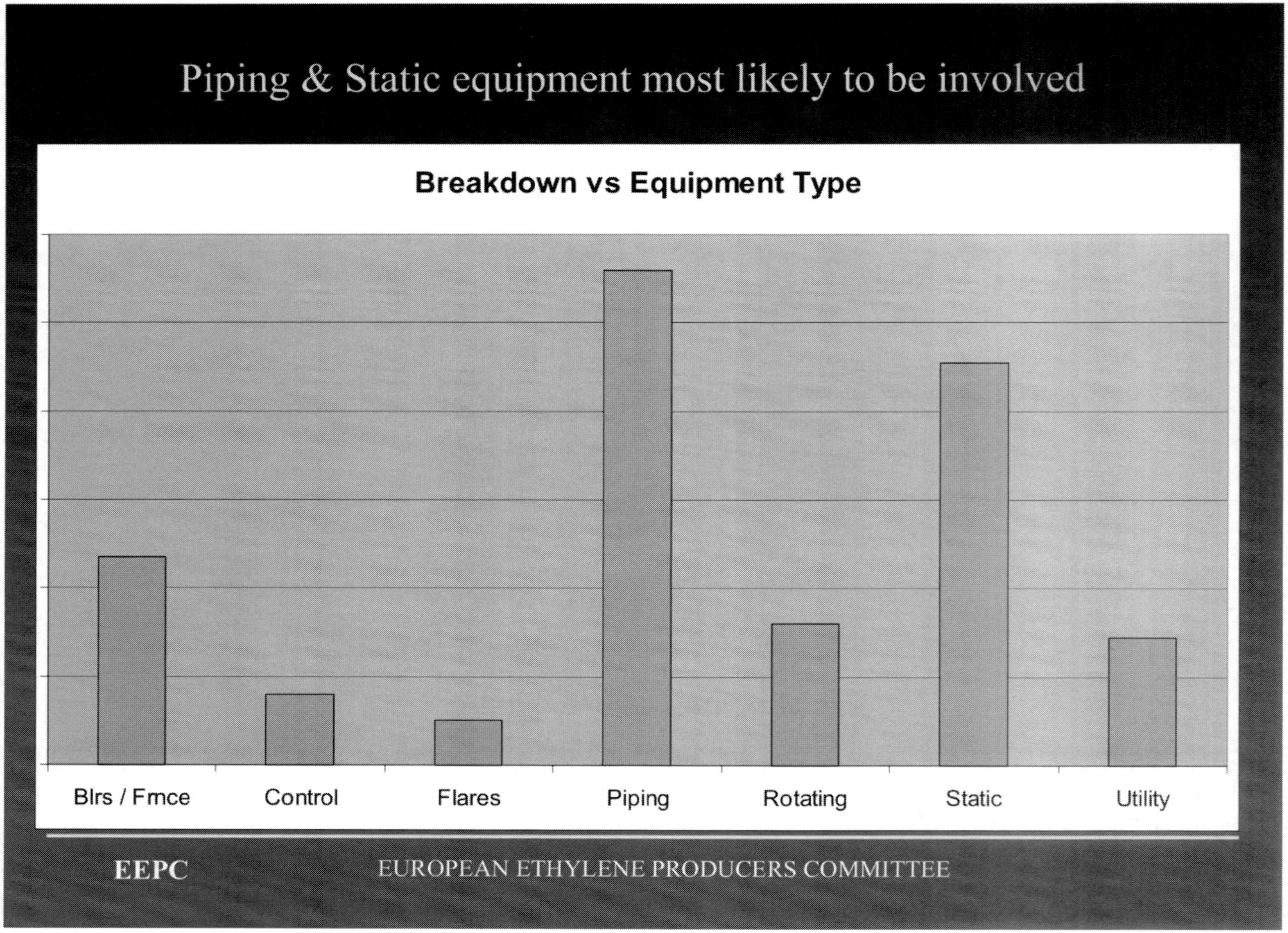
Piping & Static equipment most likely to be involved
Breakdown vs Equipment Type
Blrs / Frnce
Control
Flares
Piping
Rotating
Static
Utility
EEPC
EUROPEAN ETHYLENE PRODUCERS COMMITTEE

Of these, no clear classification emerges

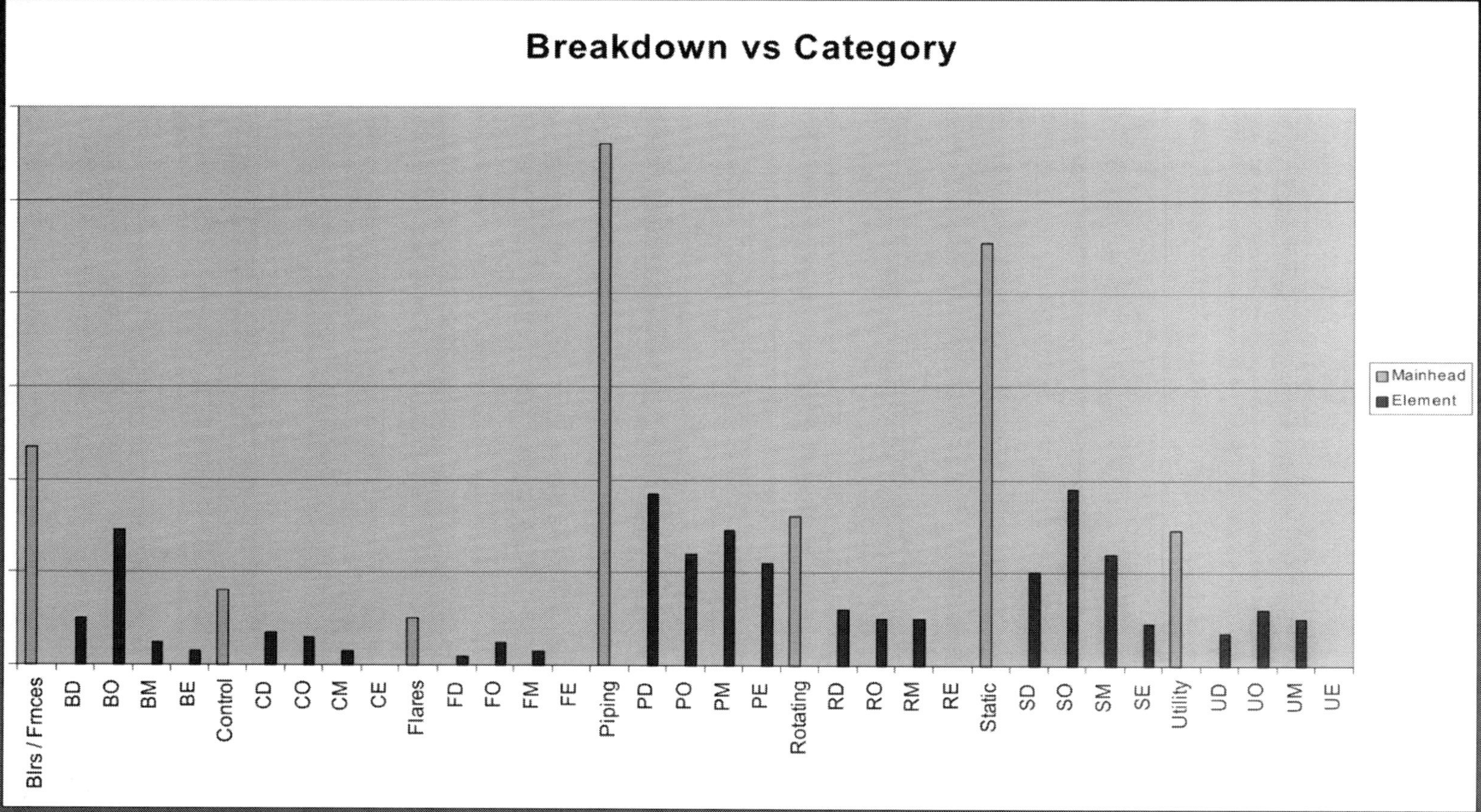

EEPC EUROPEAN ETHYLENE PRODUCERS COMMITTEE

If account is taken of multiple causes, the picture changes with Design & Procedures prominent

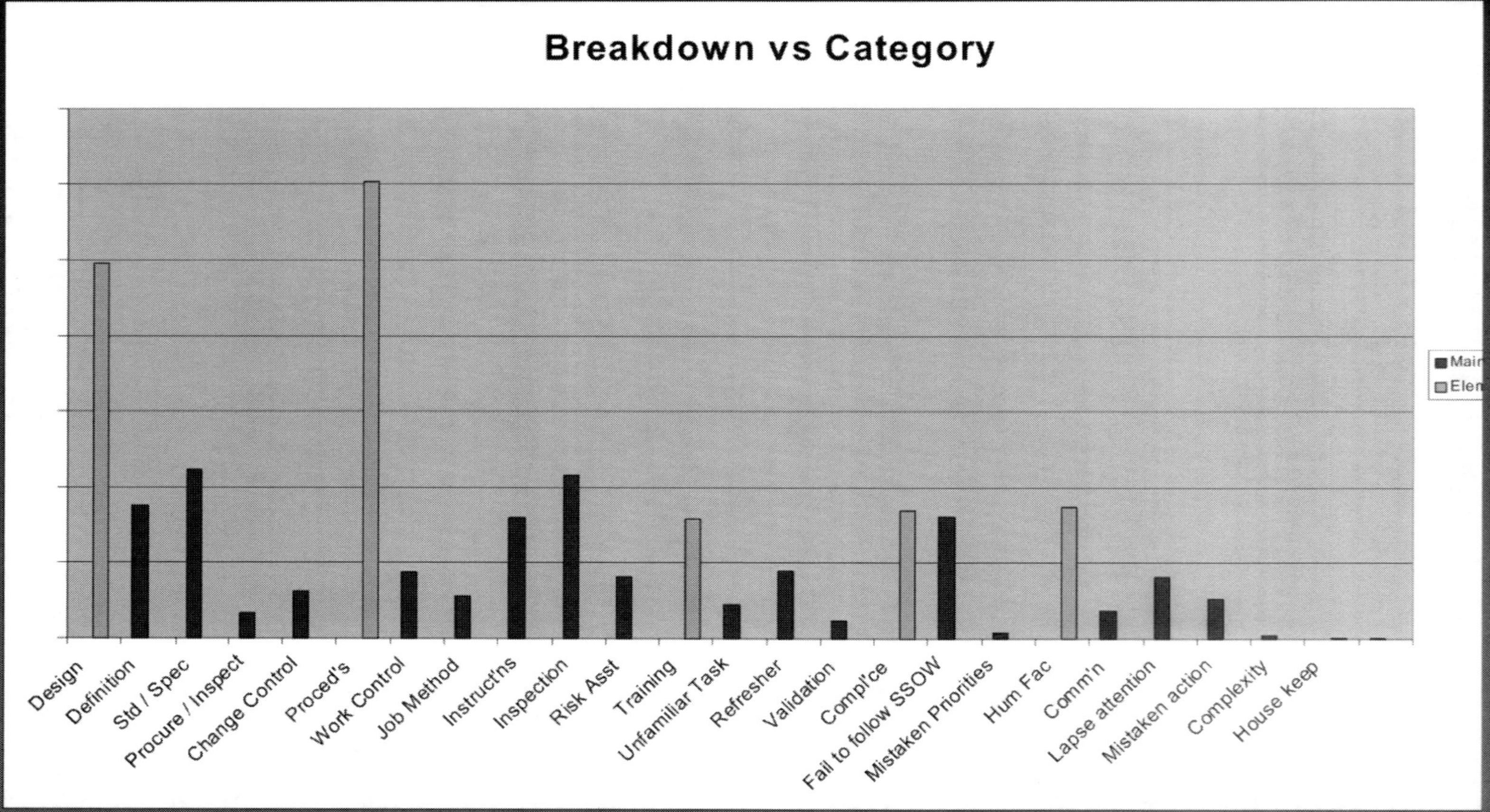

EEPC EUROPEAN ETHYLENE PRODUCERS COMMITTEE

Recurring themes

- Many incidents linked to
  - Inadequate physical checks (before / after routine work, pre-commissioning etc)
  - Training / validation related to new or unfamiliar tasks or situations
  - Failure to follow instructions and adopt Safe Systems of Work
  - Inadequately targeted maintenance / inspection procedures
- Rewrite / review operating instructions is a common reaction
- Relatively few incidents attributed to Human Factors as a primary cause

EEPC EUROPEAN ETHYLENE PRODUCERS COMMITTEE

Analysis

- More work is needed……………………………
- However, this is a valuable source of data which can be used to drive improvement
- Drill-down is possible, as is different classification approaches

Summary

- Operated assets of this type for many years
- Regulators are right to expect improvement
- Cannot afford to learn by own experience
- Sharing / learning critically important
- EEPC is providing an effective vehicle for this within Europe

Best Practices of MPC/RTO in Ethylene

Mark Darby
University of Houston
Dept. of Chemical Engineering
4800 Calhoun Ave.
Houston, TX 77204-4004

Elizabeth Teeter
Landmark Graphics
2101 City West Blvd.
Houston, TX 77042-3021

Prepared for Presentation at the 2005 Spring National Meeting, Atlanta GA, Apr 10-14
17th Annual Ethylene Producers' Conference
Session TA006 - Ethylene Plant Process Control

April 2005

UNPUBLISHED

Abstract

Multivariable Predictive Control (MPC) and Real-Time Optimization (RTO) are established technologies in Ethylene Plants today. As a result, there is a significant amount of experience in the industry resulting from applying these technologies. In this paper, we highlight the approaches that we believe represent the best practices for ensuring that these solutions are successful – both during the implementation of the project, and the subsequent support and maintenance of the applications over time. The recommendations are based on the experience of the authors as well as feedback from multiple EPC companies.

1. Positioning / Stage Setting

The application of MPC and RTO in ethylene originated in the mid to late 1980s. Since then numerous projects have been implemented in plants around the world, with benefits that have been well documented. During this time we have witnessed significant change - both with the technology itself and in the operating companies. On the technology side, computer processing capabilities have increased a thousand-fold, which has led to larger controllers and optimizer models, while allowing more frequent execution of the controls and optimizers. Just as significant, the software has evolved, from custom code to fully supported, configurable software products. With respect to the ethylene producers themselves, there have been significant reductions in manpower, which have impacted operations staff as well as the support groups for process control and RTO. Budgets are tighter and there is increased competition for investment monies (e.g., environmental and IT), while there are more frequent changes to the plant and operations as companies respond to market conditions. The result today, then, is that while the technology and software are proven, there are increased challenges to keeping these solutions performing well.

Figure 1 below provides an overview of the solution hierarchy:

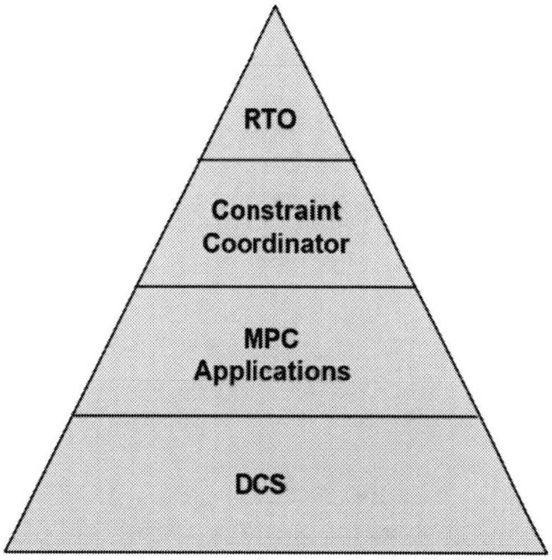

The MPC layer often includes multiple MPC controllers, one for each major unit, or groupings of major units, which output to the Distributed Control System (DCS) on a one to two minute frequency. Typically, there is a plant constraint coordinator, which pushes the plant to constraints and coordinates the individual MPCs. The constraint coordinator typically executes at the same frequency as the individual MPCs. The uppermost layer is the real-time optimization layer, which may or not be present in a given application. The RTO is based on a fundamental, nonlinear steady-state model of the plant, and typically executes multiple times a day – as often as "near steady-state" operation allows. This hierarchy is consistent with the offerings of the major technology providers, although there may be some differences in the specific functionality provided at each layer.

It is important that a company deploying MPC and RTO have a consistent view of the value these technologies provide, and to use them as tools for sustained added benefit. Obviously, these tools are justified in terms of increased profitability. But they do not achieve their benefits by themselves, without input and support from the organization. We believe it is imperative to view MPC and RTO as tools to facilitate continual improvement, increased process understanding, and improved asset reliability.

One of the key points that is often cited in articles and conferences is the increased knowledge of the plant's operation (plant capability, constraints, trade-offs) that is gained from a control or optimization project. However, this gain should not be viewed as a one-time event. The goal after the project should be to make this an ongoing, continual improvement program by the entire organization: e.g., identifying profitable bottlenecks, relieving them, identifying the next bottleneck, etc. We believe that an under-reported benefit of MPC and RTO is in driving organizational change.

A successful MPC and RTO project involves more than selecting the right technology and executing the project well (although these are absolutely necessary). Successful MPC and RTO projects are the result of getting many things right. This is captured in the following expression for the probability of success:

$$\begin{aligned}\text{Prob}\{\text{Success}\} = \\ & \text{Prob}\{\text{RightTechnology}\} \\ & \times\text{Prob}\{\text{RightProjectTeam}\} \\ & \times\text{Prob}\{\text{Good Design and Implementation}\} \\ & \times\text{Prob}\{\text{Organizational Support and Communication}\} \\ & \times\text{Prob}\{\text{Benefit Visibility}\} \\ & \times\text{Prob}\{\text{Maintenance and Improvement}\}\end{aligned}$$

Achieving ultimate success in MPC and RTO projects requires maximizing the individual probabilities. In the following, we discuss best practices for maximizing each of these probabilities.

2. Technology

MPC and RTO are, in general, mature technologies. In ethylene plants today, DMCplus and RMPCT are the MPC controllers used most often, with DMCplus cited more frequently. For the RTO layer, AspenPlus is the most often cited real-time optimizer. Ethylene RTO offerings are also provided by PAS (NOVA) and SimSci-Esscor's (ROMeo). Technip's SPYRO furnace model is frequently a component in ethylene MPC and RTO projects.

Given the reported successes in Ethylene, we conclude that the above-mentioned technologies are viable and not impediments to success, and other factors, in addition to specific technology features, impact the technology decision. This includes company standards, cost of ownership, and availability of knowledgeable third-party resources.

In addition to the basic technology of the MPC and RTO, there are other technologies that contribute to the overall success of the solution. This includes, for example, plant testing and model identification technology (including inferential predictors) and information and monitoring systems, which may come from other vendors.

It is interesting to point out that the closed-loop testing tools, Tai-Ji Online and AspenTech's SmartStep, were cited by several ethylene users as significant enhancements to MPC technology.

Users reported overall satisfaction with MPC and RTO technologies; however, the learning curve for RTO is considered to be much higher, compared to MPC, and a greater challenge to maintain. Several areas were identified for improvement. These include:

- RTO Diagnostics – Better tools to facilitate RTO troubleshooting and diagnosis when the optimization does not converge.
- Tools to facilitate understanding MPC and RTO decisions
- System Issues – Improved documentation and troubleshooting guidelines for dealing with upgrades of the operating system and clashes with other software such as security patches and upgrades.
- Tools to aid in diagnosing which MPC models need to be updated.
- Easier to use MPC and DCS monitoring tools.

3. Project Team

Putting together a good project team is an obvious best practice. The team includes members who are brought in for the duration of the project, and should include those who will be responsible for the maintenance and success of the system after the project. The key team members include:

- Lead Operator
- Control Engineer
- RTO Engineer (if RTO included)
- Consultants/Vendors

It is important that the key team members be dedicated to the project, so that their primary task is to ensure the overall success of the project. For example, many companies we spoke with reassign a lead operator from shift duty to a full-time role on the project team.

When there is a corporate function for MPC and RTO across multiple sites, personnel from the corporate group should be assigned to the team in order to leverage experience and provide consistency across the various sites. Depending on the needs and the company philosophy, the involvement by corporate staff in the project could vary from part-time to full-time. For example, it may depend on the extent to which external consultants are used.

It should go without saying that the team consultant(s) must have extensive experience applying the MPC and RTO technology in ethylene. Even when there is significant experience in-house, we believe there is significant value in using an external consultant, one who is familiar with the current technology and the latest approaches/techniques.

Due to the interaction with other departments in the company, contacts or liaisons need to be identified. This includes the following departments:

- Planning/scheduling
- Instrumentation

- DCS
- Analyzer
- Process Engineering

Personnel from these groups need to follow the overall progress of the project, getting involved at particular stages of the project, and as issues or problems arise. It is important that their time be budgeted up front in order to secure their involvement.

4. Design and Implementation

Before a project is actually started, a benefits analysis should be done to identify the source of benefits and the benefits split between MPC and RTO. This provides a base-line against which to measure and evaluate actual benefits after the project is completed (i.e., with a post audit analysis).

The key tasks for MPC in this phase of work include:

- Pre-Test and Preliminary Design
- Fix and Resolve Pretest Items
- Plant Testing
- Model Identification and Controller Simulation
- Commissioning and Training

For MPC, project implementation tends to get the most attention in articles and conferences, so we will only highlight a few key points.

The importance of the pre-test phase cannot be over emphasized, and was stressed by all producers with which we spoke. In the pretest, the instrumentation is checked and DCS controls tuned. Invariably control valves need attention – e.g., those that are sticking or need positioners. As a result it is a good idea to have a valve expert from the instrumentation departments assigned to work with the control engineers during this phase of work.

Benefits are often seen after the pre-test. We believe it is important to document and communicate these benefits to the organization. This is particularly important for companies that do not have significant experience deploying MPC and/or RTO, in order to reinforce a key advantage of deploying these technologies - that benefits are realized throughout the project cycle.

Potential constraints need to be identified early, with input from operators and the process engineers. The DCS/MPC strategy – i.e., what are manipulated variables for the controller and which DCS loops are kept versus which are "opened" is a key design issue and one where the outside consultant may be key. This will in turn lead into the specification for failure modes and back-up cascades in the DCS.

There are many specification options in the MPC that are used to specify failure modes. This includes such items as critical variables, minimum number of MVs and CVs, critical sub-controllers, and communication failures. The specification of these parameters deserves special attention as they will directly affect the robustness of the system. We have seem a range of philosophies on how to set these, which varies from company to company. These kinds of issues are good to discuss at user group meetings, to see how others approach the problem. We would say that the best practice considers instrument failure modes and looks for ways to isolate failure modes. A typical example is analyzers. Ways to isolate analyzer failures include the use of inferentials, where the inferential update from the analyzer is suspended when analyzer validity checks fail, or by including a temperature on the controller which correlates well with the analyzer and reducing the temperature limits upon analyzer failure.

The MPC and RTO support personnel who will be responsible for the system once the project is complete must be involved in the details. This includes involvement in the testing, modeling/simulation and commissioning.

The key tasks for RTO project execution are:

- Design
- Modeling
- Commissioning and Training

It is important that information necessary to execute the project be gathered up front, to avoid disruptions later. This includes such items as P&IDs, compressor curves, and any existing models.

The plant RTO engineer needs to be involved in the project details, starting if possible with the modeling phase, but at least taking responsibility for model changes during the commissioning phase of the project.

As mentioned previously, personnel from the planning and scheduling department must be involved in the project, and this holds whether the project is MPC or MPC with RTO, in order to work out details associated with pricing, constraints, and degrees of freedom. A frequent side benefit of this coordination is that it results in improvements to the planning model. A best practice item for an RTO project is to also perform a review and update of the planning model (e.g., model coefficients and constraints) based on the RTO plant model.

For companies with multiple sites, it is a good idea to provide project and implementation standards to ensure consistency and best practice across the various sites. One company we spoke with has designated an individual responsible for the corporate MPC standards document that covers a wide range of guidelines, ranging from the relationship with the contractor or consultant, project phases, and implementation specifics such as the plant testing approach, calculations, controller/optimizer pricing, etc. For these companies with an installed based of RTO, it also makes sense to develop a similar document for RTO.

Operator training – both formal and informal - should occur at multiple points across the design and implementation phase. Many of the companies we spoke with start operating training during the kick-off phase of the project. This up-front training is introductory in nature and provides an overview of both the project and the technology. Prior to commissioning, it is common to provide detailed training on the technology. During the commissioning phase, additional hands-on training occurs as the MPC and RTO is commissioned. The lead operator assigned to the project provides a critical role in the training during this phase by working closely with the operators to explain and train, as well to solicit concerns and suggestions for the project team to address.

A potential problem during the commissioning phase is in information overload of the operators. As a result, it makes more sense to plan on distributing the training over a longer period of time, and to provide best-practice training on the combined use MPC and RTO system after the commissioning phase, with periodic refresher courses thereafter.

It is common for the operators to be nervous when the MPC is initially turned on and is making frequent adjustments to the furnace feed rates, particularly when the controller is responding to a downstream constraint associated with a unit assigned to a different operator. To address this issue, many of the producers we spoke with have set-up simplified, dedicated displays that provide insight into what constraints are active and limiting plant throughput. In addition, some producers have provided electronic documentation that addresses plant constraints, and typical scenarios that lead to plant throughput reductions.

5. Organizational Support and Communication

Creating the proper support organization for the installed MPC and RTO systems is obviously of critical importance to sustaining their long-term utilization and associated benefits. As mentioned earlier, increasing economic pressures within the ethylene industry, along with a general shift from project implementation mode to maintenance mode have caused staffing levels to reduce overall. Most sites still recognize the importance of having at least one dedicated control engineer to support the MPC applications, plus a dedicated optimization engineer at sites where RTO is deployed.

Some key characteristics of the individuals who fill the control engineer roles are that they have a good understanding of both the process and underlying control systems (DCS), significant operating experience, and an understanding of dynamics and steady-state effects. Once the right individual is in place, it is critical for the control engineer to be leveraged effectively, focusing on the more difficult or strategic issues rather than on the routine trouble-shooting or maintenance activities that can be handled by others. One very effective way to accomplish this is through the best practice of using process control technicians as the first line of support for MPC. Typically, these process control technicians are not engineers, but rather former operators who have shown a special aptitude or interest in the MPC systems. In the process control technician role, they are given varying degrees of MPC-specific training and assigned, on a normal "day-job" basis, to providing the primary support for MPC. This has the benefit of relieving the control engineers of some of the more routine tasks, such as insuring that limits are set properly and addressing operator questions or concerns. Many of the sites included in this paper have used the process control technician to great advantage, and one site has even successfully expanded this role to develop an RTO technician, who similarly provides the front-line support for the RTO system.

Similarly, the RTO support engineer should be a dedicated role within the organization, not a part-time job. The main key characteristic of a good RTO engineer is the enthusiasm for modeling along with a good understanding of the process. Ideally, the RTO engineer should also be a good control engineer, but this combination often proves very difficult to find within most organizations so in the end what matters most is that the support team have the right mix of skills. There is general agreement that the RTO engineer should stay in this role several years because of the long learning curve (one to two years to develop basic proficiency in this role). For some organizations, the specialized nature of this role is at odds with the desire for well-rounded, generalist engineers.

MPC and RTO staffing levels among the sites surveyed for this paper varied fairly substantially. On one end of the spectrum is a plant in the throes of a corporate re-organization with only a single engineer supporting two RTO systems and the underlying MPC applications. On the other end of the spectrum is a site with the luxury of four process control engineers to support just its MPC applications. Admittedly, the site with only one engineer supporting all of its MPC and RTO infrastructure is struggling to properly maintain these applications and is planning to bring on additional headcount to support the RTO portion of the system, while the site with four control engineers is actually gearing up for a big plant revamp which will include substantial effort for re-tooling the MPC applications to match the new plant configuration, hence the justification for the extra personnel. On average, staffing levels in olefins seem to be one to two control engineers supporting the MPC applications for a plant, plus a separate engineer supporting the RTO system.

Some of the variability of the staffing levels is probably due to specifics of the company (e.g., frequency of revamps or operation, age of the plant). We would suggest the following practical measure for determining if a given plant's MPC and RTO staffing level are sufficient: do current staffing levels permit time for the MPC and RTO engineers to closely monitor the applications and improve them, or are they typically in catch-up or fire-fighting mode?

Changing industry demographics have had an interesting impact on role of the control and optimization staff in olefins. First off is the general trend for the unit process or production engineer to be much younger than in the past; this relative inexperience has caused the control and optimization engineers in the unit to bear more of the responsibility for basic unit troubleshooting. Another interesting change is that as the older, more experienced operators in the industry are retiring, their replacements are learning ethylene plant operations in an age where MPC is installed and heavily relied upon. One positive impact of this is that today's operators seem more accepting of the MPC as a "fact of life" in a modern olefin complex, and more apt to take steps to keep it on. The downside, however, is that many of the newer operators have never seen the plant run without the stabilizing influence of the MPC and consequently have not experienced many of the operating problems, with their inherent insight into the process, that the older operators may have faced.

It is of critical importance for the control and optimization engineers to maintain very close contacts to the production operations team. At some sites, this happens naturally when the control and optimization engineers actually report up through the operations chain of command, thereby automatically attending all production meetings etc. At other sites where this is not the case, it is important to schedule periodic meetings between the control and optimization support staff and the production operations group to insure a thorough understanding of the MPC and RTO performance, and consistency in the operating direction.

Because of the close interaction between the plant real-time optimizer and the more global LP model used for feedstock scheduling and planning, one of the key organizational issues is that of the interface to the planning group. At most plants using an on-line RTO system, the engineer who is responsible for its maintenance is often the primary plant contact to the planning group. The typical working arrangement for this interface is that the planning group sets the pricing, the feed slate and either the "priority" of each feed (such that the plant optimizer knows which one to cut first when necessary) or the range of feed rates, with the plant RTO responsible for choosing furnace severities. Among the plant sites interviewed, there seemed to be good overall coordination between the planning and plant optimization groups, and relatively little interaction required as long at the LP and on-line RTO models both seemed to be pushing the same constraints. Most sites did not feel the need to schedule periodic meetings with the planning group, instead preferring contact to be on an ad-hoc basis as required to resolve constraint coordination issues and ensure consistency in operating decisions.

Another important organizational interface for insuring the sustained performance of the MPC and RTO systems is between their support staff and the instrument/analyzer department. At most modern ethylene plants, uptime of these systems is tracked very closely and instrumentation issues are often automatically given a high priority when they start impacting this key performance measure. It is nonetheless very important for the control and optimization engineers to maintain key contacts within the instrumentation and DCS departments, and most particularly with the analyzer technicians. Among the plant sites surveyed for this paper, we found wide differences in analyzer performance. Although most sites reported high uptime and accuracy of most of the column analyzers, some plants are really struggling to keep their furnace effluent analyzers working reliably—so much so in fact that two of the sites are actually considering completely removing these analyzers and relying exclusively in inferential models!

A key component to successfully managing the various departmental interfaces required for MPC and RTO is to establish effective communication methods. We've already discussed the need for scheduled, periodic meetings between the MPC and RTO support staff and the production operations groups, but several other communication tools can help foster understanding and utilization of the systems. On the front lines, on-going verbal communications and on-the-job training with the operators is essential, especially in providing explanations for some of the less intuitive moves that the RTO and/or MPC might be driving, e.g. "why is the optimizer cutting feed?" is an often asked question. In order to insure that both the RTO and MPC have high utilization and are given the freedom of movement to truly optimize plant operations, operators must have the confidence in the results and a keen understanding of some of the underlying economic or constraint drivers for them. Therefore, the control and optimization engineers must continuously look for opportunities to educate the operators on the functionality and operation of their systems.

One best practice activity which, although not done extensively at any of the sites surveyed for this paper, can go a long way in helping to insure that the MPC and RTO have as much leeway as possible in optimizing plant operations: that is the coordination with manual operations to improve control performance. For instance, on a furnace whose feed rate may be limited by the RTO or MPC due to perceived firing constraints, the judicious manipulation of fuel gas burners and / or air registers may allow feed to be added up to the back end constraints. Similarly, the careful use of valve by-passes can sometimes help return a critical MPC variable to its control range. One ethylene site has actually taken this concept one step further and automated the coordination with manual operations by proactively sending an operator alert when a manual activity may be required to alleviate an impending MPC constraint (Arrieta and Basteris, 2004).

Beyond on-going verbal communications with the operations staff, many sites have taken the added steps of formalizing and publishing information about the MPC and RTO performance. Several sites maintain on-line documentation detailing typical constraint modes for the plant, for instance some typical causes of feed reductions, so that plant staff can determine what might be limiting throughput, and perhaps take action to alleviate the constraints. Other sites have taken steps to develop summary reports that include information from both the MPC and RTO systems, and tailored them for viewing by different departments. This type of formalized, centrally published documentation of MPC and RTO performance can foster periodic review of the constraints included in these applications, and be used as a basis for comparisons against actual plant constraints as well as the constraints built into the longer range planning models.

6. Benefit Visibility

Benefits achieved by MPC and RTO in Ethylene are well documented, and for typical Ethylene plant are in the multi-millions of dollars per year, with optimization benefits, on average, roughly 50% of the overall benefits[1]. As an industry, ethylene producers seem to have accepted the benefits that can be derived from these systems, as evidenced by the fact that many of the sites we spoke to automatically include the cost of updating the MPC/RTO as part of physical plant revamps. Furthermore, many of these sites do not even try to track specific on-going economic benefits from MPC/RTO once the systems are installed. Nonetheless, the importance of a thorough post audit exercise on completion of an MPC or RTO project, in order to quantify and document the value of the installation, is widely recognized by all as a best practice. This is particularly useful for justifying future projects, as oftentimes the benefits derived from MPC and RTO actually come from different sources than initially anticipated.

It is also important during the study or justification phase of a project that the benefits from RTO be accurately calculated and split-out from the MPC. For example, if the coordinator layer pushes feed and does not change severity, then the optimizer benefits need to account for the benefits attributable to furnace severity and non-linear effects, and relationships not present in

[1] Averaged from articles and conference presentations

the MPC layer in order to avoid double counting. To correctly justify the RTO layer it is important that the optimal point change over time as a function of operation, economics, or seasonal/environmental effects, to an extent that it is economically justified to implement and maintain. Traditionally, this has been true mainly in liquid cracking facilities, and particularly those that belong to companies with multiple ethylene plants.

There are additional benefits that can be realized using the RTO model off-line. These include:

- What if runs
- Improvement studies
- Performance monitoring and equipment health
- Update of other models (e.g., planning models)

These benefits are harder to justify up front, but they can be significant. For example, some companies assign off-line benefits in the same range as those from the on-line. In practice, however, many RTO sites find it quite difficult to use these models effectively for off-line studies. One potential issue is the size of the integrated plant model, and the inherent difficulty in exporting it and running it on a standard desktop. A work-around for this is to export and run only the portion of the plant model that is required for the study. On the other hand, the RTO model can be too detailed for many of the studies that engineers would like to do off-line, and often a simpler model is sufficient and better suited to the task. An interesting observation is that it is often engineers outside the MPC and RTO support organizations that use the models most for off-line studies, namely the process engineering group for evaluating plant improvements/de-bottlenecks, or the planning and scheduling group to evaluate new feedstocks.

There are many more intangible benefits of MPC and RTO projects. For instance, several of the sites we spoke to mentioned the step testing stage of an MPC projects as a great opportunity to dispel plant "urban myths", where both the process engineers and operators have a chance to observe first hand the cause and effect relationship between unit variables. There is also the "consistency" factor of implementing MPC, which helps to insure the plant stability day in and day out. This is also generally believed to increase plant equipment reliability, although few, if any, ethylene producers attempt to quantify this.

As mentioned previously, many ethylene producers do not attempt to track the benefits of MPC or RTO on an on-going basis, although most admit that this would be desirable. All plant sites make an effort to track truly meaningful uptime and utilization statistics, e.g. insuring that the key MV's for the applications are in service and not artificially constrained before considering the MPC application "in service". Similarly, for the RTO applications, the number of cases solved and the percentage of the solutions implemented are typical metrics tracked. Of those sites that do track economic benefits, no effort is made to separate RTO benefits from those of MPC. However, a best practice would be to periodically track benefits from MPC and RTO separately. One method for separating RTO benefits from those of the MPC could involve creating a separate, linear model within the RTO framework that executes side by side with the full non-linear model, and comparing the added benefits of having the non-linear rigorous RTO.

It is important for the long term success of these applications to maintain visibility of the performance after a project is complete. In an environment of shrinking budgets and competition for limited funds, publicizing the performance of MPC and RTO throughout the organization helps keep focus on the value and need to continue to support these applications. Some plant sites update these performance metrics on a real-time basis and post the results on the company intranet, while others choose to compile and publish only periodically in a scheduled report. Whatever the method, there is certainly a best practice to be cited for keeping the benefits of MPC and RTO front and center, and maintaining the enthusiasm for maximizing the value of these tools.

7. Maintenance and Improvement

This task is one of sustaining and hopefully, increasing benefits over time – i.e., continual improvement. Key to maintaining uptime, and therefore the benefits, is prompt resolution of problems as they arise, e.g. instrumentation, analyzer, and DCS problems. All of the companies we spoke with said they had the support of their organization, and therefore did not have problems in this area.

The MPC and RTO engineers need to follow the applications daily (and across shifts) to ensure that they are performing well and being used correctly by operations. Based on what the engineers observe, they can provide feedback, advice, or perhaps additional training.

Part of the ongoing maintenance may require revisiting decisions made during the implementation phase. This includes such things as MPC failure modes and techniques and settings for isolating instrument failures and recalibrations.

The loss of benefits due to setting limit ranges too tightly in the MPC and RTO has often been cited as a problem area, although not specifically mentioned as a major problem area with the ethylene producers to whom we spoke. A best practice in this area is implementing some type of limit management. One producer tracks limit changes via an electronic log that includes operator comments as to why the change was made (e.g., temporary change or not). Changes are reviewed by a senior operator and MPC engineer (or RTO engineer) and a decision made as to whether the default limit values should be updated. Another producer maintains a master document of limit values and is periodically updated based on input from operations, planning and MPC/RTO.

We believe it is important to develop monitoring and troubleshooting guidelines. With the possible exception of incorrectly set limits, MPC or RTO performance deterioration is often the result of problems outside the envelope of the MPC or RTO: DCS, instrumentation, or disturbance variables. Having guidelines also has the benefit of facilitating the first support function and freeing up the senior MPC and RTO engineers.

For companies with multiple sites: a best practice is having regular review meetings with other sites to share experiences and discuss solutions to specific problems. One company schedules these meeting monthly, alternating between MPC and RTO topics. User group meetings and technical conferences represent other avenues for discussing ideas and approaches.

A best practice for the RTO engineer is to look for opportunities across the organization to use the RTO model in provide additional benefit and consistency. One company encourages this by having the RTO engineer a member of a plant-wide profitability team

Finally, we believe best practice in sustaining and improving the applications involves periodic reviews or audits by outside consultants (at least outside the specific plant site). This has the benefit of bringing it a “fresh set of eyes” and possibly new ideas for improvement or for dealing with specific problems.

8. Conclusions

In conclusion, the ultimate success of MPC and RTO in the ethylene industry depends on doing many things well. We have discussed several best-practice approaches that maximize the probability of success. Ultimately, it is a matter of determining what approaches works best and tailoring them to a given organization. We hope that the ideas and opinions expressed here would be of benefit to Ethylene Producers, regardless of their experience level with these technologies.

References

Arrieta, E. and Basteris, J., "AC&O Best Practices: A Control Room Perspective", *AspenWorld 2004, Orlando.*

Development of Simplified Cracker Yield Model Correlation based on SPYRO® Rigorous Simulator for Various Operational Optimization Applications

G. D. C. Moraes, D. J. Cruse, É. R. P. Claro, P. B. Rosinha and G. Müller
COPESUL - CIA PETROQUIMICA DO SUL
Porto Alegre, RS, Brazil

J. A. DeFazio and B. Tjoa[†]
Optience Corporation
Campbell, CA 95008, USA

Omar A. Hamid
Technip USA Corporation
Claremont, CA 91711, USA

Prepared for Presentation at the 2005 Spring National Meeting, Atlanta GA, Apr 10-14
17th Annual Ethylene Producers' Conference
Session TA006 - Ethylene Plant Process Control

[†] Author to whom correspondence should be addressed. Email address: tjoa@optience.com

Development of Simplified Cracker Yield Model Correlation based on SPYRO® Rigorous Simulator for Various Operational Optimization Applications

G. D. C. Moraes, D. J. Cruse, É. R. P. Claro, P. B. Rosinha and G. Muller
COPESUL - CIA PETROQUIMICA DO SUL
Porto Alegre, RS, Brazil

J. A. DeFazio and B. Tjoa[†]
Optience Corporation
Campbell, CA 95008, USA

Omar A. Hamid
Technip USA Corporation
Claremont, CA 91711, USA

Abstract

Ethylene producers around the world face increasingly stiff competition in the market place. One of the keys to its competitiveness is the ability of a producer to evaluate and secure the most desirable feedstock and to process it at optimal operating conditions. To achieve this goal, advanced optimization planning and scheduling applications that represent the real business problem are required. Optience SCMart has been selected by COPESUL as the application platform for building nonlinear and mixed integer models. Without the limitation on the optimization technology for solving a realistic model, the key in matching the real operation data is in estimating the yield of the crackers.

SPYRO®, a rigorous cracker yield model from Technip, has been selected as the basis for developing a simplified low order nonlinear polynomial correlation model that is suitable to be embedded in the SCMart planning and scheduling model. SPYRO® is the industry standard for pyrolysis yield prediction. Its fundamental kinetic scheme of more than 3000 total reactions, representing the chain reaction mechanism of 128 molecular species and 20 radical species is well known for its accuracy in predicting pyrolysis yields for feedstock types, ranging from gas to naphthas/condensates and heavy gasoils.

There have been reports of utilizing SPYRO® in real-time optimization (RTO) and planning applications. These applications are typically either a rigorous equation based RTO or simpler LP based planning. There are basically two approaches that have been reported in the literature: (1) a direct call to the SPYRO® simulator as a 'black box' during the optimization for a function evaluation and first order numerical derivative, needed for the nonlinear optimization algorithm in the RTO application or (2) a simple 'look-up' table or first order Taylor's series expansion that has been generated a priori for the LP based optimization planning application.

In this paper, we propose a low order nonlinear polynomial correlation yield model based on SPYRO® results. We discuss our experience in obtaining a good model over the desirable range of operating conditions and feedstock types considered in the business operation policy. The model was developed using Optience's SCMart Estimator. In addition to its main function as a nonlinear parameter estimator for the optimization software, its database is designed to manage user feedstock data and SPYRO® yield results in an efficient way. SCMart Estimator is a module in Optience's SCMart Suite for Planning and Scheduling software, where its optimization solution approach is based on a mathematical programming framework that can solve a wide range of model types, from simple LP to more complex MINLP models.

Introduction

In the face of increasingly stiff competition, Ethylene producers around the world have been innovative to succeed in the market place. Critical to competitiveness is the ability of a producer to maximize its profit for a given unique opportunity in the market by evaluating and securing the most desirable feedstock. For liquid based ethylene producers, the decision on financial trade-off is much more complex than that of gas based producers. Liquid cracking producers must take into account the significant amounts of key byproducts, such as Propylene, C4's (Butadiene and Butene-1) and Aromatics (Benzene, Toluene and Xylene). It is very important to have an accurate yield evaluation tool for different feedstock types.

Furthermore, the economic evaluation process is not straightforward due to variability of the feedstock composition at the charging tanks over the planning/scheduling time horizon. This variability can be caused by many factors, such as the source of feedstock supplies, the feedstock blending occurring in the tank yard and the co-cracking of the feed with recycle materials.

The critical properties, including PINA and density of the blended feedstock in the charging tanks, can vary significantly as a result of the manufacturing policy, resources and constraints. These properties have significant influence on the effluent composition of the crackers. The challenge is to predict the yield for a variety of operating decisions (such as COT, feed rate and recycle materials) and analyze its impact on profitability.

Clearly, a simulator that can predict the yield accurately is necessary. A reliable yield simulator that gives us a point solution for each set of independent decisions is required. However, after the yield information is available, the complex trade-off with respect to process constraints and market conditions must be evaluated to study their impact on the business operations. A yield predictor or estimator must be integrated into a larger optimization application that considers the overall economic performance.

The ethylene industry has used equation based optimization applications for various operations, such as Advanced Process Control (APC), Real-Time Optimization (RTO), Production Scheduling, Production Planning and Corporate Strategic Planning (CSP). Each of these applications is designed to address a specific area of operation. The application scope can range from a unit operation, to the whole plant and up to the entire business operation level. As the application scope increases, its operation coverage (typically the time horizon) also varies: from APC in minutes, RTO in hours, scheduling in days, planning in months to CSP in years.

APC's main objective is to achieve a stable operation; the control target is often guided by decisions that take into account the economic trade-off. This target can be obtained from a higher level application, such as RTO or a scheduling system that optimizes the profit objective function.

APC and RTO applications are targeted for improving real-time plant operation, while scheduling and planning applications are designed for managing overall business operation, not only for the present but also for a future time horizon. Therefore, the nature of scheduling and planning applications is optimizing a multi-period problem. Figure 1 shows typical layers of business operation applications, model complexity and their potential economic impacts.

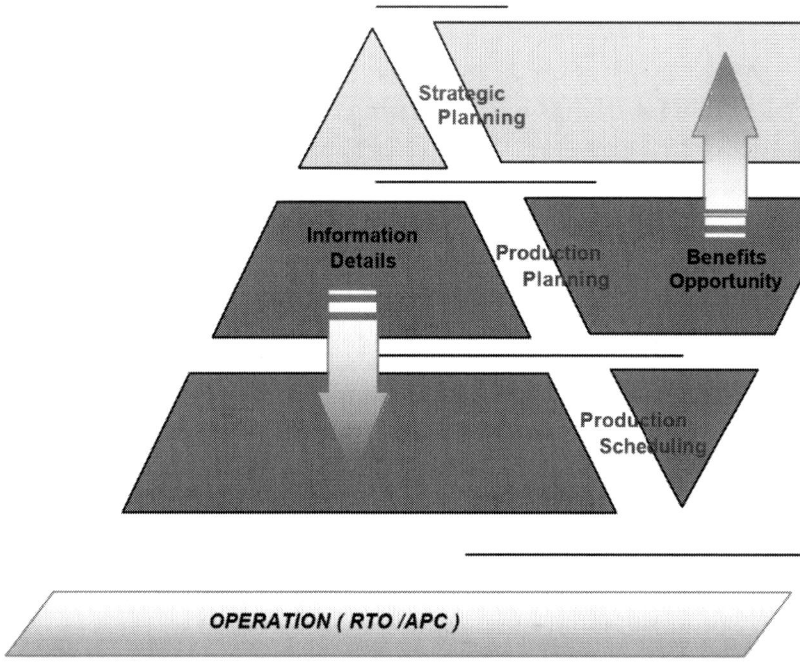

Figure 1. Hierarchy information details and potential benefits in business optimization applications

Unlike RTO, which covers optimization at each plant level, the scheduling application often has a much larger scope. In the case of COPESUL, the application covers its entire business operation in the complex that consists of two ethylene plants, its derivatives chains, the utilities plant and the tank yard. Therefore, the optimization at the scheduling level closely represents the overall business operations.

The accuracy of the model can be achieved by creating a nonlinear model that represents key operation units without rigorous thermodynamics-based models, as in a traditional RTO application. Here, the critical issue is how to create a model with enough accuracy to be used in a business optimization application.

In this paper, the development of a yield correlation model that is suited for advanced optimization based planning and scheduling applications is discussed.

Planning and Scheduling Optimization Applications

As shown in Figure 1, the link between production planning and scheduling applications is extremely tight. The main difference is in addressing different workflows in the business.

The planning application is targeted for managing monthly decisions such as 'production targets' and 'feedstock types and amount to buy'. These decisions are made by relying on market information, resource availability and the business strategy for a particular month/ period. Typical planning periods are three monthly periods with a moving time horizon. Once the monthly target has been decided, this target needs to be executed.

The scheduling application is designed for optimizing daily operation targets, based on the monthly planning targets. Here, the key difference is in the period definition. Since the scheduling is optimizing daily targets, there are many more operational constraints that need to be honored, requiring a more detailed model. Furthermore, in the scheduling it is required to optimize more periods simultaneously in order to honor the monthly planning target. For these reasons, the scheduling application should be capable of optimizing up to 31 days of scheduling decisions by considering daily-based constraints.

Although the model size for each scheduling period is much smaller than in a traditional RTO model, the number of degrees of freedom in each period is very similar (typically less than 100 in each period). In addition, in the scheduling application, the decisions between periods are also optimized. As a result, the total number of degrees of freedom in the scheduling optimization problem is much larger than in a traditional or in a simplified RTO implementation (Shindo et al. 2004).

Also, the initial condition in scheduling can be significantly different from that of RTO. In RTO, the initial point can be obtained from the previous solution that was implemented in the past few hours. Given a target that can be executed by APC, the movement from one run to the next one in RTO is extremely small compared to that of scheduling, where the period definition is daily.

Due to the nature of the problem, the scheduling optimization problem can be very challenging. The degrees of difficulty in solving a nonlinear optimization problem are not only measured by the size of the problem, but also by the number of degrees of freedom and how close the initial point is from the optimal solution. It is also typical that the scheduling problem faces discrete (logical) decisions in executing certain operations (on/off).

Clearly, a rigorous scheduling optimization application requires a system that has a strong optimization solution technology for solving large scale nonlinear (NLP) and mixed integer nonlinear programming (MINLP) model.

Considering the above explained reasons, selecting a business optimization application platform that can address not only the planning, but also the scheduling problem in the same environment is very critical to ensure coherent modeling approach across the application levels. Besides ensuring consistent decisions across application levels, it also reduces software maintenance and training cost.

After a comprehensive evaluation of commercially available systems that meet such strong business requirements, COPESUL selected Optience's SCMart Suite software for its planning and scheduling application platform. This platform utilizes a similar approach reported by Mitsubishi Chemical (Ota et al., 2002).

The SCMart Suite is a state-of-the-art planning and scheduling application platform that addresses requirements in the optimization technology (from a simple LP to the most complex MINLP model). It also has a customizable user interface and complex data management interaction with the company's internal workflow. Furthermore, one of the key advantages in the SCMart software environment is that it allows building a more rigorous (nonlinear) equation based yield model that can be embedded in both planning and scheduling models. This feature addresses both the need of building a more accurate yield model and the robustness of optimization solution technology. In order to achieve high model fidelity, the model correlation is developed based on the SPYRO® cracker yield simulator results.

SPYRO® Pyrolysis Yield Simulator

The evaluation of the potential value of each available feedstock type is critical to planning and scheduling optimization of olefins units. A model is necessary to provide yields for the various feed types. The relative values for each feed will help determine the optimal schedule, while an accurate yield prediction will ensure that the plant actually meets production targets. This model must predict yields for any type of feed; accept limited feed characteristics (such as distillation curve, PINA and density) as feed definition for liquid feeds; and reflect the impact of fouling.

COPESUL selected Technip's SPYRO® rigorous kinetic model, the industry standard for ethylene pyrolysis furnace simulation, for yield prediction. It is the most widely used ethylene yield model and has been continuously validated over the last 25 years by its industrial users. SPYRO®'s accurate and reliable yield prediction for the complex pyrolysis reactions has made it an integral part of applications in all aspects of furnace and plant operations.

At the heart of SPYRO® is a mechanistic kinetic model comprising of 3288 reactions, 128 molecular species and 20 radicals. The free radical kinetics is critical to develop a single model with sufficient flexibility to handle all feed types (e.g. ethane, propane, naphtha, gas oils and gas/liquid mixtures) and enough capability to process a wide range of operating conditions. The free radical mechanism simultaneously incorporates both theoretical kinetic background and all available industrial data. The kinetic rate equations along with the mass, momentum and energy balances are integrated along the length of the coil to obtain effluent yield predictions. Barendregt et al. (2002) provides further information on SPYRO®.

SPYRO® includes a feed characterization module that allows yield prediction for liquid feeds, such as naphtha and gas oils that only have limited feed characteristics available. It converts distillation curves, PINA and density information into the detailed feed composition breakdown required for kinetic modeling of liquid feed pyrolysis. The feed characterization algorithm in SPYRO® takes advantage of observed similarities in naphtha and gas oils from a variety of very different origins, and the relative abundance of particular isomers in such feeds. This module is able to adequately define these liquid feeds using as small number of individual hydrocarbons that can be correlated to the feed characteristics.

Coking in radiant coils affects performance by reducing runtime while impacting yields. The fouling model in SPYRO® is described by several competing carbon deposition mechanisms, including catalytic coking in the exposed coil surface, addition of unsaturated radicals and molecules, and trapping of high boiling compounds. The coking model is sensitive to specific coking precursors in the bulk fluid and the inside skin temperature of the coil. The fouling mechanisms in SPYRO® allow analysis of the impact of feedstock and operating conditions on run length.

Yield Model Correlation Development for Planning and Scheduling Applications

SPYRO® is used extensively for providing cracker yield information in ethylene planning and scheduling optimization systems. The manner on how the yield solution from SPYRO® is used in the optimization model depends on the architecture of each system.

Most state-of–the-art commercial nonlinear optimization solvers are developed using Newton based methods. This approach requires first order and second orders derivative (Jacobian and Hessian) information. The performance of this algorithm highly depends on the accuracy of this information for providing good search direction during the iterations for finding an optimal solution. For most advanced systems, accurate (analytical) Jacobian information is required in order to improve its robustness. SCMart optimization solution technology uses this approach.

SPYRO® is an extremely rigorous, but closed simulation model. For reasons related to speed and stability, it has not yet been practical to open up and simultaneously solve it together with other planning/scheduling constraints. Therefore, the choices are either embed SPYRO® directly as a 'black box' together with other 'open form' constraints in the model or develop 'an open equation based correlation model' based on SPYRO®. The 'black box' approach can compromise our ability to utilize the most advanced optimization technology for ensuring strong robustness. Here we selected the latter approach. Using SPYRO® generated yield data, we developed an open equation form yield model representation for each combination of cracker coil configuration, feedstock and plant operating range. With this procedure, it is possible to create a good model that matches SPYRO® results. This approach allows us to derive a highly accurate, open-form, application specific model from SPYRO®.

In most planning legacy systems that were built based on classical LP based algorithms, the method of choice for building a yield model is a Taylor's series expansion to first order. In this type of model, base points are selected within the operating region of a furnace coil and feed stock type. Each base point corresponds to one set of conditions; severity, feed rate, S/HC (Steam/Hydrocarbon) ratio, as well as feed properties or composition. SPYRO® calculations are done at each base point. Next, one at a time, each condition is perturbed from its base point value, with all other conditions unchanged. SPYRO® calculations are also done at these perturbed points. The difference between the solution at the perturbed point and the base point is used to calculate 'numerical' partial first order derivatives with respect to that condition. The base points and derivative vectors can be used in an LP table lookup to approximate the SPYRO® calculation within the operating region.

The reason for choosing this approach is often due to limitations of the optimization algorithm. Clearly, this is not the most rigorous way of building correlation models for the planning/scheduling purposes. There are several basic limitations of this approach:

1. During the optimization iterations of recursive solution method, interpolation of the first order function is needed. Depending on how the base points are selected (sampling points) and the size of the perturbation, the yield prediction based on this linearization 'extrapolation' from the base point can result in significant inaccuracy.

2. Due to its structure, it is difficult to develop a meaningful interpretation of a parameter value. It is only possible to interpret the parameter sign to indicate whether it has a positive or negative direction (gradient).
3. This type of model is inherently a linearization technique. Therefore, it is difficult to expand the model to a higher order polynomial equation form that can estimate the yields more accurately.

The SCMart modeling environment for building planning and scheduling applications is a task ('unit operation') based approach. The material chains in the scope of the application are tracked in a series of unit operation blocks that indicate the transformation from input to output. In the cracker task model, the feedstock that enters a reactor model is transformed into products. The yield for each product can be defined as a function of a set of feedstock types, including co-cracking, that enters the cracker, its properties and the cracker operating conditions. Since it is a nonlinear polynomial function, there is no limitation on how the user defines the function to best represent the yield estimation. Therefore, the ability to define an accurate model is no longer limited by the restriction in the optimization technology, but on user knowledge and experience.

Here, we would like to share our experience in developing a good model. There are several key steps that are important:

1. Data analysis on feedstock:
Understanding the history of feedstock purchases, and how they are blended before reaching the cracker. The feedstock properties have a strong influence on the yield; what properties should be included as independent variables in the correlation depends on this analysis.
2. Data analysis on process operation data:
This analysis can provide a better understanding of the operation constraints and possible range of operating variables that effect the yield.
3. Selecting independent variables:
Based on the results of steps 1 and 2, it is possible to establish what to select as independent variables.
4. Sampling points:
The design of the sampling points will have an impact on the size of the sampling point population, as well as in the quality of the correlation model. Figure 2 illustrates a classical factorial design for generating sample points, assuming that all of them are truly independent variables. In this application, implicit operational constraints and their possible correlations need to be taken into account in generating the sampling points. As one can image, the number of data points that need to be generated is easily in the thousands, considering the possible range of feedstock properties and co-cracking conditions per furnace coil type.
5. Yield data analysis:
Analyzing the yield data against the independent variables enables another level of checking when validating the results. This step is very important for validating against known facts about the cracker design.
6. Equation form and function of the correlation model:
The selection of independent variables to be included in the correlation model depends largely on the analysis of the previous steps. A lot of judgment needs to be made in this selection because the goal is to develop a good model with predictive value. Here, experience in modeling is

useful. Our goal is not doing parameter fitting only; the goal is to find a simple correlation model that can capture the structure of the functionality without sacrificing the model quality.

When the number of selected independent variables is not sufficient, the model cannot provide a unique response to the changes in the independent variable sets, as illustrated in Figure 3. Over parameterization of the model often gives a better fit, but this kind of model has little predictive value. It will be difficult to have clearly interpreted the physical meaning of the parameters. Therefore, it would be difficult to make adjustments when it becomes necessary to match the overall production data in the scheduling and planning real results. Figure 4 illustrates a good model. The accuracy of the model is sufficient for estimating the yield in the planning and scheduling application.

A typical ethylene plant can have several types of crackers with different coil configuration and feedstock alignment. The development of correlation models for all the crackers requires a large number of SPYRO® simulation runs. These results are stored in the SCMart Estimator database. The data then can be analyzed further and used for estimating the model parameters. Most of the steps in the yield model development process have been automated in SCMart Estimator module.

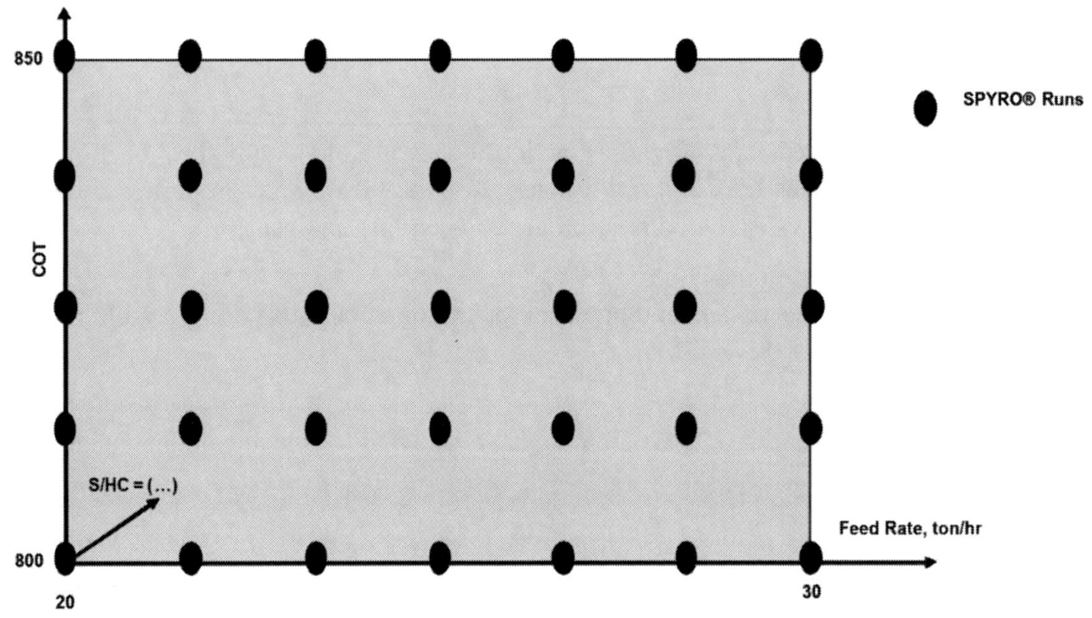

Figure 2. Illustration of a classical factorial design.

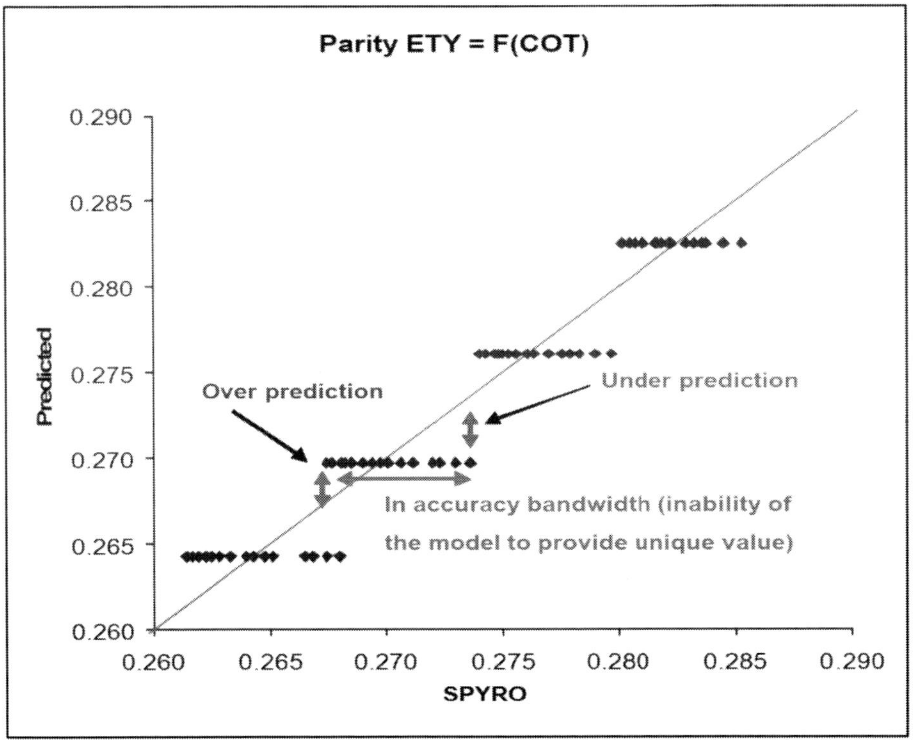

Figure 3. A parity plot illustrating a poor model due to insufficient independent variables

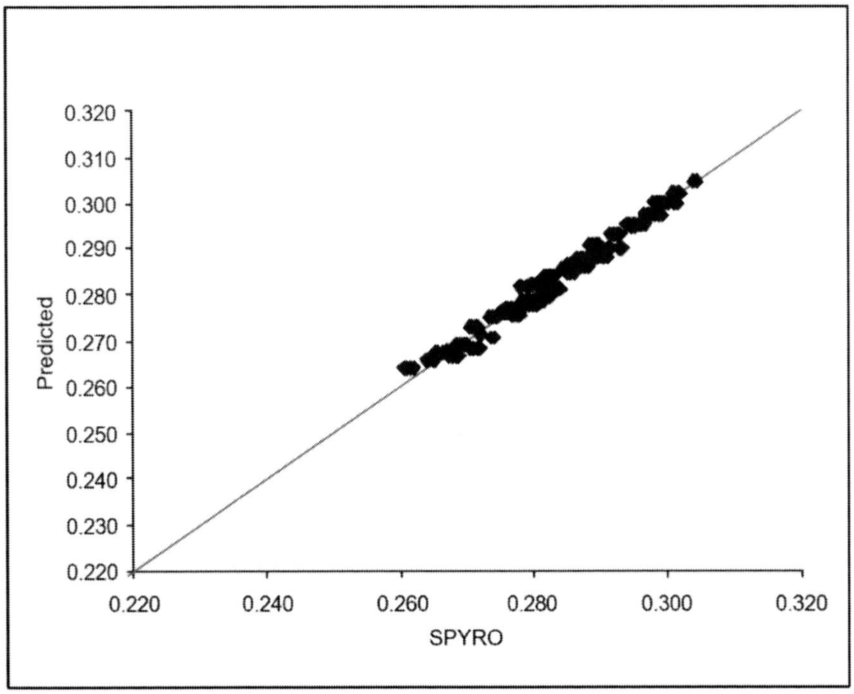

Figure 4. A parity plot illustrating a good model.

Summary

Ethylene producers have been known to be very innovative in applying optimization technology for improving profit margin. One of the major challenges in achieving a better performance is the purchase of the most appropriate feedstock and cracking it at optimal operating conditions. This objective can be reached by employing an optimization-based planning and scheduling system integrated with a high fidelity pyrolysis yield model to accurately represent the real business problem.

COPESUL selected Optience's SCMart software as the application platform. In the SCMart platform, applications that require nonlinear and mixed integer model can be formulated without a linear approximation model and optimized. Without limitation on the optimization technology and model representation, the accuracy of the solution is highly dependent on how well we can represent the problem.

Technip's SPYRO® rigorous cracker model was selected for yield prediction. Since the cracker is where feed is converted to products, a reliable yield model is the key factor in matching operation results.

In SCMart, the development of simplified yield model of the crackers was done by using an open form lower order polynomial. The parameters in the model are optimized using yield data generated from SPYRO®. Since the yield model has an open equation form, the robustness of the optimization solution method can be significantly improved. The effectiveness of this approach has been proven in solving large scale planning and scheduling applications at COPESUL.

References

Barendregt, Valkenburg, Wagner, Dente, Ranzi, 'History and Recent Development in SPYRO®, a Review', presented at the 14th Annual Ethylene Producers' Conference, AICHE Spring National Meeting, New Orleans, Louisiana, 2002

Ota, Y.; Namatame, K.; Hamataka, H.; Nakagawa, K.; Abe, H; Cervantes, A.; Tjoa, I.B. and Valli, F., 'Optimization of Naphtha Feedstock Blending for Integrated Olefins-Aromatics Plant Production Scheduling', presented at ESCAPE-12, The Hague, The Nederlands, May 2002.

Shindo, H.; Lau, K.; Kaneko, R. and Ayala, J. S., 'Furnace Optimizer in a Naphtha Cracker', presented at the 16th Annual Ethylene Producers' Conference, AIChE Spring National Meeting, New Orleans, Louisiana, 2004.

Paper TA006c

New AspenTech DMC Controller design For the C3 Hydrogenation Reactors in an Ethylene Plant

By
Alexandre Mota, André Vilelas, Pedro Braga and Rui Chaves

of
Repsol Polímeros Lda
Sines, Portugal

Prepared for Presentation at the 2005 Spring National Meeting, Atlanta GA, Apr 10-14
17th Annual Ethylene Producers' Conference
Session TA006 - Ethylene Plant Process Control

Abstract

The Repsol Steam Cracker, in Sines, is modelled with the AspenTech PIMS and it was found that the propylene production was not optimized, i.e. losses of near 18 t/day.

The C_3 hydrogenation reactors were one of the systems to be studied, since typical steam cracker C_3 cuts contain 92% propylene and 4% to 6% methyl acetylene and propadiene (MAPD) which must be removed to meet propylene and propane product specifications. This hydrogenation gas reaction consists of four adiabatic catalyst beds, in series, of which three beds are used in normal operation. The palladium (0.03% to 0.04%) base catalyst must be selective to the hydrogenation of C_3H_4 to C_3H_6.

On this study, it was concluded that the C_3 reactors had been operated with a very low selectivity (45%).

Repsol Polímeros Lda has designed and developed a new DMC controller in order to optimize MAPD reaction conversion and minimize the losses of propylene to propane, due to over hydrogenation. This controller manipulates the total hydrogen flow rate to the reactors maximizing the propadiene concentration at the outlet.

The results achieved are outstanding since the propylene production has increased about 5.5 kta, corresponding to 1.27 M€ gain in the plant operation profit.

Introduction

The Repsol YPF Sines Petrochemical Site has six plants: *Steam Cracker*, Butadiene, MTBE/ETBE, HDPE, LDPE and a Power Plant.

The *Steam Cracker* produce ethylene and propylene *polymer grade* from different hydrocarbons feedstock as propane, butane and naphtha by the Linde AG process. As co-products there are: a C_4 cut (rich in 1, 3 - Butadiene), aromatic mixture and fuel oil.

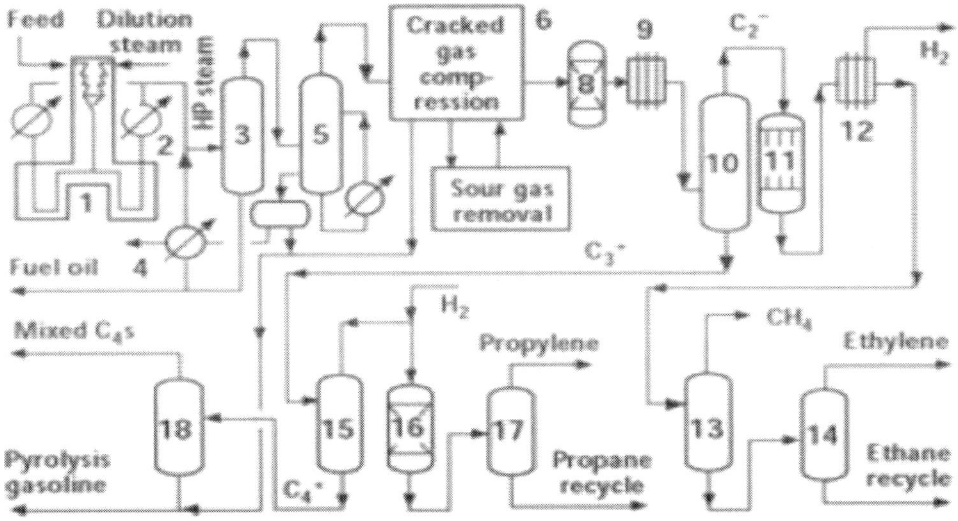

Figure 1 – *Steam Cracking* Process Flow sheet (Linde Process)

The feedstock and the recycle streams are pre-heated and cracked with dilution steam in the furnaces that are highly selective (1). The furnaces are optimize in the residence time (0,3-0,5 s), temperature and pressure profiles for the actual feedstock in order to have high yields in ethylene and propylene.

The furnace effluent is cooled in the transfer line exchangers (TLX) (2), generating high pressure steam, and injecting quench oil.

The raw gas is further cooled and purified in the primary fractionator (3) and in the quench water tower (5). The heat is recovery in the quench oil train, generating steam and pre-heating the naphtha to furnaces (4). The raw gas is compressed in a 5 stages compressor (6) driven by a high pressure steam turbine and then is dried in gas and liquid molecular sieves adsorbent beds (8). The CO_2 e H_2S are removed in the caustic scrubber tower located between the 4th and 5th stages of the compressor.

The compressed gas is pre-cooled (9) and fed to the separation section: Deethanizer *front-end* (10), C_2 *front-end* hydrogenation (11), cold box (12), Demethanizer (13) and C_2 splitter (14) that is operated as a heat pump and is integrated in the ethylene refrigeration cycle.

The C_3^+ stream in the Deethanizer bottoms is routed to the Depropanizer (15) and the top product is hydrogenated in *tail-end* C_3 reactors (16) to remove the methyl acetylene and propadiene, and then is fractionated in the C_3 splitters (17) to produce *polymer grade* propylene. The components on the C_4^+ stream are fractionated in the Debutanizer (18) recovering a C_4 cut, rich in 1,3 – Butadiene, and a C_5 stream that together with the condensate hydrocarbons in the hot section, are the rich aromatic gasoline.

Problem Description

The *Steam Cracking* process of the Repsol Polímeros is modelled with the AspenTech PIMS® (*Process Industry Modelling System*), which objective is to planning and economically optimizes the production. Comparing the model with the actual production it was noticed an 18 t/day negative deviation in the propylene production, about 5% of the demonstrate capacity.

Troubleshooting

The C_3 stream hydrogenation for the methyl acetylene and propadiene (MAPD) removal is carried in 3 gas phase catalytic adiabatic reactors in series. Tail-end means that hydrogen is fed to the reactor in a separate stream, the opposite of the C_2 hydrogenation (front-end). The catalyst active metals are Palladium and Chromium (0.03%) supported in alumina.

The main reactions in this system are:

Methyl acetylene to Propylene Hydrogenation: $CH_3{-}C{\equiv}CH + H_2 \rightarrow CH_3{-}CH{=}CH_2$

Propadiene to Propylene Hydrogenation: $CH_2{=}C{=}CH_2 + H_2 \rightarrow CH_3{-}CH{=}CH_2$

Propylene to Propane Hydrogenation: $CH_3{-}CH{=}CH_2 + H_2 \rightarrow CH_3{-}CH_2{-}CH_3$

The last one should be minimized, optimizing the system, in order to maximize the propylene production, more valuable product than propane.

Besides these, other oligomerization reactions take place yielding hexenes and even nonenes, but in a reduced scale as well more unsaturated species isomerization. These secondary reactions are responsible for the green oil formation, selectivity loss, pressure drop increase and even to catalytic beds deactivation.

The reactions are exothermic, in which the temperature control is done by inter-reactors cooling. Thermodynamically, the first two chemical reactions are quite equivalent because the ΔH_r is similar. In a kinetic way, methyl acetylene hydrogenation is faster, reason why the downstream control variable is the propadiene concentration.

Based on the reactions exothermicity, optimal temperature differentials were calculated for the MAPD reactors inlet concentrations considering the excess ΔT method.

When high propylene losses were found and the investigation and system optimization started, the main reason that was found was over-hydrogenation induced by:

- High H_2 flow rate to the reactors
- High temperature differential and, sometimes, wrong profile
- Wrong residence times and space velocity (VHSV)
- Low catalyst selectivity

High hydrogen flow rate and excessive propylene to propane hydrogenation decreases the catalyst selectivity to the MAPD reaction, inducing higher inlet temperatures as well higher DT for the three reactors.

It is now possible to see the descendent temperature profile and that still MAPD conversion occurs in the last bed, which was not happening before, minimizing the propylene loss with a difference less than 5°C between the ideal and real reaction ΔT.

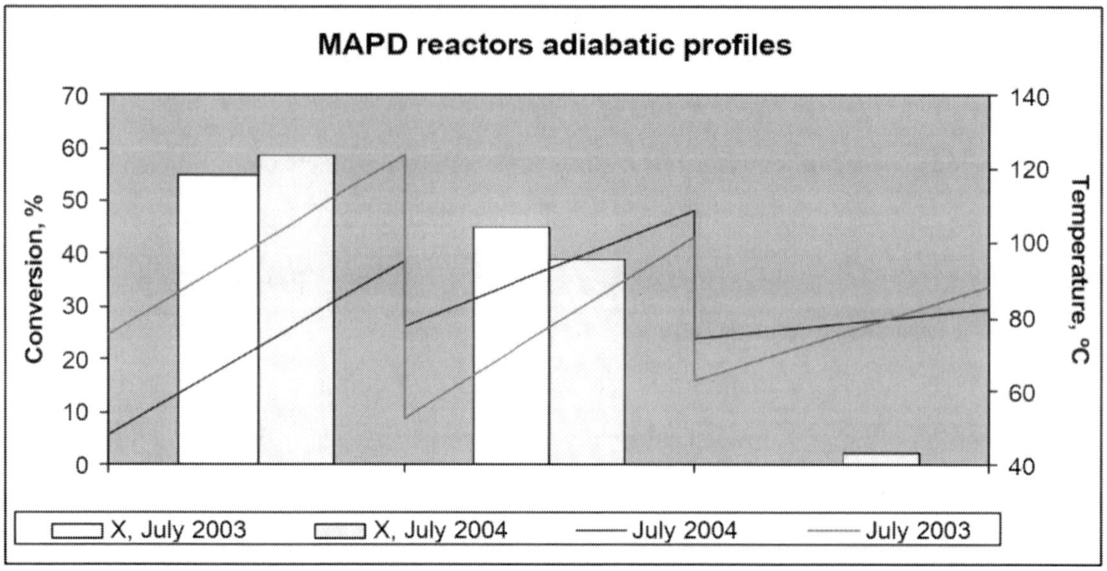

Figure 2 – Adiabatic temperature profiles and MAPD conversion between July 2003 and July 2004

The temperature profile is difficult to optimize because hydrogen is controlled with manual valves, which means that is necessary a permanent cooperation between control room and plant operation.

Catalytic beds disposition between regenerations is also crucial because the more active bed is not always working as the first bed, giving bad temperature profiles in the system.

Another important factor is the C_4 olefinic components concentration in the reactors feed C_3 stream. These have a high influence by a decrease in the C_3/H_2 ratio as well in the selectivity and catalytic activity of the first bed. Although in small quantities, with hydrogen, they react among them exothermically being responsible by green oil formation and pressure drop increase in the beds.

Optimization

After the findings about the problems concerning the reactors operation, it was necessary to optimize and improve the operation conditions, having for this been designed and implemented a new APC (*Advanced Process Control*) controller, utilizing the AspenTech DMCplus software tool.

The DMC (*Dynamic Matrix Control*) is a control technique characterized by this features [6]:

1. Multivariable

Different type of variables used:

- Manipulated (MV): independents – things that the operator can move
- Feed forward (FF): independents - things that impact the process, yet can not be moved directly
- Controlled (CV): dependents – dynamic behaviour can be described totally in terms of specific independent variable changes over time

2. Model Predictive

The response of a DMC controller to the change of one determined been stationary state for a new future state, it is based on the existing linear dynamic models between the independents and dependents variables.

3. Constrained

All variables have low and high limits, that in the case of the independents ones never are violated, and that for the dependents can be violated, depending on the importance of the same one.

4. Optimizing

Steady State:

- Maximize profit
- Robust handling of infeasibility

Dynamic State:

- Minimize the predictive error
- Minimize the manipulated variable movement
- Follow the optimum dynamic path to economically optimum steady-state targets

5. Rigorous

The DMC controller solution must:

- Solve rigorous optimization problems
- Have the process models based on the actual process behaviour
- Calculate a detailed series of future moves for each MV

The APC controller strategy for MAPD reactors is showed in the next figure:

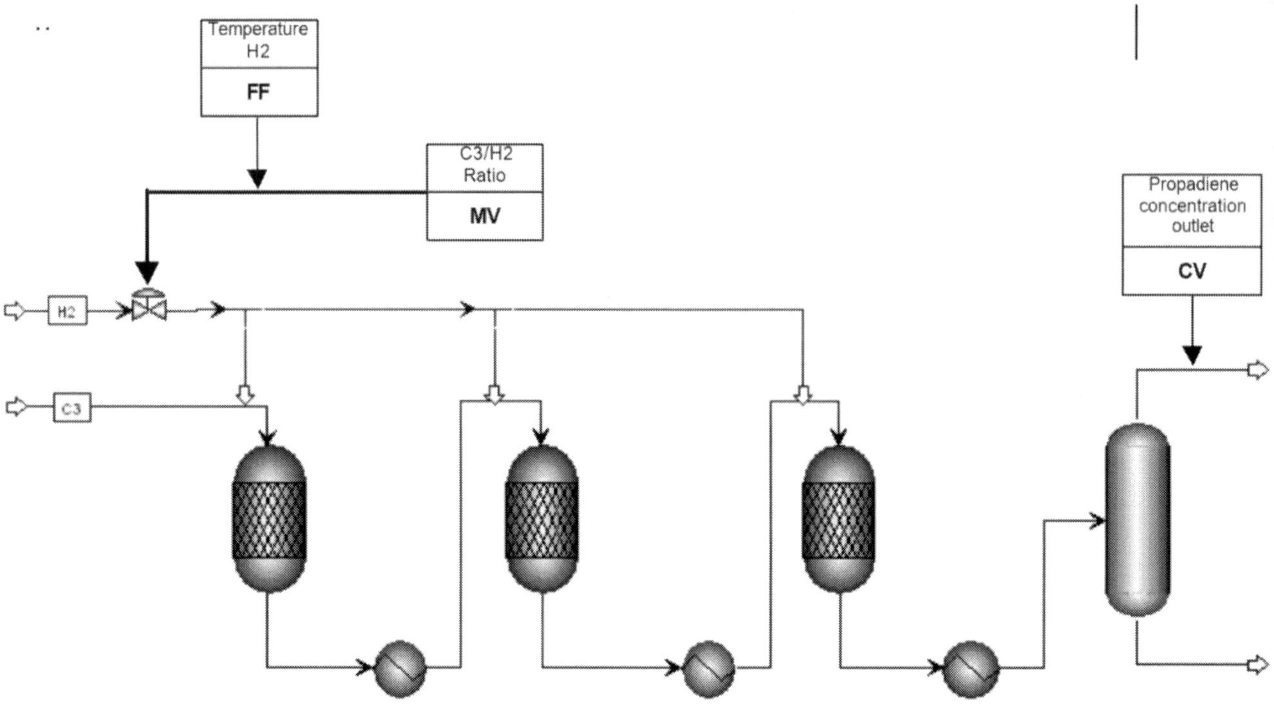

Figure 3 – MAPD Hydrogenation schematic with the DMC control strategy

The steps to develop and to implement this controller APC had been the following ones:

1. <u>Model Identification</u>

To create the linear dynamic models between the independent and dependent variables, it was used the DMCplus Model, graphical based modelling tool, that utilize the actual process data to identify the models. Plant tests were conducted during two weeks and all necessary data was collected, together with some improving in the PID controllers and HPLC analyser.

The utilize method during the tests was the FIR (*Finite Impulse Response*), that allows to identify the response of a dependent variable to a unit step in a manipulated variable.

For this APC controller the manipulated variable is the ratio between the C_3 stream volume flow and the total hydrogen volume flow to the reactors (C_3/H_2 ratio), whiles the hydrogen temperature, i.e. the purity, is a feed-forward variable that can not be controlled by the DMC controller.

The propadiene reactor outlet concentration is the controlled variable.

The linear dynamic models that relate the two independent variables, one manipulated (C_3/H_2 ratio) and other feed-forward (hydrogen temperature), with the dependent or controlled variable (propadiene outlet concentration) are presented in the next figure:

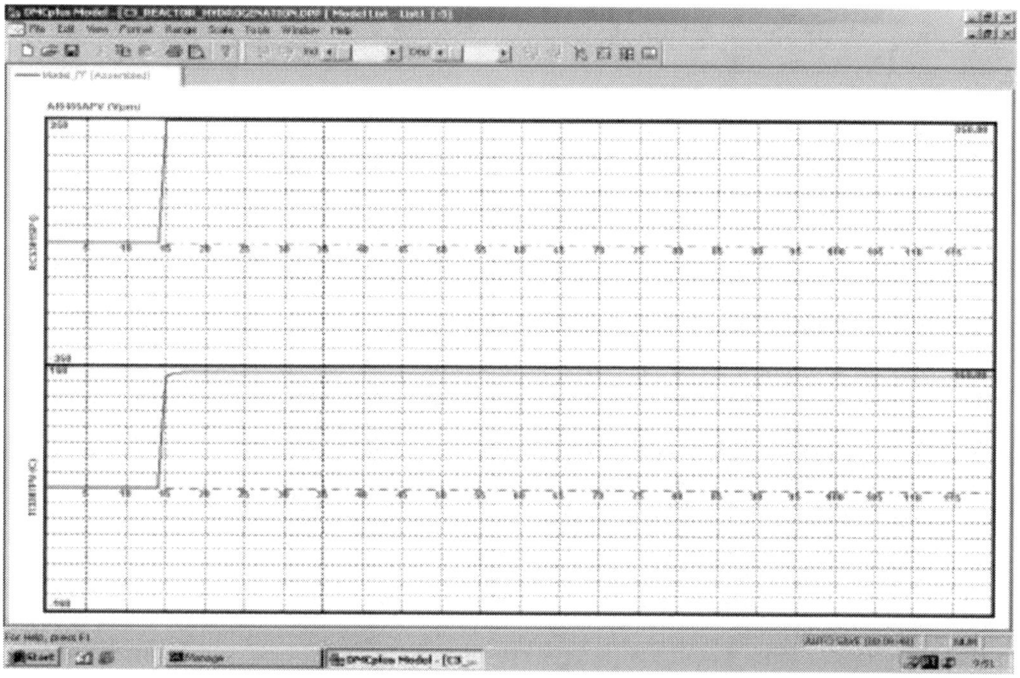

Figure 4 – Linear Dynamic Models

It is possible to conclude that for an increase of a unit in the C_3/H_2 ratio the propadiene outlet concentration increases 350 vpm with a delay of 14 minutes, i.e., increasing the C_3 stream flow and maintaining the H_2 stream flow, the propadiene conversion to propylene will be lesser. The delay in the response is due to the fact that the propadiene analyser is not situated in the reactors outlet (to avoid sampling contamination), it is located on the top of the polymers washing column.

The hydrogen purity (about 80%) has a great impact in the hydrogenation reactions, being therefore necessary to use the temperature as feed-forward variable.

2. Controller Configuration

On this phase it was used the DMCplus Build software to configure the controller parameters using the validated dynamic models.

3. Controller Simulation

The controller configuration was tested and validated using the DMCplus Simulate software, allowing simulating the controller in different operating process conditions, and thus, to optimize the parameters previously configured.

Results

In the figure 5 it is possible to visualize the propadiene outlet during concentration July 2003, when there was no process control optimization.

It is important to notice that during several days the propadiene concentration was zero and the C_3/H_2 ratio was constant, since the control was done by the room control operator. In this case is easy to conclude that the reactors were operating with over hydrogenation, converting propylene to propane. On this period the catalyst selectivity was about 45%.

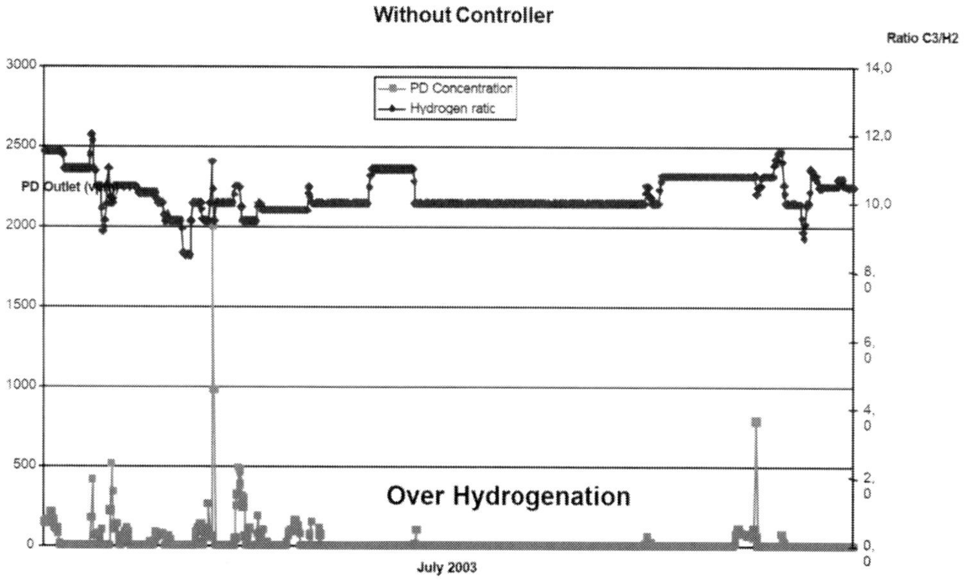

Figure 5 – C_3/H_2 ration and propadiene outlet concentration in July 2003, without controller

In the next figure 6 it is possible to see the propadiene outlet concentration and the C_3/H_2 ratio in July 2004, when the APC controller was already commissioned and in operation.

The propadiene outlet concentration is stable and within the set point value (250 vpm), while the C_3/H_2 ratio was manipulated by the controller. In this situation there was no over hydrogenation, that was confirm by doing laboratory analysis to the gas composition at the reactors inlet and outlet. The catalyst selectivity increased to about 90%.

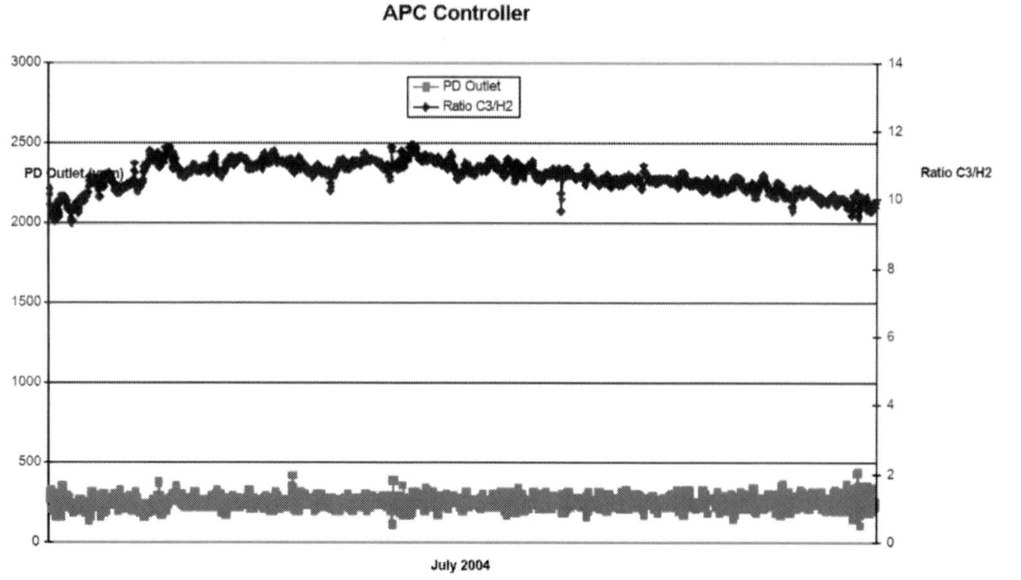

Figure 6 – C_3/H_2 ration and propadiene outlet concentration in July 2003, with APC controller

The increase on the selectivity has an immediate repercussion in the propylene production, and being the propylene price 25% higher than the propane, the increase on the operation profit is very significant.

The reactors optimization allowed doubling the selectivity value, from 45 to 90%. As consequence of this alteration is the decrease, about 650%, of the propane quantity produced by the hydrogenation reaction, as showed in figure 7.

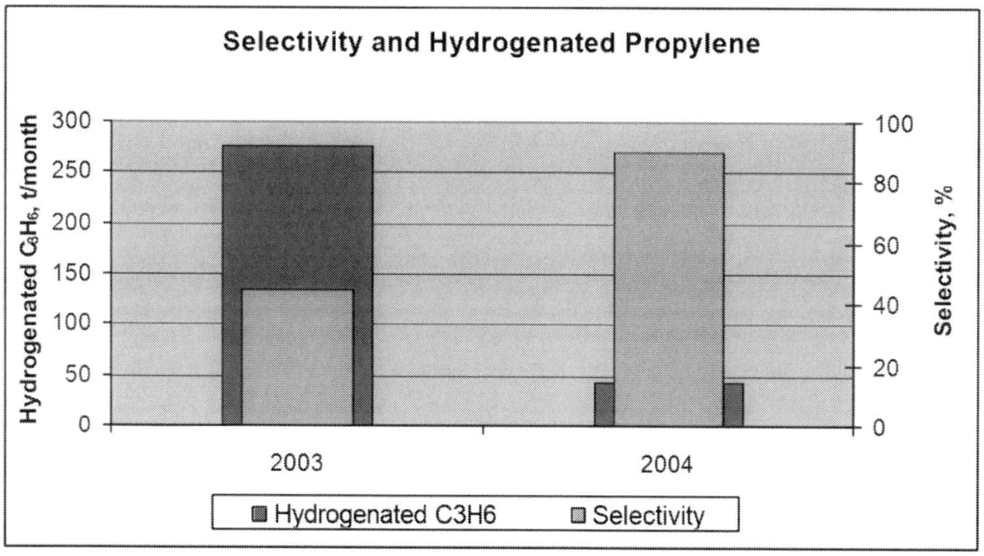

Figure 7 – Selectivity and propane production decrease, by propylene hydrogenation.

Final Considerations

It was successfully developed and implemented an advanced controller that enables these hydrogenation reactors optimization, minimizing propylene to propane hydrogenation. Outlet propadiene concentration is controlled by manipulation of the C_3/H_2 ratio, responsible for the hydrogen flow rate adjustment to the system, maintaining the process value around the set point.

The major advantage is to be unnecessary to operate in the limit in order to fulfil the specification, without over hydrogenation. The controller's response time and robustness allows an immediate action in the new steady state lookup, in case of an upset. The selectivity increased significantly and the hydrogenated propylene decreased around 650%.

It is done regularly a selectivity follow-up and the temperature differentials adjustment in the reactors as the feed to system changes with the plant feedstock, with direct influence in the reaction exotermicity.

This innovation, a teamwork result, actually allows a better operation and reactional system comprehension. There were recovered around 16 t/day of propylene, which is an extra profit to Repsol of approximately 1.27 M€ per year.

Bibliography

1. Cabanes, A.L., Cano, G.V., Gomes, J.C.F., Analisis Termodinâmico de la hidrogenacion de la fraccion C_3 de "Steam-Cracking", Ingenieria Química, Maio 1990

2. Mohundro, E.L., Overview on C_2 and C_3 Selective Hydrogenation in Ethylene Plants, 15th Ethylene Producers Conference, Abril 2003

3. Seborg, D.E., Edgar, T.F., Mellichamp, D.A., Process Dynamics and Control, John Wiley & Sons

4. Selective hydrogenation of Methyl acetylene + Propadiene in Steam Cracked C_3 cut, Institute Français du Pétrole

5. Smith, J.M., Chemical Engineering Kinetics, 3rd Edition, McGraw-Hill International Editions

6. Training Manual, Introduction to Multivariable Predictive Control with DMCPlus, AspenTech, 2002

7. Westerterp, K.R., Swaaij, W.P.M., Beenackers, A.A.C.M., Chemical Reactor Design and Operation, John Wiley & Sons

Paper TA006d

Supporting Key Console Operator Interactions through the Control System Interface

Jamie Errington
Human Centered Solutions

Tim DeMaere[1], Eric Wade
NOVA Chemicals

Prepared for Presentation at the 2005 Spring National Meeting, Atlanta GA, Apr 10-14
17th Annual Ethylene Producers' Conference
Session TA006 – Ethylene Plant Process Control

[1] Mr. DeMaere is currently a partner in Human Centered Solutions, LLP

Abstract

In today's world of rapidly developing control systems, supporting the key interactions between the console operator and their control system is a difficult task that continues to grow as more process data from the field and more information from outside the control system becomes available. The control system interface forms a core part of a decision support system that can aid the operators in preventing or minimizing the impact of unplanned plant disturbances and improving plant performance.

This paper will discuss how plant data and external information have been combined in a single interface to meet the key console operator interaction requirements. The resulting interface includes integrated trending, yoked multi-windows, and simultaneous increasingly detailed views of process equipment.

A recent case study that compared the performance of console operators using this interface to operators who used a traditional control system interface is reviewed.

Introduction

In 1996, NOVA Chemicals Corporation (NOVA Chemicals) launched a major project in Alberta to expand its Joffre production facility. At the time, Joffre was the home of two ethylene plants, a polyethylene plant and a large utilities unit. In total, the site produced over three billion pounds of product each year. The expansion project was to add new ethylene, polyethylene and cogeneration plants as well as expand the utilities plant to support the expanded facilities. When completed in the summer of 2000, the new facilities would increase the site production to over 6 billion pounds per year making it one of the largest ethylene / polyethylene sites in the world.

At this time, NOVA Chemicals was a very active member of the Abnormal Situation Management Consortium® (ASM)[2]. The ASM Consortium was developing a novel approach to operator interface design for its Abnormal Event and Guidance Information System (AEGIS) prototype. The ASM Consortium member companies were funding the AEGIS prototype as a multi-year, multi-million dollar project with additional support from NIST (National Institute of Standards and Technology). The goal of the AEGIS project was to demonstrate that with current technologies, a significant improvement in petrochemical plant performance and reliability could be achieved.

The ASM interface approach[3] used a structured multi-window format to improve the console operator's awareness to changing plant situations so that abnormal situations were identified early and averted or mitigated. The structured format included explicit use of colour, navigational strategies, windows management, and coordinated display associations.

With NOVA Chemicals' knowledge of the AEGIS work, NOVA Chemicals applied and further enhanced the approach that had been pioneered by the Consortium and moved it from the lab to the field to work with a plant control system.

NOVA's ASM style of operator interface was in place for all of the new Joffre plant startups in mid-2000. Operator acceptance of the new interface design approach was wide ranging. Some operators welcomed the new approaches while others were more reluctant during the project. Extensive training on dynamic simulators in the major units prior to the plant startup significantly aided the adoption of the new approaches. Following the plant startups, operators reported that they found many of the elements of the new interface invaluable and would not go back to the older more traditional interface designs.

---

[2] ®ASM and Abnormal Situation Management are U.S. registered trademarks of Honeywell International, Inc.

[3] Bullemer, P. T. (1998) *User interface design framework: Abnormal Situation Management concepts and features*. Invited paper at Honeywell IAC User Group. Phoenix, AZ.

Bullemer, P. T., Cochran, E., Harp, S., & Miller, C. (1999). *Collaborative decision support for operations personnel.* Paper presented at the INTERKAMMA ISA Technical Conference, Düsseldorf, Germany.

As a follow-up case study[4] in 2004, NOVA Chemicals and the ASM Consortium wanted to quantify the response performance difference of operators of similar experience levels due to the use of either the ASM or a Traditional style. A set of simulated plant malfunctions were developed that could be run on two of the ethylene plant simulators where one of the simulators used the new ASM style of interface while the other used a more Traditional style. Two of the key features of the ASM style of interface that differentiate it from the Traditional style are the number of simultaneous views available and the direct integration of process trends.

This paper will describe the design approach of the ASM style of interface and present a summary of the case study.

Interface Design Elements

The objective of NOVA Chemicals' ASM style of operator interface is to improve the operator's situational awareness of changing plant conditions so that abnormal events can be avoided or mitigated. To achieve this objective the operator interface must address how the operators interact with the control system and it must support their decision-making.

Through the ASM research, it was determined that a good operator interface design needed to integrate process alarms, process and trend information cohesively together. To create a new interface approach, the strengths and weaknesses of the earlier approaches (such as large panel boards, strip chart recorders, alarm annunciators, off-normal indicators and former DCS display formats) were used in conjunction with an understanding of human factors.

The earlier approaches showed that people worked best with a fixed structured format that emphasized key physical and process interrelationships. The ability to see strip charts or trends of key variables immediately alongside a controller was also very useful in making decisions on what should be changed. Pulling controllers out of the panel to indicate that they were in an "off-normal" mode helped people remember this during the shift and at shift-change when they needed to relay experiences to the next crew. The location of the alarm annunciators, showing the critical but very limited number of plant alarms, in association with the corresponding section of the plant gave people an immediate localization of where a problem was occurring. These were all examples of approaches that were taken from the pre-control system interface formats.

The human factors aspect of the interface design approach is best described by a definition of human factors: "Human factors discovers and applies information about human behavior, abilities, limitations, and other characteristics to the design of tools, machines, systems, tasks, jobs, and environments for productive, safe, comfortable and effective human use."[5] An interface design therefore should enhance people's attention and perception of the plant.

---

[4] Errington, J. et al. Establishing human performance improvements and economic benefits for a human-centered operator interface: An industrial evaluation., (2005). Paper to appear in *Proceedings of the Human Factors and Ergonomics Society 49th Annual Meeting*, Orlando, FL, 26-30 September.

[5] Sanders, M.S, and McCormick, E.J., "Human Factors in Engineering and Design", (1993) McGraw-Hill, New York

From the historical practices and the human factors appreciation, the major design principles of the NOVA ASM style of interface included:

- Support the operator's scope of work
  - Situational awareness
  - Routine operations
  - Mode changes
- Take advantage of human and computer strengths
- Present information through the use of data in context
- Maximize focus on task at hand
  - Minimize distractions such as desktop and windows management

Problems that can be experienced by operators using a poorly designed operator interface can be wide ranging. When trying to locate the source of a plant upset, operators can miss important information, be overloaded with too much information or be distracted by an aspect in the interface. When evaluating a problem to understand what is occurring, operators can often not get the information that they need immediately from the interface. For example access to trend information is often several awkward steps away in most interfaces. When the operator is ready to take action, the poor interface is often prone to entry errors, may provide limited feedback to the operator, or require additional actions to make the changes needed. After making a corrective action, an operator will need to assess if the correction is having the desired impact, and will need a "big picture" of the plant so that they can see the broad impacts of the changes and not get focused on a very small section.

The major design elements of the NOVA ASM style of interface included:

- Single, Integrated View
- Window Management and Layout
- Navigation Scheme
- Visual Coding Scheme – use of colour & shapes
- Interaction Objects
- Contextual Menus
- Task View Organization

The NOVA Chemicals' ASM style of interface is illustrated in Figure 1. The interface uses a set of eight - 21" screens. The screens are operated from four dual-headed workstations. The four screens across the top of the console show an area-wide overview summarizing the status of the console operator's scope of responsibility. It is made up of 4 different displays. In the top left corner a Type 1 graphic shows a Status View. Beside that is an Alarm Summary with two related trends. Further to the right are two screens showing key area variables in either a standard trend format or a newer specialized event detection application format.

The operator interactions are typically performed though one pair of lower screens in Figure 1 labeled as Operations View. Each pair is identical and allows for a second operator to support the console during upset or planned high workload periods.

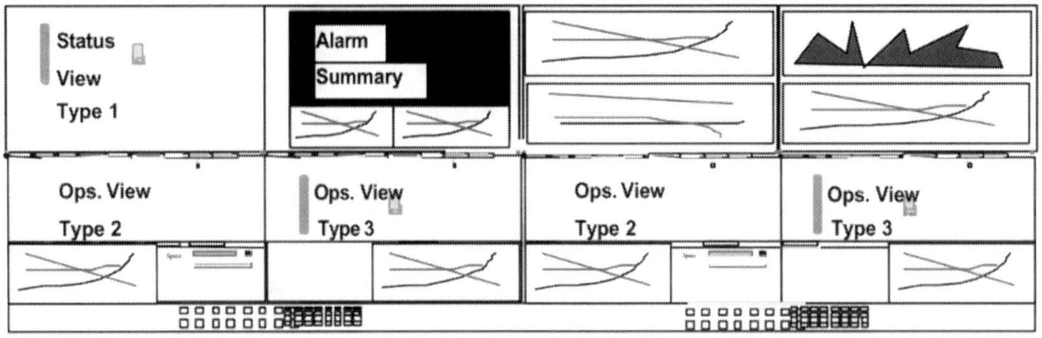

Figure 1: NOVA Chemicals' ASM Style Interface

The Operations View is comprised of four levels of displays with two fixed trends as shown in more detail in Figure 2. The Type 2 display (upper left) provides a high-level view of an individual process unit. The Type 3 display shows a portion of that unit in higher detail. Below the Type 3 display is a region used to provide detailed control loop information and a trend object. Below the Type 2 display there is another trend plus a detail view showing the selected point.

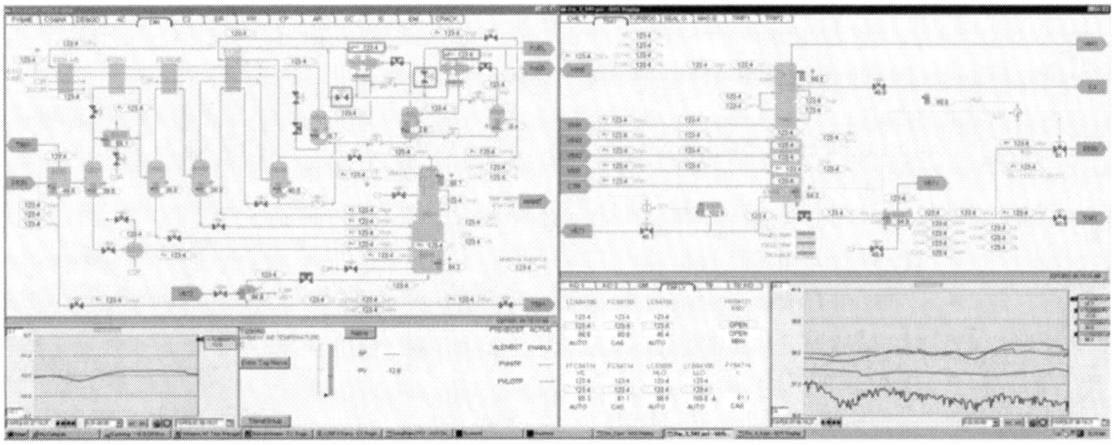

Figure 2: NOVA Chemicals' ASM Style Operator View

Key features of the Operations View include

- On-screen navigation via display tabs with integrated alarm information
- Restricted use of colour, reserving colour to highlight off normal conditions
- Windows management minimizing display overlays
- Automated display invocation through pre-configured display associations (i.e., assisted, task-relevant navigation)
- Access to online information
- Integration of trend information

The ASM style is intended to provide a continuous broad overview of process conditions while simultaneously giving access to the needed detail information to reduce the potential for "tunnel visioning" when operators are solving complex process problems.

Interaction Requirements

"The design of an operator interface that enables efficient and effective job performance requires the anticipation of the operators' needs in understanding the plant's process interactions, their interactions with computers and with other individuals in the workplace."[6] Often an operator interface is described as merely presenting information to people however this description is about a one-way communication flow. The operator is an active and integral element to good plant performance and their working with the operator interface is a two-way process. A well designed operator interface will provide for not only an effective display of process information but will also meet the operator's interaction needs that are critical to job performance.

Aspects of the NOVA ASM style interface explicitly addressed the interaction requirements of the console operator in working with the control system. The key element of this portion of the design was the integration of trend information, the use of context sensitive menus and the development of the detail display.

Figure 3 shows a picture of a detail view in the NOVA ASM style of interface for a regulatory controller. Its key features with respect to meeting operator interaction requirements are items such as:

- Highlighted dialog boxes
- Highlighted operator entry points (white background)
- Control valve status message box (STSMSG)
- Instrument maintenance status (REDTAG)
- Alarm limits displayed with philosophy matching highlights
- Access to the setpoint target value (SPTV) and ramp controls (RAMPTIME, TVPROC)
- Access to an alarm acknowledge button and alarm summary controls
- Access to group trend control menu (TrendControl)
- Off-normal indicators (magenta)
- Ramp boxes (displayed on selection of setpoint or output)

[6] Bullemer, P., Reising D.V., Hajdukiewicz, J., Errington, J, *Interaction Requirements Methods for Effective Operator Interfaces*, Abnormal Situation Management Consortium internal publication, (2004)

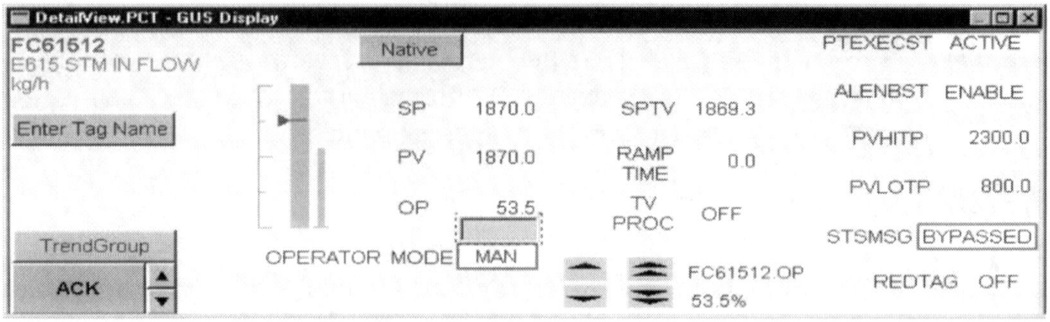

Figure 3: NOVA ASM Style Detail View

Anytime an operator selects an object on any of the displays, it is immediately placed in the Detail View (Figure 3). As part of this action, trend information about this tag is also immediately created in a detail trend (Figure 4) adjacent to the Detail View. The trend used in the interface integrated both historical and real-time data about each point to give the operator a strip-chart like component to see how the tag had changed and project the tags trajectory into the future.

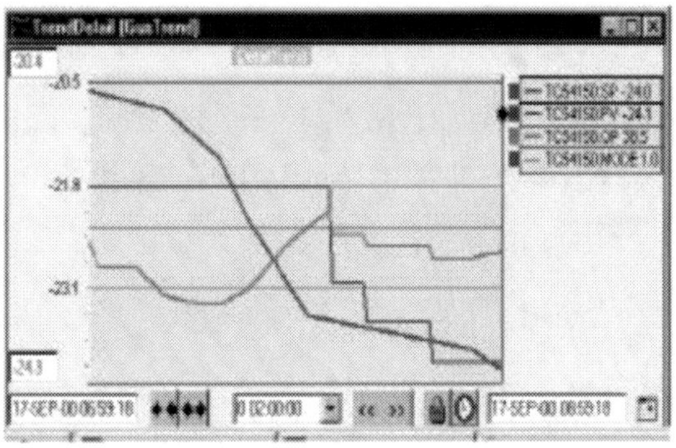

Figure 4: NOVA ASM Style Detail Trend

The operator has a wide range of interactions that they can do with the trend in terms of changing time scales, scrolling back in time, tag ranges, changing from a trend to a tabular format, and adding scooter markers so that they can get the information that they need. For regulatory control tags, the process value, setpoint, output and controller mode were all simultaneously trended. For other tag types, the traces are configurable to any available parameters.

Online Support Information

To further support the operator's interactions, three additional types of information are provided and examples are listed in the following table. Point specific information is provided through context sensitive menus for the tag that the operator is focusing on.

Table 1: Online Support Information

Single DCS Point (Context Sensitive Search)	DCS System/Console Alarm Information	Desktop Tools Access
• Alarm Objective Analysis (AOA) Information • Alarm /Operator Change History • APC Cross Reference • Document/Procedure/Logic Search	• Disabled/Inhibited Alarm Summary • Alarm System Metrics (Bad Actors, KPI's)	• SAP Access for Maintenance Requests • Electronic Operator Logbook • Desktop – Internet, Intranet, MS Office Tools

Figure 5 shows an example of a context sensitive menu that has been used to recall alarm configuration information. The menu is displayed from a "right-mouse" click on any dynamic object in the displays. What the menu offers is both a direct access to additional information about the point in question for the operator or control of the group trend information. If a group trend change is requested by the operator, then that addition or deletion of a trend element is made immediately on the station's group trend.

If the operator requires additional information it is presented in a separate window either on an adjacent workstation display or on the screen that they are working from. The choice of display location is made so that on primary workstations, the additional information does not cover or obscure any critical process information that might be needed by the operator.

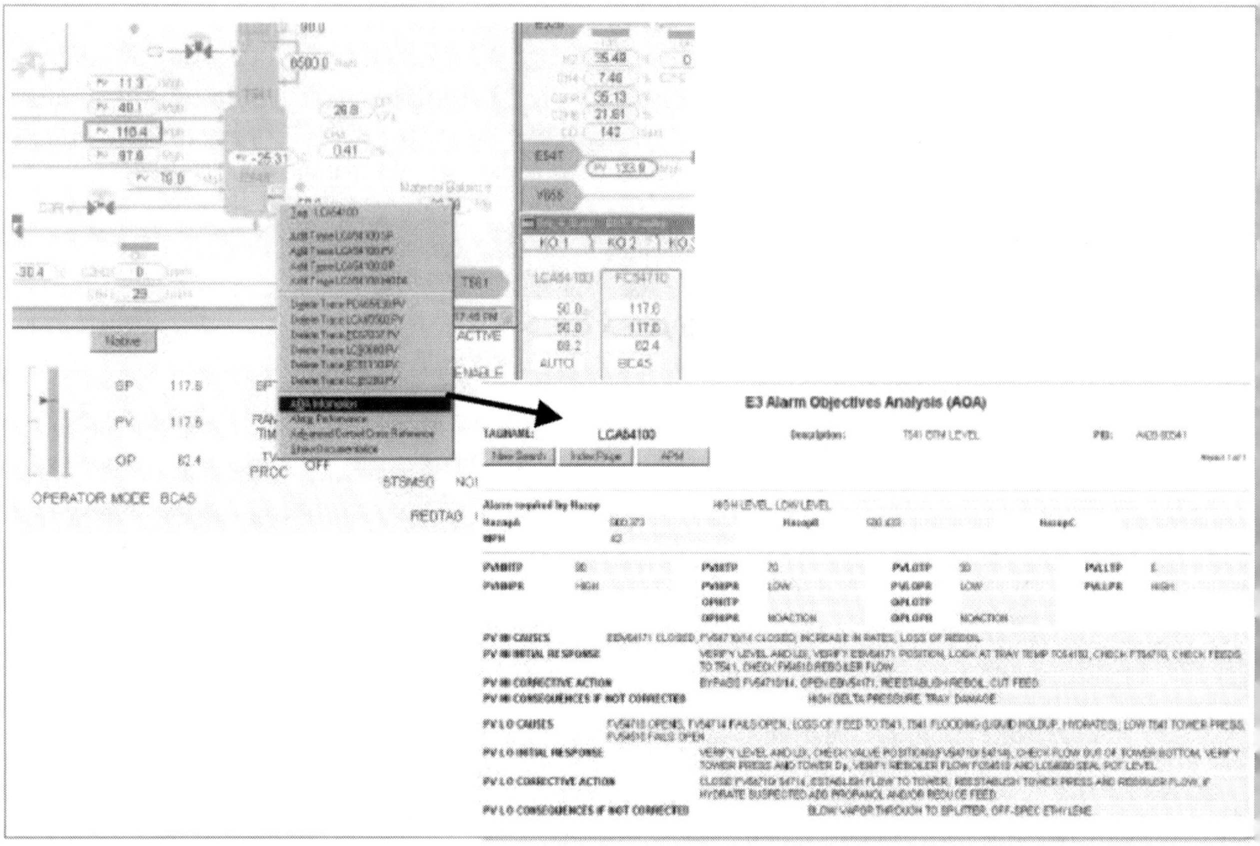

Figure 5: Context Sensitive Menus

Providing a user interface to support operations with the procedural step information enables a more consistent execution of routine procedures such as a furnace swing. Process control tuning and monitoring can best be setup when the procedure is executed the same way each time. Figure 6 details an example of a furnace swing application that utilizes the integrated displays and trending to allow operations to monitor this task while executing the swing application. In this case additional trending displays are invoked for monitoring purposes.

Figure 7 details an example of a semi-automated compressor trip application to expedite execution of immediate response items while flagging those steps still incomplete. Access to all emergency procedures is via a single click through this interface.

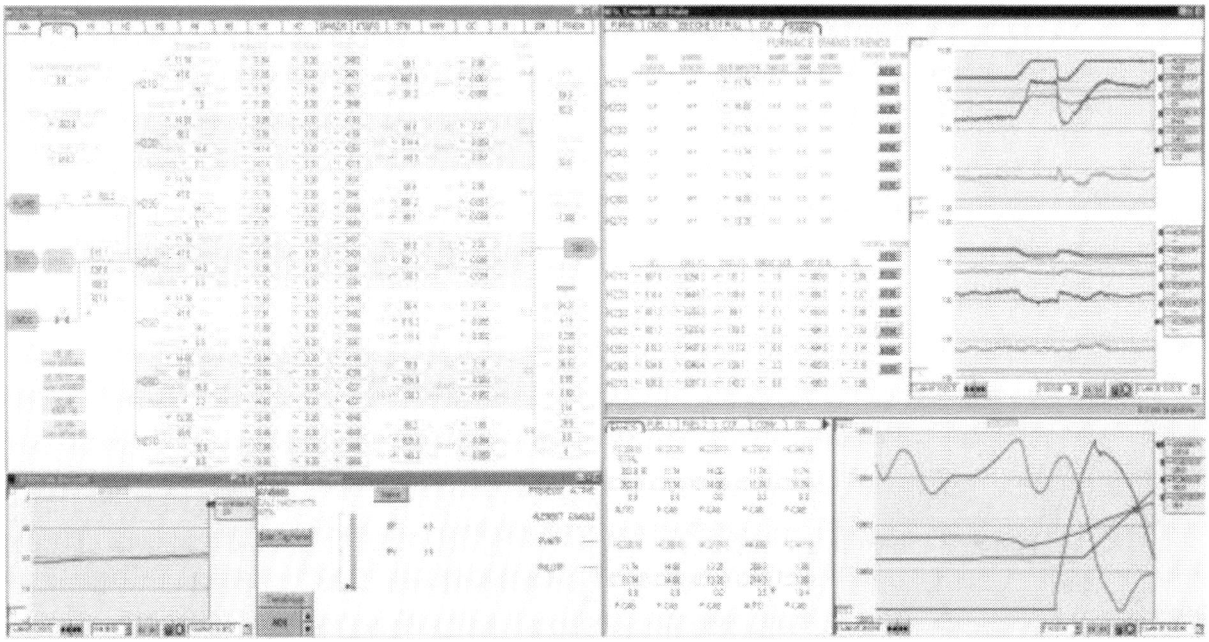

Figure 6: Furnace Swing Procedure Support

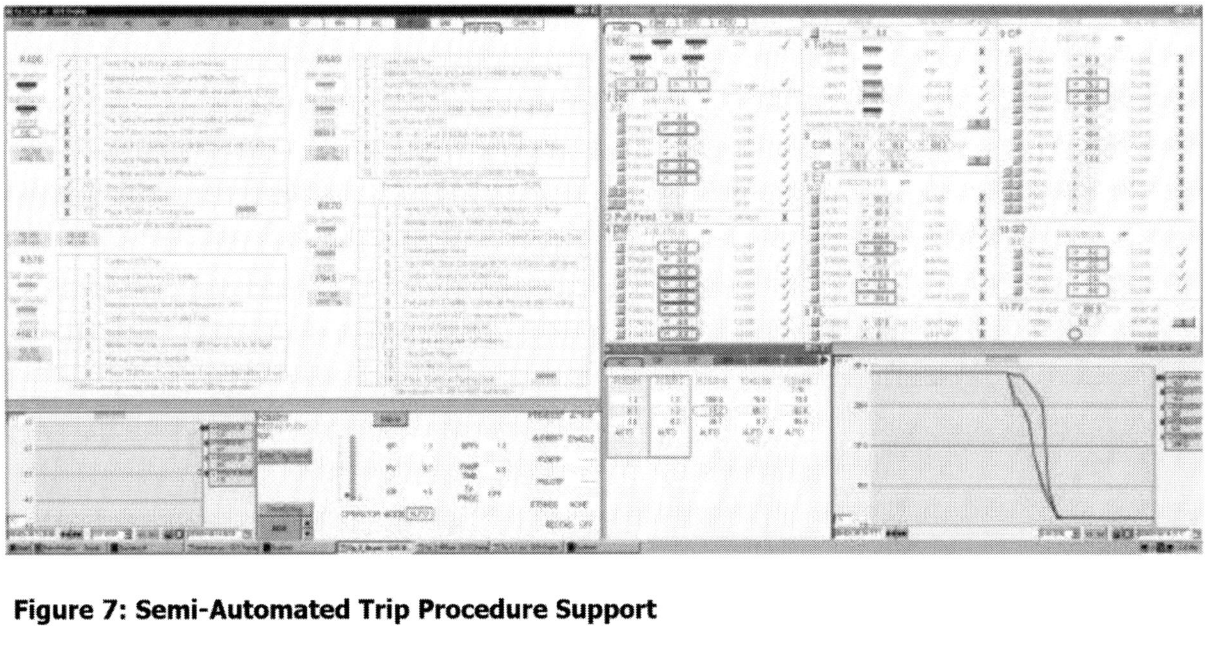

Figure 7: Semi-Automated Trip Procedure Support

Case Study Background

The objective of this case study was to evaluate an ASM style of operator interface used in one unit at the NOVA Chemicals Joffre site against a more traditional style of interface used in another unit at the same site. The key features of the ASM style of interface that differentiate it from the Traditional style are the number of simultaneous views available and the direct integration of process trends.

The project was part of a larger overall project to understand the business case values of designing for ASM. The ASM Consortium and NOVA Chemicals funded this interface case study portion of the project.

The interface case study compared the response performance of two groups of operators using either the ASM style of interface described earlier in this paper, or a more Traditional style of control system interface. Each group of operators worked on matching problems using two of NOVA's high fidelity plant simulators. The two simulators were configured to match each group's respective plant and operator interface. The time for each operator to orient to the problem and the total time to solve the problems was recorded. A third element, identified as early event detection was also recorded during the trials and this captured the point that the operators first identified an upset.

Other components of the study included tests of plant knowledge, individual experience records and other information so that a comparison of the two groups could be examined to see if differences other than the interface design were impacting any observed upset performance differences.

The NOVA Joffre site has three ethylene plants operating on it. The first two ethylene plants were commissioned in 1979 and 1984 while the third plant went into production in 2000. The original two plants have gone through several control system changes over the years but both are currently using Honeywell Total Plant control systems. The operator interfaces in these plants are single window interfaces developed on the proprietary Honeywell workstations. An example of a typical traditional interface display is shown in Figure 8.

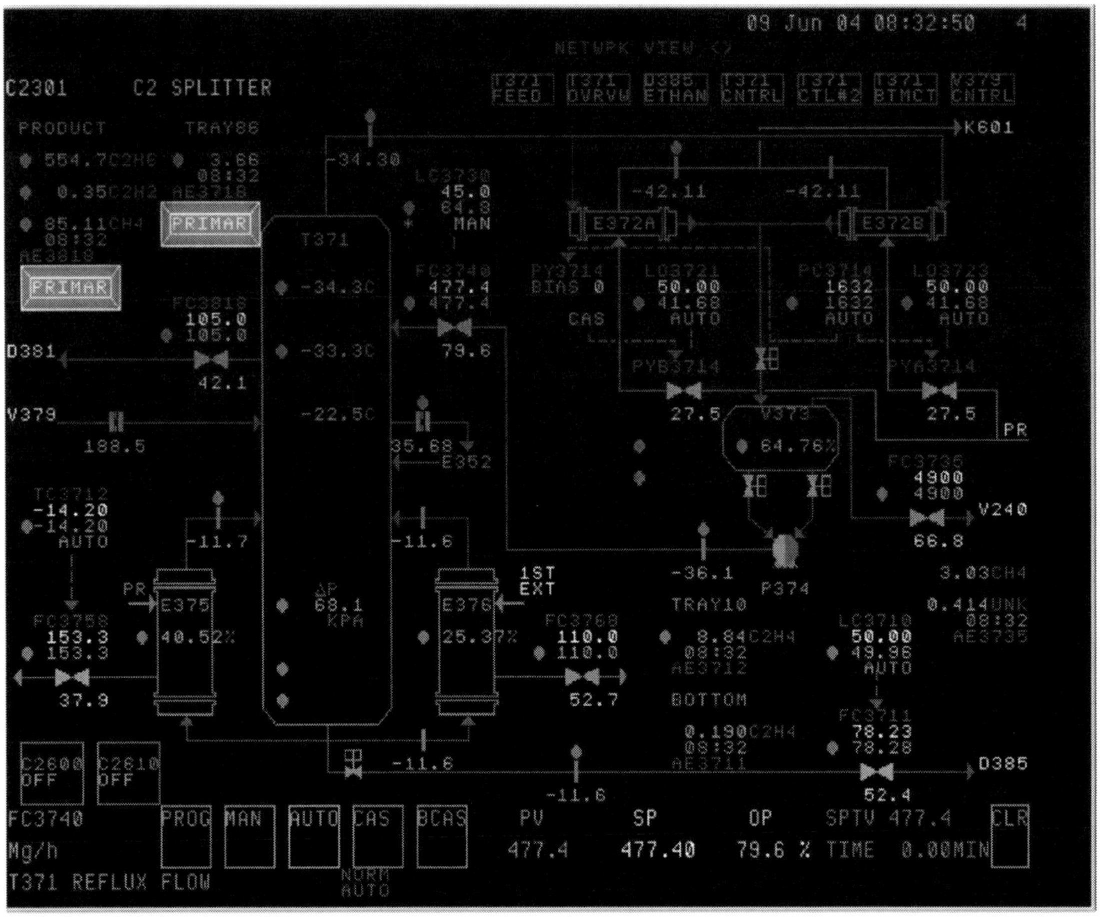

Figure 8: Example of Traditional Operator Interface Display

Navigational aids are in boxes across the top of the display and links to other areas of the plant are accessible from the text boxes at the ends of the process lines. Process variables are generally shown as a group of text where the tag name is in half-intensity green, setpoints are white, process values are cyan and valve outputs are yellow. Control valves are also colour coded to show the mode of the controller (yellow is manual, green is auto and blue is cascade). Motor-operated valves (MOVs) are displayed with the valve positioners on top of the controllers. Values in alarm are in red along with the sensor object that will flash until the alarm has been acknowledged by the operating technician. Process lines are colour-coded to distinguish between utility (half-intensity white) and hydrocarbon lines (half-intensity green). The pieces of equipment, such as the compressor and steam turbine in the centre of the display, are outlined in the display with half-intensity white lines. The half-intensity dashed lines represent the instrument connections from the controllers to the control valves.

When an operator selects a tag on the display to make changes or to get additional information (i.e. historical or trend), a horizontal change zone is displayed at the bottom of the display highlighting that tag with the required access points (setpoint, output or mode).

To test the hypothesis that the ASM style interface provided an advantage in diagnosing and responding to abnormal situations, a between-subjects experiment (ASM vs. Traditional) was designed that consisted of three phases: pre-test, scenario performance evaluation, and post-test.

High fidelity simulators are used extensively in the Joffre ethylene plants for operator training, refresher and qualification testing platform for all of the console operators. These platforms became the ideal testing ground for the scenario performance portion of the case study. The training coordinators in the two ethylene units developed a series of matching plant malfunctions. The operators participating in the study were asked to monitor the plant and respond to any upsets that occurred. Operator actions such as display changes, alarm responses, trend requests and setpoint changes were recorded during each scenario. In all, the participants' performance was examined across 4 different plant scenarios.

Case Study Results

One of the key study results was the total time to solve the simulated problems. Total time data was analyzed in a Repeated Measures Analysis of Variance with Interface type (ASM vs. Traditional) as the between-subjects factor, and Scenario as the within-subjects (repeated) measure. Using an 80% confidence level[7], the results of the Repeated Measures analysis of variance indicated a significant difference for Interface Type, $F(1,19) = 11.689$, $p < .001$ only, and no significant effect of Scenario or Scenario by Interface Type interaction. In general, operators using the ASM interface spent significantly less total time solving each scenario than did operators using the Traditional interface.

In addition, an examination of the results shown in Figure 9 shows not only did the operators require less time, but also that the variability for the operators using the ASM interface was much smaller than those of the operators using the Traditional interface. A difference between scenarios was expected, as no attempts were made to ensure overall equality between scenarios, and the fact that the interaction between Scenario and Interface Type was not significant indicates that any differences in Scenario had no effect on the time performance of either operator group.

---

[7] Wickens, C.D. (1998, Oct.), Commonsense Statistics. *Ergonomics in Design*, 6(4), pp.19-22

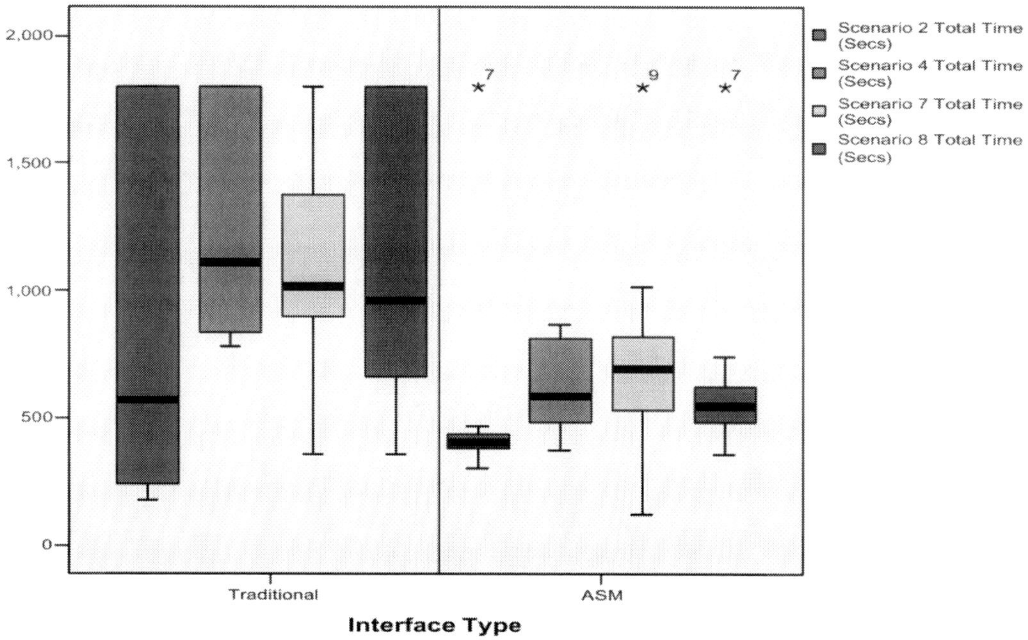

Figure 9: Total Time to Solve Scenarios (Traditional vs. ASM Interface) – Box Plot

In summary, the case study results (Table 2) showed that the operators using the ASM multi-level, multi-view interface rather than a Traditional style of interface:

- Recognized problems faster and more consistently
- Responded and solved problems faster and more consistently
  - 6.5 – 9.7 min improvement (35 – 48 %) on four scenarios of 15 – 20 minute durations
- Were more likely to successfully solve the problems
  - 25 % higher success rate for participants using the ASM interface versus Traditional interface

Table 2: Case Study Key Results Summary

	Traditional	ASM
Scenario Times to Complete - Mean (min)	18.1	10.6
Scenario Completion (%)	70.0	95.5
Early Event Detection (%)	10.0	47.7

Overall, the ASM style interface has demonstrated in this case study to be superior to the Traditional interface in terms of its impact on operator performance in handling abnormal situations.

Advanced Process Control Computer-based Operator Training

Edward Chang
Equistar Chemicals, a Lyondell Company
Corpus Christi, TX

Chandra Gannavarapu
Aspen Technology, Inc.
Houston, TX

Prepared for Presentation at the 2005 Spring National Meeting, Atlanta GA, Apr 10-14
17th Annual Ethylene Producers' Conference
Session TA006 - Ethylene Plant Process Control

Abstract

The success of any advanced process control (APC) application is highly dependent on the operator's understanding of the application. A high on-stream factor does not necessarily translate to a high return in benefits. Proper operator training and application knowledge play key roles in APC sustained performance and associated benefits. APC training can also be considered an important requirement for board operator qualification mandated by the Occupational Safety and Health Administration (OSHA). At the Lyondell's Corpus Christi Complex in Corpus Christi, Texas, a project was initiated to implement Aspen Technology's Computer Based Operator Support Training (CBOST™) modules on the olefins unit. This paper discusses project justification, scope, issues and results.

Lyondell's Corpus Christi Complex completed an advanced process control (APC) and real-time optimization (RTOpt) project on the olefins unit in 2003. During project commissioning, operator training was performed via traditional methods – presentations, user manuals and testing. It was always expected that operator training would be an on-going and continuous process after commissioning – to ensure high controller on-stream time, effective use of the controllers, address operator turnover and handle controller changes and enhancements.

To provide a key component of this post commissioning training program, Aspen Technology, Inc. was contracted to develop a series of computer-based training (CBT) modules in 2004. These CBT modules cover all the APC controllers in use at the plant, and to a lesser extent, the real-time optimizer. Although the CBTs are still in the rollout phase, the response from the operators and the in-house training department has been very positive.

Background. Lyondell's Corpus Christi Complex was originally constructed in 1980. One of the main units of the plant is the olefins unit, which produces ethylene as its primary product. The nominal ethylene production rate is 1.7 billion pounds per year.

Lyondell's Equistar Chemicals business initiated a project to install APC and RTOpt on the olefins unit in 1999. Advanced controls in the form of DMCplus™ controllers were commissioned on the cracking furnaces, quench tower, cracked gas compressors, cold fractionation towers, hot fractionation towers, acetylene reactors and refrigeration system. A composite linear program (CLP) was also commissioned to maximize unit feed rate. The project was successfully completed in 2003 and the project has achieved its expected benefits.

Advanced process control was new to the plant. Like many other sites that have commissioned advanced controls, the plant used a combination of presentations, documents and hands-on training to initially train the board operators. Plant personnel and the consulting engineers who commissioned the advanced controls performed this training jointly. Follow-up training has generally been on an as needed basis and has consisted of additional documentation, question and answer sessions, simulations and training on the board on a one-on-one or small group basis by plant personnel (controls engineers and specialists).

A more structured training program sought. Although the initial APC training served to get the controllers up and running, follow-up training has been a necessity to ensure high controller on-stream time and effective use of the controllers. Other important reasons for follow-up training include:

- Notification of controller changes
- Regulatory (OSHA) requirements for training programs
- The plant's internal operator certification/qualification program (for both new and current board operators)
- A need for consistent use of the controllers (shift-to-shift), leading to consistent operation of the process unit
- Minimization of the risk of process upsets due to the incorrect use of the controls

The shortcomings of the plant's existing training program include:

- Time constraints on the controls engineers and specialists to prepare and present information
- Availability of the controls engineer and specialists (not on site 24/7 and not always available while on site)
- Occasional conflicting information coming from different training personnel
- No formal testing
- No regular review
- Operators did not always seek out controls personnel to get their questions answered (sometimes relied on other operators, which may or may not have yielded the correct information)
- Many operators expressed a desire for additional training
- No consistent system to track training progress

Due to the shortcomings of the existing training program, improvements to the existing system were sought. CBTs were identified as a potential tool to address some of the training issues and improve the current system.

CBT Justification and Benefits. While no formal benefits analysis was performed to justify the CBT project, the plant felt that there were sufficient intangible benefits to justify the project. APC CBTs do not address all training issues and requirements, but do provide benefits such as:

- Consistent training material and information
- Availability 24/7
- Troubleshooting capability. Operators can interactively search a database of questions and answers.
- Frees up process control engineers and specialist for other activities
- Provides training material for process engineers and process specialists
- Retains in-house knowledge of APC applications
- Provides a means to track training progress

CBT Scope. The scope of the CBTs includes general APC modules as well as unit specific APC modules. Since the general APC modules are not unit specific, they are applicable to all operators that interface with APC. The general APC modules include:

- DMCplus General Information
- Frequently Asked Questions (FAQ)
- APC Glossary

Plant operation is divided into five areas, consolidated into one main control room. Three of these operating areas contain APC applications - furnaces, cold fractionation and hot fractionation. Unit specific modules were developed for each of these three operating areas to ensure that all the operators that interface with APC had relevant CBTs to train on. Unit specific modules were developed for:

- Furnaces (2 furnace types - 1 feed type each)
- Quench tower
- Cracked gas compression and cold fractionation (4 towers and 2 refrigeration systems)
- Acetylene reactors
- Hot fractionation (4 towers)
- Composite linear program (feed maximizer)

In order to maximize training effectiveness and CBT utilization, the unit specific modules are customized for the plant and utilize the plant's distributed control system (DCS) operating displays. The CBTs explain details of each DMCplus controller application in simple and effective ways. This accelerates the learning process and reduces time required for the operators to become familiar with the controllers. Lessons within each unit specific module include:

- Process overviews
- Controller (or sub-controller if applicable) objectives
- Process response relationships
- Controller variables prioritization
- Common operating scenarios (normal and abnormal)
- Review questions
- Final exam

Although there is no specific RTOpt CBT, some basic RTOpt interface information is included in the CBTs to explain how the RTOpt application works and how it interacts with the DMCplus controllers.

The CBT modules reside on a network server and the modules can be accessed by any workstations connected to the network. The software used to build the CBTs is Authorware™, a product from Macromedia, Inc. The plant did not need to purchase a license for Authorware since the plant only receives the executable files. The CBTs are maintained as off-line modules and are in no way connected to the DCS, thus there is no live data exchange.

Sample Screenshots. Figures 1 through 7 are screenshots from the CBTs. The DMCplus operating screens are actual DCS screenshots incorporated into the CBTs.

Figure 1 is a screenshot from the DMCplus General Information CBT module. It describes how a DMCplus controller uses process models to improve process control. The information in this General Information module is not specific to a particular operating unit or controller and is applicable to all operators. This module serves to provide a foundation for understanding how multivariable controllers work and why they are an improvement over traditional regulatory controls.

In Figures 2(a) and 2(b), the CBT demonstrates one of the recommended operator actions prior to turning a DMCplus controller on – the review of the manipulated variables. The required steps are listed in blue as steps 1 through 4. As the user passes the cursor over each one of these 4 steps, the particular item is highlighted in yellow on the DCS operating display. In Figure 2(a), the highlighted step is Step 1. In Figure 2(b), the highlighted step is Step 3.

The CBTs are highly interactive and include simulations to mimic the actions and responses when operating a DMCplus controller. The CBT user performs tasks on a DCS operating display and the simulation mimics the response that the operator would normally see on the DCS console. Figures 3(a) through 3(d) simulate the switching of a manipulated variable to a feed-forward/disturbance variable. The user is required to perform all the steps necessary to perform this task. This includes selecting the variable service switch (Figure 3(a)), selecting the feed-forward option (Figure 3(b)) and pressing the enter key (Figure 3(c)). Figure 3(d) shows the end result.

Figure 4 is a screenshot from a unit specific module. It shows a simplified process flow diagram and the variables included in the DMCplus controller. This is an interactive screen in that as the user passes the cursor over each controlled variable (CV), manipulated variables (MVs) that have a DMCplus model relationship with the selected CV are highlighted. In Figure 4, the cursor is placed over a CV, which becomes highlighted by a blue box. Immediately, yellow circles highlight all MVs that have a model to the selected CV. An equivalent lesson page is also available for MVs, in which the user passes the cursor over MVs and the CVs affected by the selected MVs are highlighted.

Figure 5 shows a typical review question. Review questions are comprised of a mix of multiple choice and true/false questions. Some of these review questions incorporate operating graphics. Figure 5(a) is a screenshot of a question and Figure 5(b) is the resulting screenshot when the question is answered correctly. If the user does not select the correct answer, the answer is provided for immediate feedback. A random sample of review questions is chosen for the final exam. The user does not receive a score (in the form of percent correct) in either the review or the final exam. However, in the final exam, the user does receive a pass/fail grade, which is logged in a file that can be accessed by the CBT administrator.

A glossary of APC terms is available to the CBT user at any time during a lesson. A call-up button at the bottom of the CBTs allows the user to access the glossary without having to exit the module. The glossary is in layman terms to facilitate the understanding of complex concepts. Figure 6 shows a typical glossary entry.

Figures 7(a) and 7(b) show the frequently asked questions (FAQ). Figure 7(a) is the opening page and Figure 7(b) is the resulting screenshot when one of the FAQ is selected. The CBT administrator has the capability to add entries to this module.

Project schedule and issues. The project to develop the APC CBTs was kicked off in February 2004 and completed in December 2004. The project turned out to take longer than originally anticipated. The primary reason for this was plant resource constraints, specifically process control engineer and specialist time. The amount of review required was extensive and underestimated. In all, the CBTs contained approximately 3000 pages, all of which required review. In addition, detailed controller information had to be verified for accuracy. A total of three review passes were required. Plant resources were also limited by a planned facility turn-around, which resulted in a suspension of CBT development activities by plant personnel prior to, during and immediately after the turn-around.

Operator feedback. The amount of operator feedback is very limited due to the timing of the initial CBT rollout and the deadline of this paper. The initial operator response is very favorable. The plant's in-house training department, which handles the majority of operator training, was not directly involved in the development of the modules, but was kept abreast of project progress. The training department is impressed with the results.

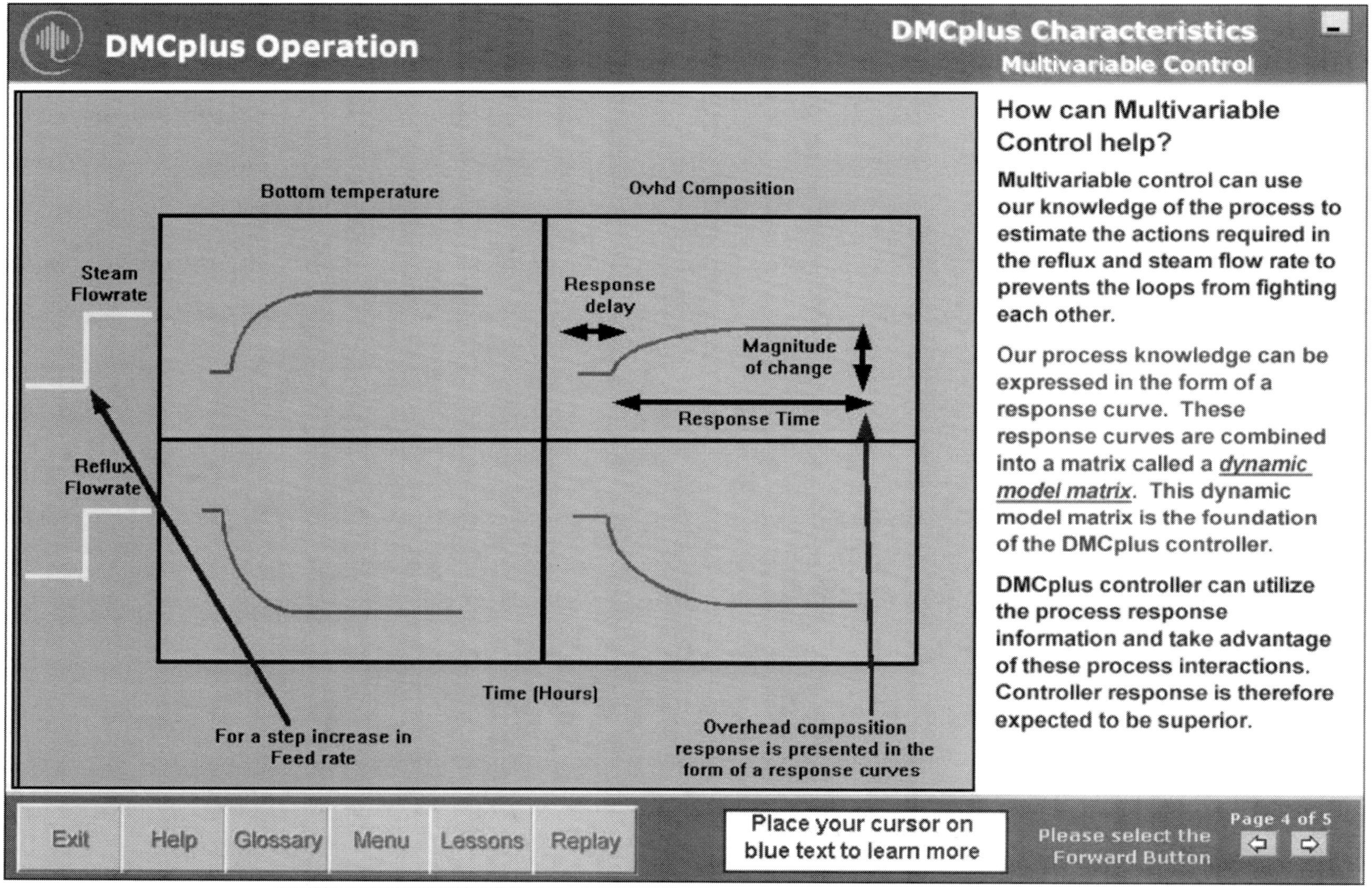

Figure 1: General DMCplus™ Concepts

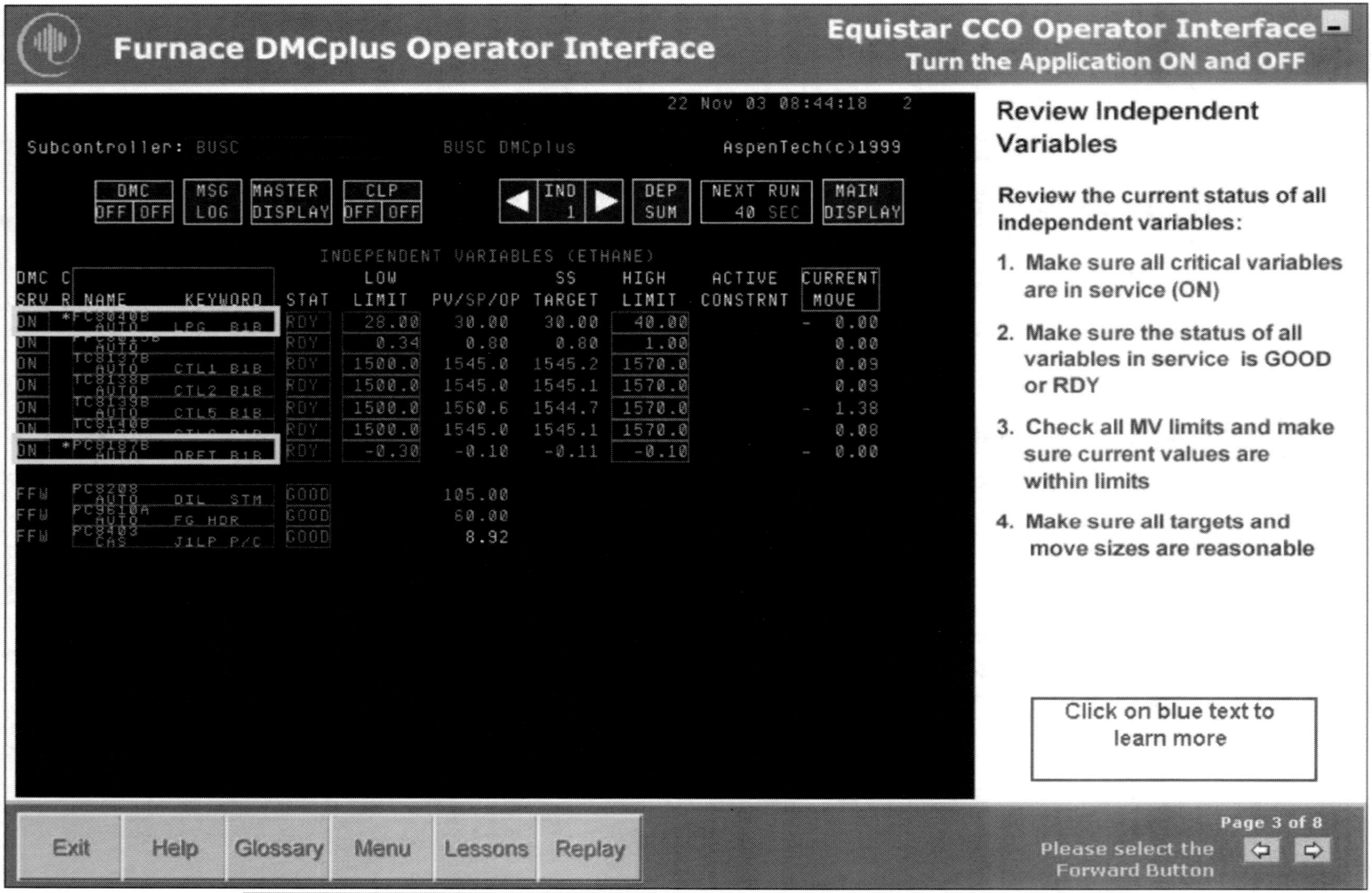

Figure 2(a): Steps in Turning a Controller ON
Step 1 Highlighted

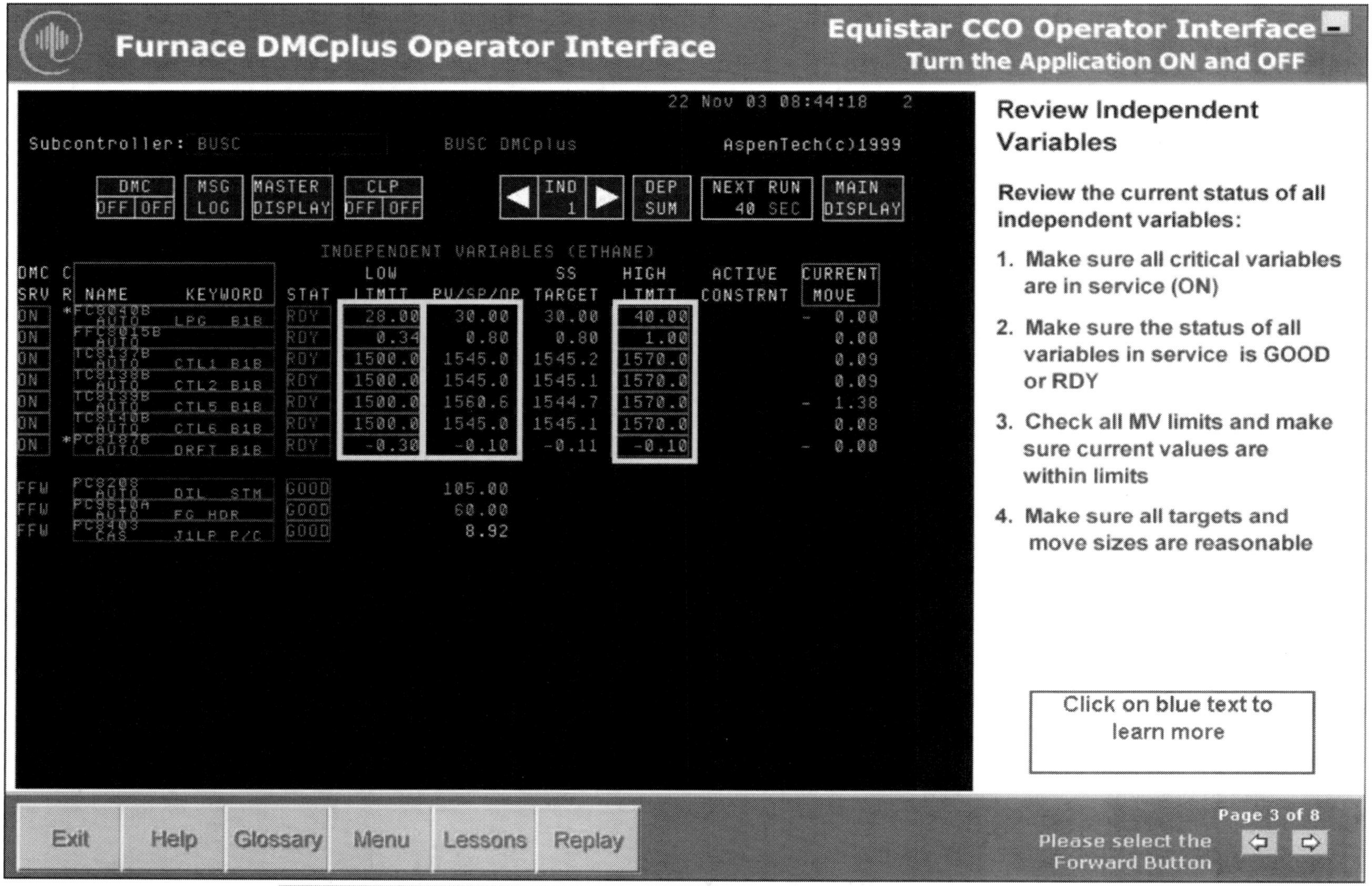

<u>Figure 2(b)</u>: **Steps in Turning a Controller ON**
Step 3 Highlighted

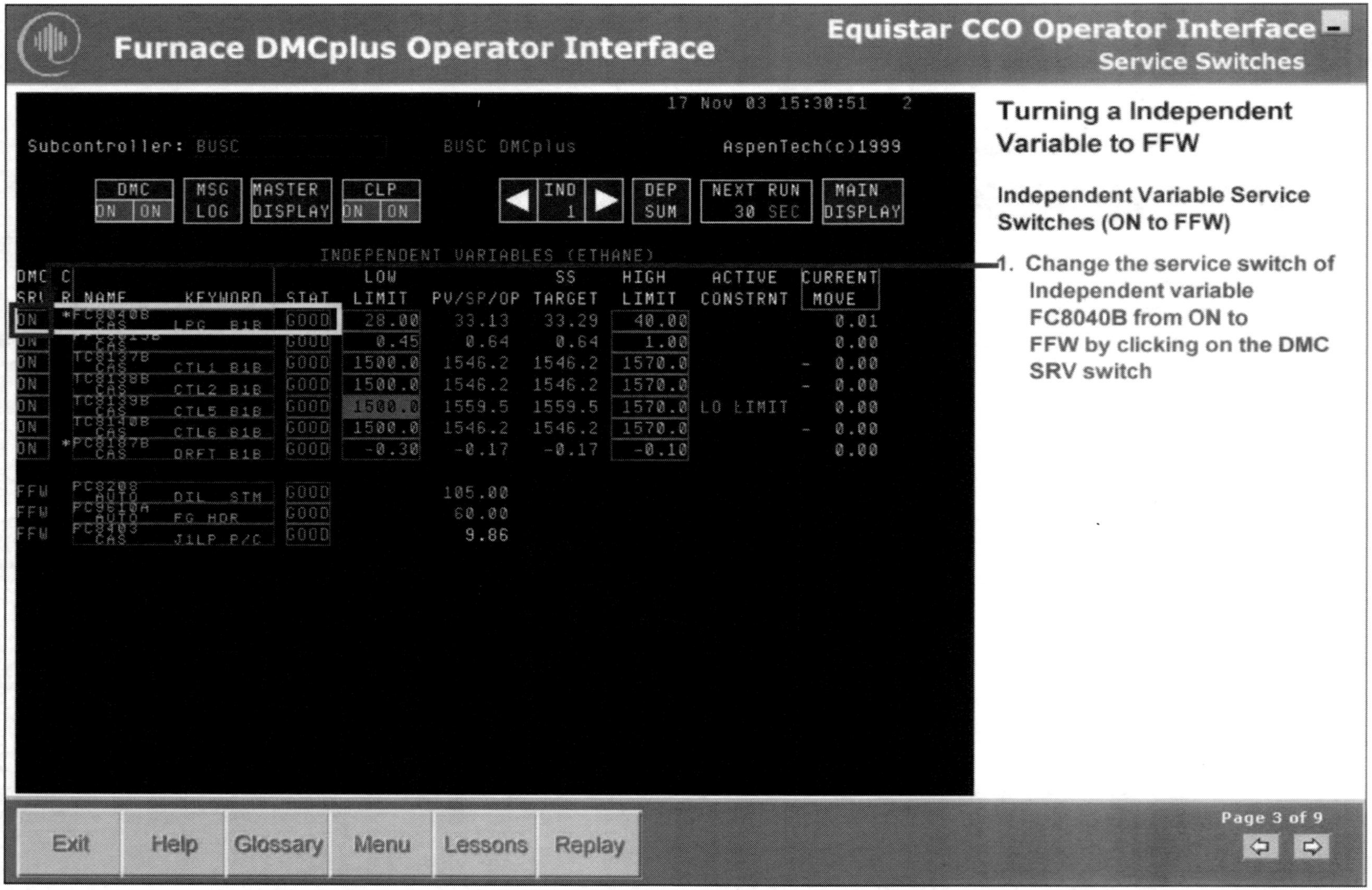

Figure 3(a): Changing an MV to a DV
Interactive Step 1

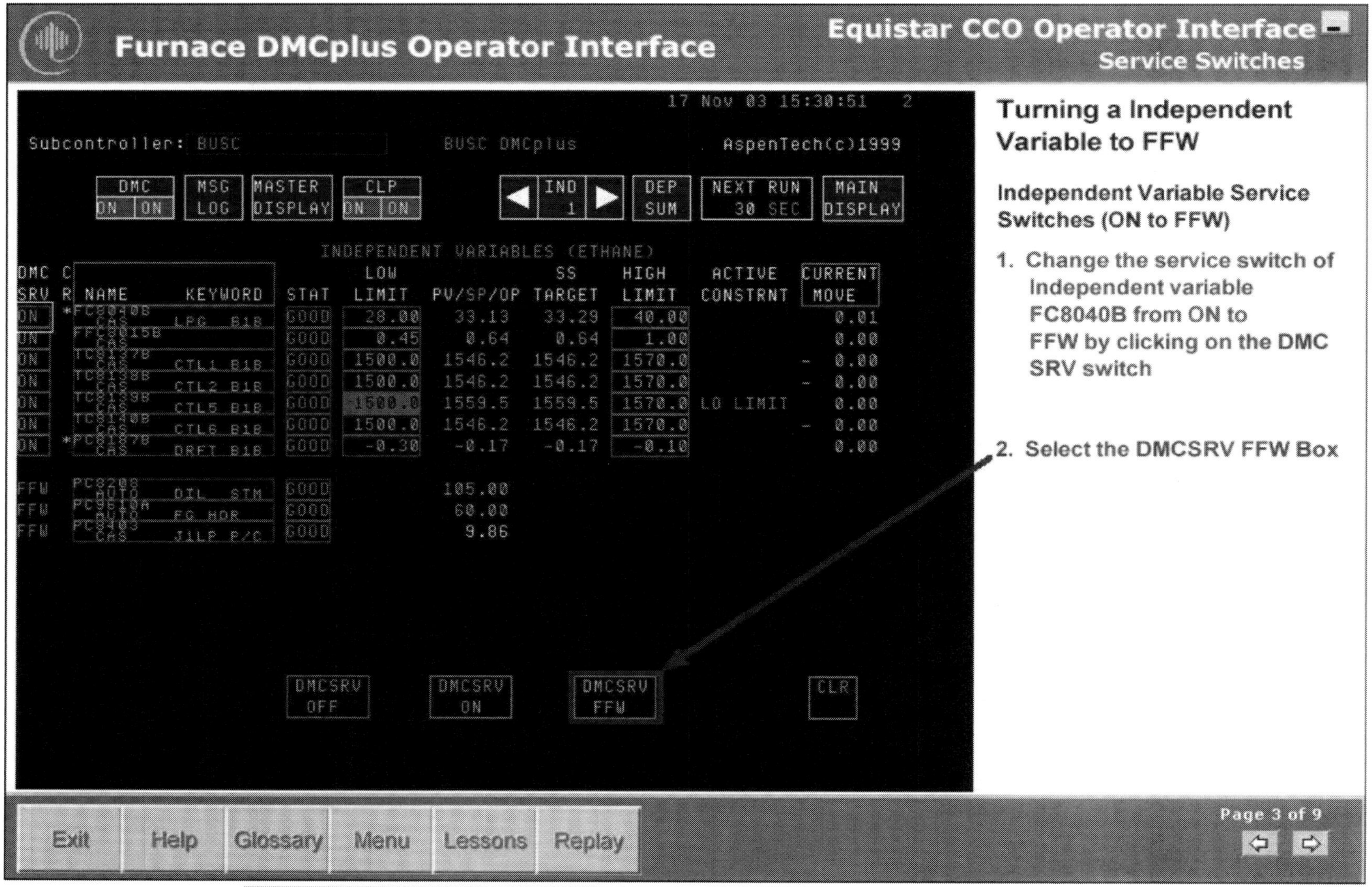

Figure 3(b): Changing an MV to a DV
Interactive Step 2

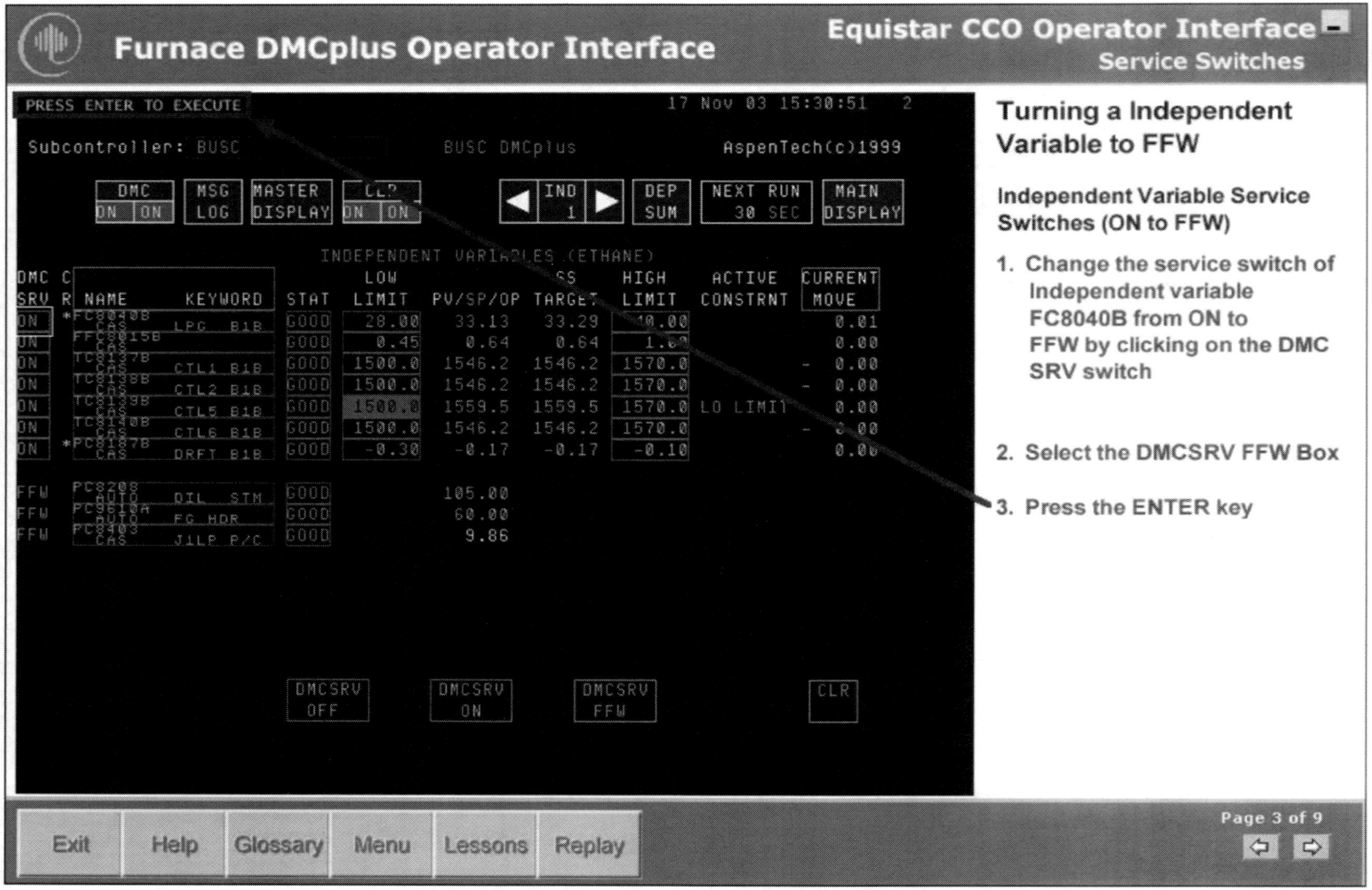

Figure 3(c): Changing an MV to a DV Interactive Step 3

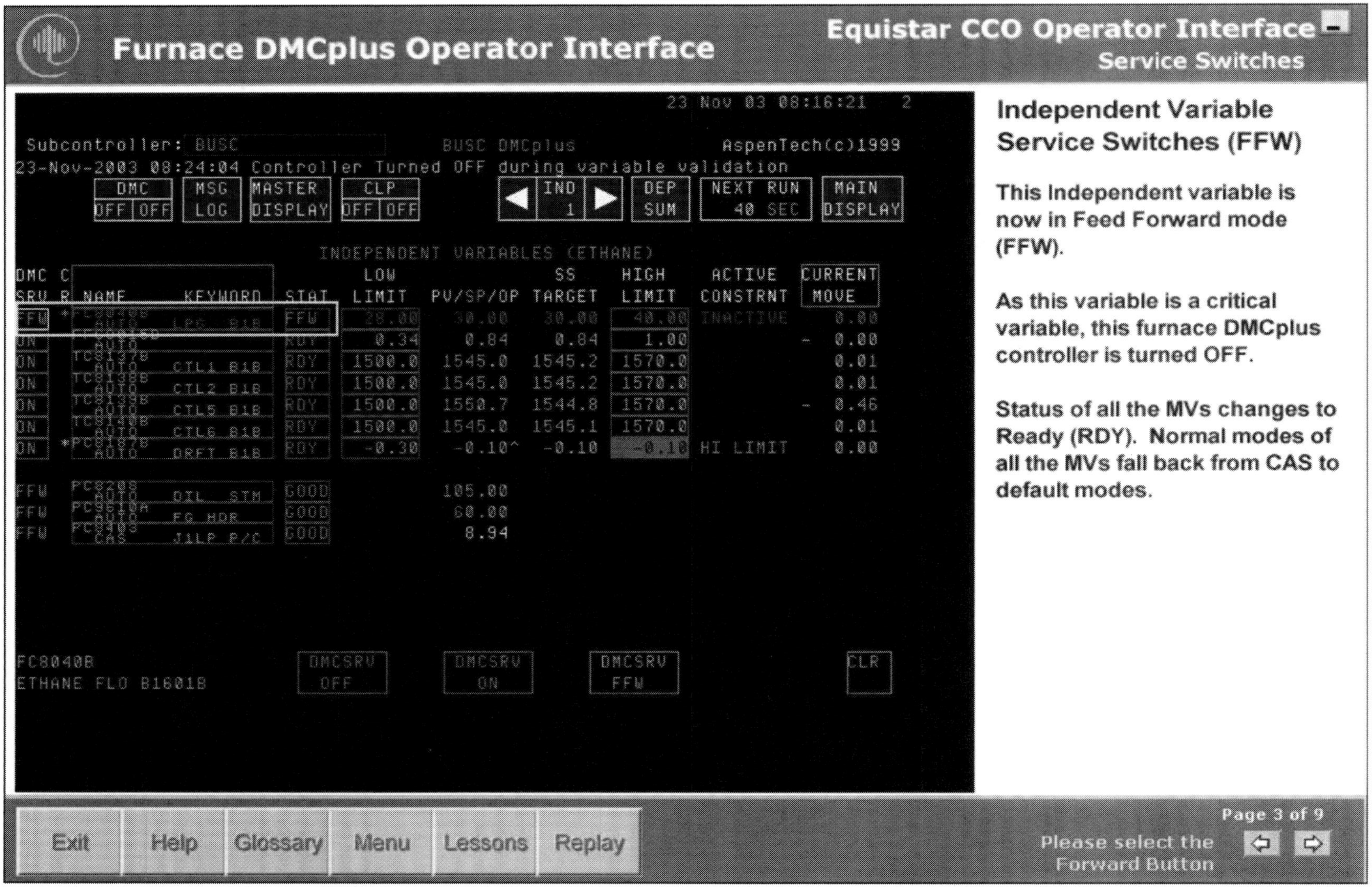

Figure 3(d): Changing an MV to a DV
Interactive Step 4

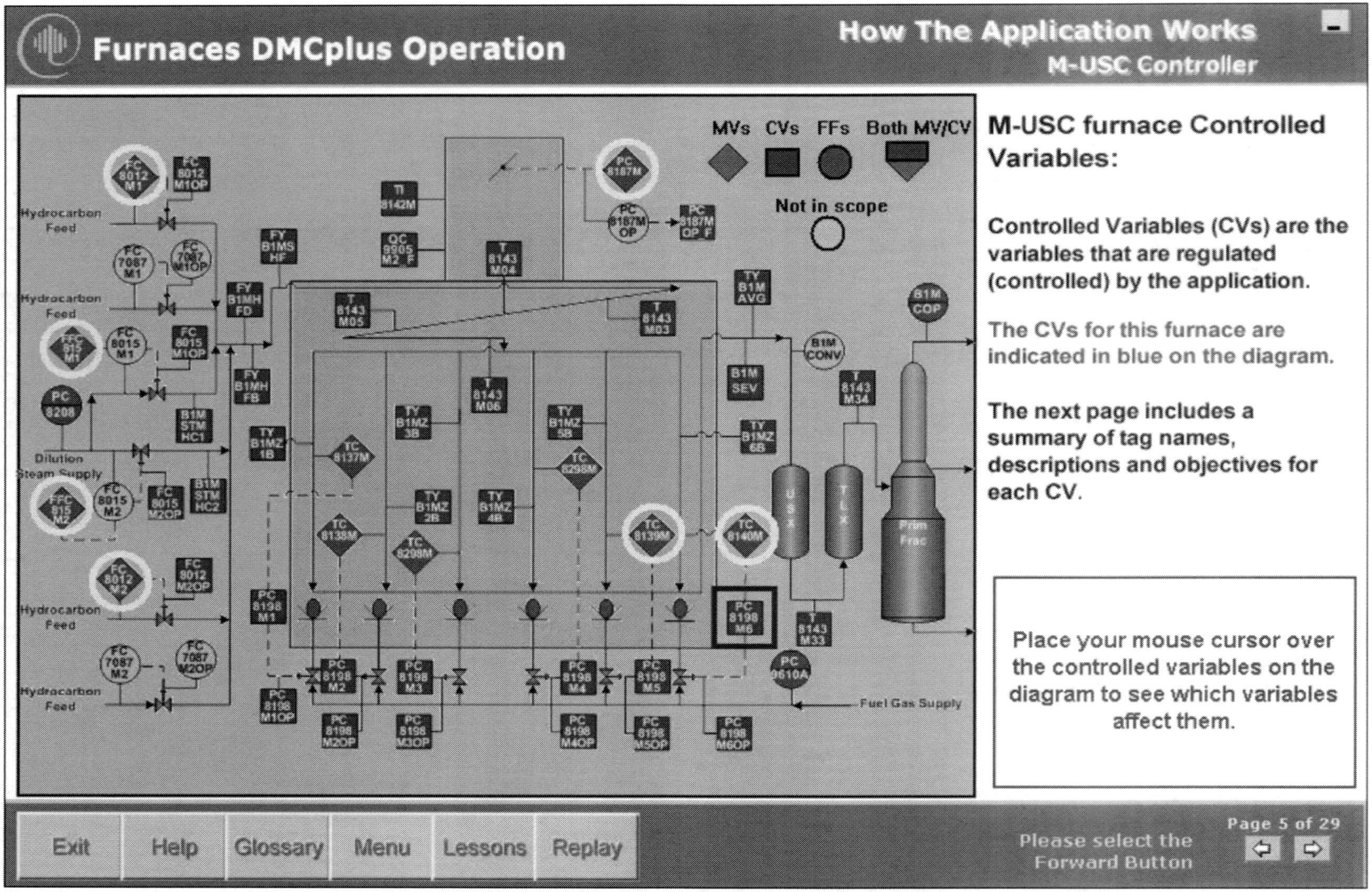

Figure 4: Process Flow Diagram with APC Variables
MVs (Yellow Circles) that Affect Selected CV (Blue Box)

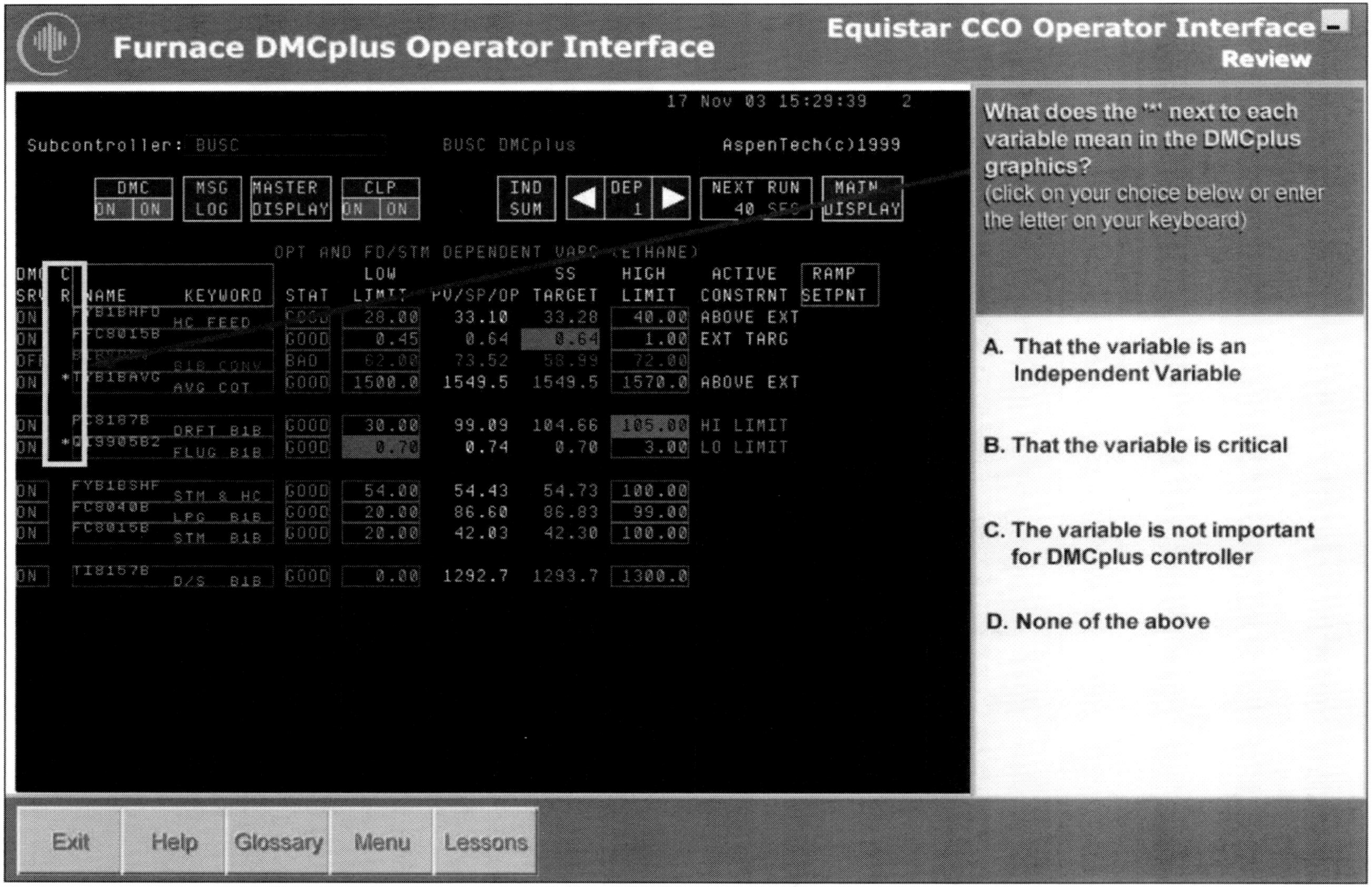

Figure 5(a): Typical Review Question

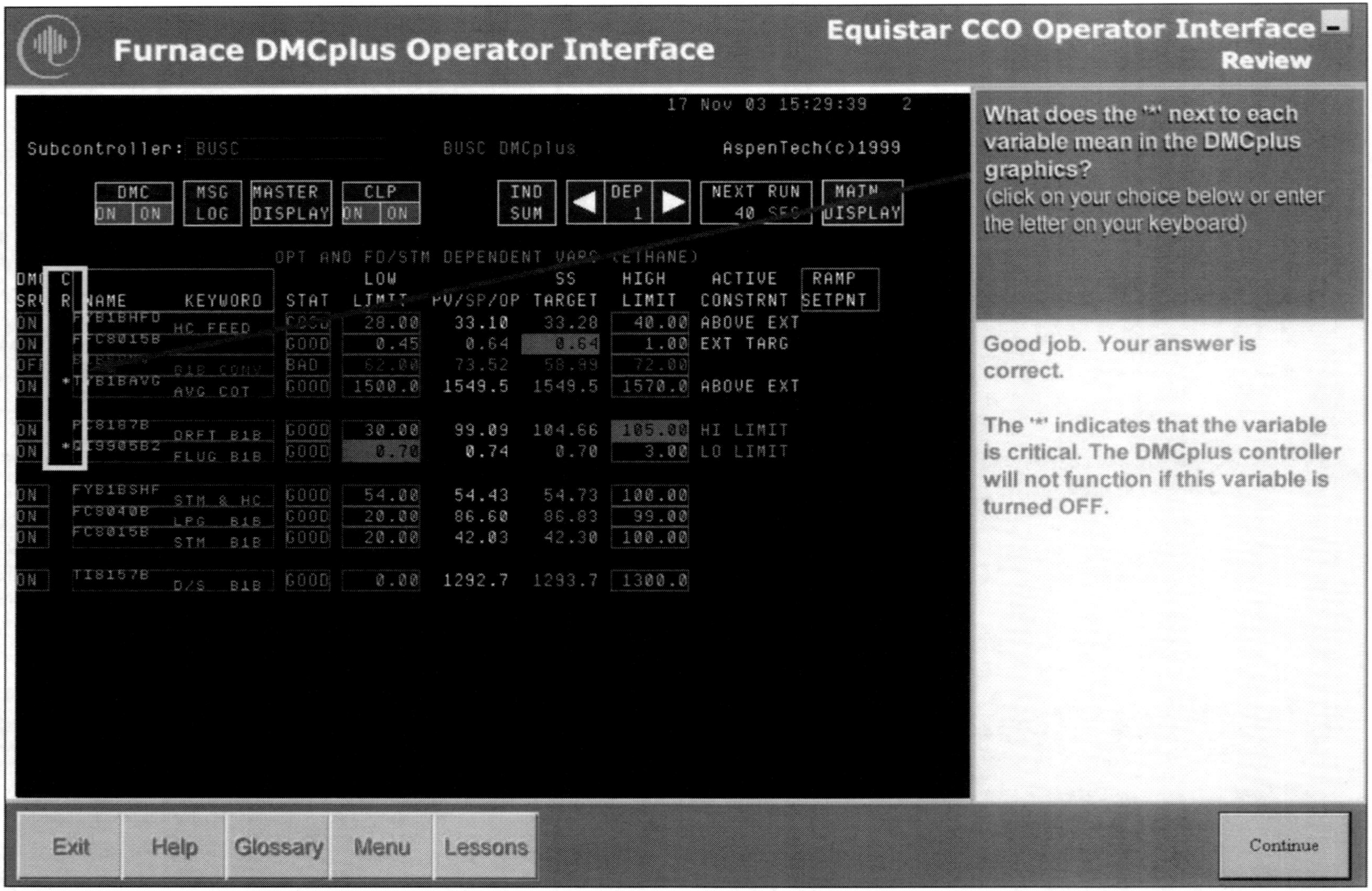

Figure 5(b): Typical Review Answer

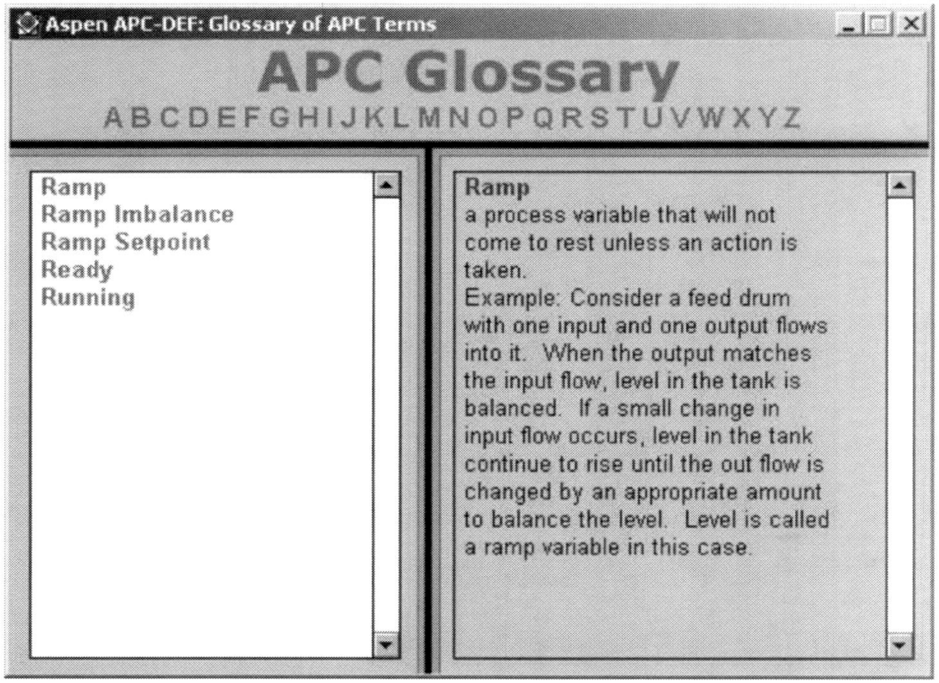

Figure 6: Glossary

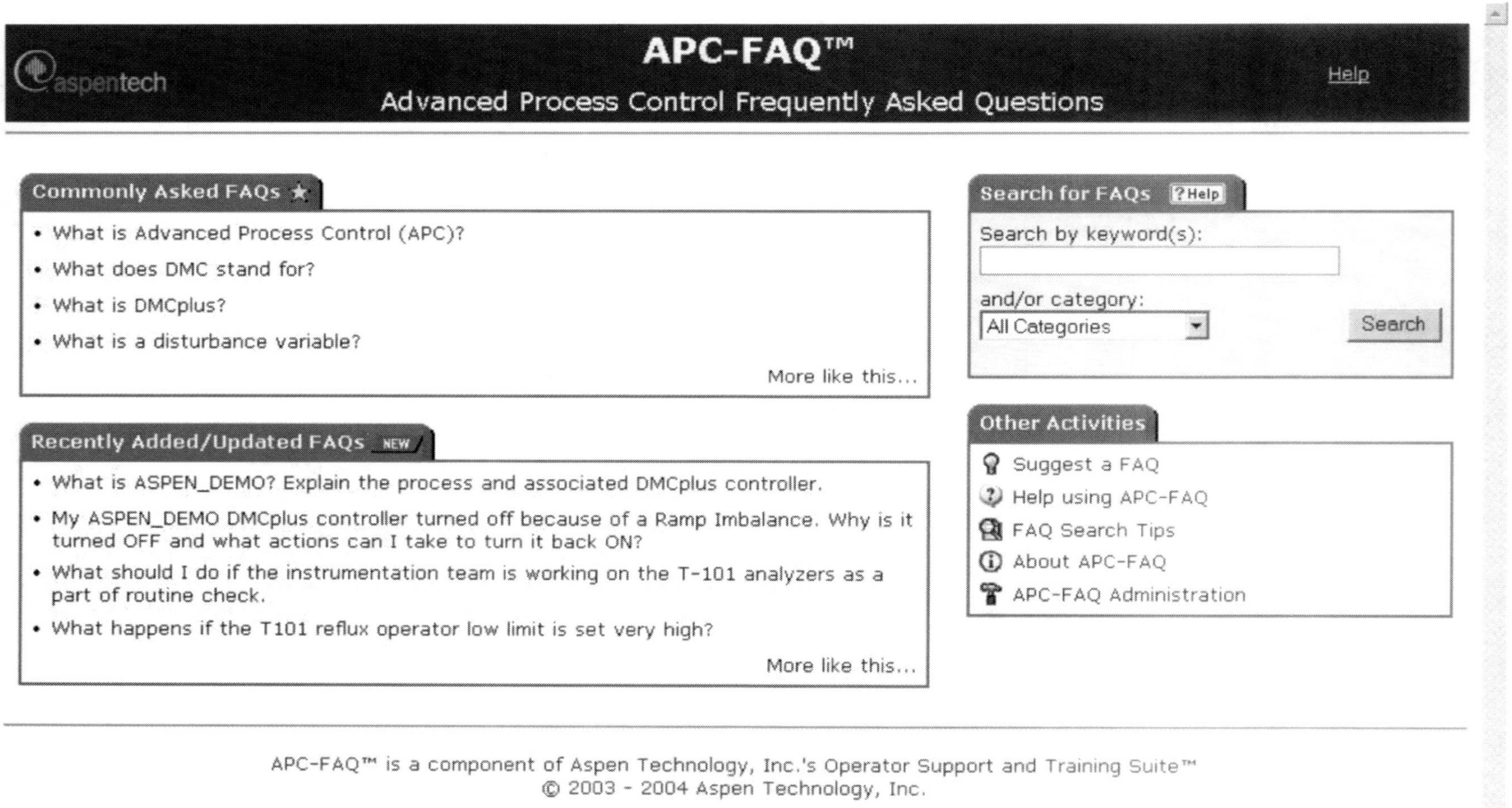

Figure 7(a): FAQ Opening Page

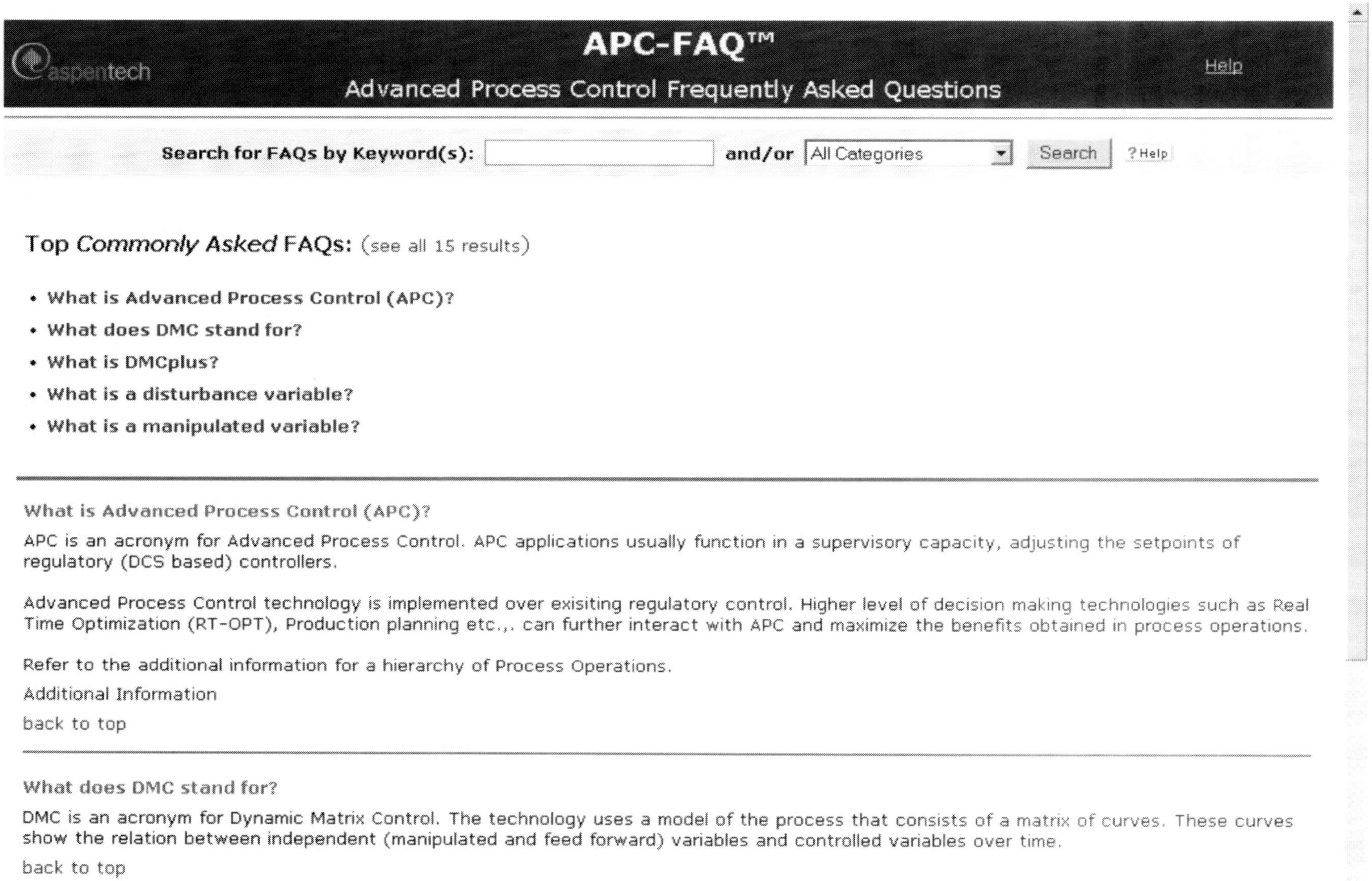

Figure 7(b): FAQ Answer

Paper TA006f

Early Event Detection – Results From A Prototype Implementation

Michael B. Bell, MASc, P. Eng.
NOVA Chemicals

Wendy K. Foslien
Honeywell

Prepared for Presentation at the 2005 Spring National Meeting, Atlanta GA, Apr 10-14
17th Annual Ethylene Producers' Conference
Session TA006 - Ethylene Plant Process Control

Abstract

NOVA Chemicals Corporation and Honeywell Laboratories have been working together to evaluate the effectiveness of an Early Event Detection (*EED*) prototype in a typical manufacturing environment. The role of an *EED* application is to augment the console operators monitoring behavior and situation awareness as well as provide them with a tool to reduce or avoid process upsets. The *EED* application uses robust mathematical algorithms to monitor the correlations in key process measurements and estimate the state or condition of the plant unit. This paper is the second part of a paper presented at the Honeywell Users Group in 2003, Early Event Detection – A Prototype Implementation [Bell, Errington, Reising, & Mylaraswamy, 2003]. This continuation paper will describe approximately one year's worth of runtime results of the model. Examples of successes and operational issues will be highlighted.

1. Introduction

An *Early Event Detection (EED)* system [Mylaraswamy, Bullemer, & Emigholz, 2000] is designed to assist a console operator in better maintaining their situation awareness, providing early indications of potential process excursions, and therefore help the operator avoid or better manage abnormal situations. An *EED* application delivers multivariate monitoring of large process units and detects the preliminary patterns that forecast upcoming events that could drive the process into abnormal situations. Unlike an equipment monitoring application that is focused on specific mechanical failures, *EED* takes a high-level process view. *EED* acts as an intelligent assistant to the operations team for monitoring process functions, rather than to the maintenance team for monitoring equipment health. As an assistant, once *EED* has identified a potential problem, it provides diagnostic clues that orient the operator in the direction of the potential causes of the approaching problem. The overall modeling principle is to model on "normal" data and then detect when the process is not behaving "normally". Another advantage to using and *EED* solution is that is can condense many measurements down to only a handful so that monitoring a process unit can be performed more easily by the panel operator.

There are two types of detection models available. The first is a general detection model. It is designed to detect any process abnormality. The second is a focused detector, which is designed to pickup a specific event. In our prototype we combined the two models.

NOVA Chemicals Corp. (NOVA), as a member of the Abnormal Situation Management (ASM®)[1] Joint Research and Development Consortium [Nimmo 1998], is participating with Honeywell in the testing of an Early Event Detection Prototype in its Ethylene 3 facility in Joffre, Alberta. The project began in 2001 and has been running online for over one and a half years.

The *EED* system described in this paper is based on earlier work by the ASM® Consortium, and uses principal component analysis models (PCA) to provide the correlation description of a process state [Mylaraswamy 2000]. The use of PCA models for process monitoring is well established [Kourti1996, Kourti1995, Kresta 1991, Wold 1987, Piovoso 1992]. An example of the online application of PCA models for process monitoring is given in [Dudzic 2000]. Background information on PCA modeling can be found in [Eriksson].

The objective of this paper is to describe each of the four major steps involved in this project as illustrated in Figure 1. Particular detail will be given to the Phase 3 – Online Implementation and Results and Phase 4 - Learnings and Benefits.

---

[1] ®ASM and Abnormal Situation Management are U.S. registered trademarks of Honeywell International, Inc.

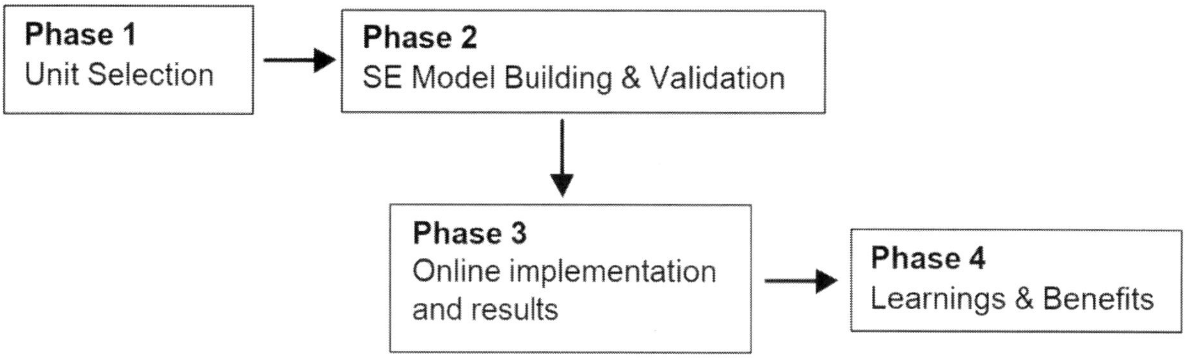

Figure 1: ***EED*** **Project Steps**

2. Unit Selection

Selection of the appropriate process unit to apply the *EED* technology can be a difficult task. The creation of an *EED* system does require some significant investment in both time and money; therefore selecting a unit that will warrant the investment must have a reasonable chance of success. There are three major categories, which are used in the unit selection:

- Events which are hard to detect
- Events which are expensive
- Events that can be detected by the *EED* technology early enough that appropriate action can be taken.

At Nova's Joffre site there are three ethylene units, two polyethylene units, a hydrogen offgas unit as well as on offsite utility process. From these five potential candidate processes were compared. Due to the low maturity of the modeling tool being used batch or semi-batch processes were not chosen – eliminating the polyethylene processes.

The chosen process was the Ethylene-Ethane (C2) Splitter in the Ethylene 3 plant. A simplified process diagram is shown in of the Ethylene 3 plant is shown in Figure 2. The C2 Splitter is the final separation column where the overhead stream from the tower is the product ethylene that is fed into a pipeline for distribution to NOVA's customers. The bottoms stream is unconverted ethane that is recycled back to the cracking furnaces.

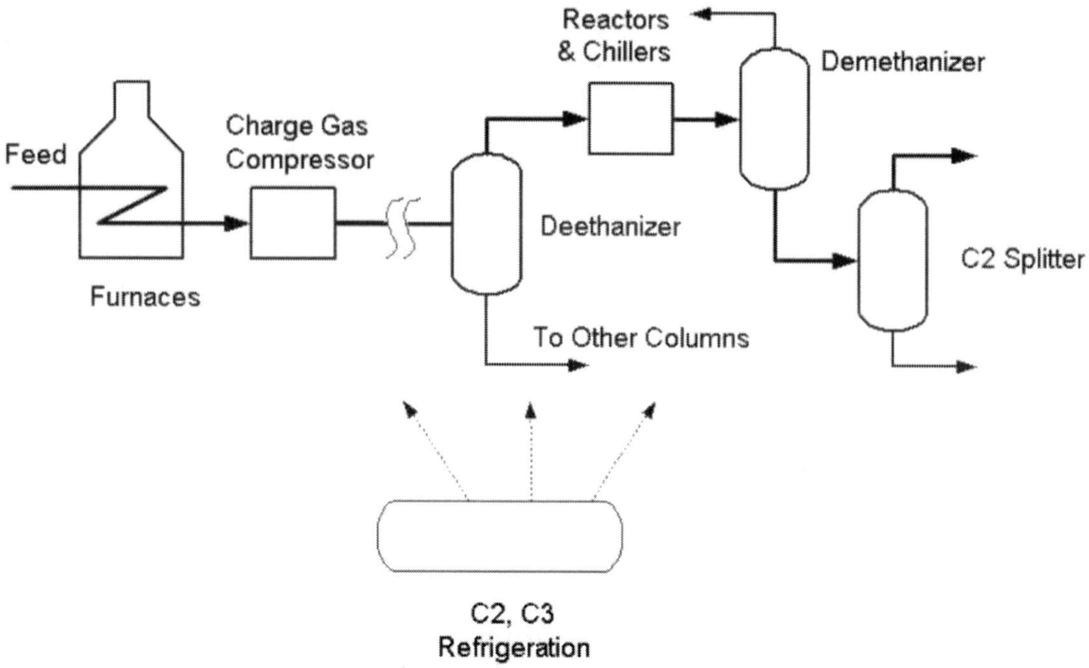

Figure 2: Ethylene 3 – Simplified Process Layout

Quality control of the product ethylene stream is the ultimate function of the C2 Splitter and any major disturbance in this tower can represent significant cost impacts. The tower is also difficult for the Operating Technicians to monitor and control due to a long process time constants and its tight heat integration as it employs a heat-pump compressor that supplies both the overheads cooling and bottoms heat supply. Due to its complexity and high-energy use, the tower has also recently had a new multi-variable controller application added to it. A process diagram illustrating the heat integration is shown in Figure 3.

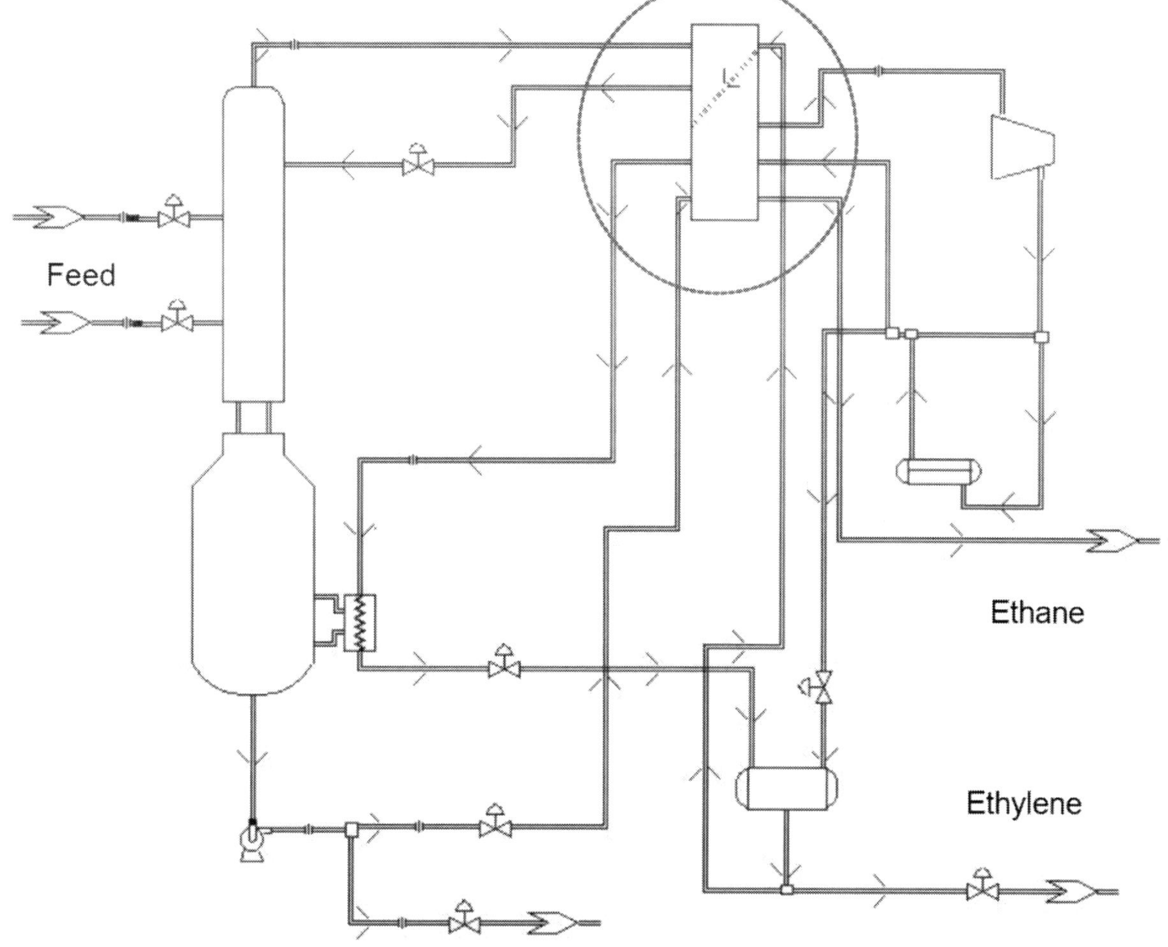

Figure 3: C2 Splitter Simplified Process Flowsheet Showing Heat Integration

3. Model Development

The *EED* technology that was developed by the ASM Consortium and used in this project has four basic parts as shown in Figure 4. The first element is the data processing component that lets the user view and select "normal" data from which to build the model. The second element that is applied to the data is an Exponentially Weighted Moving Average (EWMA) filter to remove long-term trends. The third piece creates the actual model using Principal Component Analysis techniques. Finally a fuzzy logic block is used as a tuning parameter to adjust the sensitivity of the results and to avoid chattering.

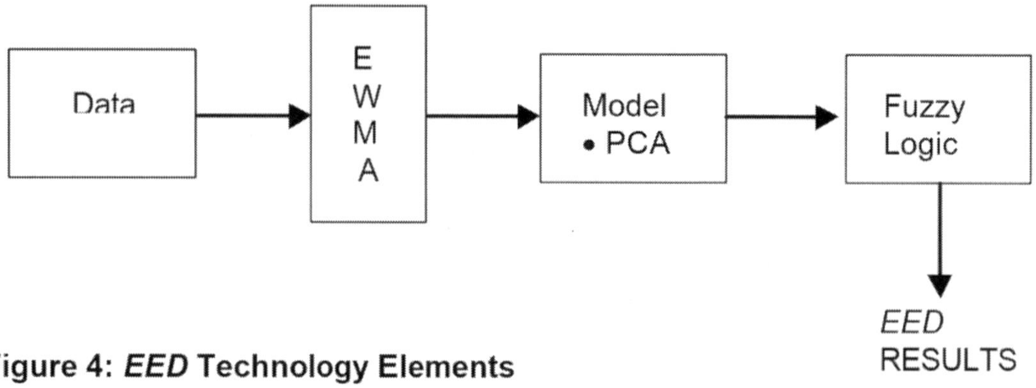

Figure 4: *EED* Technology Elements

The initial *EED* tool was constructed in Matlab using the ASM Consortium's SETools package. Data was extracted from the plant historian using a spreadsheet package.

Modeling With "Normal" Data

In order to detect abnormal situations the *EED* model must be designed around "normal" data. During the development of the *EED* model at Nova a few different approaches were taken to model on "normal" data.

The first approach was to select approximately 60 one day, (1 minute sampling) data sets from two years worth of process data. This data was not contiguous. The original modeling was performed by simple concatenation of the 60 data sets without any mean shifting of the data (there were large discontinuities in most of the measurements). Using standard PCA analysis techniques it was found that the standard deviation calculated did not represent the actual plant standard deviation – it was much too large. This was due to the large discontinuities in the data sets.

The second approach was to then take only one of the data sets. Unfortunately the standard deviation was too small in most of the variables in the data set making the model perform poorly with random samples of process data.

The third approach was to take data from the step tests used to generate models for a new multi-variable controller. This data set contained 15 days of data where significant but not unacceptable moves were. There were no rate increases or compositional changes in the feed. Two other data sets were also added to include data that showed plant variation due to feed rate and compositional changes, as this data was not in the original step tests. A new method of concatenating data sets was also implemented so that discontinuities would be minimized when adding more data. Note that the variance used for scaling individual variables calculated with the step test data tended to be quite high. For this reason, the scale factor for individual variables was calculated from a data set without the step test data.

Solving the PCA model

Once the "best" data sets were selected PCA modeling was started. Allowing for more or less Principal Components (PCs) determines the amount of variance that is explained by the model. If there are too many PCs the model starts modeling noise in the variables themselves. Since the objective of the model is to only model process changes due to true variance in the process itself, the number of principle components must not be too large. In the C2 Splitter problem there were 83 process variables being modeled. When the modeling was complete 10 Principal Components were calculated with 70% of the variance in the data sets being explained.

An attempt was made to name the Principal Components. It was believed that by doing this and by monitoring the values of the PCs the operator could be directed to the area of the splitter where the abnormality was. The naming proved to be quite difficult, as some of the Principal Components seemed to be the same. It is believed that the heat integration in the actual process, may have contributed to the difficulty.

Focused Detectors

With the *EED* model being a general detection model it was decided that if focused detection models around two specific events could be developed as it would aid the operator in identifying what the abnormality was. The first known event was the rapid decrease in level in a liquid seal pot. The level normally sits at 100% with a low level alarm at 80%. This vessel is quite small and if allowed to drain totally, serious operational issues would occur. The second event occurs when the variation of the ethane (C_2H_6) in the splitter overhead increases significantly in size (4-5 times). The benefits of identifying this event would be running tighter to the overhead specification, which would result in less energy use. New training sets were found with the happening events and new focused PCA identifier models were created.

Using the *EED* Model

For the *EED* model to be effective appropriate operator intervention is needed. Unlike multi-variable controllers, which automatically move the plant to the desired targets, *EED* technology generates diagnostic clues to the operator as to where an abnormality exists. Two key parameters are used to determine the status of the plant compared to the model: Q-Statistic (also called the Squared Prediction Error or DModX) and the Bad Actors or key contributors [Eriksson, Johansson, Kettaneh-Wold, Wold]. The Q-Statistic is the amount of unexplained error between the model and the process. In general, the Q statistic is a positive and unbounded value. The post processing described earlier scales the Q-statistic between 0 and 1, which is useful for display and alarming purposes. A threshold was placed on the scaled Q-Statistic value so that when it reached 0.6 the operator would receive an indication of model mismatch. The Bad Actors are a list of individual variables, ranked by their contributions to the Q-Statistic.

When the Q-Statistic gets large, the operator then looks at the Bad Actors to help determine where the model-plant mismatch is highest. To reduce chattering in the Q-Statistic some simple fuzzy logic and temporal filtering was used.

Once an event was triggered and the problem had gone away, the operator was requested to enter an annotation into the online operator log. The log was to be used to help determine what action should be taken, if any, by the operator for a similar typed event. Trending the Bad Actors was also a useful tool to determine what was happening in the plant during the abnormal event.

4.0 Results, Learnings and Benefits

Effect of Multi-variable Controller

Approximately 2 months into the trial an MVC controller was implemented on the C2 Splitter. Its effect on the *EED* model was pronounced. This was due to the MVC control strategy being significantly different than the standard strategy. The MVC controller had shifted major variances in the process to different places. One example of this was the bottoms level in the stripper. Under PID level control the level remained quite close to its setpoint. With the new MVC controller the level was used as surge increasing it variance dramatically. The underlying mathematics behind the *EED* is PCA, which is dependent on the variance of the input variables. With a significant change in the variance of some of the variables it was noticed that the Bad Actors rarely changed. Also the MVC controller now began to control the ΔT across a heat exchanger so that the level in the seal pot could be controlled better. The focused *EED* now no longer represented the level decrease problem in the seal pot and never indicated an abnormality.

Using post MVC data, new variances were calculated for each of the variables. These variances were combined with previous data set variances so that the largest variance was used as the scaling factor in the PCA model. This modeling step removed the Bad Actors problem.

Two *EED* Detection Examples

On October 31, 2003 the *EED* detection model successfully identified an unknown event. Figure 5 shows how the event unfolded. For the first hour in the plot there is no indication of anything abnormal happening. Slowly the Q-Statistic begins to rise as well as the two Bad Actors become constant. At 3:00pm the Q-Statistic indication alerted the operator. By trending the two Bad Actors with a down stream pressure it was determined by the operator that plugging was occurring. Within 30 minutes, propanol was injected upstream of the pluggage removing the blockage and the process went back to normal. This event propagated itself in about 5 hours. The pressure variable is not routinely monitored by the operator and if the situation progressed to a more serious nature a major upset would have occurred. It is somewhat difficult to place a dollar value on an upset that was avoided due to generally increasing the events value the longer it is ignored. With this in mind the accepted value of this avoided incident was about 130 k$.

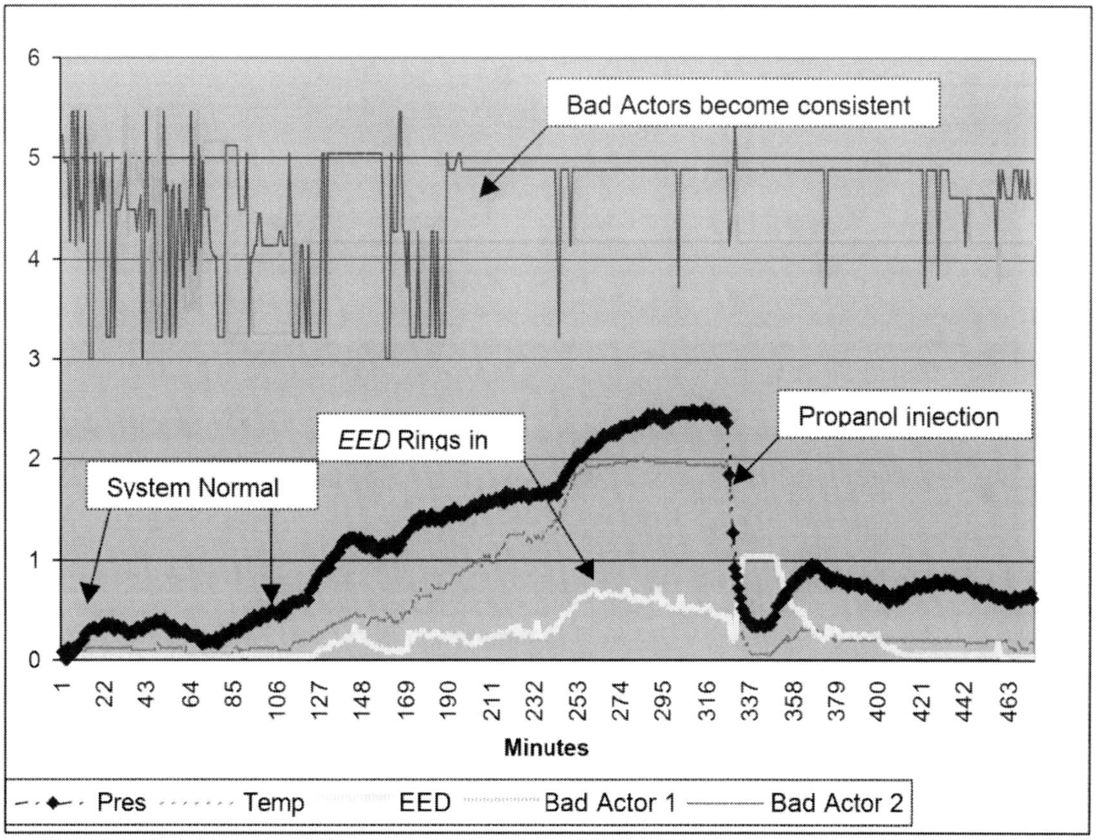

Figure 5: October 2003 Hydration Event

On January 19, 2004 another hydration event occurred. This time it developed more slowly. The *EED* detection model recognized that something was happening but never reached the threshold. Figure 6 shows how the event developed. This event propagated itself in about 5 hours also but much more gradually than the previous case. What is interesting to note is the effect of the EWMA filter. Once the input began to flatten out the *EED* calculated Q-Statistic begins to reduce in size. Therefore you would expect the very slow and smooth occurring events would be filtered out, which is shown in Figure 6. This illustrates the tradeoff in filtering the inputs. Common events such as slow feed rate changes or effects from ambient are filtered out, but absolute limits on variables or very slow developing events are not detected as easily.

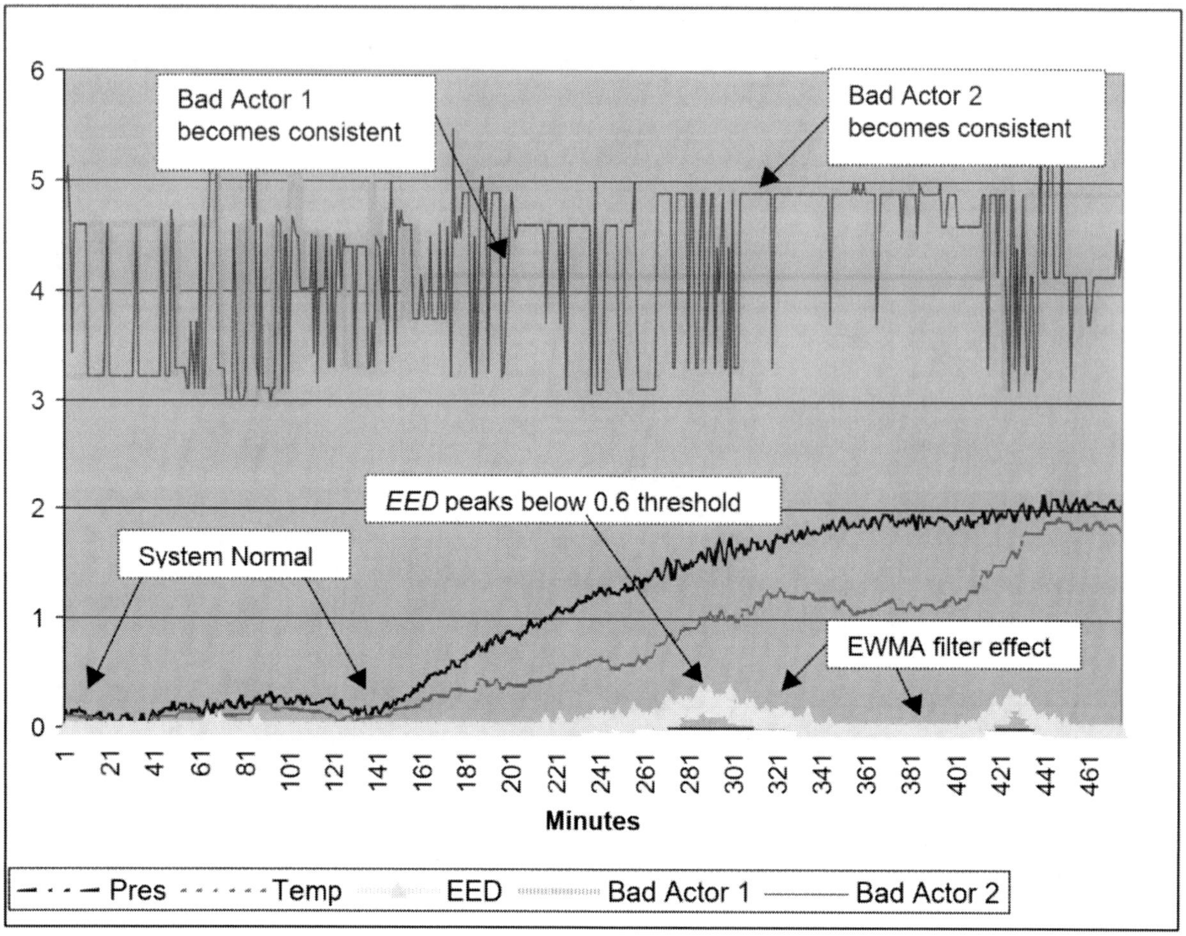

Figure 6: January 2004 Hydration Event

Operator and *EED* Saturation

During the first year of *EED* online time it was found that many events that were detected were already known to the operator. These events include furnace swings in the cracking section of the plant, rate increases, changes in furnace conversions, etc. A typical furnace swing is shown in Figure 7. Even though some of our "normal" data included rate and conversion changes *EED* typically would ring in. This poses a problem in that the operator becomes desensitized to the detection system. Similarly if a single tag becomes erratic in its indication *EED* will become a "nuisance alarm". In an attempt to reduce the nuisance events in *EED*, scaling factors for the offending variable was modified so that the model became more insensitive to that variable's variance. Care should be taken with this approach, as the scaling should be put back to its original values once the problem has been fixed.

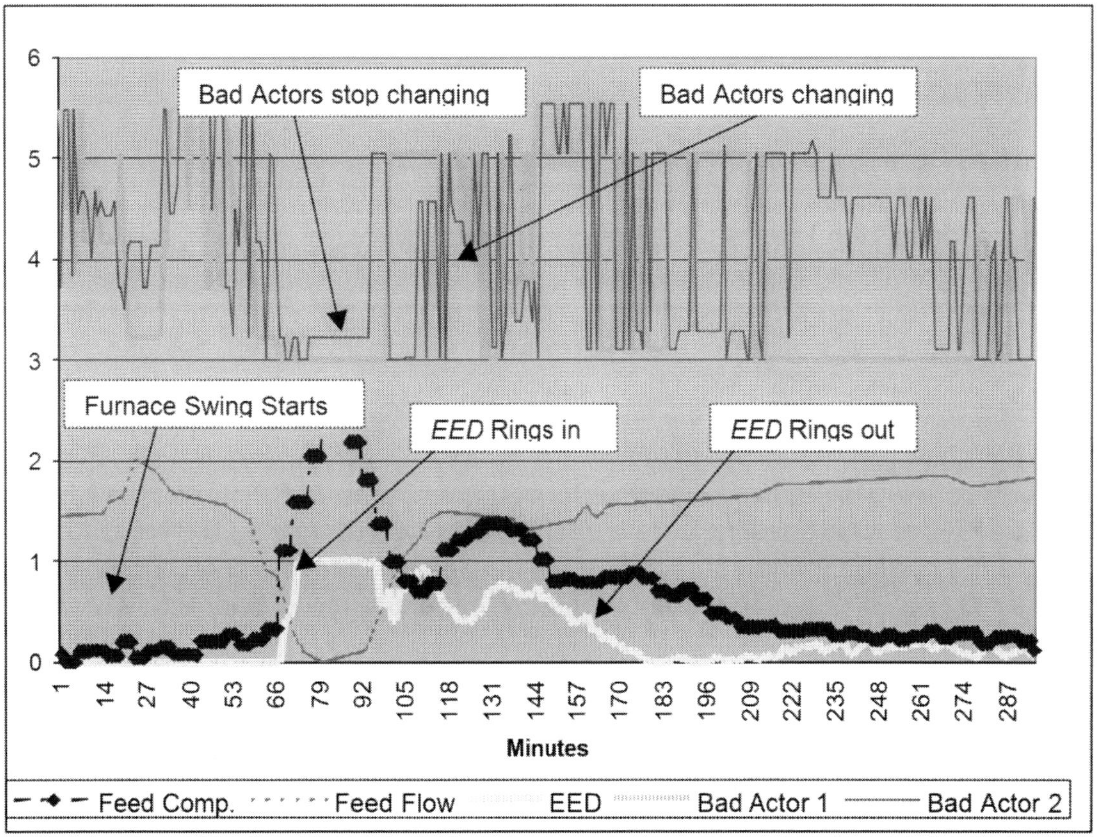

Figure 7: Typical Furnace Swing

The *EED* indication was scaled from 0 to 1 so that plotting would be made easier. It does cause some issues when an event is progressing. Significant events will quite often move from one part of the plant to the subsequent unit. If the event is large enough the *EED* indication will saturate at a peak value of 1, even though the Bad Actors are changing. Because the *EED* interface indication peaks and stays above the threshold value of 0.6, the detector never re-indicates a problem. From the furnace swing event in Figure 7, the *EED* indicator remains above its threshold for about 1 hour. As the furnace swing progresses over time the Bad Actors change indicating a potentially different event.

Plant Configuration

Many C2 Splitter *EED* events where characterized with 5 or more significant Bad Actors appearing in a 10 minute period. This was generally in the reboiler, reflux and heat exchange variables in the plant. This made it difficult to determine what was happening abnormally in as the tags composed more that one or two pieces of equipment. Fixing the abnormality is even more difficult as it is hard to pin down exactly where the problem is. It is believed that this is an artifact of the heat integration of the plant and not necessarily and inherent problem with the *EED*.

Even with a high degree of recycles in a plant the *EED* performs very well in detecting bad sensors and single/double Bad Actor variable events. These events are easy to decipher from the *EED* diagnostic clues. An operator action is also quite easily determined.

Focused Detectors

The two focused detectors had limited success. With the implementation of the MVC controller the seal pot level problems had pretty much been eliminated. Of note, was that even though scaling factors were changed in the model, when the MVC controller was off, the model performed back to its original design. The C2H6 variance detector performed reasonably well but generally did not add more information than what the Bad Actors were highlighting, which is to say the general detection model was sufficient. Figure 8 shows the same furnace swing event as shown in Figure 7 with its effect on the overhead C2H6 concentration. The C2H6 detector does indicate a deviation in approximately 20-30 minutes before the change in the composition but this information is also available from the general *EED* detector in combination with the change in Bad Actors.

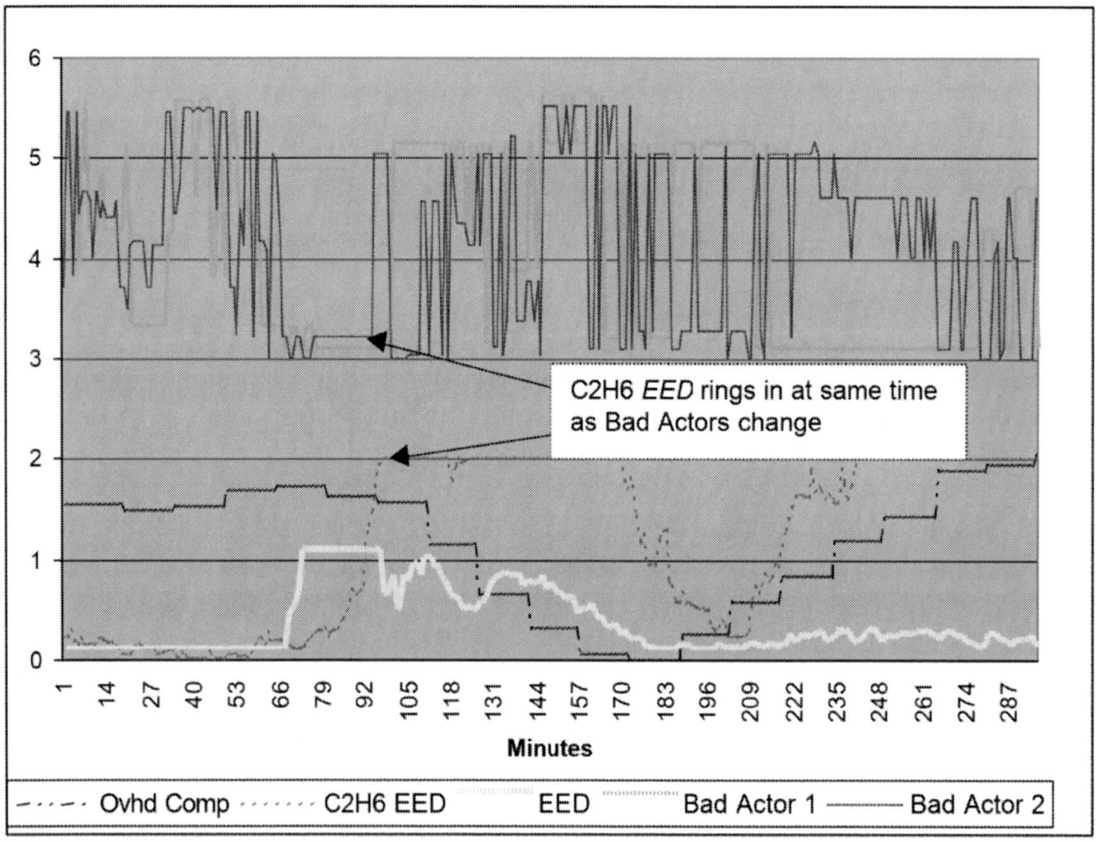

Figure 8: C2H6 Focused *EED*

5.0 Summary and Conclusions

A summary of our findings from the prototype are:

- The selection of "normal" data is critical in determining the effectiveness of the Early Event Detection model.
- Overall during the online test, the ability to pick up abnormalities with a general detection model works quite well, especially with only a few Bad Actors. Sorting out the actual process issue was not always simple, especially with a highly heat integrated process.
- Operator saturation, illustrated by failure to respond, can occur if detection indications become nuisance types.
- The degree of EWMA filtering can trend out slow occurring events.
- If changes to the process variance occur, the *EED* model must be changed to reflect the new conditions. The *EED* model can be used to evaluate the significance of the process changes, and modified if significant process/model mismatch is present. In our experience, the modification included the basic steps of adding process data and retraining the model.

Future enhancements to the *EED* will include:

- Method to suppress detections when known events are occurring
- Increase the scope of the model to include more upstream units
- Improve the *EED* diagnostic clues

6.0 References

1. Bell, M., Errington, J., Reising, D., Mylaraswamy, D., 2003, *Early Event Detection – A Prototype Implementation*, Honeywell Users Group Meeting, Phoenix, AZ.
2. Dudzic, M., Vaculik, V., Miletic, I., *On-line applications of multivariate statistics at Dofasco,* in Proc. Int. Fed. Automat. Contr. Workshop on Future Trends in Automation in Mineral and Metal Processing, Helsinki, Finland, 2000, pp. 438-443.
3. Eriksson, L., Johansson, E., Kettaneh-Wold, N., Wold, S., *Multi- and Megavariate Data Analysis – Principles and Applications*, Umetrics, P.O.B. 7960, SE-90719 Umeå, Sweden
4. Kourti, T., MacGregor, J.F., (1995). *Process Analysis, Monitoring, and Diagnosis using Multivariate Projection Methods - A Tutorial.* Chemometrics and Intelligent Laboratory Systems, 28, 3-21.
5. Kourti, T., MacGregor, J.F., *Recent developments in multivariate SPC methods for monitoring and diagnosing process and product performance,* J. Quality Technol., vol. 28, no. 4, pp. 409-428, 1996.
6. Kresta, J.,V., MacGregor, J.F., Marlin, T.E., (1991). *Multivariate Statistical Monitoring of Process Operating Performance.* Can.J.Chem Eng., 69, 35-47.
7. Mylaraswamy, D., Bullemer, P.T. and Emigholz, K., 2000, *Fielding a Multiple State Estimator Platform.* Proceedings of NPRA Computer Conference, Chicago, IL.
8. Nimmo, I., 1998, The ASM Story, *Hydrocarbon Processing*
9. Piovoso, M.J., Kosanovich, K.A., Yuk, J.P., *Process data chemometrics,* IEEE Trans. Instrument. Measure., vol. 41, pp. 262-268, 1992.
10. Wold, S., Esbensen, K., Geladi, P., 1987, *Principal Components Analysis.* Chemometrics and Intelligent Laboratory Systems. 2, 37-55.

Paper TA007a

Roundable Discussion on Abnormal Event Management

Eric Gildea (ABB), Don Bartusiak (ExxonMobil Chemical) moderators

Michael Bell (Nova Chemicals), Mike Clark (BP Chemicals),
Ken Emigholz (ExxonMobil Research & Engineering), Don Morrison (ASM Consortium),
Ian Nimmo (User Centered Design Services), Ramesh Vaidhyanathan (Plant Automation Services),
Venkat Venkatasubramanian (Purdue University) panelists

Prepared for Presentation at the 2005 Spring National Meeting, Atlanta GA, Apr 10-14
17th Annual Ethylene Producers' Conference
Session TA007 - Ethylene Plant Process Control

UNPUBLISHED

Abstract

This session will consist of a moderated roundtable discussion among a panel of operating company, vendors, and academic engineers with expertise and experience in the following topics:

- Process Fault Detection
- Early Event Detection

The session will begin with a concise primer to define terms, followed by opening statements from the panelists on their areas of expertise, current R&D efforts, or project implementation. This will be followed by a moderated open forum among the panelists and the audience.

The focus of the discussion will be on the application of various model-based technologies to detect the onset of abnormal plant operating states, and possibly to isolate problem signals more precisely. Alarm management technologies are outside of the scope of this roundtable. Examples of model-based abnormal event detection technologies include (1) empirical statistical approaches such as principal component analysis and partial least squares, and (2) fundamental model approaches such as observers and optimal state estimators. Complementary technologies for interpreting the results of the model-based event detectors are within the scope of this roundtable.

DEVELOPMENT OF A PARKING MODE AT SHELL CHEMICAL'S DEER PARK PLANT OLEFIN UNIT OP-III

Authors

Nicolas Genty, Bryce Kagay, Shell Chemical LP, Houston, Texas

Abstract

Shell Chemical LP's OP-III olefins unit located at its Deer Park site has defined a set of guidelines for the unit operators to consistently reduce feed rates to the unit when an unanticipated process upset results in flaring of material. The use of these guidelines will minimize flaring and thereby the potential impact on the environment and the adjacent community. The so-called "parking mode" allows the operator to move the plant to a more stable position and to have longer response times for process changes. It also allows going back to the normal operating mode with minimum impact on the hardware. Please know that reference to operating conditions, such as process upsets, is not made in the context of any potentially applicable regulatory definition.

The paper and presentation discuss not only the philosophical aspects of the parking mode, but also the practical implications to an olefins unit like OP-III. It also describes in more details the different application tools provided to the operators to help them in the decision making process, namely (1) the operating guideline proper, (2) a decision making chart allowing the proper selection of the furnaces to shutdown and (3) an Excel-based decision making tool providing the operator with the sequence of furnace shutdowns according to the philosophy developed.

Introduction

The olefins unit OP-III at Shell Chemical LP's Deer Park site was designed and built at the end of the 1970's. Its furnace section consists of liquid-feed furnaces and gas-feed furnaces. Some of the liquid furnaces are capable of cracking different types of feedstock, including ethane, and butane. The recovery section of the unit is a typical front-end demethanizer scheme with conversion of acetylene and of methyl-acetylene/propadiene. The main products are hydrogen, tail gas, ethylene, propylene, a C4-cut, pygas and pitch. Ethane is recycled and cracked to extinction.

The Deer Park site includes a second ethylene unit, OP-II, which may export ethane to OP-III as furnace feed. The OP-II unit may also export process gas from the cracked gas compressor (PGC) directly to the fractionation section of OP-III.

Both olefins units have no ability to store off-spec ethylene (either on-site or off-site) and have limited ability to store off-spec propylene onsite. Flaring is the only other option in case of a unit process upset.

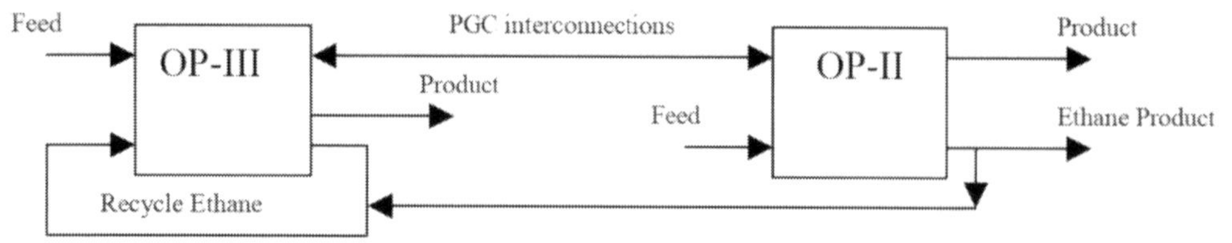

Fig. 1: block scheme of OP-II and OP-III interconnections

Driven by its HSE Commitment and Policy Statement, Shell Chemical LP is committed to being a good neighbor and to continuously improve its environmental performance. Following this commitment, it became clear, that olefins flaring should be minimized to reduce the potential impacts of additional light, noise, odors and emissions associated with flaring on the adjacent community of Deer Park and the local environment. Since Shell Chemical LP's Deer Park site has a limited ability to store off-spec olefins material, focus was put on flaring minimization and reduction of the time required for the OP-III olefins unit to recover after an off-specification incident.

Shell Chemical LP has developed a set of guidelines to define the furnace shutdown sequence when an unanticipated process upset causes a need to reduce feed rates to the unit to minimize flaring. Some operating companies may call this type of guideline a "parking mode" for the unit. The concept of a "parking mode" is therefore to define consistent responses that operations should take for a number of pre-defined events thereby developing consistent responses to process upsets across shifts.

By defining the response for certain events operations will:

- Reduce total flaring
- Move the plant to a more stable position
- Allow operators to have longer response times for process changes

Philosophy applied in OP-III

Due to the large number of potential feedstock-related cases and types of furnaces, a philosophy for shedding feed had to be developed. The following guidelines were therefore established:

Safety First!

Safety is the major concern during any process upset and the operators must know in advance the action that must be taken to safely secure the plant. The response should be planned according to applicable regulations and should proceed in an orderly manner, and care should be taken to work at a steady pace to avoid injury. With this in mind, the operational response always refers to emergency and normal shut down procedures that pertain to shutting down the plant, and refers to start up procedures prior to starting up the affected equipment.

Environmental Impact

With respect to environmental impacts, reducing and eventually eliminating flaring as quickly as reasonably possible is the highest priority and requires immediate action by all operators.

The expectation of the parking mode is to minimize the potential impact on the environment by immediately reducing the feed rates to the cracking furnaces to the minimum flow allowed. Furnaces can then be diverted from the process in a planned sequence by shutting down, decoking or putting furnaces on hot steam standby. The guideline is designed to quickly reduce all flaring and to protect the integrity of the furnace tubes.

While developing the concept of "parking mode" for OP-III, it was observed that the time required to recover from significant process upsets historically has been much longer than initial estimates. However, at any given time during the upset and subsequent recovery, "parking the plant" may be put on hold or aborted to resume production.

First Priority: Reduce flaring – this is the main purpose of this "parking mode." The guidelines defined simply derive from basic logic and relative impact on flaring, as shown in Table 1 below, giving typical approximated expected flare contribution (as % of feed) by feed type, providing the flaring takes place upstream of the process gas compressor.

- Shut down large-capacity furnaces before small furnaces
- Shut down ethane furnaces (95% throughput to flare) prior to butane furnaces (80% to flare), prior to liquid furnaces
- Shut down liquid furnaces according to heaviness (e.g. specific gravity) of the feed - light liquid feeds first, heavy liquid feeds last

Ethane	95%
Butane	80%
C5-Type feed	75%
Raffinate Type	60%
Gas Oil	50%
Naphtha Range	50%
Condensate type Material	40%
Hot Steam Standby	0%
Decoke	0%
Stop	0%

Table 1: relative flare load "Flare Yields" for different feedstock types

<u>Second Priority</u>: Minimize impact of parking mode on hardware; because the objective of the ethylene unit is ultimately to produce valuable products and therefore to recover as soon as possible from an off-spec flaring event, this second priority takes into account the impact on the hardware of a brusque or expedited furnace shutdown. With this in mind, an assessment was made resulting in the following guidelines:

- Shutdown furnaces based on runtime, "youngest" furnaces (defined by run length) and / or lowest measured tube skin temperatures first
- Put furnaces on hot steam standby or decoke when applicable/possible

<u>Third Priority</u>: Minimize economic impact of the parking mode by positioning the plant for a swift restart of production when the issue is solved.

<u>Other Considerations</u>:

- Since ethane from OP-II is typically processed by OP-III, care must be exercised while "parking the OP-III plant" to prevent a process upset in OP-II; no value is gained by reducing emissions from OP-III and increasing them by an equal or greater amount at OP-II. OP-II ethane continues to be cracked until operational personnel start the off-site export pump.

- Watch for possible dilution steam system bottlenecks; due to plant feed reduction and furnaces going to decoke and/or hot-steam standby, dilution steam system and blow down balance may become very difficult. Steam make-up usage will be at a high rate, therefore blowdown will be increased to balance the system. It is important to monitor this system closely to avoid possible overloading of the water handling system that may lead to elevated levels of benzene in the wastewater.

Modes of Response

A set of expected modes of response were specifically defined for the OP-III unit. Table 2 indicates below what the expected "parked" unit would look like, in terms of furnace configuration.

Table 2: Modes of response OP-III

Incident	Immediate Response	Final Furnace Configuration
PGC trip, PRC trip, ERC Trip, AC Converter trip, Hydrogen System (both) trip, Core plugging, Cold side leak, Steam Failure, Instrument Air Failure, Cooling Water Tower Failure, Total Power Failure	Take all furnaces to minimum feed as soon as possible during/after the incident. Begin moving to target alignment	1 liquid furnace at min. on each transfer line - 2 Ethane at min.
C3 System Issues - MAP Offspec, Water in C3=	All liquid furnaces to minimum feed Use off-spec storage facilities first, then move toward Target Alignment	1 liquid furnace at min. on each transfer line - 2 Ethane remain at capacity
C2 Offspec	All furnaces to minimum feed If flaring event is expected to be prolonged (if C2 Splitter is contaminated with AC), move to target alignment	1 liquid furnace at min. on each transfer line - 2 Ethane at min.
Reboiler failures in cold fractionation	All furnaces to minimum feed	1 liquid furnace at min. on each transfer line - 2 Ethane at min.

Initiation of the Parking Mode

Two techniques are available to the operational personnel to make a decision on how to reach the predefined parking mode:

- A spreadsheet
- A decision tree

The spreadsheet

A spreadsheet using Visual Basic code was developed to assist the operational personnel in determining how OP-III furnaces should be "parked" in the event of an unanticipated process upset (e.g., Process Gas Compressor or Propylene Refrigeration Compressor trip).

The code assigns penalties and credits to each furnace as it "decides" when a furnace should be shut down. The furnace with the highest objective function is recommended for shutdown. Another portion of code determines which off-line furnaces should be decoked based on constraints in the pyrolysis fractionator.

Initial Conditions:
The computer determines the plant's initial conditions by linking to the process computer. Gathered data includes feed type, feed rate, and days online for each furnace immediately prior to olefins process upset.

Constraints:
The user may identify special constraints facing the plant. These include the ethane production that must be cracked, import ethane heater's capacity, and the pyrolysis fractionator's design constraints.

Furnace capabilities and typical decoking steam flows are entered as constraints. Also, the expected flare contribution (as % of feed) is entered for each feed type. Constraint values are infrequently modified.

The code:
The code follows the following objectives:

- First, if ethane furnace feed is greater than the available ethane (assumes no long-term recycle ethane from OP-III is available), identify an ethane furnace to shut down. Remove feed from the largest flare flow contributor that won't create other ethane balance problems. Try to avoid back-to-back shutdowns of large ethane furnaces. Furnaces with short runs are preferentially taken off-line. Observe all guidelines about how many furnaces are required on each transfer line.

- Second, remove feed from furnaces with large flare flow contribution. Avoid ethane furnace shutdown if it causes material balance problems in the ethane system. Avoid back-to-back shutdowns of large ethane furnaces. Furnaces with short runs are preferentially taken off-line. Observe all guidelines about how many furnaces are required on each transfer line.

- Third, prioritize decoking needs. Furnaces with long run lengths have priority over recently decoked furnaces. Frequently, the user will observe a furnace go to decoke for a short (30 minute) time. The logic is that a short steam-out is better than no decoke, but when a furnace with higher decoke priority becomes available, it should go to decoke.

Results:
The results of the problem are displayed on the Interface worksheet as shown on Figure 2 below.

Furnaces at Minimum Feed

Drop one furnace at a time to minimize flaring

November 24, 2004														
Elapsed Time:	0	0:10	0:15	0:30	1:00	1:30	2:00	2:30	3:00	3:30	4:00	4:30	5:00	5:30
Current Time:	0:11	0:21	0:26	0:41	1:11	1:41	2:11	2:41	3:11	3:41	4:11	4:41	5:11	5:41
F-P3-11 - (30 days)	N	N	N	N	N	N	N	N	N	N	N	N	N	N
F-P3-12 - (0 days)	Stop	Stop	Stop	Stop	Stop	Stop	Stop	Stop	Stop	Stop	Stop	Stop	Stop	Stop
F-P3-13 - (13 days)	N	N	N	N	N	Decoke	Stop	Stop	Stop	Stop	Stop	Stop	Stop	Stop
F-P3-14 - (39 days)	Decoke	Decoke	Decoke	Decoke	Decoke	Decoke	Decoke	Decoke	Decoke	Decoke	Decoke	Decoke	Decoke	Decoke
F-P3-15 - (0 days)	HSSB	HSSB	HSSB	HSSB	HSSB	HSSB	HSSB	HSSB	HSSB	HSSB	HSSB	HSSB	HSSB	HSSB
F-P3-16 - (43 days)	Decoke	Decoke	Decoke	Decoke	Decoke	Decoke	Decoke	Decoke	Decoke	Decoke	Decoke	Decoke	Decoke	Decoke
F-P3-17 - (19 days)	N	N	N	N	N	N	N	N	N	N	N	N	N	N
F-P3-18 - (22 days)	N	N	N	N	N	N	Decoke	Decoke	Decoke	Decoke	Decoke	Decoke	Decoke	Decoke
F-P3-19 - (1 days)	ET	ET	ET	ET	ET	ET	ET	ET	ET	ET	ET	ET	ET	ET
F-P3-20 - (3 days)	ET	ET	ET	ET	ET	ET	ET	ET	ET	ET	ET	ET	ET	ET
F-P3-21 - (0 days)	GO	GO	GO	Decoke	Decoke	Stop	Stop	Stop	Stop	Stop	Stop	Stop	Stop	Stop
F-P3-23 - (6 days)	N	N	N	N	N	N	N	N	N	N	N	N	N	N
F-P3-24 - (0 days)	Stop	Stop	Stop	Stop	Stop	Stop	Stop	Stop	Stop	Stop	Stop	Stop	Stop	Stop
F-P3-25 - (43 days)	GO	GO	GO	GO	Decoke	Decoke	Decoke	Decoke	Decoke	Decoke	Decoke	Decoke	Decoke	Decoke
F-P3-26 - (0 days)	Stop	Stop	Stop	Stop	Stop	Stop	Stop	Stop	Stop	Stop	Stop	Stop	Stop	Stop
F-P3-27 - (61 days)	ET	ET	Decoke	Decoke	Decoke	Decoke	Decoke	Decoke	Decoke	Decoke	Decoke	Decoke	Decoke	Decoke
Flare Flow, Mlb/hr	255	228	209	179	149	127	105	105	105	105	105	105	105	105

Current Conditions

Max OP-II Ethane Export, Mlb/hr	50
Max Import Ethane, Mlb/hr	70
Max Decoking Steam to Pyro Frac, Mlb/hr	130

User describes the current plant constraints

Figure 2: Parking Mode Result on the Interface Worksheet

The decision tree

Reference is made to the decision chart in Figure 5.

The following describes the logic in the worst-case scenario (i.e., most major events). In case of C3 system issue, the ethane furnaces remain on line.

1. The immediate response to any significant olefins process upset is to reduce the feed rates to all the furnaces to the minimum allowed by the existing furnace operating guidelines and to minimize the interconnecting flows(s) between the two units that contribute to flaring at OP-III.

2. The Furnace operator on duty at the time of the event contacts OP-II to determine if OP-II can export ethane. If OP-II cannot export, determine how many/which furnaces to keep on-line to secure OP-II ethane.

3. Two furnaces will be shut down within the first 30 minutes of an unanticipated process upset

First step: A large-capacity furnace operating on ethane shall be put on decoke.
Second Step: A small furnace operating on ethane shall be put to decoke

4. One furnace should be shut down every 30 minutes thereafter until the target alignment is achieved. Focus efforts on furnaces with greatest contribution to flaring.

Results

The implementation of a "parking mode" at OP-III has allowed the Shell Chemical LP's Deer Park site to reduce flaring by some 50% between 2002 and 2003.

Acceptance level of this "parking mode" by the operational personnel has been high. Future work anticipated at the site is to integrate further the two units and maximize the use of the synergies between them, taking into account the guidelines laid out in this document.

Case Study #1:

During a heavy thunderstorm, the Process Gas compressor of OP-III tripped because of a faulty axial displacement indication. The process gases generated by twelve furnaces on line at the time of the incident were routed to flare. The ethylene and propylene products were off specification due to the loss of hydrogen supply to the hydrogenation reactors. Because the situation could not be alleviated in the first minutes of the incident, Operations decided to "park" the unit.

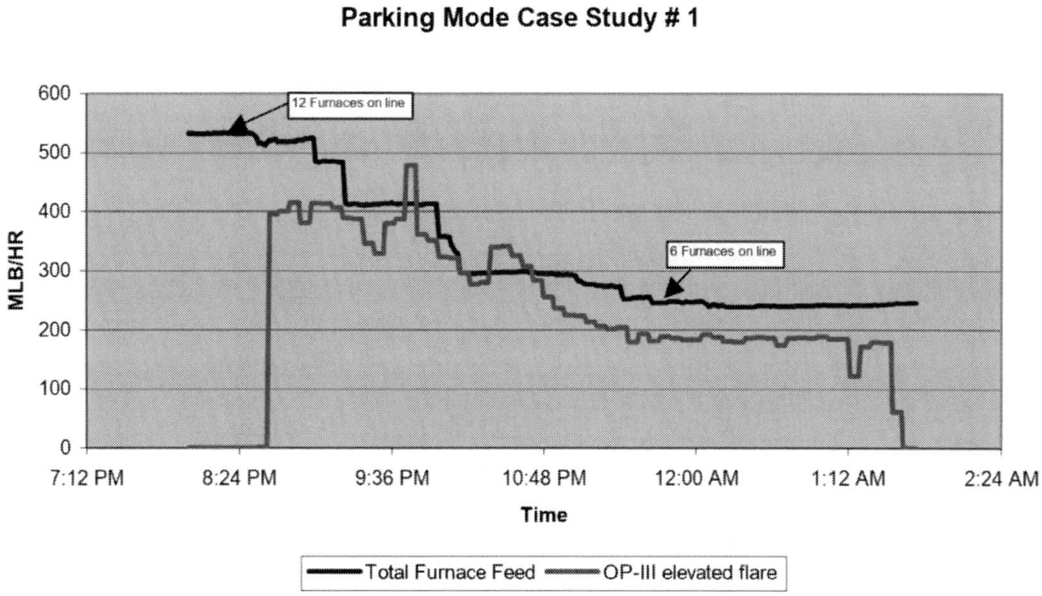

Figure 3: Case Study # 1: PGC Trip at OP-III

Case Study #2:

OP-III experienced a sudden loss of the 1250 psig steam supply due to an equipment failure upstream in the steam network. The ethylene refrigeration compressor (ERC) tripped due to lack of steam supply. Consequently, the downstream refrigeration and cold fractionation were significantly impacted. Ethylene, propylene, and C4 products were off-specification and were routed to flare. Because the situation could not recover without major flaring, Operations decided to "park" the unit.

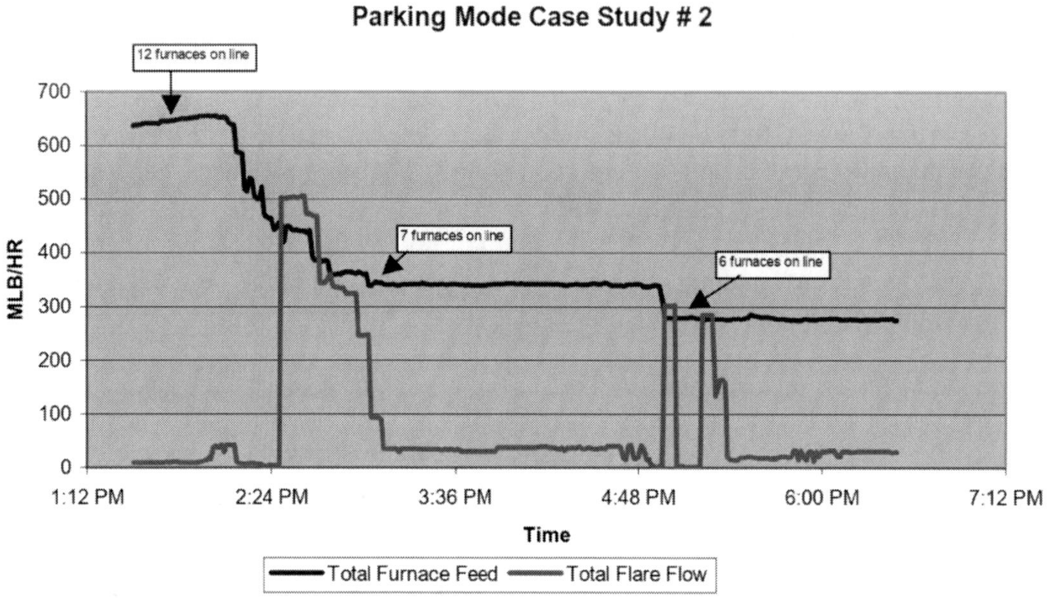

Figure 4: Case Study # 2: ERC Trip at OP-III

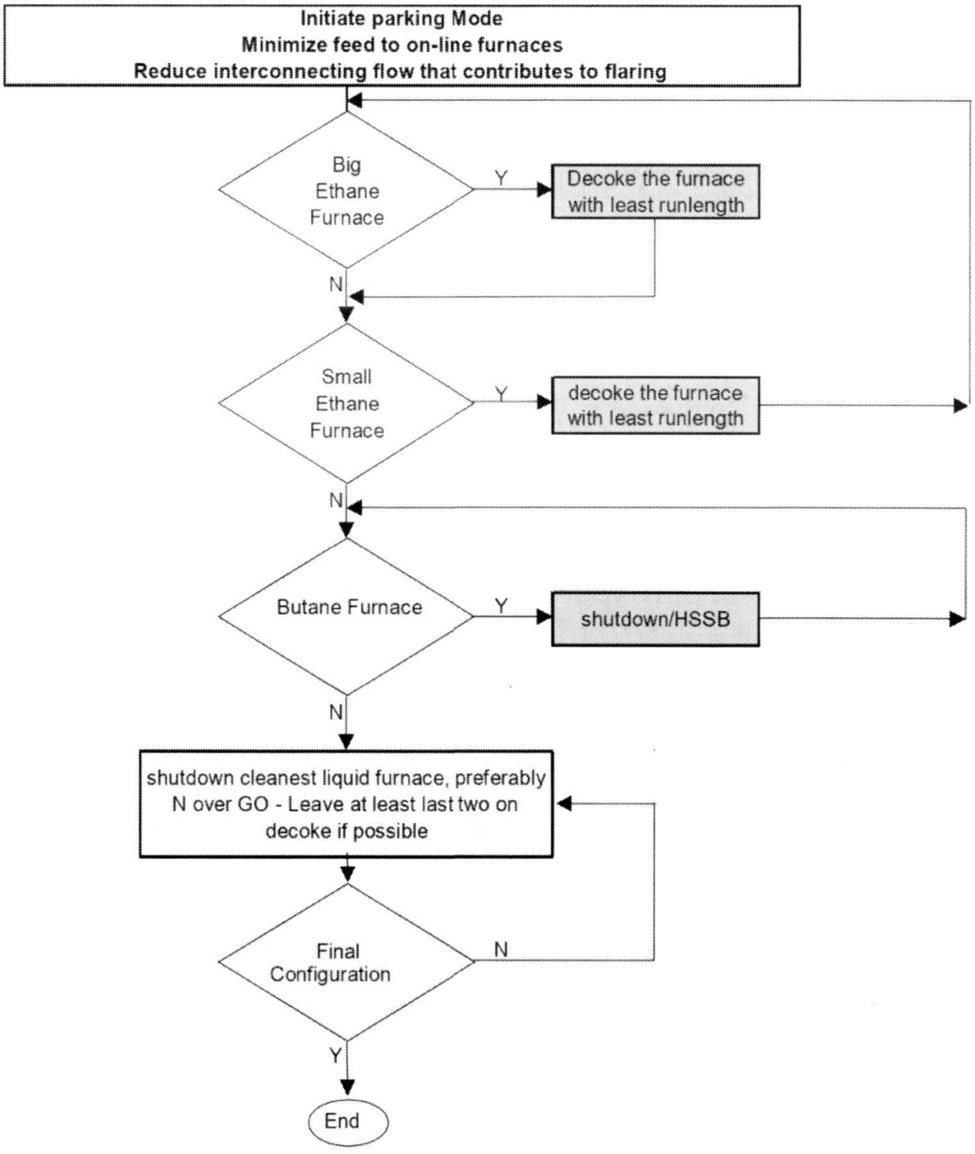

Figure 5: Parking Mode Decision

Analyses of a Rapid loss of a Steam System During a Power Failure

By

André Bernard, Process Design Engineer, NOVA Chemicals (Canada) Ltd
P.O. Box 3060, Sarnia, ON Canada N7T 8C7
Bernara@novachem.com

Abstract

NOVA Chemicals' Corunna Ethylene facility has originally been designed to ride through power failure of short duration. Operation of critical rotating equipments with steam drivers along with auto starts on other key equipment was to allow quick recovery of the plant once power supply re-established. The steam system was designed and has always been operated so that it would be unaffected by a loss of power. Power supply to the site has been proven reliable until the year 2000, where only 1 incident had been recorded over 25 years of operations. Since then, 4 occurrences of power losses have been experienced. On 2 of those 4 events, the steam system was lost on low deaerator level and a third event, which could be considered a near miss. Premature losses of the steam system caused a crash shut down of the site leading to damage on a number of process units. Analyses of the water, steam and condensate systems reveals that excessive losses of water along with the duration of the outage are the key contributor to failure of the steam system. This paper presents the key findings from the incidents analyses. Water conservation, steam shedding and proper coordination of emergency shut down activities consistent with reduced capabilities of water supply to the steam system are also discussed.

Introduction

NOVA Chemicals' Corunna Ethylene facility is integrated with a front end Crude Unit and back end Aromatics Unit. A variety of feedstock is being processed in 14 cracking furnaces, 4 operated with induce draft (ID) fans driven by electrical motors and all others on natural draft. The furnaces produced 2/3 of the total site steam generation with the remaining 1/3 supplemented by 3 high pressure boilers. The boilers are operated with force draft (FD) fans, which could be driven by electrical motors or steam turbines. The typical plant configuration is to operate the boilers with 2 FD fans on electric motors and 1 on steam turbine. On a power failure, the loss of electric driver trips 4 furnaces and 2 boilers. Despite this, the steam system can remain on stream depending on the length of the power loss and the effectiveness of operation in responding to the upset. Keeping the steam system on stream minimizes equipment damage and allows a quick recovery if the power loss is of short duration or provides favorable conditions for idle state or safe shut down of the unit if the interruption is to last for an extended period.

Analyses of power failure incidents at NOVA Chemicals Corunna site reveal that the likelihood of keeping the steam system on stream depends on the extent and duration of the power interruption. Boiler water conservation and efficient steam shedding are key elements in keeping the steam system on stream.

Power Distribution

Dual 230 kV lines supply electrical power to the site. Each incoming line and transformer is sized to carry the ultimate electric load of the entire plant. The lines are also interconnected by tiebreaker. During normal operation, both lines are energized so that the plant stays energized by the remaining line if one line is lost. The 230 KV supply is transformed to 13.8 KV through two 25 MVA transformers and then carried out at 13.8 KV to the individual area sub-station. Two separate feeders from the main sub-station switchgear to each area sub-station provide reliability and flexibility through the use of tiebreakers at the 600 Volt level of the areas sub-station. Power is reestablished by automatic re-closure of the lines during power failure lasting less then 30 seconds. It has to be manually reset once the fault is cleared for interruption exceeding 30 seconds.

A standby diesel driven generator provides emergency power in case of power failure. The generator supplies emergency power to critical lighting, control and UPS system.

A certain number of duplicated electric motors are also designed to re-arm automatically upon a failure lasting not more than 30 seconds.

Critical rotating equipment drivers (electric motors/steam turbines) are strategically configured to maximize protection against power or steam system failure.

Site Description and Vulnerability to Power Failure

A simplified block diagram of the site operation is illustrated on Figure 1. Loss of fin fans, reduction in cooling water flows and loss of critical pumps electrically driven makes the Crude Unit vulnerable to power failure. The unit cannot idle and is therefore slumped when power is loss. Turbine pump operations with electric backups ensure that the feed system supplying liquid cracking furnaces in the Olefins area remains on stream if only power is lost. NGL furnaces are supplied from pressurized wells. Depending on the duration of the power interruption, the loss of electrical brine pumps will eventually affect the ability of the wells to supply NGL feed to the furnaces. 4 of the 14 cracking furnaces are driven by electrical ID fans and are lost during a power failure. Because of multiple equipment loss in the recovery area, the recycle ethane stream supplying one furnace is stopped. Strategic combination of turbine/electric drivers makes the quench area less sensitive to power failure. It is however disturbed by the furnaces through feed rate reduction and incremental dilution steam requirements. The charge gas compressor (CGC) is driven by a steam turbine and is not directly affected by power but rather impacted by the disturbance caused on the steam and cooling water system. The CGC turbine extracts steam at 200 psig and exhaust at 4 in Hg_a in a surface condenser. Reduction in cooling water capacity caused by a power failure limits the ability of the surface condenser and can cause the CGC to trip on high condenser pressure. Steam pressure fluctuation on the jet ejector system could also cause high condenser pressure. Electrical motors drive all the pumps in the recovery area. A power interruption would therefore cause loss of reflux on all towers and stop fluid flow on streams involving pumps. The propylene refrigeration compressor (PRC) situation is similar to the CGC where the compressor is driven by a steam turbine that extracts at 500 psig and exhaust at 4 in Hg_a in a surface condenser. As for the CGC the compressor could trip on high condenser pressure. Loss of reflux on towers and other key streams significantly reduce the power required to drive the PRC during a power loss. The ethylene refrigeration compressor (ERC) is driven by a 1500/500 psig steam turbine. Its stability during a power failure is affected by the disturbance caused on the steam system and the PRC. Excessive steam header pressure loss could also limit the power supply to the CGC, PRC and ERC turbines, which would cause the compressors to slow down and trip on high vibration as the speed approach the first critical speed. Electrical motors drive all the rotating equipment in the Aromatic unit, which cause all the process flow to stop during a power failure. Steam to exchangers is however regulated on flow control and is still free to flow during a power failure.

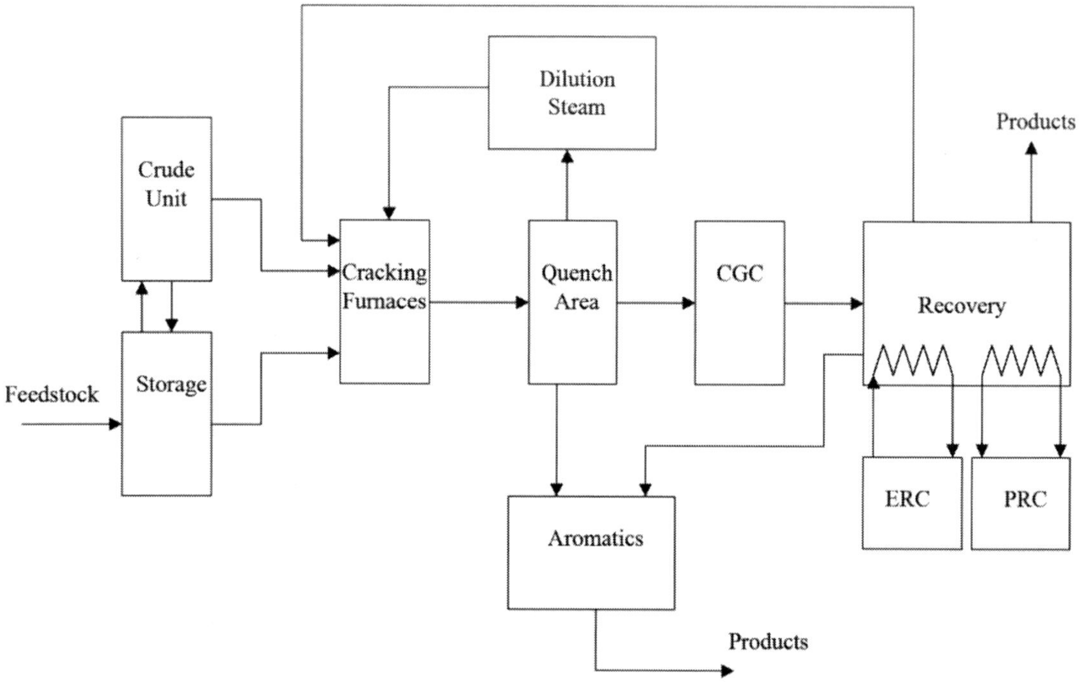

Figure 1: Process Block Diagram

Water/Steam/Condensate System Description and Vulnerability to Power Failure

As shown on Figure 2, raw water from the St-Clair River is pumped to the site from a pump house located beside the river (approximately 3 miles from the plant site). The pump house is equipped with 3 pumps, 2 operating and 1 spare all driven by electrical motors. Given its location, the pump house is not powered from the plant electrical system but directly from the grid. Raw water is first processed in a clarifier and then pumped to the cooling tower for makeup requirements, to the utility water system and to a filtration system. Filtered water is further purified to meet the quality required for high pressure boiler feed water. This process is achieved in an ion exchange unit. Demineralized water from that unit is stored in a concrete vessel from where it is pumped to the deaerator as makeup water to the high pressure boiler feed water system. The movement of demineralized water to the deaerator is ensured by a steam turbine driven pump and an electric driven standby pump. All other pumps involved between the clarifier and the demineralized water storage tank are all electrically driven. Approximately 81.5% of the raw water is used for cooling tower makeup, 15% for boiler feed water makeup an 3.5% for the utility water system.

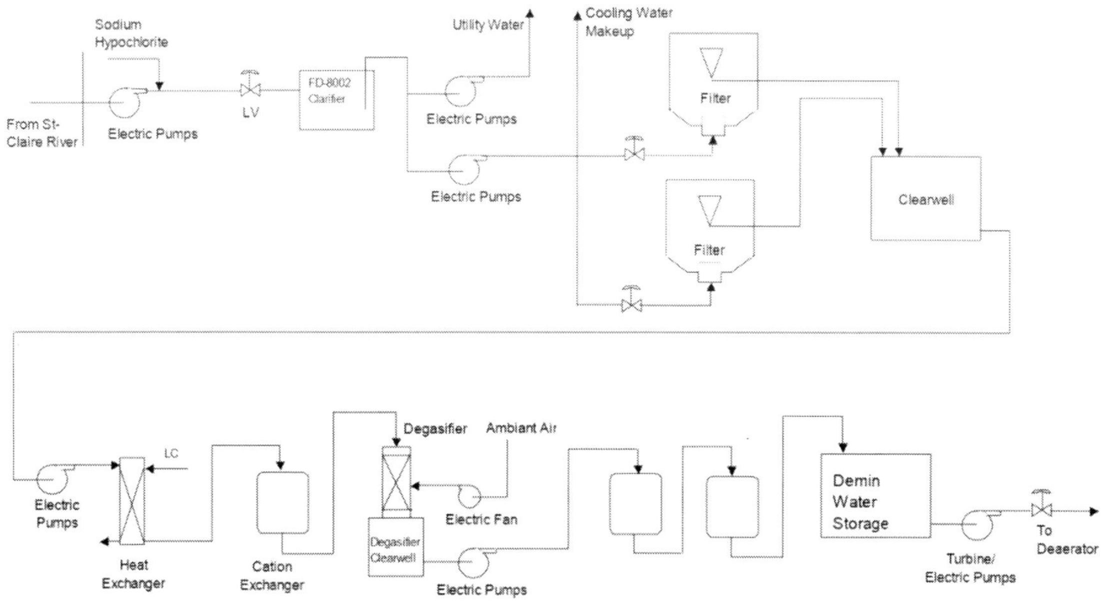

Figure 2: Water Treatment

Demineralized water mixed with condensate returns streams in the deaerator. As illustrated on Figure 3, the deaerated boiler feed water flow by gravity to a vessel underneath, which is referred as the deaerator storage vessel. The deaerator storage vessel provides inventory and static head to satisfy the high pressure boiler feed water pumps NPSH requirements. Its inventory is on a level/flow cascade control loop with the demineralized water makeup stream. 2 turbine driven boiler feed water pumps normally supply 3 high pressure steam boilers and 14 cracking furnaces. 1 electric driven standby pump is available in case of a turbine pump trip or during maintenance activities on one of the turbine pumps. A deaerator storage vessel low level trip protects the boiler feed water pump from running dry.

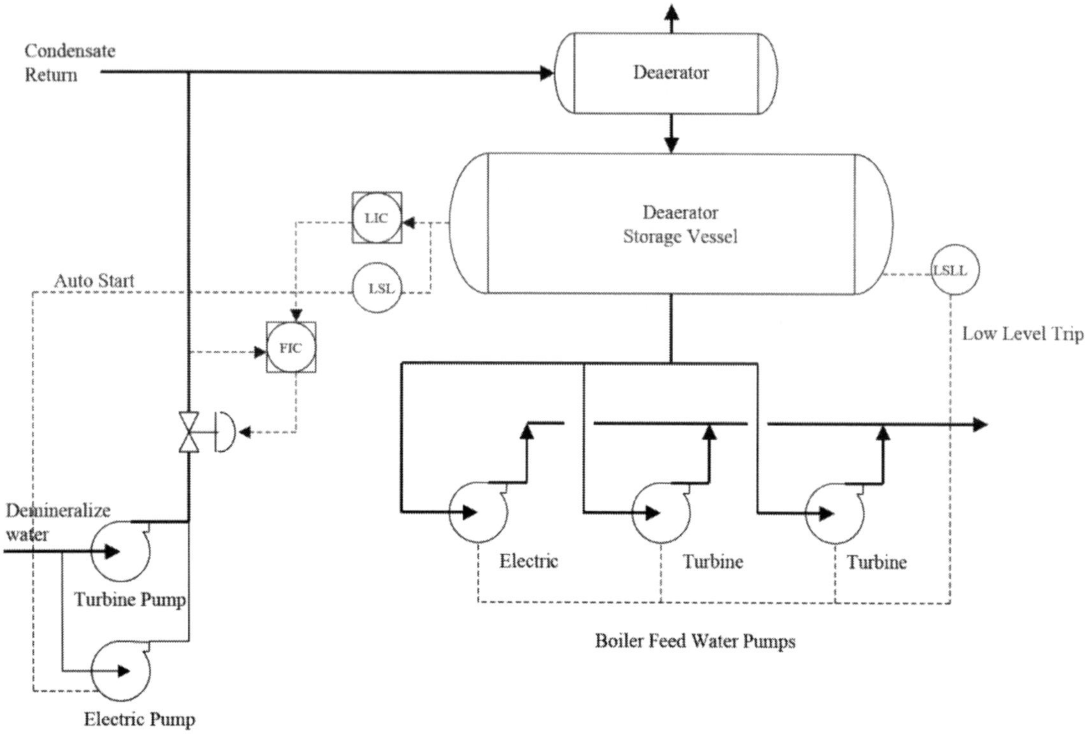

Figure 3: Deaerator

A simplified diagram of the steam system, which includes 4 different levels of steam is illustrated on Figure 4. The very high pressure steam (YS) header, controlled at 1550 psig through boiler firing supply steam required to drive the CGC, PRC and the ERC. The high pressure steam (XS) header, controlled at 500 psig through steam let down; is used to drive 500/200 psig, 500/50 psig turbines and a small number of process units. The medium pressure steam (MS) header is controlled at 200 psig through steam let down and drive 200/50 psig turbines as well as supplying a significant number of process units. Approximately 47% of the MS steam used by the process occurs in the Aromatics unit and 40% in the Olefins dilution steam area. The low pressure steam (LS) header is controlled at 50 psig through steam let down and is entirely used by process units. Dumping of LS steam through a steam dump condenser is required when extracted steam from turbines exceeds the LS steam process demands. This state can be avoided by optimum selection of steam vs. electric driver. However, consideration to plant vulnerability with regards to turbine or power failure leaves little flexibility in selection. Steam balance optimization is also done through manipulation of steam extraction on the CGC (200 psig extraction) and the PRC (500 psig extraction) turbines.

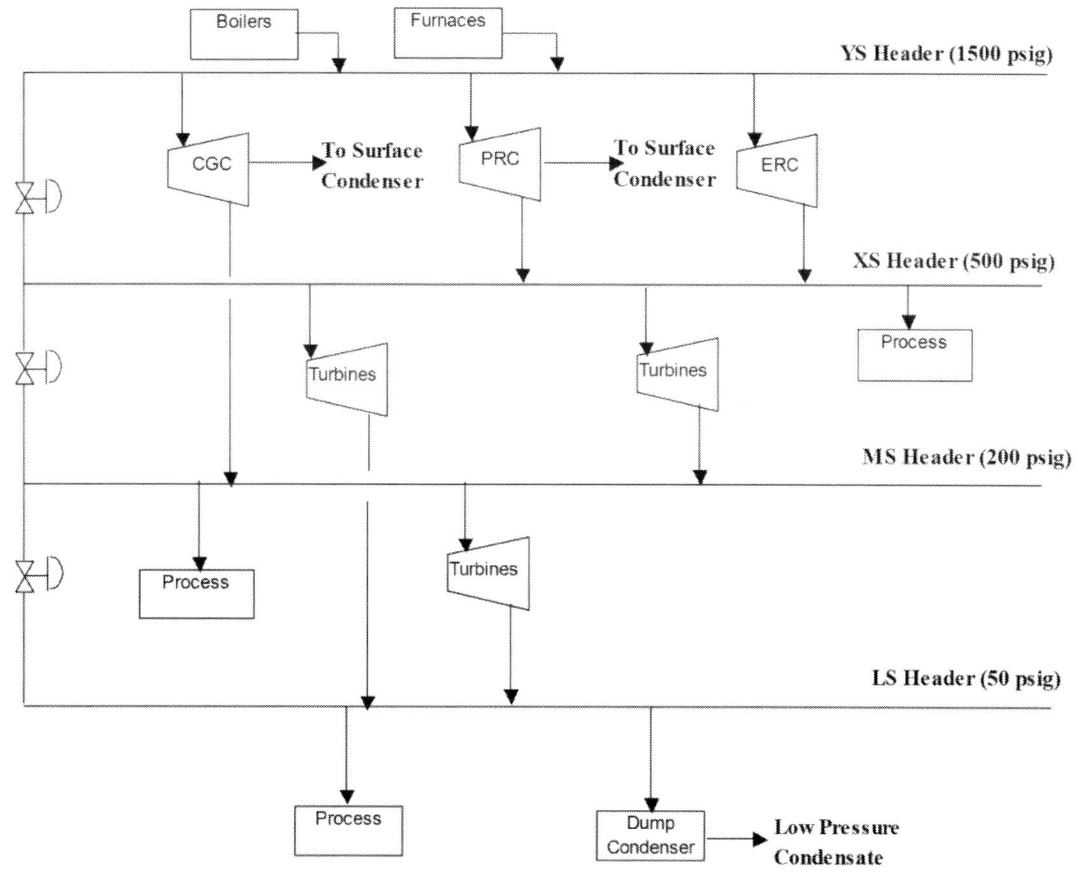

Figure 4: Steam System

Condensate from steam exchangers is recovered in condensate drums located in the different process areas. It is pumped back to the deaerator after treatment in a mix bed polisher. High and medium pressure condensate from the olefins area flows directly back to deaerator by pressure differential. Figure 5 illustrates the condensate recovery system with the combination of pump drivers at each condensate collection drum. The absence of turbine drivers on the Aromatics and one Olefins drums make condensate recovery from those areas vulnerable to power failure. An electric booster pump downstream the mix bed polisher supplies additional head require by condensate to flow to the deaerator. The pump and polisher is bypassed during power failure. The reduction in condensate flow during a power failure allows condensate return without the need of the booster pump. Condensate from the CGC and PRC surface condenser is collected and treated in a similar system. Combinations of turbine/electric drivers are available to pump condensate from the surface condensers back to the deaerator with the turbine drive normally operating.

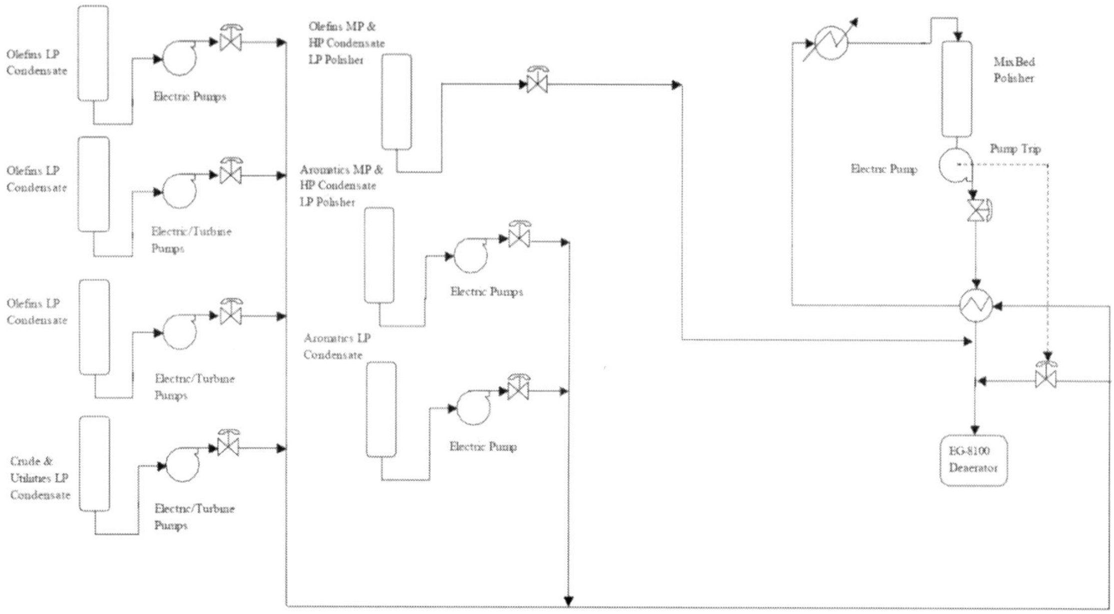

Figure 5: Condensate Recovery System

Power Failure Incidents

Table 1 summarized the power failure incidents that occurred at the site during the last 4 years. The incident of April 2001 did not cause a total site failure as it only involved a drop in voltage. The steam system was not affected while many other process areas were. Given the minor impact, it is excluded from the analyses. The deaerator storage vessel level rate of decay is illustrated on Figure 6 for the incidents of July 2000, April 2003 and August 2003. Time 0 minutes correspond to the loss of power. The boiler feed water pumps are tripped by low level switches, which are positioned at a location equivalent to a 28% reading on the drum level transmitter as illustrated by a horizontal line on Figure 6. Trip of the boiler feed water pump results in a total failure of the steam system, which then cause the boilers and heaters to trip.

Table 1: Power Failure Incidents

Incidents	Duration	Description
July 2000	Less then 10 sec	An equipment failure caused the two 230 kV line to trip
April 2001	Less then 10 sec	Power dip that dragged the system voltage to 50% of its nominal rating.
April 2003	26 sec	Total loss of Power in the Sarnia area
August 2003	> 12 hrs	Blackout

Power was reestablished within 30 seconds on the July 2000 and April 2003 incidents. The disruption was much longer on August 2003 where power was lost during the blackout that affected part of Canada and the US. The level rate of decay on the deaerator storage vessel is relatively similar for all the cases during the first 3 minutes that follow the power loss. After that period the vessel level struggle to re-establish itself on 2 cases but reached the trip point after 65 minutes on the July 2000 event and came within less than 5% from the trip point on the April 2003 incident. During the blackout of August 2003, the duration of the disruption was such that the dematerialized water standby pump was unavailable, which explains that the level rate of decay remained constant and caused a rapid loss of the steam system. The demineralized water standby makeup pump is driven by an electric motor and is automatically started when the deaerator storage vessel reached a low level of 48%. It took approximately 3 minutes for the level to reach 48%, at which time power was already re-established on the July 2000 and April 2003 events.

As illustrated on Figure 7 Significant pressure fluctuations on the steam headers were observed for the July 2000 and April 2003 incident. As expected a total loss of pressure followed the deaerator low level trip that occurred during the blackout of August 2003. The time scale on those figures only covers the first 2 hours of the upset, YS steam pressure was reestablished within 16 to 20 hours and 3 to 4 hours on the XS and MS headers for the July 2000 and April 2003 incident. The rapid drop in header pressure is caused by the loss of 2 boilers and at least 4 furnaces, which reduces dramatically the amount of steam generated. The loss of feed from the trip furnaces reduce the power require to drive the CGC, PRC and ERC. However, heavy flaring and dilution steam requirements cause a significant increase in MS steam demand. In summary the overall demand significantly increases the steam generation capabilities.

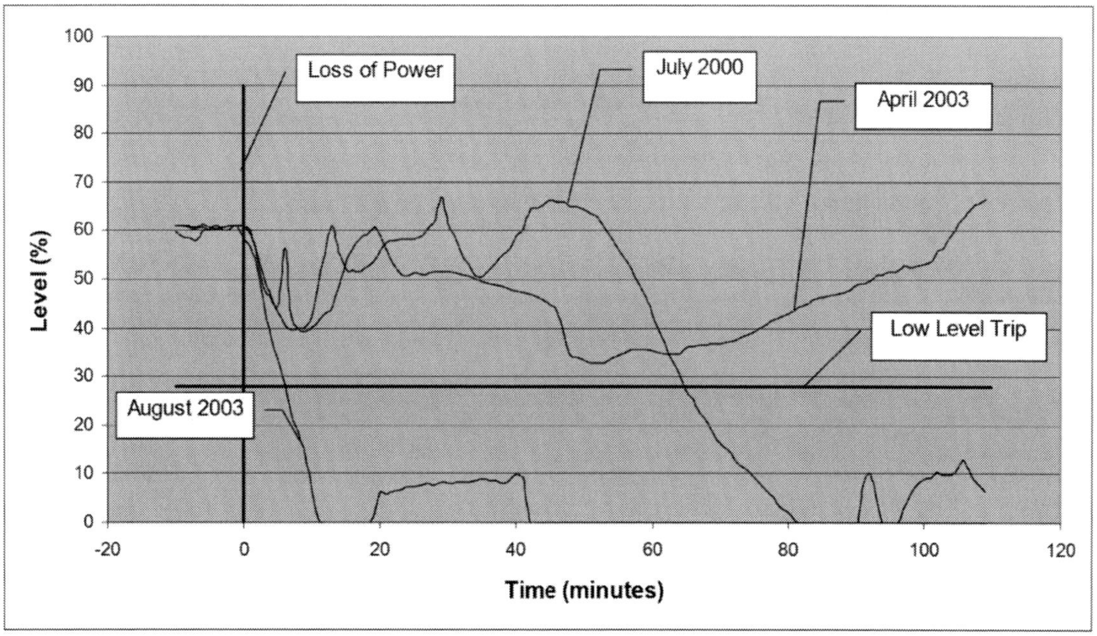

Figure 6: Deaerator Storage Level Trend

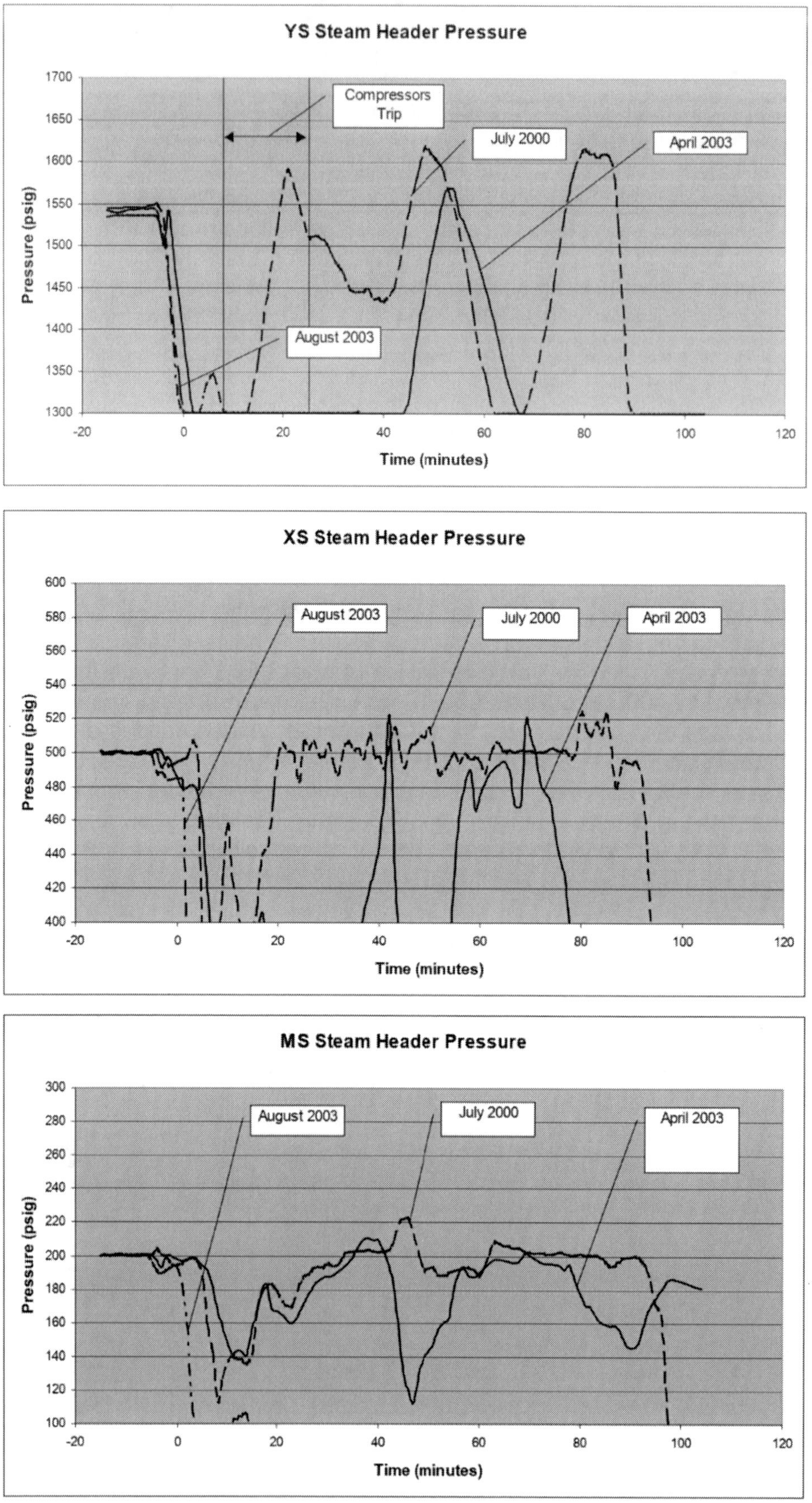

Figure 7: Steam Headers Pressure Trend

Water & Condensate Balance

The blackout of August 2003 has been more thoroughly analyzed, as its consequences are likely to be equivalent to any power disruption exceeding a 30 seconds period. The manual interventions required when an interruption exceeds 30 seconds is expected to make the demineralized water standby makeup pump unavailable once the deaerator storage vessel reach a critical low level.

The water shortage to the deaerator has been estimated from the vessel storage section rate of decay for the time interval included between the loss of power and the low level trip. The following figure illustrates how the level reading has been correlated to liquid height in the vessel.

Deaerator Level Decay

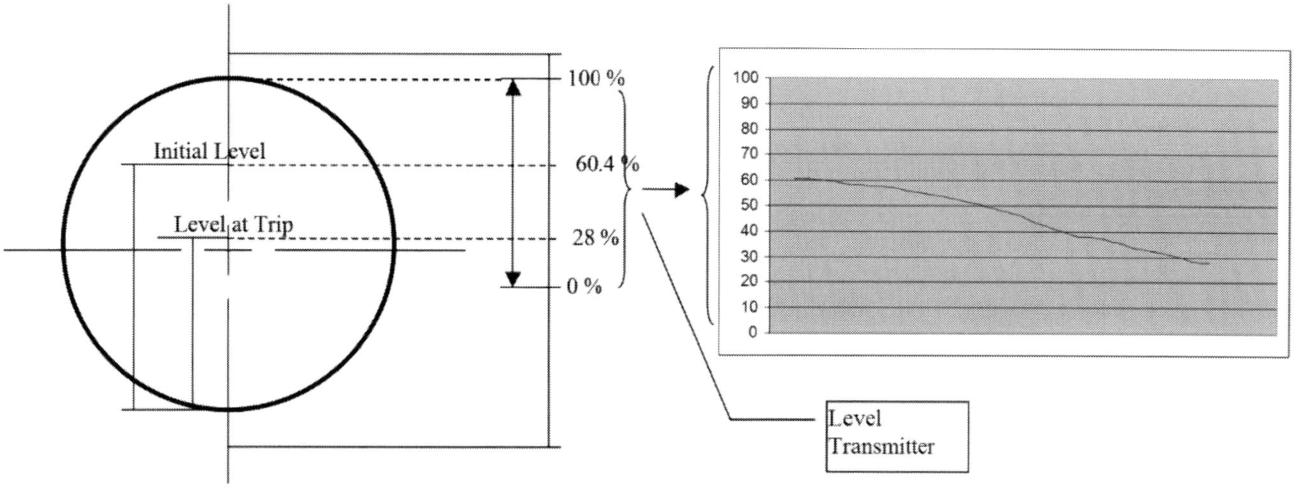

The rate of decay has been estimated from the volume loss over the 6 minutes period it took to reach the trip level. The estimated rate of decay was found equivalent to 88% of one deaerator water makeup pump. In other words, it would require 188% of the capacity of one pump to keep the level steady under those conditions. Condensate losses caused by drums operated by electric pumps is reported on Table 2 based on a 100% condensate recovery assumption (1lb steam = 1 lb condensate).

A high integrity trip system intended to minimize flare loading is activated by the loss of cooling water pumps. The interlock cut steam to the deethanizer reboiler and therefore helps the water balance by eliminating a condensate stream that would otherwise be lost as it is collected in drum operated by an electrical pump. 100% condensate recovery has been assumed from drums operated by turbine pumps.

Table 2: LP Condensate Losses

Condensate Drum	Losses (%) (Rate/Deaerator Rate of Decay)*100
Olefins	30
Aromatics Medium/High Pressure	61
Aromatics Low Pressure	0.3
Total	91.3

As power was lost, flare releases increased causing additional water losses through steam injection at the flare tip. Average losses of 17% of the rate of decay were estimated for the period prior to the low level trip.

15% of the rate of decay is also observed in MS steam makeup to the dilution steam system, which is explained by the reduction of quench oil circulation. The loss of an electric pump on an electric/turbine pump combination affects the quench oil circulation capacity. The circulation is however suspected to be more affected by the quench fitting temperature control loop backing out the quench oil flow on the cracking furnaces that tripped or lost feed (ethane recycle). The dilution steam requirements increase on the furnace affected by the power failure to prevent coke formation in the coils. The heat required to compensate for the reduction in quench oil circulation along with the additional dilution steam requirements is transferred to the steam system. 2 MS steam exchangers normally supplement the heat shortage in the dilution steam system. Under those upset conditions, the 2 exchangers reached their maximum capacity and live MS steam must be supplied in the dilution steam system to achieve the incremental heat requirements. Condensate from the MS steam exchangers is recycled by pressure differential to the deaerator. However, live steam makeup is lost in the process and contributes to the water losses.

Another source of water losses occur on the cracking furnaces steam drums. This was not observed during the August 2003 blackout given the short time it took for the deaerator storage vessel to trip the boiler feed water pumps. However, they were identified as a major source of water loss and largely responsible for the boiler feed water trip during the July 2000 incident. Those loses were caused by a minimum bypass orifice on the cracking furnaces steam drums intended to protect the economizer in the event of boiler feed water valve failure. Removing the feed and cutting the heat flux on the furnace affected by the loss of power cause steam generation to stop, hence stopping boiler feed water to the furnace steam drum. Boiler feed water does however still flow across the minimum bypass orifice causing high level in the drum. A high level control loop purges the excess water into a continuous blow down drum from where it is drained to the sewer. As power is recovered the affected furnaces are restarted on hot steam standby. The boiler feed water requirement under this mode of operation is less then the minimum flow orifice. During normal operation, the boiler feed water stream is normally pinched back to prevent excessive boiler feed water flow. Given the stress and the workload of the furnace field operator, a significant amount of time elapsed before those streams could be pinched back causing significant loss of boiler feed water.

Discussion

Excessive water losses are responsible for the rapid drop of the deaerator storage vessel level on the August 2003 blackout. As indicated above 91% of the water demand exceeding the deaerator makeup capability was caused by the inability to recover condensate from the different process areas. The Aromatics unit alone contributes to 61% of the unbalance. Since the Aromatics unit cannot operate without power as it looses all its rotating equipment, there is no reason why steam should still flow across reboilers and exchangers. Stopping the steam flow to all the process users in the Aromatic unit would have extended the window before the boiler feed water trip from 6 to 55 minutes. This rationale could be extended to the recovery section of the Olefins area where all reflux pumps are driven by electrical motors.

Another significant source of water losses occurs in the furnace's boiler feed water minimum bypass orifice. Removing or blocking the bypass should be considered, as it is not included in many furnace's design.

Limiting steam to flare could also be considered if the boiler feed water supply is in jeopardy. A compromise on a smoky flare is well justified if the intent is to prevent failure of the steam system. The steam flow should not be stopped but rather reduced as a total stop may damage the steam injection hardware at the flare tip.

Rapid loss of the YS header pressure is known to cause damage on furnace TLE's. A leak on one furnace TLE had caused a process fire during the plant restart on the July 2000 incident. Dynamic modeling of the YS steam header has been used to evaluate actions that could limit the pressure fall and is illustrated on Figure 8. A base case that use a steam supply shortage equivalent to the April 2003 incident assumed no action after the power loss. 3 different interventions intended to stop pressure from falling were also tested. Case 1 requires immediate shut down of the CGC with immediate interruption of the steam supply to the Aromatics unit. Case 2 is an extension of the first one by adding immediate shut down of the ERC to the sequence. Case 3 extends on the first 2 cases where the CGC, ERC and PRC are immediately shut down along with steam supply interruption to the Aromatics unit. Even with immediate implementation following the loss of power, the pressure fall is reduced but is still significant. It is however interesting to observe that shutting down only the CGC with steam supply interruption to the Aromatics unit provides the best response among the 3 cases for the first 8 minutes following the loss of power. Adding shut down of the ERC in case 2 makes the pressure decay worse then the first case. The power savings of shutting down the ERC are offset by the loss of extraction steam from the ERC backpressure turbine. The third case, which requires shut down of the three compressors show a recovery in header pressure. Dynamics of controller and flows ramping down caused the initial fall in pressure. The overall steam balance gets within the capability of the remaining boiler once flows are stabilizing. Operators will normally stop the CGC, ERC and PRC when the pressure reaches approximately 1250 psig. If not stopped by the operator, the compressors would likely trip on high vibration as steam can no longer supply the required power, which causes the compressor to slowdown and eventually reach the first critical speed. Based on the simulation, the compressors would have been shut down or tripped within 2 minutes. Operation trends indicate that the compressors stopped within 10 to 20 minutes. The simulation predicts slightly steeper pressure trend then those actually observed during upsets. No credit was taken in the simulation for the loss of feed from 4 to 5 furnaces, which will cause a reduction in power requirements to the 3 compressors. Some other operating interventions are hard to predict and were disregarded by the simulation. The simulation however, shows that an immediate shut down of the CGC along with immediate interruption of the steam flow to the Aromatics unit provide 6 additional minutes before the pressure falls below 1250 psig when compared to the base case. Therefore, implementing those actions could have extended the period before the compressors were brought down from 10 – 20 minutes to 16 – 26 minutes. The extension of this time period is important as it could allow sufficient time for restarting one of the boilers that tripped and reestablished the steam header pressure. This window could potentially be extended further through extraction increase on the PRC turbine and other interventions.

There is no logical reason to keep the CGC on stream if a power failure is to exceed 30 seconds, as all reflux pumps in the recovery area are lost. Even though the CGC does not affect the water balance, it does contribute to the rapid fall of the YS steam header pressure. Shut down of the CGC has to be coordinated with stopping the steam flow to the Aromatics unit. Not coordinating the two activities could deteriorate the situation rather then improving it because of the CGC steam extraction, which supplies the steam level used by the Aromatics unit. Those actions need to occur quickly given the rate of decay of the YS header pressure. Whether it needs to occur immediately after the loss of power or 30 seconds latter still needs to be evaluated. There are good incentives in keeping the refrigeration systems on stream as it reduces flaring and accelerates the plant restart.

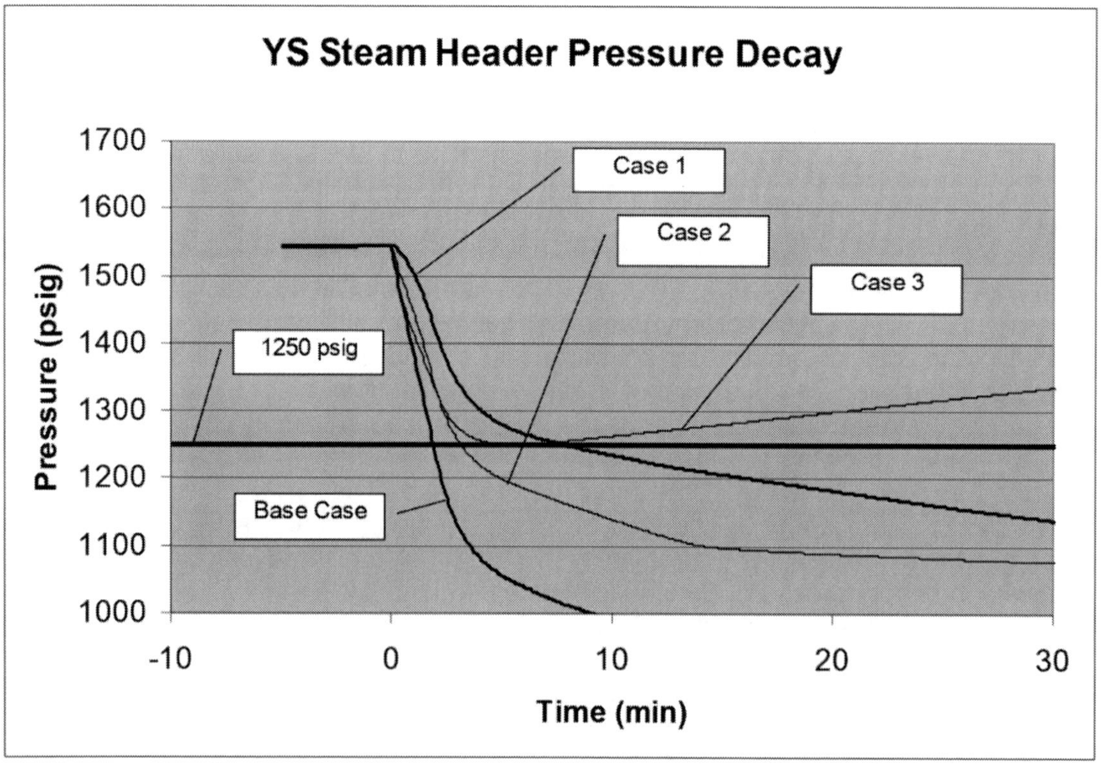

Figure 8: YS Steam Header Dynamic Simulation

Without power the site does not have the capability of processing demineralized water, as most of the pumps in the water plant are motor driven. Demin water supply to the deaerator then becomes constrained by the demin water storage tank reserve. Water supply could be sustained for 3.25 to 3.5 hours without power based on 80% of the full tank content available and high demineralized water makeup flow. The water inventory required to bring the plant down will then dictated how long an idle state can be maintained before implementing a total shut down of the facility. Again, water conservation is a key element in extending that time window.

Power Failure Strategy

A clear understanding of priorities should be defined prior to developing any plan or procedure intended to address power failure upsets. Obviously the first priority is to prevent any unsafe conditions and minimize equipment damage. Understanding of system limitation is key to successfully achieving the desired idle or control shut down state. Table 3 provides priorities that should be given to the key system affected by a power interruption in an attempt to reach a safe idle state.

Table 3: Power Failure Strategy

Systems	Priority	Action	Objectives
Water & Condensate	High	Stop Steam to the Aromatic Units. Stop Steam to any unnecessary unit in Olefins area. Design Change on Furnaces (min bypass orifices).	Prevent the boiler feed water to trip on low deaerator storage vessel level.
Steam System	High	Steam Shedding – Shut Down CGC (coordinated with the action of stopping steam to the Aromatics Unit) Boiler Restart	Prevent steam header pressure from falling. Keep the PRC & ERC on stream.
Protect Furnaces	High	Provide incremental dilution steam to trip furnaces.	Prevent coking.
PRC	Medium	Keep the PRC on stream	Speeds up plant re-start & reduce flaring.
ERC	Medium	Keep the ERC on stream	Speeds up plant re-start.

There may not be sufficient dilution steam available depending on the capability of the steam system and the number of furnaces affected by the power interruption. In such case, some furnaces would need to be sacrificed by cutting the dilution steam flow. The furnaces on stream time, type of feed, vulnerability of the coil and configuration for plant restart (minimum number of furnaces, feed types) have to be considered in selecting those to be sacrificed.

The actions listed on Table 3 require many tasks to be initiated by a number of field and board operators. The sequence of implementing those actions is also critical. The duration of the power interruption is another key element to consider. Interruptions lasting less then 30 seconds are of less-consequence than those exceeding it. A short interruption may require a more-limited response then a longer one. Given all the variables to consider and the coordination required, responding to a power failure by a procedure that relies strictly on human intervention has a poor likelihood of succeeding. Automation or a combination of automation and human intervention needs to be considered for effectively implementing a procedure intended to handle a power failure event.

Conclusion

The North American blackout of August 2003 caused a power interruption never experienced at the site. The analyses of the incident provided valuable information for identifying the weak areas that caused a total steam system failure. The steps required to minimize equipment damage and bring the plant to a safe idle state were identified. Full effectiveness of the response would however require the combination of human intervention and a certain degree of automation. Procedures relying on human intervention are still the only available tool for responding to such an upset. Even though not expected to achieve all the required steps a significant reduction in damage is still expected. The idea of automation is being entertained but is at a very early stage of development. It is now understood that a number of systems need to come down during those upsets. The assumption that the plant could ride through a power failure is no longer true. Procedures and communication to operators have been updated accordingly.

Nomenclature

TLE	Transfer Line Exchanger
FD fan	Force Draft Fan
ID fan	Induce Draft Fan
UPS	Uninterrupted Power Supply
kV	Kilo Volts
MVA	Mega Volts Amps
NGL	Natural Gas Liquid
CGC	Charge Gas Compressor
PRC	Propylene Refrigeration Compressor
ERC	Ethylene Refrigeration Compressor
NPSH	Net Positive Suction Head
YS	Very High Pressure Steam
XS	High Pressure Steam

Nomenclature

MS	Medium Pressure Steam
LS	Low Pressure Steam

References:

1. Donald F Schneider, Plant power failure and its indirect effects: a case study, Petrochemicals and Gas Processing Winter 1998/99.

2. D.P. Wallace, P.J. Nightingale, A.P. Walker, Failure of Boiler Feedwater Pump Turbine following Site Power Failure, International Journal of Hydrocarbon Engineering October 1997.

3. Harry Forbes, Blackout remedies may come from the HPI, Hydrocarbon Processing October 2003.

4. Donald F Schneider, Improving plant reliability: Look Beyond the Usual Suspects, Chemical Engineering October 2003.

Shell Chemical's Deer Park Plant

OP-III KC3-3001 PGC 70M Vibration

Authors

Mia Zager and Bryce Kagay, Shell Chemical LP, Houston, Texas

Contributors

Charles D'Angelo III, C.J. Lloyd and Axel Muzerie, Shell Chemical LP, Houston, Texas

Abstract

Shell Chemical LP's OP-III olefins unit, located in Deer Park, Texas, has a process gas train consisting of a 60,500 hp GE turbine driving three Elliott M-Line centrifugal compressors in series. Approximately 5.5 years after the last overhaul, the middle compressor developed large amplitude vibration that would come and go. A technical team was chartered to conduct a root-cause analysis of the vibrations without taking the machine off-line. The team theorized that the most probable cause of the vibration was a rub at the balance piston seal exacerbated by polymer build-up and thermal expansion. The vibration signature and other supporting data were consistent with the theory. Recommendations were made to increase the flow of wash oil, to change wash oil chemistry, and to limit discharge temperatures. Operation was able to continue with the vibration through 2004.

Introduction

Shell Chemical LP's OP-III olefins unit located at its Deer Park, Texas site has a process gas train consisting of a 60,500 hp GE turbine driving three Elliott M-line centrifugal compressors in series: 88M, 70M, and 46M (See Figure 1 – Plot plan of OP-III PGC). Approximately 5.5 years after the last overhaul, the middle compressor developed a vibration. A technical team was chartered to investigate the phenomena while the machine remained on-line. The investigation immediately noted a repetitive pattern to the vibration, which typically peaked during mid-day and repeated every twenty-four hours. Further analysis of the vibration signature noted vibration frequency at two times running speed, continuously increasing phase lag, higher vibration magnitude at the suction-end of the compressor, and increased bearing temperature at the suction-end of the compressor. High temperature in the balance line led investigators to conclude there was a rub at the balance piston. The balance piston is located at the discharge end of the compressor and a rub at that location was consistent with the observed vibration signature and other supporting data.

By moderating process gas temperature, operation has continued throughout 2003 and 2004. The ever-present rub led investigators to conclude that a metal-to-metal rub was unlikely to be sustainable and that only a rub against polymer could last in excess of one year. Polymer has been observed in the compressor train during past overhauls and in other compressors in similar service in the industry. During warmer times of the day, it is theorized that as process gas temperatures increase, thermal expansion of the compressor rotor relative to the stationary gas path components causes tighter clearances in the machine, which increases the severity of the polymer rub.

One of the recommendations from the investigation was to increase the flow of wash gasoline to combat polymer formation. Guidance was also given to operations to limit process gas temperatures. The intent was to limit thermal expansion, which could cause the rub to be severe enough to impact the operation of the machine and of the unit.

This paper describes the most probable cause to the vibration, the supporting data, and recommendations made to mitigate the risk of a vibration trip until the next planned outage. The findings are potentially applicable to process gas compressors in all olefins plants.

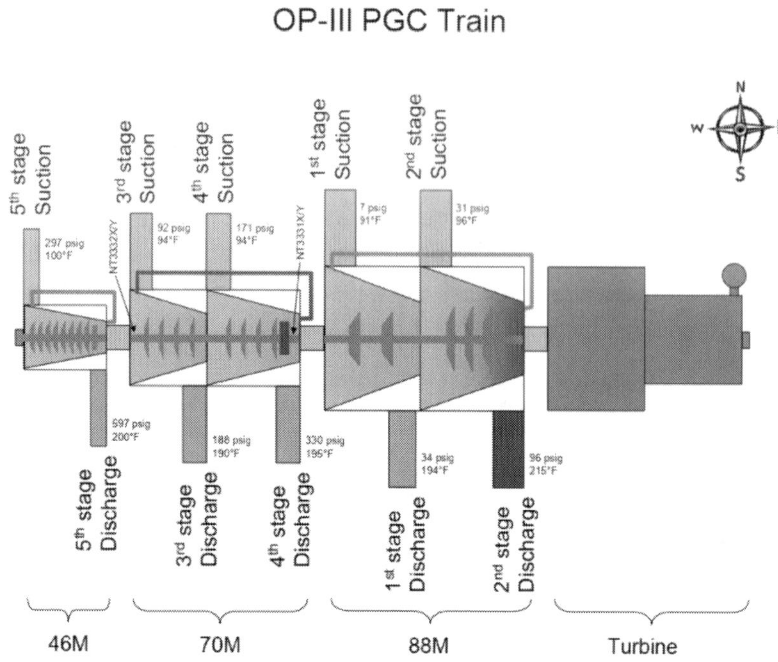

Figure 1. Plot plan of OP-III PGC

Most Probable Cause Theory

Photo 1. Illustration of PGC 70M balance piston

Balance Piston Rub Theory

The balance piston is located on the discharge end of the rotor (See Photo 1). The differential pressure across the balance piston and seal acts against the balance piston area to control the majority of the thrust developed through compression. The radial clearances between the balance piston and the balance piston seal are extremely tight with a specification of 0.004 inch to 0.0065 inch. This is in the same range of the specified radial bearing clearance of 0.0042 inch to 0.0052 inch. The accumulation of polymer (See Photo 2) between the balance piston and the balance piston seal makes the clearances even tighter (See Figure 2). In fact, there is likely a continuous small rub at the balance piston. This rub at the balance piston creates a localized hot-spot that further promotes hard-polymer formation. This hot spot creates the heat source that causes the gas in the balance line to be hotter than the discharge gas.

During the hottest periods of the day, process gas temperatures rise by about 10°F. Increase of the 4th stage discharge temperature causes the balance piston to grow faster than the case can grow. Also, if the balance piston grows, so will the thickness of the balance piston seal. Clearance at the balance piston may close by 1.6 mils or more ($6x10^{-6}$ in/in/°F * 27 in diameter *10°F = 1.6 mils).

As the clearance closes, the rub will become more severe. The rub will likely be on the bottom half of the balance piston seal because that is the tightest clearance due to normal bearing wear. The deflection shape of the rotor will start to change and the bearing nearest the balance piston will become less loaded.

The rub excites vibration; it will be most severe on the west end of the rotor, since the balance piston rub constrains vibration on the east end. The west-end vibration will increase friction at the bearing and raise the bearing temperature.

As the vibration continues, the rub will clear as the abradable material wears away. Also, as the transient temperature excursion reaches new equilibrium, the balance piston seal will grow away from balance piston. Both of these actions will relieve the rub and reduce vibration.

As the process gas temperature cools, clearances open and fill in with additional polymer. This sets up the next vibration excursion.

Photo 2. Illustration of polymer on Balance Piston Seal

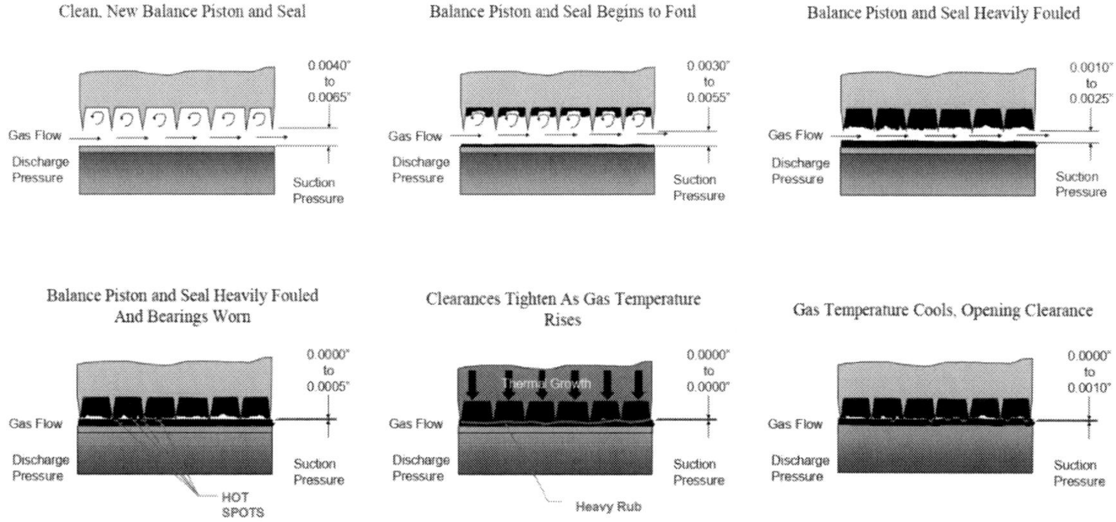

Figure 2. Accumulation of polymer at balance piston

Most Probable Cause Theory Data

An extensive amount of data was reviewed during the investigation. The following categorizes the data into supporting, consistent and inconsistent with the most probable cause theory of a rub at the balance piston:

Supporting Data:

A. Have observed polymer build-up in balance piston grooves at other Shell Chemical LP's ethylene units.
B. Other manufacturers report vibration in Elliott M-line compressors because of polymer build-up in balance piston area.
C. High balance line temperature (~50°F higher than discharge temperature).
D. Increased bearing clearance would increase likelihood of a rub at the balance piston.
E. Orbits develop flat spots as the vibration grows.
F. Vibration vector phase lag increases continuously in a "loop" on 1X & 2X on east and west bearings as the vibration grows and subsides.
G. Vibration phase angle continuously increases in lag from vibration min to max to min, again and again, even when speed is constant.
H. Vibration comes and goes.
I. Observe reverse precession during most vibration episodes, but some vibration episodes have forward precession. Partial rubs create reverse precession, but full annular rubs can create forward or reverse precession.

Consistent data:

A. Periodic polymer growth into balance piston seal could cause vibration (rub) to come and go.
B. Higher amplitude vibration on compressor suction end - opposite balance piston, because rotor is less constrained there.
C. Polymer can be created & may be removed & can come again. Polymer formation can come and go.
D. Hot, contaminated seal oil drain temperature is consistent with rub at the balance piston (210°F seal drain which is greater than discharge temperature of 197°F).
E. Dominant 2X vibration during episodes (rub could cause 2X, but so could other mechanisms).
F. Shaft centerline plot during episode moves towards center (instead of toward bearing wall). This suggests that the rotor is against the rub and begins to vibrate more.
G. Bearing temperature increase could be the result of increased bearing load (from shaft being forced toward bearing or from higher vibration)
H. An increase in radial gap voltage indicating the shaft has moved down & closer to stationary object.
I. 1.5 to 3 hour periodicity is explained by the amount of time it takes to abrade away polymer and time for temperature to reach equilibrium in compressor case.
J. Daily periodicity is explained by daily process temperature cycles.

Inconsistent data:

A. Steel-to-lead rub is not sustainable over a day. Even steel-to-steel rub is doubtfully sustainable over very many days.
B. Polymer was found in the Shell Chemical LP's Norco OL-5 ethylene unit balance piston seal with some signs of a minor rub but elevated vibration was not observed.

Recommendations

The investigation team recommended the following short term and long term mitigation plan:

1. Maintain discharge temperature limit at 215°F and re-evaluate if there are repeated alerts. Figure 3 shows the relationship between vibration and temperature.

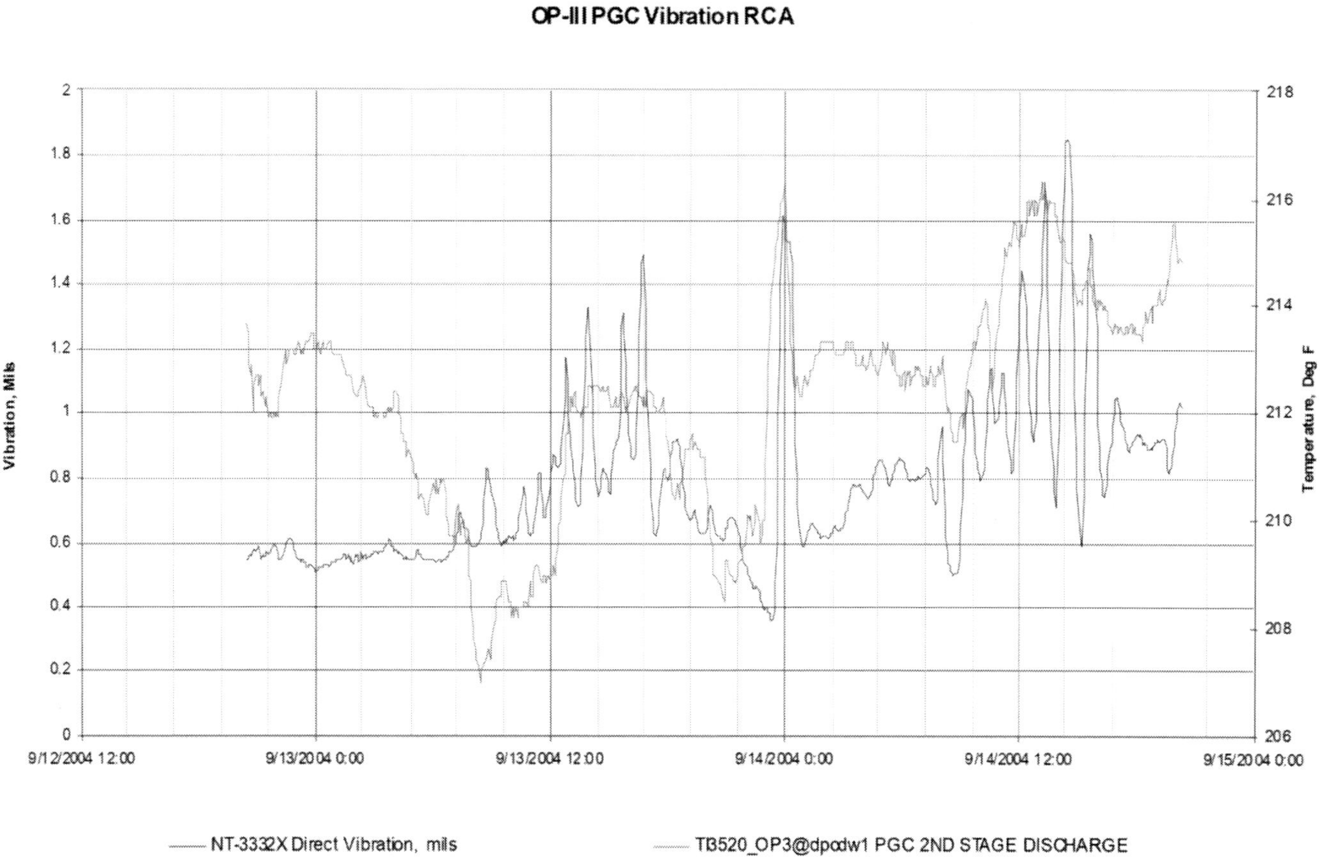

Figure 3. Process gas temperature and vibration

2. Consider adding water quench to the 2nd, 3rd, and 4th stage suction lines to reduce process gas discharge temperatures.

3. Consider adding specialty chemical to on-line CLGO wash to inhibit and remove polymer growth more effectively.

4. Consider coating spare rotor, including balance piston, to reduce polymer accumulation (See Photo 3).

Photo 3. Illustration of 70M coated rotor

Conclusion

This paper describes the vibration problem with the Elliott 70M compressor of the process gas compressor train at Shell Chemical LP's Deer Park plant. The problem was analyzed without the benefit of a machine outage and inspection. The most probable cause theory was polymer accumulation in a tight fitting balance piston seal that causes high compressor vibration. This may be applicable to any process gas compressor with sufficient fouling in tight areas. Thermal expansion exacerbates the problem. The recommendations proposed are intended to reduce the vibration and should be considered by any olefins plant operator with a process gas compressor vibration problem that seems to be related to polymer formation.

FLARE MINIMIZATION STRATEGY DURING PLANT UPSETS: FREEPORT

By
Steven Krietenstein
The Dow Chemical Company
Freeport, Texas

Prepared for Presentation at the 2005 AIChE Spring National Meeting
17th Annual Ethylene Producers' Conference
Session TA009 – Ethylene Plant Operations
Atlanta, Georgia
April 12, 2005

UNPUBLISHED

FLARE MINIMIZATION STRATEGY DURING PLANT UPSETS

ABSTRACT

Dow owns and operates an E/P cracker in Freeport, Texas. The cracker was started up in the early 1970's and has a nameplate capacity of 1.3 billion lbs/year ethylene. During plant startups, shutdowns, and off-spec incidents, hydrocarbon streams within the plant are diverted to the elevated flare. In 2001, a project utilizing Six Sigma methodology was initiated to minimize the amount of hydrocarbons flared during plant startups, shutdowns, and off-spec incidents, in order to reduce the impact to the environment and the local community. The reduction in hydrocarbons flared was achieved by minimizing startup / shutdown times, by utilizing existing recycle lines, and by maximizing recovery of products. Problems which occurred during previous startups and shutdowns were also identified and a plan for overcoming those problems was developed. From the start of the project in 2001 to the end of 2003, the project reduced emissions during startups, shutdowns, and off-spec incidents by 54% and resulted in a documented savings of $2.5 MM. The paper will describe the Six Sigma methodology used to execute the project and the strategies that were implemented in the plant.

INTRODUCTION

The Light Hydrocarbons 7 (LHC-7) plant's flare serves a number of beneficial functions. It prevents the release of off-spec hydrocarbons, a potentially volatile emission, directly to the environment. In an upset condition, such as a plant start-up or shutdown, the flare ensures system stability by burning excess hydrocarbon vapor down to an environmentally safe state.

Despite these benefits, the community has a different perception of flaring. Noise, light, and smoke emitted from the flare during plant upsets draws concerns from the increasingly sensitive community. Stricter environmental laws have also limited the amounts of hydrocarbons legally emitted from the plants including flare emissions.

Prior to 2000, nearly all off-spec hydrocarbons at LHC-7 were sent to the flare for incineration. In 2001, the "LHC-7 Flare Minimization Project" was completed to optimize the use of equipment that would re-circulate out-of-spec hydrocarbons to the front of the plant to be reprocessed again and brought into specification. The analysis was broken into three distinct categories:

- Plant Start-up
- Off-Spec Upsets
- Plant Trips/Shutdowns

The procedures and controls were optimized to reduce flaring in each of these events. The system impacts throughout the plant were considered and optimized.

PROCESS OVERVIEW

The plant is an ethane/propane cracker with a name plate capacity of 1.3 billion lbs/year ethylene. Figure 1 shows the front-end de-methanizer / back-end acetylene reactor configuration. The ethane/propane feedstock are split into their pure components in the LPG splitter and fed to the furnaces separately. The furnace effluent is then cooled in the quench system and compressed to 500 psig by the cracked gas compressor. After compression, the gas is chilled through a series of heat exchangers in the chill train and then de-methanized. Then the C2s are split from the C3+ in the de-ethanizer from which the acetylene is reacted out in the acetylene reactors. Finally the ethylene is split from the ethane in the C2 splitter and sold as product from one of the two "bullet" storage vessels.

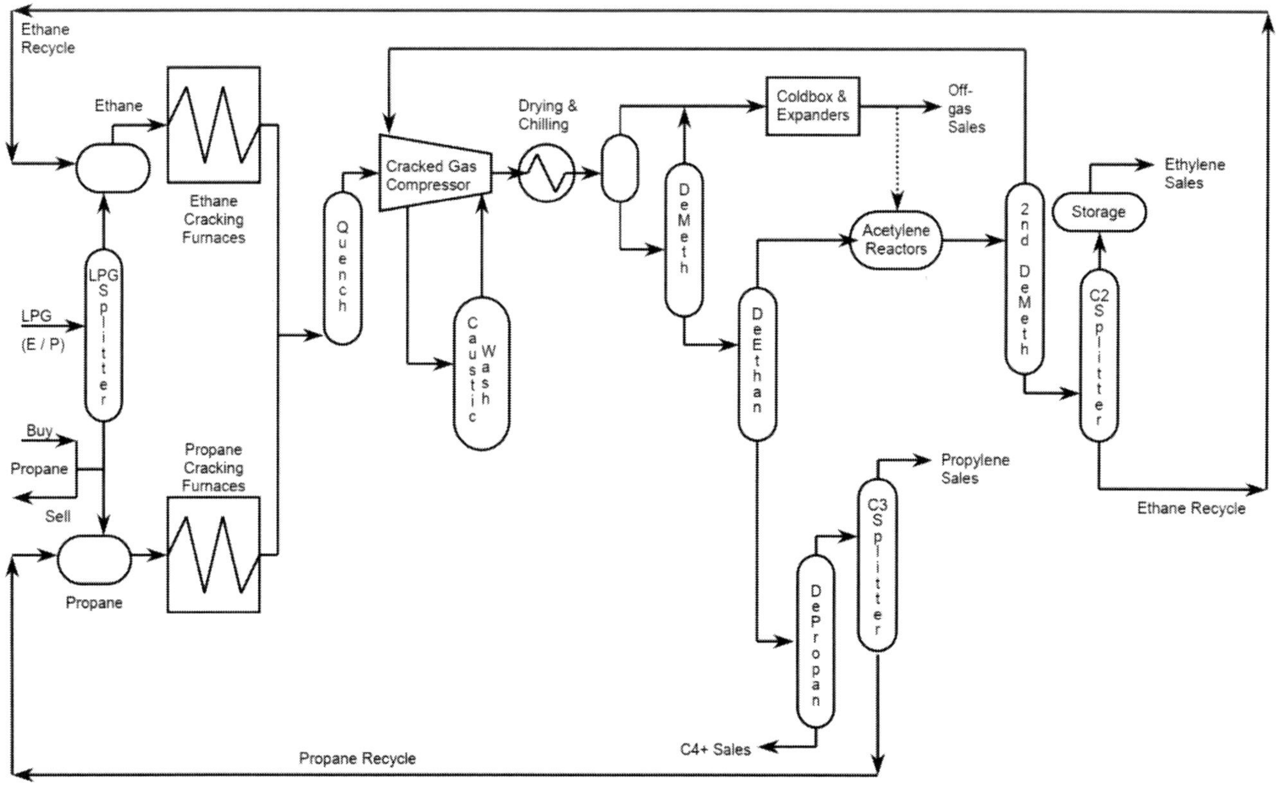

Figure 1: LHC-7 Simplified Ethane/Propane Cracking Process Diagram

The product storage was originally designed to store about 45 minutes worth of off-spec ethylene at which the product must be returned to the product header or the material will have to be flared. This limited storage capacity resulted in high quantities of off-spec material flared. A better way of handling this off-spec stream was needed.

RECYCLE HARDWARE – PIPING / CONTROLS

The two process locations from which off-spec flaring is most common are the ethane/ethylene mixture downstream of the reactors and the ethylene product from the storage drum. Piping was installed in the late 1990s to enable operations to recycle off-spec ethylene from these locations to the front end of the plant (see Figure 2). These locations could be lined up to the inlet of the furnaces or to the suction of the Cracked Gas Compressor. A control valve and flowmeter were added to control this flow.

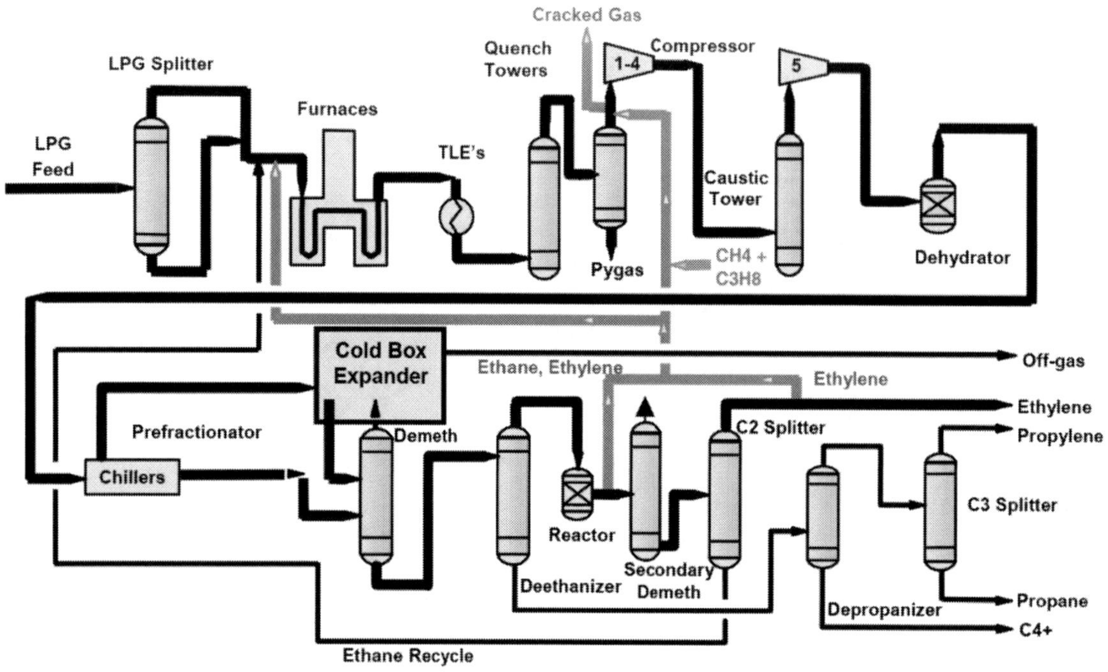

Figure 2: Recycle Piping

To help mitigate flaring during start-up situations, steam piping was also installed from the adjacent power plant to help minimize the amount of self-produced steam required. This external supply helps reduce the amount of furnace feed gas required to be flared in order to produce the steam required to start-up the steam turbines that drive the three compressors (550,000 lb/hr steam). In addition, piping was added to introduce propane and methane directly to the suction of the Cracked Gas Compressor to facilitate a "flareless start-up". In theory, this would allow the compressor to be started-up on total recycle (false load) with propane/methane, avoiding the flaring of furnace effluent gas.

PLANT START-UPS

LHC-7 had some particular challenges with minimizing Start-up flaring: shortages of high pressure steam, smooth Cracked Gas Compressor start-ups, pressure safety valve relieving, furnace start-up operating discipline, and back-end operating discipline. To reach the goal of minimizing flaring, each of these aspects was analyzed and optimized using the Six Sigma methodologies.

The Six Sigma approach requires the team to define an appropriate "defect" that, with the appropriate use of the Six Sigma tools, will be measured and analyzed before any improvements are made. Based on this detailed analysis, improvements are prioritized and implemented to most effectively reduce the defect's occurrence. In the "Start-up" case, the defect was defined as a "pound of hydrocarbons flared". A general philosophy used in Six Sigma is to track the defects per "opportunity" (to create the defect).

$$\frac{\text{Defect}}{\text{Opportunity}}$$

In this way, tracking the improvements is not confused with other improvements in the plant that might reduce the opportunity. For the start-up analysis, a

"pound of HC flared per start-up"

was used for measuring and tracking the improvement. By using the "per Start-up" as the "opportunity" to produce a pound of flared material, this rids the defect of getting credit for other improvements made such as plant reliability. In other words, if there were less start-ups, the data would show that there had been a reduction in start-up flaring, but not because of this project.

LHC-7's normal start-up procedure can be summarized as:
(1) Ramp up furnace feed which creates the feed for the cracked gas compressor while helping to make the high pressure steam to help drive the compressor's steam turbine;
(2) Start the cracked gas compressor;
(3) Start-up the chill train;
(4) Start-up tower operations;
(5) Bring acetylene reactors on-spec; and
(6) Bring the product on spec.

As mentioned earlier, high pressure steam piping was installed from the power plant to minimize the amount of self-generated steam required for start-up. Unfortunately, this pipeline was proven to be too small for a complete "flareless start-up". The start-up data showed that we were lacking about 160,000 lbs/hr of steam for the start-up of the cracked gas compressor even with use of the imported steam. Therefore, furnace hydrocarbon load had to be used to help generated the required steam and nullified the use of the Propane/Methane "flareless start-up" piping. Every effort was made to maximize steam production without adding hydrocarbon load such as running excess air in the furnaces i.e. excess draft in all running furnaces (online and hot steam standby) to promote steam production. Even more effort was done to optimize the amount of steam pushed through the steam import line which required direct communications with the external suppliers of high pressure steam.

When analyzing past start-ups, a smooth start-up of the Cracked Gas Compressor proved to be essential in avoiding unnecessary flaring. Many past start-ups have resulted in surging this machine upon start-up as the discharge of the machine slowly builds discharge pressure. To combat this issue, LHC-7 utilized a high pressure ethylene stream available from our downstream customers to pressure up the back-end of the compressor before it is brought on-line. This new operating discipline coupled with better false load controls has significantly reduced the risk of harming the mechanical internals of the compressor during start-up.

In several start-up incidents, the inability to adequately control tower pressures on cold service distillation columns caused excessive flaring. This was due to pressure safety valves (PSVs) relieving and not reseating when the pressure returned to normal. Data and experience showed that the de-ethanizer is the most prone to relieve its PSV. During start-up conditions, the de-ethanizer experiences stability problems due to a lack of sub-cooling of the propylene refrigerant in the overhead condenser. This sub-cooling is absent during start-up because this refrigerant stream is normally cross exchanged with the product ethylene (See Figure 3).

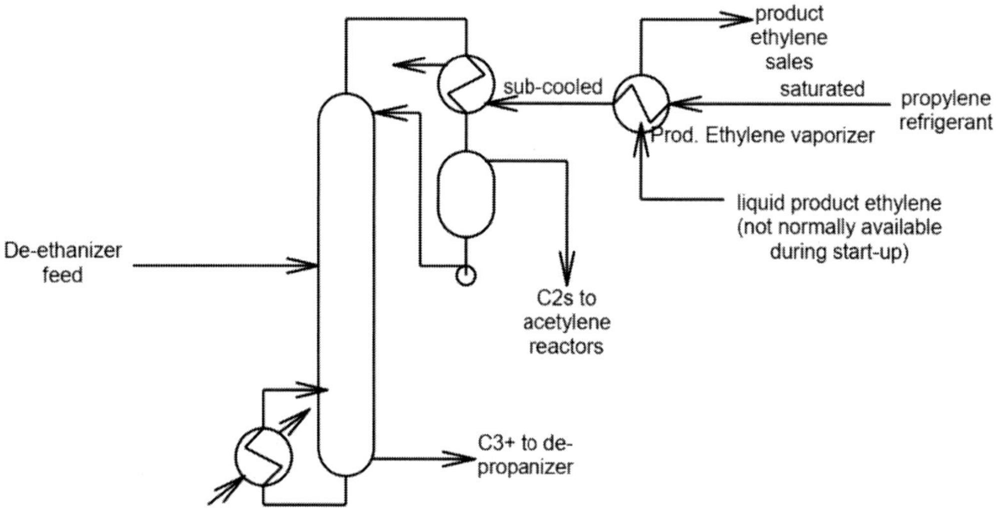

Figure 3: De-ethanizer Unit Operation

This issue causes several subsequent effects such as instability in the acetylene reactors in addition to the pressure problems within the unit operation itself. Team members implemented a novel approach by re-circulating pure ethylene through this cross-exchanger to get the necessary sub-cooling. Use of this procedure has decreased the start-up time to stabilize this column and subsequent reactor operations which enabled the plant to reduce the amount of time to create on-spec product. There is a drawback to this operation, however. This can delay the start-up timing significantly if the product ethylene storage drums and the C2 splitter column were de-inventoried during the shutdown or trip.

Other operating discipline items were also studied. One of the more critical start-up disciplines is furnace load management. Speeding up the furnace load ramp-in rates helps to minimize the flaring time while building to the minimum load for the start-up of the cracked gas compressor. In addition, the timing when furnace ramps are begun is important so that excess furnace gas is not flared while preparations to the compressor are still being completed. Other back-end operating discipline improvements consisted of using the recycle piping mentioned earlier. This discipline will be described under "Off-spec Upsets".

Even though some compromises were made (e.g. self produced steam), the overall start-up philosophy and operating discipline were optimized. From a defect perspective, a 64% improvement was made from the baseline "pound of HC flared / start-up".

OFF-SPEC UPSETS

The most typical, and by far, the most significant flaring potential while the plant is running is due to off-spec acetylene contamination. Some upsets are significant and many are minor. So a way of judging project improvements had to be devised so that big events were not penalized for their severity. Again, a Six Sigma "defect per opportunity" was chosen to judiciously critique this type of flaring:

"pound of HC flared per pound of off-spec produced"

As mentioned earlier, the piping was added to recycle off-spec material to the front end of the plant. The first attempt to use this new piping was somewhat unsuccessful. As the plant recognized that off-spec material was being produced during a minor upset, the recycle piping was lined up to the front end of the plant i.e. to the cracked gas compressor suction (see Figure 2). This additional flow through the compressor caused it to significantly slowdown and began to upset other parts of the plant. The recycle flow was quickly stopped and the upset was handled in the normal fashion. This lesson helped the team to understand the term "getting fat" when recycling. At full load, our systems have little capacitance to carry higher loads especially flows as high as 25% more than the normal flow. This lesson was also extrapolated to some of the new recycling piping that allows the plant to recycle product ethylene back to the front of the acetylene reactors. A quick calculation shows that when the recycle flows and the plant flows are added together, the sensitive acetylene reactors will not handle the flow swings nor the sum of these flows. Hence, this recycle piping back to the acetylene reactors should never be used.

To avoid the "getting fat" syndrome, the recycle flow had to be routed to the furnace feed. Using this piping, the feed to the cracked gas compressor was kept constant and helped maintain stable operation of the back-end of the plant. However, as this flow is introduced to the furnace ethane feed, the LPG raw material flow must be backed down to make room for the recycled flow, otherwise, the ethane feed system will pressure up. Much attention was given to optimizing and automating the ramping out of LPG (Ethane) feedstock. This affected the operation of the LPG splitter, ethane feed vaporizers, ethane storage vessel, and ethane charge drum operations. Each set of process controls had to be redesigned.

Other process control changes had to occur as well. A side effect of recycling ethylene to the furnaces is that ethylene as a furnace feed doesn't re-crack. In other words, ethylene feed doesn't result in the endothermic reaction like ethane or propane creates. If prompt action isn't taken to reduce furnace heat duty proactively before recycling ethylene, the furnaces tend to automatically trip on high coil outlet temperature due to the absence of the endothermic reaction.

Recycling ethylene also affects the back-end. When the ethylene flows through the furnaces, no hydrogen is produced as when ethane is cracked. Therefore, hydrogen concentrations drop proportionately to the recycle flows in the back-end of the plant. This reduction in "lights" production causes the turbo-expanders to slowdown. Consequently the cooling capacity of the turbo-expanders can limit the amount of off-spec ethylene recycled. At full rates, the plant must limit recycling to 50% of the total product flow i.e. the other half must be flared. At lower plant rates, such as start-up conditions, this percentage of recycle flow can be as much as 90% without turbo-expander problems due to the extra capacity of the ethylene refrigeration system under those reduced rate scenarios.

As with the "Start-up" case, operating discipline was updated and process controls were installed to help operations minimize off-spec flaring. Better management practices of off-spec bullet storage were employed and vent lines from these vessels that used to connect to the flare were rerouted to the furnace ethane feed system.

The results of these enhancements reduced the average defect of "pound HC flared / pound of off-spec produced" by 57%.

PLANT TRIPS / SHUTDOWNS

Most unplanned shutdowns are due to the cracked gas compressor or the propylene refrigeration compressor tripping. Regardless of the reason for the shutdown, this project studied possible actions for minimizing flaring under all trip conditions.

To track project progress, a defect was selected similar to the Start-up case:

"<u>pound of HC flared</u> per <u>plant trip</u>"

This metric is slightly biased to how much load the plant is carrying at the time of the trip, but other more complicated metrics didn't address this issue much better.

The plant trip scenario is the simplest of the flaring events to analyze. The most significant parameter to flare reduction during trips is reducing furnace load as quickly as possible. The challenge with cutting furnace load quickly is that fast ramp rates can cause furnace damage such as coke sloughing and coil ruptures due to uneven/over heating issues. LHC-7's furnaces also require the manual cutting of burners as the heat duty is reduced. This is a particular challenge for one outside furnace operator considering that there are 8 furnaces on-line each having 48 burners (total of 384 burners). Furnace quick ramp-down rate parameters were increased with the expectation that some furnace damage could occur.

Other improvements were process control related. The back-end distillation columns (primary de-methanizer, de-ethanizer, secondary de-methanizer, and the C2 splitter) have a tendency to vent the pressure safety valves during refrigeration compressor trips (either propylene or ethylene refrigeration).

This increase in pressure is a result from the loss of refrigerant in the overhead condenser. To address this issue, an automated reboiler duty reduction was implemented to proactively reduce the chance of over pressure in each of these towers.

Conversely, a different approach was taken with the C2 splitter column during a cracked gas compressor trip. In this case, specific operating discipline and process controls were created to minimize the loss of C2 splitter and product ethylene storage inventory. The C2 splitter is now automatically moved to a "total reflux" mode. Although these actions do not contribute to the reductions in flaring from the trip scenario, they can reduce the amount of time spent to re-inventory these unit operations in preparation for start-up (see de-ethanizer stability issue under the "Plant Start-up" section).

These improvements have reduced the average defect of a "pound HC flared / plant trip" by 43%.

OVERALL UPSET FLARING PERFORMANCE

Regardless of the improvements made as described previously, the major factor in reducing upset emission flaring is to prevent the incident from occurring – plant reliability. Using the Six Sigma methodology of defining an appropriate defect per opportunity should compensate for this discrepancy. This gets it on an even playing field - apples to apples comparison. However, the improvements in flare minimization, coupled with the improvements that LHC-7 has made with plant reliability, have shown drastic reductions in upset flaring. The plant has shown an 89% reduction in overall upset flaring – using a two year running average (see Figure 4).

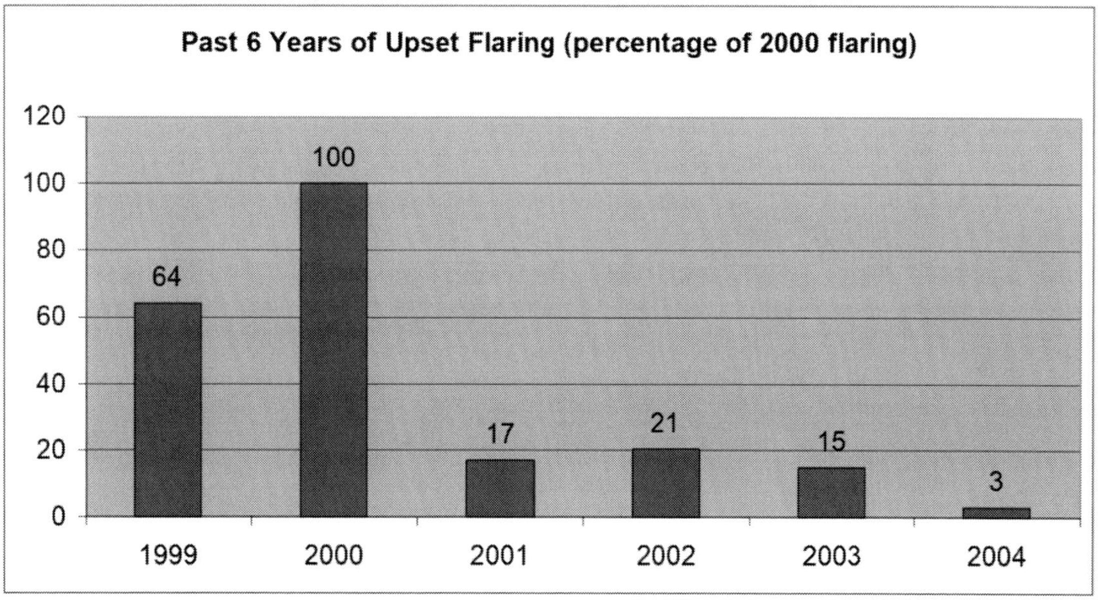

Figure 4: LHC-7's Upset Flaring History

CONCLUSION

One of the main contributors to the overall success of this project was the adoption of a philosophy that it is an unacceptable practice to flare. This philosophy had to be accepted at all levels. Management (Leadership) supported the project efforts by allowing operations the liberty to slow down the process of upset recovery if the efforts would result in less emissions. Under these conditions, production rates were third in priority – safety first, flare minimization second, and finally production rates. This allowed operations more freedom to think through upset situations and take more appropriate action in an effort to reduce flaring. This mindset also opened the doors to a larger group to generate further improvements. Even the smallest upset events are now managed to reduce flaring.

ACKNOWLEDGEMENTS

The project success is credited to the Flare Minimization Green Belts: Edward Boomhower, Kenny Coe, Brian Funk, DeWayne Gerber, John Holderness, and Robert Trevino. Other valuable contributors were Dennis Ganstine for process control and Blayne Friehauf, Doug Sheldon, and the whole LHC-7 team for their operational support.

Overspeed Incident at Refinery Results in Destruction of Turbine Casing

Lloyd Filemon
Superintendent Process II
Operations department
TRN
Vlissingen
The Netherlands
lloyd.filemon@trn.nl
+31 113 619 351

Peter van Limmen
Rotating Equipment
Engineer
Dow Benelux B.V.
Terneuzen
The Netherlands
pfvanlimmen@dow.com
+31 115 672 043

Prepared for Presentation at 2005 AICHE Spring National Meeting
Atlanta, Georgia - April 10 - 14, 2005
16th Ethylene Producer's Conference
Session TA009 - Ethylene Plant Operations

Unpublished

Overspeed Incident at Refinery Results in Destruction of Turbine Casing

Background

At the Vlissingen Refinery in the Netherlands an incident occurred on February 28, 2004 which led to an overspeed of a steam turbine. During the incident both the trip and throttle valve and the steam valve failed to close. The incident proceeded until the coupling failed and blades were thrown off the rotor destroying the casing and sending metal through the roof of the compressor house. This paper describes the conditions and causes of the incident.

Plant process

The Vlissingen Refinery operates a Distillate Hydrocracker (DHC) unit. The Distillate Hydrocracker includes three reciprocating hydrogen make-up compressors (302C01, 302C02 and 302C05) which operate in parallel. Two of these reciprocating compressors are driven by high pressure steam turbines (302CT01 and 302CT02 resp.) and are equipped with mechanical governors. The 3rd make-up compressor (302C05) is E-motor driven. Since there is excess Hydrogen make-up gas compressor capacity, speed is typically held constant at minimum governor speed (5400 rpm), and the process is controlled by changing the compressor recycle flow (kickback). A fourth compressor in the DHC unit is a steam turbine driven centrifugal recycle gas compressor. All HP steam turbines are installed on a common steam supply line.

Conditions leading up to the incident

Eleven days prior to the incident (February 17, 2004), there was an upset at the plant's demin water system. The upset resulted in demin water that was both high in pH and high in conductivity. The demin unit supplies boilerfeedwater (BFW) to all of the refinery's boilers including the boiler of the Hydrogen Production Unit (HPU). The HPU ultimately supplies Hydrogen and steam to the Hydrogen make-up compressors at the DHC. The supply of off-spec boilerfeedwater to the HPU caused a serious foaming incident, leading to carryover of water into the high pressure steam system. The carryover would have entrained any salts that were contained in the boiler water. The foaming lasted about two hours and was corrected by maximizing blowdown and BFW chemicals injection.

On February 25, it was reported that the Hydrogen recycle compressor (302C03), was slowly but steadily losing capacity. It was postulated that there was fouling by salt deposits on the internals of the turbine, which was a result of the boiler foaming eight days earlier in the HPU.

On February 26, problems were reported with the 302C01 turbine. At intervals the turbine speed would drop below minimum governor speed. This caused the compressor control system to switch the compressor to 50% load (at rpm $<$ 5200).

Note that in cases were the speed drops to below 5200 rpm's the compressor load is automatically switched to 50% load by the interlock system, in order to safeguard the gearbox and turbine as part of the shutdown procedure. However, the reduction to 5,200 rpm was not due to the fact that the turbine was being shut down, but due to its inability to maintain the set speed. When compressor load was reduced, the turbine would rise again above minimum governor speed and the compressor control system would restore load. Initial diagnosis was that there was a problem related to the speed governor.

On February 27 a team was formed to plan a water wash of turbine 302CT03.

On February 28, plans were made to shutdown compressor 302C01 and replace the governor. The compressor load was changed from 100% to 50% per normal shutdown procedures. The turbine soon began to gain speed. One minute later, with turbine at 6680 rpm, the compressor safety interlock system switched the compressor to unload conditions (0% load) on high compressor differential pressure across the 3rd stage. This caused the turbine to gain more speed. Six seconds later, the high speed alarm came in at 7700 rpm. Within another 45 seconds the panel operator attempted to trip the turbine via the manual control room trip switch. By this time the turbine had exceeded 8400 rpm.

Meanwhile the operators that were in the compressor building attempted to trip the turbine manually, but observed that the mechanical overspeed handle activated (8216 rpm), and that the trip and throttle valve unlatched. When the turbine began to make the "noise of a jet engine" they fled the compressor building. Within 20 seconds the turbine oversped to destruction. On the panel in the technical room speeds in excess of 14,000 rpm were recorded. The attached photographs show the destruction of the turbine.

Figure 1 - Turbine Damage

Figure 2 – Turbine Building Damage

Results of Root Cause Analysis
Various elements of all three steam turbines (302CT01, 302CT02, and 302CT03) were disassembled and analyzed. Two types of deposits were found - black deposits which proved to be magnetite, and white deposits which were determined to be salts. Most of the salts were found on turbine blades.

The T&T valve on the failed turbine (302CT01) was found to have excessive amounts of magnetite and also some sodium salts. It is believed that the upset in the steam system may have resulted in a carry-over of free magnetite particles from the steam drum and steam piping, leading to excessive deposits in the T&T valve. The clearance between the valve stem and bushing was less than normal. The last time that the T&T valve had been exercised was during the last start-up of the turbine approx. 3 months before.

The steam valve was heavily damaged in the incident; however analysis revealed higher than expected levels of magnetite and also some evidence of salts. Though damaged, investigation revealed that the valve had seized due to deposits in the balance cylinder. Contributing factors to the seizure were

- misalignment between the balance piston and the bushing of the valve stem
- the fact that the steam valve was normally always in the same position (turbine speed was typically held constant at minimum governor speed).

The T&T valve of the second turbine (302CT02) was also found not to be fully closed. After the incident this valve was manually closed and then tested several times without problems. The reason this turbine did not overspeed was the closing of the steam valve.

Actions taken to prepare plant for start-up
Prior to restart of the DHC unit the following actions were taken

- Refinery HP steam network cleaned by water washing
- TTV's and steam valves of the turbines 302CT02 and 302CT03 (not damaged) were inspected, cleaned and overhauled.
- These turbines were also cleaned by water washing
- Extensive functional testing of overspeed protection systems of 302CT02 and 302CT03

Recommended Action Items
The refinery developed a list of action items which can also be recommended to others:

<u>Maintenance strategy changes:</u>

- Steam valves, trip and throttle valves and associated controls should be thoroughly inspected and overhauled during each major turnaround.
- Investigate improved on-line testing techniques for steam- and trip-throttle valves

Operational procedure changes:
- Functional testing of trip and throttle valves by partial stroking on a weekly basis
- Avoid prolonged static position of steam valves
- Functional testing of overspeed protection system prior to each start-up of the turbines
- Stroking of trip and throttle valve and steam valve prior to each planned shutdown of the compressors
- In general: regular testing of all valves in HP steam turbines protection systems.

Design improvements:
- Search for improved design steam valves and trip and throttle valves which are less sensitive for fouling
- Search for additional lines of defense with respect to turbine overspeed protection (e.g. additional HP steam shutdown valve, speed rate of change)
- Re-evaluate trip- and unload philosophy of the compressors.

Human behavior:
- Improve risk awareness of non-routine interventions
- Extra alertness in case of upset in BFW systems

AUTHOR INDEX

AUTHOR INDEX

AUTHOR INDEX

American Institute of Chemical Engineers
3 Park Avenue
New York, New York 10016-5991

ISBN: 978-1-60423-460-2

American Institute of Chemical Engineers

17th Topical Conference on Ethylene Producers 2005

Held at the 2005 AIChE Spring National Meeting

April 10-14, 2005
Atlanta, Georgia, USA

Volume 2 of 2

American Institute of Chemical Engineers

17th Topical Conference on Ethylene Producers 2005

Held at the 2005 AIChE Spring National Meeting

April 10-14, 2005
Atlanta, Georgia, USA

Volume 2 of 2

Printed from e-media with permission by:

Curran Associates, Inc.
57 Morehouse Lane
Red Hook, NY 12571
www.proceedings.com

ISBN: 978-1-60423-460-2

Some format issues inherent in the e-media version may also appear in this print version.

American Institute of Chemical Engineers

17th Topical Conference on Ethylene Producers
2005

Table of Contents

Volume 1

Paper 42a

Refrigeration Technology as Practiced in Olefins Plants

Thomas M. Pickett

Senior Consultant

Barnes and Click, Inc.

Prepared for Presentation at the 2005 AIChE Spring National Meeting

17th Annual Ethylene Producers' Conference

Atlanta, GA, April 10-14, 2005

Session TA010 – Ethylene Plant Technology – Refrigeration Systems

April 13, 2005

Refrigeration Technology as Practiced in Olefin Plants

Thomas M. Pickett
Senior Consultant
Barnes and Click, Inc.

Introduction

Refrigeration systems in the modern olefins plant account for a large percentage of the initial capital as well as a significant portion of the ongoing plant operating costs. As much as 15-20% of the installed ISBL ethylene plant cost can be attributed to the refrigeration system CAPEX. A similar percentage of the total plant energy consumption is due to the operating costs of the refrigeration compressors. Although a common topic, and a field familiar to most in the business of manufacturing ethylene, refrigeration technology and the application thereof remains an important subject to those interested in optimal plant design and performance.

Top performing olefins producers must have reliable and efficient refrigeration systems to maintain today's competitive edge. For the producer, there is almost always room for improvement in nearly every facility whether in the area of operational efficiency, reliability or plant maintenance costs. This paper provides an overview and update on refrigeration systems in olefins plants. The paper will touch on refrigeration theory, but the focus will mostly be related to design developments, flow scheme options, design and operational considerations. The paper will be addressing ISBL issues. A later session paper will address OSBL refrigeration as it pertains to storage.

Background

For the olefins producer, there are essentially three sources of refrigeration which can be used to meet chilling or stream condensing needs. These include:

1. Isentropic Refrigeration (Turboexpander Systems)
2. Isenthalpic Refrigeration (Vapor Compression/Expansion Processes)
3. Process-Process Integration (Cross Exchange)

Each of these has associated advantages and disadvantages. To properly understand the various refrigeration alternatives, a basic understanding of refrigeration theory is required. The typical plant today may employ all of these in order to realize efficient fractionation and product recovery.

Isentropic Refrigeration

From a thermodynamic view point, expansion across a valve takes place at constant enthalpy (isenthalpically) while expansion through a theoretical turbine takes place at constant entropy (isentropically). For the expansion turbine, there is an enthalpy change in the fluid during the expansion process. Of further significance is the work which can be extracted from the turbine compared to the isenthalpic valve, or Joule-Thomson expansion. The expander therefore acts as both a producer of work (driver) and a source of low level cryogenic refrigeration. Figure 1 below illustrates the advantage of the isentropic expansion process over constant enthalpy expansion. As noted, expansion of a fluid from point A to point C through an expansion valve results in a smaller inlet to outlet ΔT than expansion from A to B through a turboexpander to the same pressure as C. The resulting stream represented by C therefore has less refrigeration capacity than B.

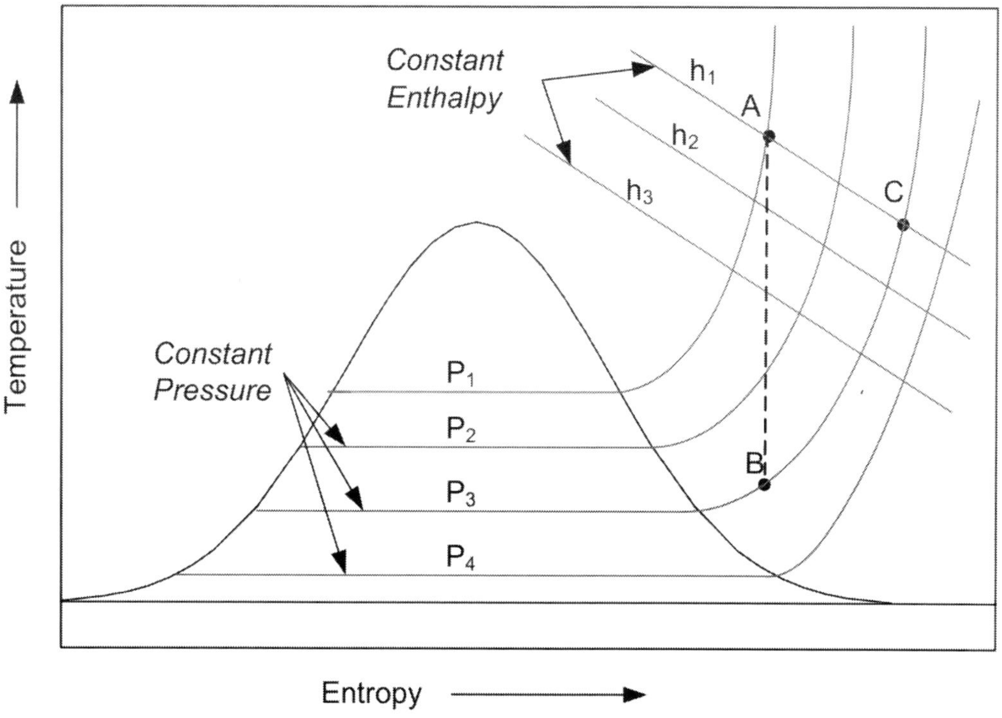

Figure 1 - Isentropic vs. Isenthalpic Cooling Comparison[1]

A more practical illustration follows:

> Suppose a hydrogen/methane stream of molecular weight 13 exists at -140°F and 450 PSIA. If the stream is reduced in pressure across a valve to 150 PSIA, the resulting temperature is -176°F. If the same stream is passed through a turboexpander, the outlet temperature of the expander would be -200°F, assuming typical machinery efficiencies. For a typical 1000 KTA plant, there is a recovery of approximately 12 MM Btu/hr realized by heating the stream to 15°F for the expander scenario compared to 10 MM Btu/hr for the valve expansion case, or a difference of around 20%. The 2 MM Btu/hr difference is at a cold temperature level unattainable via valve pressure reduction. Of more significance is the potential power recovery of 1000 HP from the stream as driver power. A 1000 KTA plant might have a two stage expander-recompressor system resulting in power recovery of around 2000 HP.

Commercial expanders today work in the vapor phase, typically on streams which have already been fractionated and refrigerated. Deep refrigeration below -150°F is therefore recovered. Cold methane liquid formed at the outlet of the turboexpander provides added latent refrigeration compared to a constant enthalpy Joule-Thomson (J-T) valve which will generally produce only superheated vapors from which sensible heat is recovered. Typical turboexpander efficiencies today are in the 80-85% (polytropic) range. As machinery efficiency increases, more power can be recovered and more liquid is formed through the expansion process, further increasing the available refrigeration from the outlet stream. In practice, actual machine operating efficiencies are difficult to determine since outlet temperatures do not vary significantly with changing efficiencies. If the machine is coupled with a compressor, the most practical method of evaluating the turboexpander performance may be via calculation of the compressor power. Coupling efficiency is generally 95% or above.

When the expander is out of service the refrigeration capability of the system reverts to an isenthalpic system as the Joule-Thomson by-pass valves are activated. Although less efficient, the system remains in service albeit generally with higher ethylene losses into the plant's residue gas.

Isenthalpic Refrigeration

As important as the turboexpander is in today's ethylene plants, the vast majority of refrigeration is still derived from the more conventional isenthalpic vapor compression-condensing process. In this case, refrigerant is compressed, condensed, expanded across a valve (resulting in a temperature reduction) and vaporized at constant temperature. The refrigerants in today's plants are commonly pure component systems made up of ethylene, propylene and sometimes methane. However, binary or multi-component systems composed of mixtures of these components are now being employed.

The conventional refrigeration process is illustrated simply as follows:

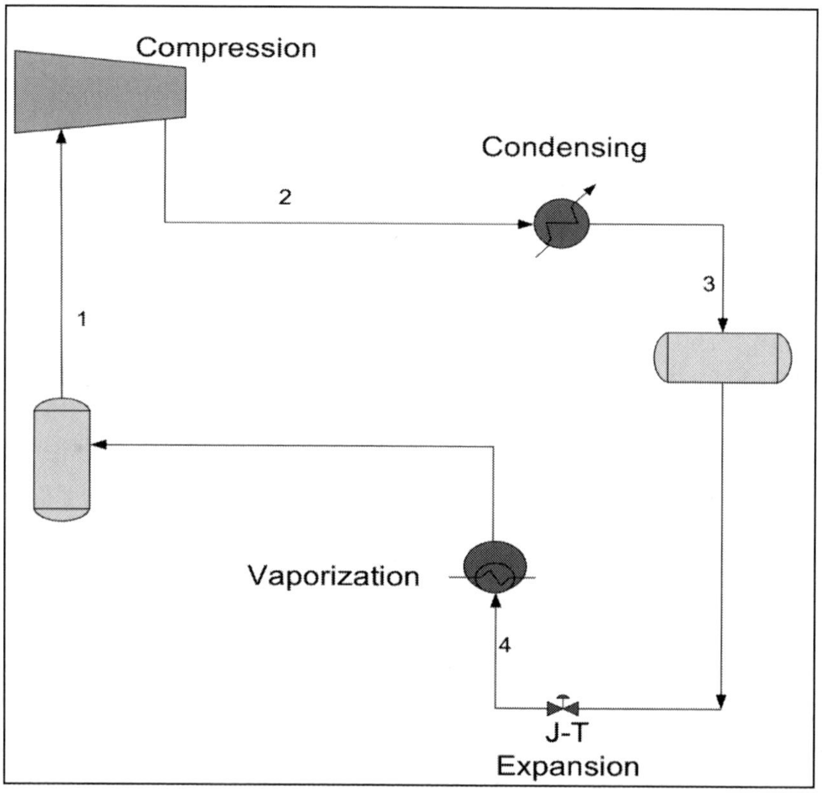

Figure 2 – Simplified Refrigeration Schematic

The well known refrigeration cycle for the above schematic is represented below in Figure 3.

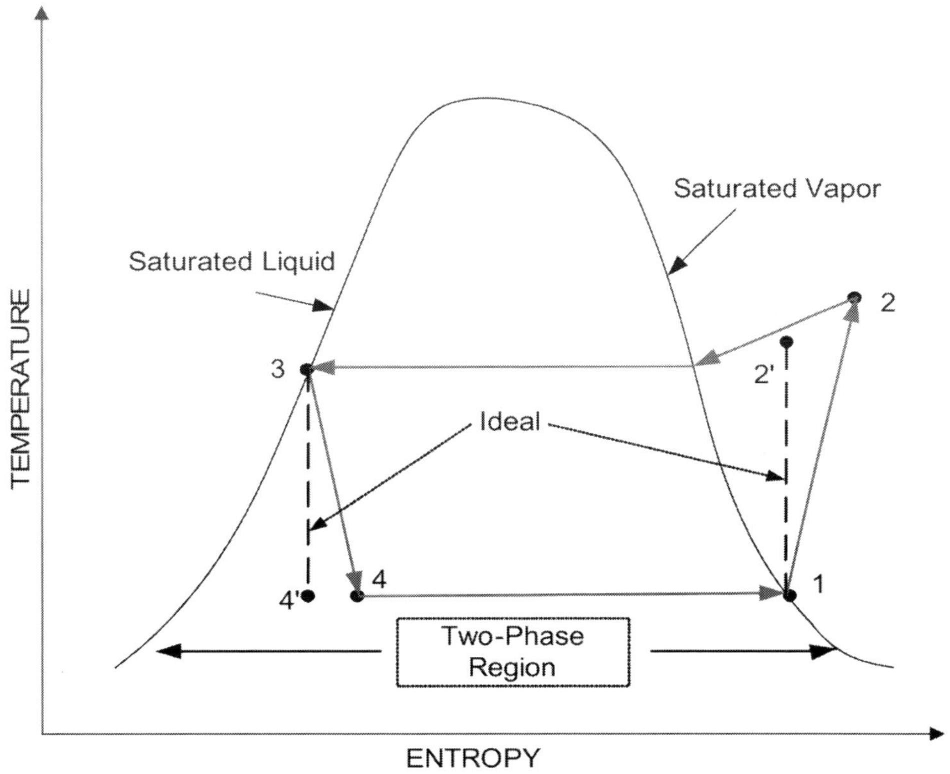

Figure 3 – Vapor Compression Cycle

These figures are an oversimplification of the refrigeration process in today's more complex plants. Nevertheless, there are several important points to be understood from the simplified cycle:

1. The process side benefits from steps 4→1. In this portion of the cycle, heat is absorbed from the process into the refrigerant as the liquid vaporizes at constant temperature. It is the phase change of the refrigeration which facilitates the heat exchange and provides chilling into the process stream.

2. One objective of the process designer and plant engineer is to design and operate the system such that Point 4 is as close to the liquid saturation line as practical. In theory, this would require the expansion to take place in an isentropic fashion. As previously stated, expansion engines are used in vapor phase only; hence, this expansion generally occurs across a Joule-Thomson valve at constant enthalpy. In practice the two-phase valve outlet represents a loss in refrigeration efficiency since it is the liquid refrigerant which is needed by the process. The flash vapors provide little useful refrigeration.

3. Energy is expended in the compression step 1→2. One of the primary goals of the compressor designer is to reduce the power requirement for a given process. The equipment supplier does this through improvements in mechanical efficiency. The

process designer and plant engineer attempt this through plant optimization, reducing the vapor flow to the compressor for a given process side refrigeration load or duty. This reduces the overall power requirement for the compression step. Examples will be shown later in this paper, but fundamentally this can be accomplished through subcooling of the refrigerant liquid prior to the expansion step. Doing so moves Point 4 closer to the saturated liquid line resulting in less vapor flash. Alternatively, for multi-stage refrigeration systems, another approach is to shift as much of the refrigeration as possible to a warmer level, reducing the magnitude of compression required on the refrigerant.

The operating range for the typical refrigeration system is from slightly above atmospheric pressure for the first stage suction to a pressure coincident with the condensing temperature for the final discharge, the condensing temperature being an approach to the cooling media. For example, for a standard propylene refrigeration system, the designer might set the first stage suction at 5 psig, the final discharge at 230 psig. The condensing temperature of propylene at 230 psig is 105°F which provides a suitable approach to 95°F cooling water. Side loads from economizers are normally selected at pressure levels which provide the dual benefit of reducing horsepower and lowering internal temperatures for a reasonable mechanical design. Users are also placed at appropriate interstage positions to supplement economizer flow and unload the duty requirement at the lower refrigeration levels.

For a typical ethylene refrigeration system, the first stage would commonly be set again at slightly above atmospheric pressure, with the final discharge set at a pressure to facilitate condensing with the low level propylene refrigeration, say 230 psig. At 230 psig, ethylene condenses at approximately -30°F, again a suitable approach to the 5 psig (-40°) low level propylene refrigerant. Refrigeration systems operating at lower than atmospheric pressure are in service today in older units. These type systems are usually avoided in new designs due to nuisance and safety concerns over potential air ingress through seals or flanges.

With the advent of computational flow dynamics, the compressor manufacturer has been able to achieve increases in polytropic efficiencies typically into the mid or upper 80's. Machine efficiencies are vital to good overall plant energy performance. Compressors with poor efficiencies not only erode earnings with increased OPEX, but may also limit plant capacity if the driver is inadequately sized or sized based on an assumed higher efficiency. All plant operators should have a regular program of monitoring compressor efficiency, watching for any sign of deterioration which may need attention in the event of a scheduled or unscheduled shutdown. As long as refrigeration flows are known, operating efficiencies should be able to be determined using most commercial plant simulators to match plant operation.

Process-Process Integration

Ethylene plants today have become highly integrated, meaning that streams from one section of the plant are used to chill or heat streams from another section. Some refrigeration systems are even self-integrated, using the cold temperatures generated from flashed liquid and vapor streams to cool and condense warmer refrigerant streams at higher pressure within the same system. Increased plant integration schemes have been the primary reason for reduction in the plant specific energy (energy expended per unit of product produced) in the past 10-15 years. At the same time, plant operability has become increasingly complex. The greater the extent of integration, the more susceptible plants are to upsets as a problem in one area of the plant can in-turn lead to upsets in other plant systems. Even so, the greatest opportunity to satisfy refrigeration needs with the least impact on energy comes from interchange within the plant's primary process streams. This is routine in nearly all plant designs as tail gas streams are reheated against incoming cracked gas streams, reducing ethylene and propylene compressor horsepower. To achieve greater reduction in plant energy, designers must now consider cross exchanging in more critical services such as tower condensers and reboilers, refrigeration condensers and other key services with process (not refrigeration) streams. Incorporating such modifications into plant expansions will not only reduce refrigeration horsepower, but may also salvage compressor casings and/or prevent costly compressor re-wheels.

For the typical olefins plant, the relationship between refrigeration temperature level and compression power is shown on Figure 4. The obvious trend is asymptotic as the temperature trends towards -150°F. This trend is reflective of the cascade affect due to power requirements from both ethylene and propylene compressors when chilling in the ethylene refrigeration temperature range. Thus the incentive to operate efficiently and where possible minimize refrigeration requirements at the coldest levels can be seen.

FIGURE 4
HORSEPOWER vs REFRIGERANT TEMPERATURE

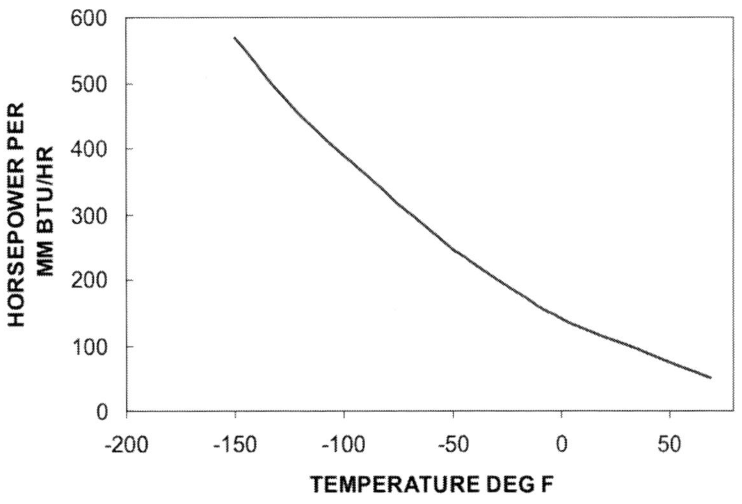

* Figure Courtesy of Shaw Stone & Webster, Inc.

(Note: The curve is a generalized correlation and values vary slightly depending upon machine efficiency, specific discharge pressure and overall system configuration.)

A variant of the Process-Process integration benefit is an optimized integration of the interacting refrigeration systems. Where propylene refrigeration is routinely used to desuperheat and condense the discharge of the ethylene refrigeration compressor, substitution with a process user instead yields a generous payback. Example 1 serves to illustrate this concept.

Example 1

A typical ethylene refrigeration system would utilize chillers in series against propylene refrigerant to desuperheat and subsequently condense the ethylene refrigerant. Such systems are commonly configured as seen in Figure 5A below.

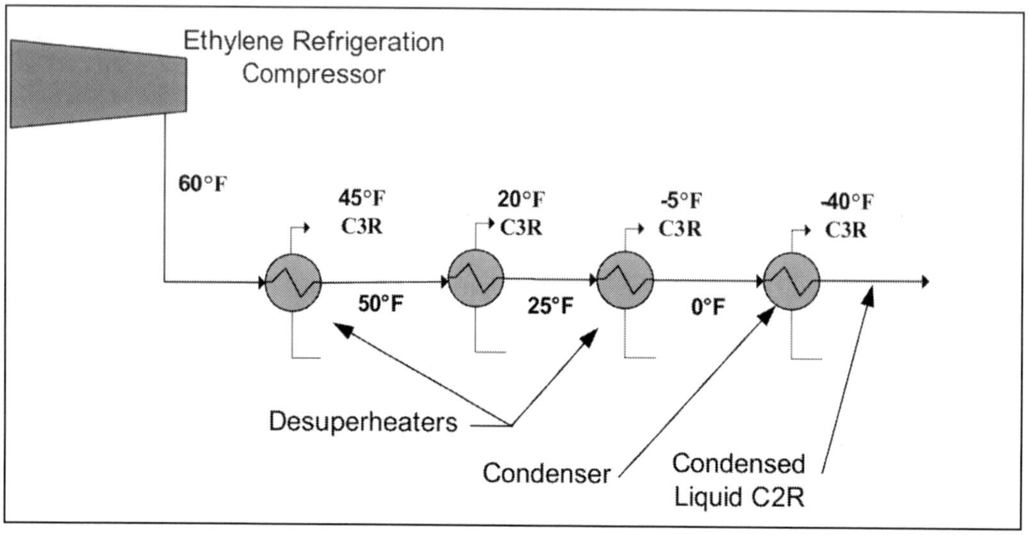

Figure 5A – Common Ethylene Refrigeration Discharge Scheme

In this example propylene refrigeration is available at 45, 20, -5 and -40°F, respectively. For a nominal 1000 KTA ethylene plant, the desuperheating duties are roughly as follows:

	DUTY (MM BTU/HR)	**C3R LEVEL (°F)**	**HORSEPOWER PER MM BTU/HR**	**HORSEPOWER**
DESUPERHEATER NO. 1	3.0	45	80	240
DESUPERHEATER NO. 2	8.5	20	115	977
DESUPERHEATER NO. 3	8.5	-5	150	1275
TOTAL	**20**			**2492**

Propylene refrigeration system horsepower is also shown in the above table using the data in the graph from Figure 4. Cold service towers such as demethanizers and C2 Splitters are commonly reboiled using vapors from refrigeration suction drums, offsetting flows to the refrigeration compressors resulting in horsepower "credits" to the system. Again, for the sample plant, the duty requirement for the demethanizer reboiler is 13 MM Btu/hr with a tower bottoms temperature of 13°F. In this case, the tower would commonly be reboiled using propylene vapors at the 45° level. With the savings of 80 HP per MM Btu/hr, a credit of 1040 horsepower is realized in the propylene refrigeration system.

Consider instead the scheme shown in Figure 5B:

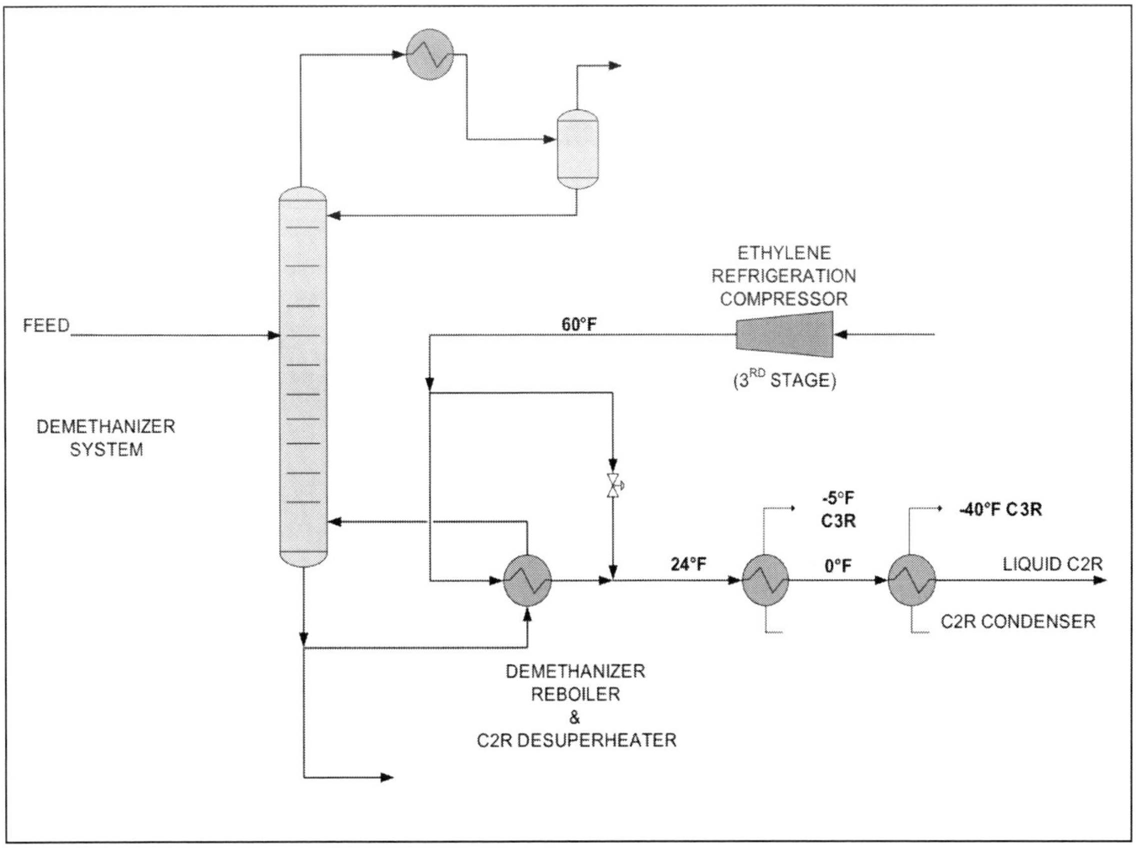

Fig 5B – Process-Process Integration
Demethanizer System with Ethylene Refrigeration System

The same 13 MM Btu/hr is this time exchanged with the ethylene refrigeration compressor discharge, eliminating the 45 and 20°F propylene desuperheating requirement as well as a portion of the -5°F C3R duty. From the table above it can be seen that only 7 MM Btu/hr of desuperheating duty is required from propylene refrigerant, an equivalent reduction of 1,442 horsepower. Comparing this with the conventional scheme using C3R to reboil the demethanizer, we are left with the following:

HP Reduction - Integrated Approach	1442 HP
HP Reduction – Conventional Scheme	1040 HP
Difference:	402 HP

The integrated approach has the obvious energy advantage compared to the conventional scheme.

Example 2

For Figure 5C, the Precut-Demethanizer Tower would typically be condensed with ethylene refrigerant at the -90°F level. Using the Deethanizer overhead liquid product as refrigerant in the condenser before feeding the C2 Splitter tower, the refrigeration load is shifted from the -90°F C2R level to the -40°F propylene refrigeration level, the typical condenser level for the C2 Splitter Tower. The relative difference in horsepower per BTU being about 50% more at the -90 C2R level compared to the -40 C3R level. Hence a savings approaching 500 horsepower for the 1000 KTA ethylene plant compared to the traditional use of ethylene refrigeration for condensing this service.

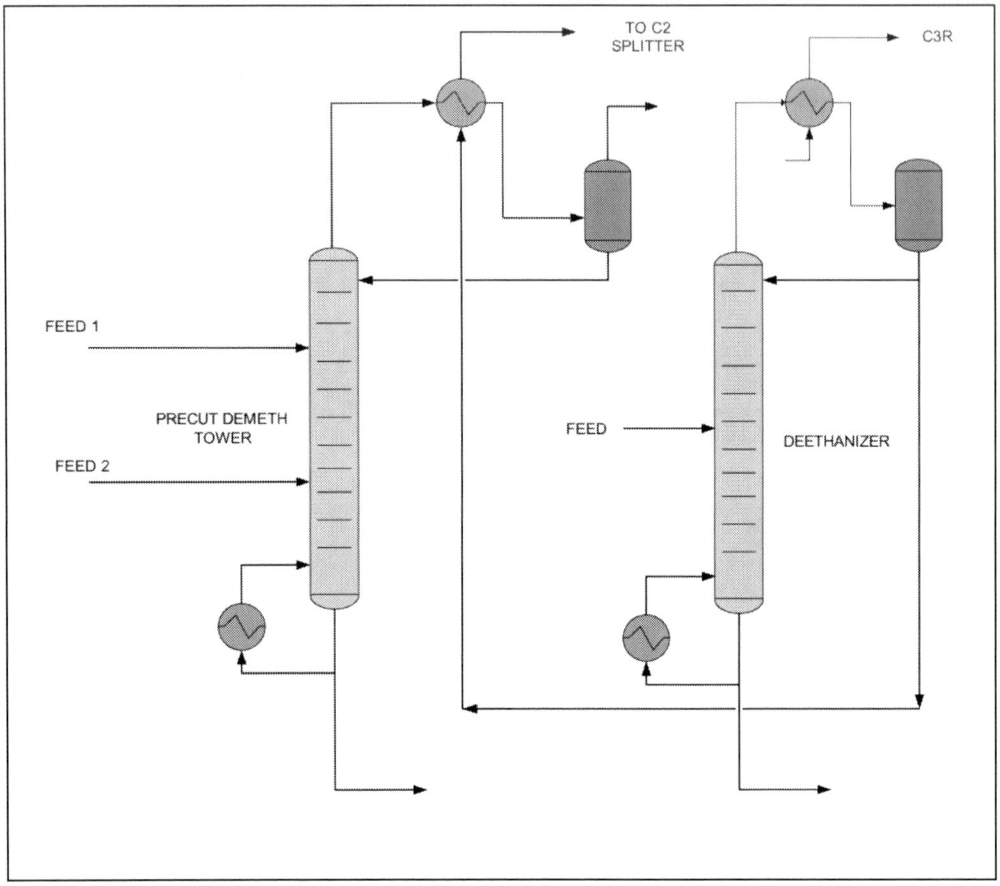

Fig 5C – Process-Process Integration
Deethanizer Overhead Product Condensing Precut Demethanizer Tower

Design Developments

Previous papers presented [2] at the Ethylene Producers Conference have demonstrated improvements through the years in machinery efficiencies. From Figure 6 below, it can be seen that these improvements have leveled out, with little room now for the substantive increases compared to the previous years. Dry gas seals were also considered relatively new 10-15 years ago, but have now become a standard feature for new equipment with most of the older machines having been retro-fitted as well. Comparable to the dry-gas seal in compressor technology, magnetic bearings have become standard for cryogenic turbo-expanders. These developments have significantly improved plant reliability as seal-oil freezing and plugging problems were eliminated. Dry-gas seals for refrigeration systems and magnetic bearings for expander systems are no longer considered new features.

FIGURE 6
EFFICIENCY TREND

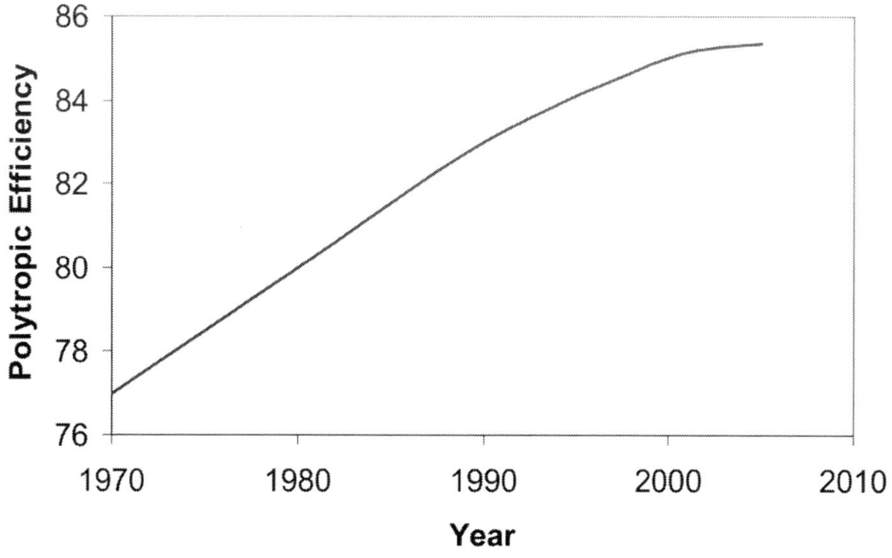

The most notable development in olefins plant refrigeration systems in the past 5 years has been the tremendous growth in machinery size. As the latest plant designs have ballooned in terms of production volume, so too have the refrigeration systems grown to keep pace. Whereas much of the plant equipment such as pumps, exchangers and even towers can be paralleled to meet capacity requirements, paralleling compressor systems for a new plant is typically avoided to hold down capital cost. Operators are also cautious of the complex controls required to operate large compressors in parallel. Ten years ago, compressors in the 50,000 horsepower range were extremely rare in olefins plants. Today these are common for the Cracked Gas Compressor service. Refrigeration machines are frequently supplied with power ratings above 30,000 HP. The casings of the refrigeration compressors have thus far not posed a design limit to plant size, the general view being that the Cracked Gas (Process Gas) Compressor casing size will reach a limitation prior to that of the refrigeration machine casings. Likewise, turboexpanders above 2000 HP are now being supplied as plants have moved into the "Mega" size.

One particular technology which is gaining popularity, though not a new concept, is that of mixed refrigerant use in olefins plants. Mixed refrigeration systems have been in use in LNG facilities for many years. A recent plant startup in China of an ABB-Lummus licensed unit incorporated the Tertiary Refrigerant design. Prior to this there have been two recent designs started up based on the Binary Refrigerant concept.

Tertiary Refrigeration Concept[3] - The primary objective of the tertiary system is to combine the traditional function of the propylene, ethylene and methane refrigeration systems into a single system, thereby minimizing capital cost and simplifying plant operations. The tertiary refrigerant is generally a mix of methane, ethylene and propylene. The process involves the separation of the tertiary refrigerant from the closed loop compressor interstage and final discharge into a methane rich portion and two propylene rich portions. As such, within the same system there exists both a light and heavy refrigerant. The lighter refrigerant is used for the coldest requirements such as the demethanizer condenser. The heavy refrigerant (propylene) is used for the large C2 Splitter condenser load, similar to the traditional propylene refrigeration system. In addition, the heavy refrigerant is used as "self refrigeration" to chill and condense the lighter refrigerant from the final discharge. The full benefit of this system is realized when applied to low pressure demethanizer designs which require refrigeration below the -150°F level typically supplied by ethylene refrigeration. The compressor would typically be a 4 or 5 stage machine with suction pressure and speed controls similar to that of a conventional propylene refrigeration system.

Binary Refrigeration Concept[4] - The binary refrigeration system, generally a mixture of methane and ethylene, combines the service of the traditional ethylene and methane compressors into a single system. The binary system is a cascade system utilizing a separate propylene refrigeration system to condense the compressor discharge. This technology is typically applied to low pressure demethanizer systems which require a reflux below that achievable with low level ethylene refrigeration, commonly below -200°F.

Flow Schemes and Design Considerations

Refrigeration compressors are specially designed machines requiring careful attention in the design and fabrication stages by both system and equipment engineer in order to meet specific process performance requirements. Although the compressor casings may come in common "off-the-shelf" sizes, rarely are two machines designed alike. Factors such as cooling water temperatures, plant configuration, plant size, feedstock and owner preferences (just to mention a few) prevent the standardization of these machines. Though the casings may be standardized, the nozzle sizes, wheels and other internals are custom made for the specific application. Therefore, a thorough understanding is required by all parties involved in the implementation of any system. Expertise and input is required from stationary equipment designers, process specialists, instrumentation engineers, plant operations personnel and field erection individuals.

Ninety percent of all plants utilize ethylene and propylene refrigeration systems in a conventional manner to chill and condense process streams and tower overhead condensers. Previous papers presented at Ethylene Producers Conference's have adequately addressed both design and operation of these systems. Following is a brief summary of the features of the conventional system followed by a more detailed discussion of alternate flow schemes.

Conventional Propylene Refrigeration System

The Propylene Refrigeration Compressor is usually the second largest compressor in the ethylene plant after the cracked gas compressor. It is commonly a single bodied machine having from 4-8 wheels.[5] Typically there are 4-5 process stages, where internal discharge flows are mixed with incoming cooler side load streams. The features most common in a conventional design are as follows:

- Closed Loop Circulating System
- Centrifugal Compressor Design
- Suction Conditions @ approximately -40°F and 5-6 psig
- Discharge Conditions @ approximately 220-250 psig
- Multiple (typically 2-3) side load and extraction nozzles
- Condensing Media – cooling water or air cooling
- Driver - Mostly steam turbine

Because the side load streams normally originate from the chilling of process users, the pressure level is carefully selected and must be met by the system designer. Otherwise the process will not achieve the intended design targets for operating temperature. Similarly, the pressure of any extraction stream withdrawn from the compressor must match that selected by the system designer as these streams are often used to drive reboilers which are constructed based on tight temperature approaches. Only slight deviations from the specified values are permitted and these should be reviewed prior to finalizing the design of the compressor and other system components.

Conventional Ethylene Refrigeration System

The Ethylene Refrigeration Compressor likewise is commonly a single bodied machine. These machines are typically supplied with from 6-10 wheels[6], but with the same or fewer process stages than the propylene compressor. The features most common in a conventional design are as follows:

- Closed Loop Circulating System
- Centrifugal Compressor Design
- Suction Conditions @ approximately -150°F and 2-3 psig
- Discharge Conditions @ approximately 230-290 psig
- Multiple (typically 1-3) side load and extraction nozzles
- Condensing Media – Propylene Refrigeration
- Driver - Mostly Steam Turbine

Having the smaller body and therefore smaller diameter wheels, the ethylene compressor will typically operate at a higher speed than the propylene compressor.

Heat Pump Systems

For superfractionator services such as that required for high purity ethylene and propylene production, the advantages of the heat pumped design should be considered. One of the chief advantages of the heat pump is the ability to perform the distillation at a lower pressure resulting in increased relative volatility between the fractionated components. Fewer fractionation stages are therefore necessary for a given R/D. For services such as propylene-propane fractionation, this may mean that the purification can be accomplished in a single large tower instead of two. When integrated into the corresponding refrigeration system, the number of critical rotating equipment items is reduced and therefore plant reliability is improved. Integrated Heat Pump systems have been in operation for many years, although they seem to have gone in and out of favor over the past 20 years. With the recent spike in energy costs, operators should consider this option in the plant conception phase.

Integration of the C2 Splitter into the plant's overall refrigeration system has become a common practice. Generally this would feature a 4 stage single-bodied compressor design with the tower overhead directed to the compressor's third stage suction. Whether an integrated or a stand-alone system, designers often debate the cost benefit due to metallurgy implications for the low pressure tower which may be operated close to or below the impact tested carbon steel-to-stainless steel breakpoint. There is no debate, however, on the energy benefit from low pressure fractionation.

Most propylene heat pump systems are thought of in the context of propylene recovery and fractionation, not refrigeration. In ethylene plants, especially liquids crackers, due to the abundance of warm quench water and the need for the cooling thereof, propylene heat pump systems are the exception and not the rule. Instead, ethylene plants will normally use high pressure C3 Splitter systems for propylene fractionation so that the quench water can be used as the heating media to satisfy the large reboiler requirements. Otherwise, a large cooling water demand would be incurred to chill the quench water. C3 Splitter heat pump systems are routinely seen however in stand-alone processes for the purification of propylene from refinery streams. Nonetheless, low pressure heat pump systems do have their application in olefins plants under the right circumstances. With the recent upward trend in propylene demand, the heat pump approach has attracted more interest. Because ethylene crackers are now integrating with other emerging technologies for propylene production, excess quench water may not always be available. More and more ethylene facilities are also receiving FCC or coker P-P streams for upgrading to chemical or polymer grade propylene. Revamps are an ideal place for heat pumps to add capacity since these can be more or less stand-alone systems, or systems designed to meet increased propylene and refrigeration capacity.

Know-how in the design and implementation of heat-pump systems cannot be over emphasized. Thorough reviews by experienced engineers and operators are essential. Among other things, the following system features should be considered:

> Reflux Subcooler – Subcooling the reflux will reduce the large reflux flash associated with these systems, and thus the system horsepower. Existing systems without reflux subcoolers may be able to increase capacity or reduce compressor power with the addition of this cooler.
>
> Light-Ends Removal – Few systems are implemented with tower pasteurizing sections. Normally, if light-ends are present, a dedicated stripper tower is positioned up-stream for removal.

Discharge Pressure Control – At reduced rates, the over-surface of the tower reboiler and/or auxiliary condenser can result in excessively low compressor discharge pressure such that reflux flow to the tower is jeopardized. Review of the controls and provisions for a hot vapor by-pass should be considered.

A simplified flow scheme of an integrated C3 Heat Pump System is shown in Figure 7.

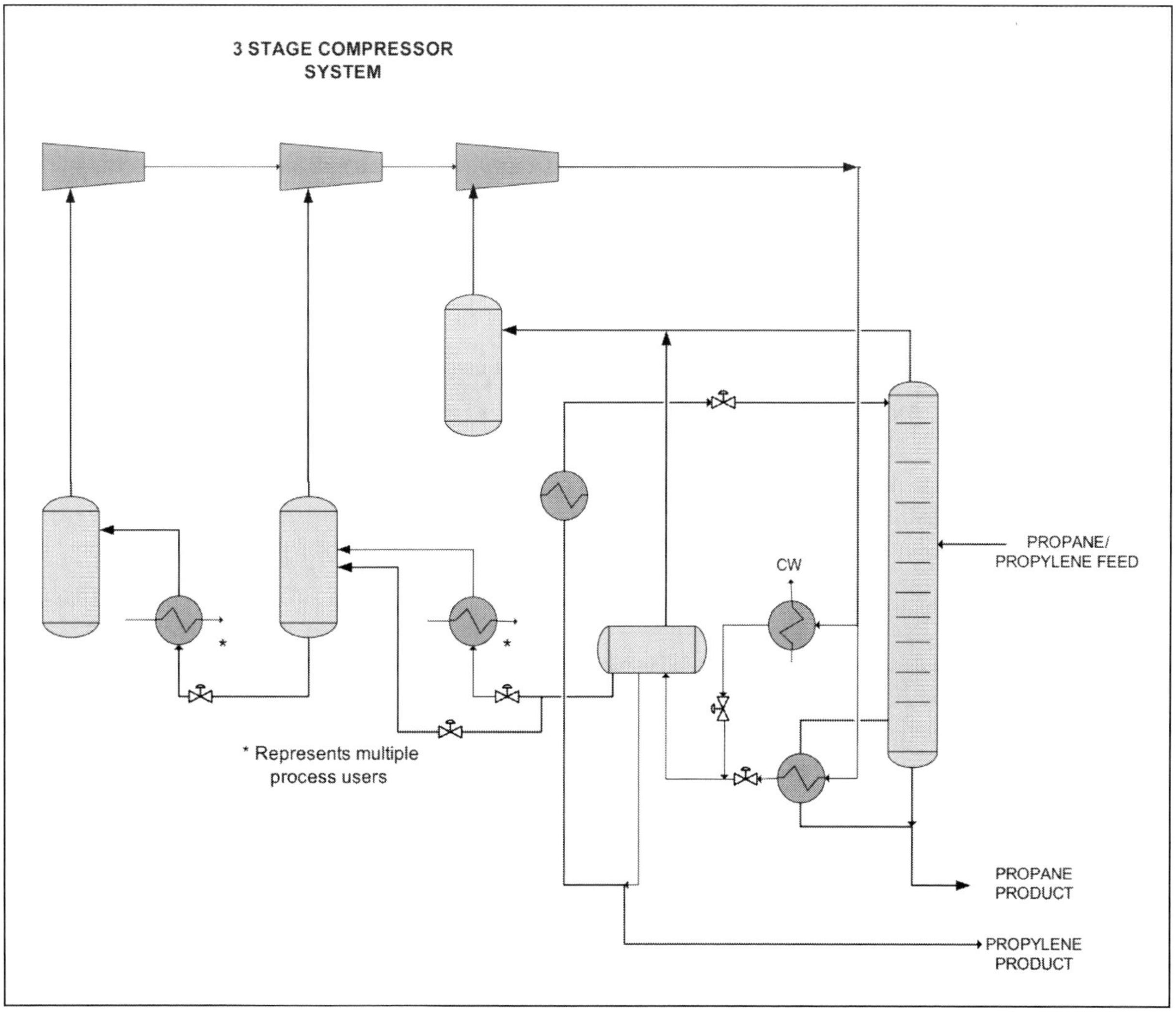

Figure 7 - Integrated C3 Splitter/Refrigeration Heat Pump System

Design Considerations

There are a number of design considerations and system features which the plant owner and design engineer should be familiar with for optimum scope of supply and performance of refrigeration systems.

System Turndown - Compression systems are limited to an operating window which must be recognized in the implementation phase. On the low side, this is normally around 75% of the normal design point before intervention controls take over to load the compressor and prevent surge. Ideally, all systems will operate at fixed flow requirements. However, in practice this is not the case. With the advent of large furnaces, where fewer and fewer furnaces are installed in the newer plants, the impact of a furnace decoke or trip can substantially influence cracked gas compressor and refrigeration system performance. Systems are required to automatically react to changing demands from the process side. Other factors such as changing market demands and corresponding plant production rates all influence compressor efficiency and therefore plant efficiency. Although selecting machinery with optimal efficiency at the expected load is the logical choice, an understanding of the efficiency drop–off at reduced rates is important. Consideration of compromises at the "normal" rates may sometimes be justified if extended reduced rate operation is foreseen. This is also where properly installed and configured anti-surge systems are important. These sophisticated compressor controls are not only valued for machine protection, but also guide compressors to operate closer to the surge-line so that over-recycle does not occur along with wasteful horsepower expenditures.

Cooling Water Variation – A frequently overlooked consideration in locations where cooling water temperatures and therefore refrigeration condensing temperatures vary is the related impact on system operating conditions. The prudent plant operator will take advantage of the reduced power resulting from the ability to operate at lower discharge pressures. Power savings on the order of 6500 HP/hr can potentially be realized seasonally for the world scale plant. The compressor will naturally seek its own operating point on the curve; however, if the system is not designed to permit the lower pressure, there will be little benefit to the operator. The common limitation in this regard is the ability to reboil services such as the demethanizer, which is commonly reboiled against warmer propylene vapors. As system interstage pressures drop, the condensing temperature of the refrigerant in the reboiler likewise drops, pinching-out the necessary exchanger ΔT driver. Thus, the machine is required to maintain operation closer to design, i.e. – higher pressure. Implementing controls and possibly minor piping with additional reboiler surface can have rapid pay back in energy savings.

Driver Selection – Factors such as electrical power and cooling water supply costs may influence the driver selection and result in different approaches for different geographical regions. Due to the abundance of high pressure steam available from the plant's furnace section, large compressor drivers are usually steam-turbine driven. Where fixed speed motor drivers are specified, suction throttling for capacity control is generally required. Unless variable speed drives are used, size limits should be recognized due to the large in-rush current required for compressor starts. Site infrastructure requirements may therefore have to be set to satisfy non-routine operating scenarios such as start-up and shutdown.

Curve Selection – Compressor curve characteristics should follow industry and machinery guidelines. Wheel selection which results in too large of a drop in head will not provide adequate flexibility for the operator. For example, at 15% overcapacity the recommended head may be a minimum of 85% of the specified (normal) head. If less than 85%, the process could fluctuate due to relatively small compressor load variations. On the other hand, if the head curve is too flat with respect to flow, the machine may oscillate wildly, seeking its operating point on the curve. This rapid oscillation could drive the machine into surge. Consultation with the manufacturer is important to understand interacting parameters and prevent over-specification. Relaxing certain guidelines may be prudent to achieve a flexible design which satisfies the most critical and likely operating requirements. Proper evaluation of the proposed curve characteristics is essential prior to the final certification of wheel design.

Core Exchangers/Design Temperature Approach – A common trait of most efficient refrigeration designs is a small temperature approach between the refrigerant and the process user. The use of aluminum style core-exchangers has enabled approaches in 2-4°F range. Once considered troublesome, fabrication techniques have improved the core exchanger's mechanical reliability. Although mechanical failures are now rare, the system designer must still be aware of inherent limitations when using these exchangers. Clean service is a must and fine mesh strainers should be installed to catch debris and avoid start-up and operating problems. Once debris enters the exchanger passages, performance may be permanently affected. Typically, provisions for bypassing cores on the initial air-blowing and dry-out are added. Contaminants such as mercury are also a known hazard with aluminum. If there are any questions regarding the presence of mercury, precautionary guard beds should be installed.

With the reduced ΔT design, however, the requirement of highly accurate physical property data is vital. Errors as small as one fourth of a degree can affect performance due to the reduced exchanger LMTD. Careful qualification of plant design simulators and selection of the appropriate thermal package is important if exchangers are to be constructed based results from these tools.

Trip Systems - As with other compression systems, refrigeration systems require trip provisions in order to protect plant equipment and personnel. These systems generally include trips due to:

- High Suction Drum Level
- High Discharge Temperature and Pressure
- High Radial Vibration or Axial Displacement
- Low Lube Oil Pressure

Inclusion of high integrity trip systems can also aid in reducing large flare loads caused by events such as cooling water failure.

Settle-Out Pressure – During compressor trip or shutdown scenarios, the higher pressure vapors of closed loop systems will "equalize" or settle-out to the lower pressure levels. Unless the lower pressure stages (drums, exchangers, piping, etc.) are upgraded in terms of the mechanical design pressure, relief to the flare and loss of refrigerant inventory will occur. Therefore, as a preventative measure, most systems are upgraded above that required for normal operation to prevent this occurrence. Settle-out pressure must also be communicated to the compressor supplier as this in some cases impacts driver size.

Liquid Drain/Vaporization Provisions – A common problem during startup and upset scenarios is the filling of the low pressure suction drum with excess liquid. Unless provisions are included for handling this liquid, a compressor shutdown due to high level will occur and rapid restart may not be possible until the inventory is removed. Routing the liquid contents to the plant's cold blow-down or flare system is generally not an option due to the low drum operating pressure. Typically a separate drain drum and/or a warm vaporizing coil internal to the drum are provided to handle this situation. The drum provides an alternate location to route the liquid. The coil is simply a small line which receives warm vapor from the compressor discharge, through the drum liquid level and back to an intermediate drum.

Suction Temperature "Quench" Provisions – During operating modes where minimum flow is required to maintain operation away from surge, suction temperatures can increase resulting in internal or discharge temperatures above manufacturers' limits. Provisions to maintain low suction temperatures are therefore essential. This is normally accomplished via auto-refrigeration from letting down liquid at the higher pressure level to the lower level, mixing with the warm minimum flow vapors. Reliable instrumentation, or preferably systems requiring no instrumentation, should be specified in order to quench the hot minimum flow vapors before entering the machine's suction.

Materials of Construction – A detailed analysis of potential upset conditions must be carried out in order to establish system-wide materials of construction. Extremely low depressurizing temperatures may dictate fabrication of vessels and piping using alloy metals. If fractionation towers are involved, as may be the case with heat pumped systems, there could be a substantial cost impact to the project. Establishing the depressuring philosophy early in the design phase may eliminate late surprises.

Conclusion

Both high installation cost and high on-going operating costs for plant refrigeration systems create challenge and opportunity for the plant owner. Whether modifying existing systems or considering new installations, designers and plant owners today must properly weigh the many available options. Evaluating and selecting the best process scheme and hardware features are key to achieving reliability, flexibility, and efficiency. These decisions are life-long decisions for the plant. Decisions based in know-how, and know-how based in experience must be applied to the evaluation and selection process.

---

[1] "Chemical Engineer's Handbook", Fifth Edition, Page 12-50, Figure 12-60.

[2] Taylor, R.A., Beck, R., Eads, K.S., "Advancements in Ethylene Plant Compressor and Steam Turbine Design", presented at the AIChE 2000 Spring Meeting, New Orleans, LA, Mar 2000.

[3] Vitus, T.W., Ma Q., Wu, T., US Patent 6,637,237 to ABB-Lummus Global Inc., April 11, 2002.

[4] Vitus, T.W., Sumner, C., Crawford, J.J., Stanley, S.J., McNab, R.J., US Patent 5,979,177 to ABB-Lummus Global Inc., Nov. 9, 1999.

[5] Taylor, R.A., Beck, R., Eads, K.S., "Advancements in Ethylene Plant Compressor and Steam Turbine Design", presented at the AIChE 2000 Spring Meeting, New Orleans, LA, Mar 2000.

[6] Taylor, R.A., Beck, R., Eads, K.S., "Advancements in Ethylene Plant Compressor and Steam Turbine Design", presented at the AIChE 2000 Spring Meeting, New Orleans, LA, Mar 2000.

Paper 42b

Design Considerations for Centrifugal Compressors in Refrigeration Duty

William Hohlweg

Senior Consulting Engineer – Advanced Technology

Elliott Turbomachinery Co., Inc.

James Blahovec

Manager, Application Engineering

Elliott Turbomachinery Co., Inc.

John Wright

Manager of Engineering

Elliott Turbomachinery Co., Inc.

Prepared for Presentation at the 2005 AIChE Spring National Meeting

17th Annual Ethylene Producers' Conference

Atlanta, GA, April 10-14, 2005

Session TA010 – Ethylene Plant Technology – Refrigeration Systems

April 13, 2005

Design Considerations for Centrifugal Compressors in Refrigeration Duty

William Hohlweg, Senior Consulting Engineer, Elliott Company
James Blahovec, Manager of Application Engineering, Elliott Company
John Wright, Manager of Engineering, Elliott Company

Abstract

This paper will explain the process that is followed to properly select centrifugal compressors for typical refrigeration duty. Refrigeration compressors, when compared to other applications, require some different approaches to the aerodynamic and mechanical design. Discussion topics will include frame size selection, aerodynamic component specification and design capabilities, sideload and extraction flow designs, materials selection and mechanical design requirements. The impact of customer-supplied operating conditions on compressor design is presented as well as the affect of variations in process conditions among different operating points. Also, special end user aerodynamic and mechanical requirements / guarantees are reviewed relative to their influence on equipment selection.

Introduction

Centrifugal compressors have been an integral part of the industrial refrigeration process in ethylene production for many decades. There are a number of versions of the process, seemingly as many as there are EPC contractors, and a number of services within the process. Within most ethylene plants, the two primary refrigeration services are propylene (sometimes propane) and ethylene. The characteristics of these compressors that differentiate them from most other process applications are the low temperature material requirements and the presence of inlet sidestream and/or extraction nozzles that introduce external flow into the compressor at an intermediate point or siphon off flow from an intermediate point in the compressor. Since it is not uncommon to have as many as three sidestreams, resulting in four compression sections in a single body, the design process becomes more complicated than for most other types of compressors. The addition of customer-imposed requirements, along with multiple operating points, can make these compressors even more specialized.

Initial Selection

The basic process for selecting a refrigeration compressor is much the same as for any other compressor. The first step is to gather all data relevant to the compressor. These include operating parameters for all specified operating points and/or a set of compressor data sheets, the requisition document for the project, the compressor specification and any other documents that define aerodynamic or mechanical requirements for the compressor. A quick but thorough review of these documents is necessary prior to selecting the compressor to tell the engineer the type of driver (fixed versus variable speed), range of operation, alternate operating modes for dryout and startup, the gas composition, special performance guarantees and special mechanical constraints. These all have an effect on how the hardware should be selected as explained by Blahovec, et al. (1998).

Getting a feel for the required range of operation helps to clarify the relationship among the different specified operating points, and defines how the engineer should select the aero components. For instance, if the operating points are at constant inlet conditions with a common discharge pressure, by looking at the required sectional flows, one can get a good idea as to how to choose components to satisfy as many of the points as possible. If flow trends are variable for the different compression sections at different operating points, it may be quite difficult to simply "eyeball" conditions and get a sense of how they relate to one another. Inlet sidestream/extraction flows may vary disproportionately from point to point or an intermediate nozzle may be a sideload for some conditions and an extraction for other conditions. Such situations can make estimating performance much more difficult and time consuming.

Refrigeration compressors present greater challenges than do straight-through or intercooled compressors since the largest volume flow in the compressor is not always at the first impeller. Inlet sidestreams/ extractions "externally" increase or decrease the flow to successive sections to upset the normal progression of volume reduction through the compressor. The compressor frame size required to meet the specified duty is determined based on the highest volume flow that occurs in any of the sections. Casing size is estimated by selecting an impeller tip diameter that provides a reasonably high flow coefficient (for the highest flow section) while maintaining good efficiency and range of operation. Impeller flow coefficients higher than 0.1 and lower than 0.2 are typical. At the same time, each section's pressure ratio is used to estimate the head required and the number of stages needed per section. The number of stages is based on a nominal head coefficient per stage, typically in the range of 0.5 to 0.6. A nominal set of stages is then selected for each section with the estimated diameters and flow coefficients, and the speed is varied to meet the required final discharge pressure.

Flow coefficient, $\phi = (700)\ (Q)\ /\ (N)\ (d_2^3)$

Where

Q = inlet volume flow (CFM)

N = rotational speed (rpm)

d_2 = Impeller tip diameter (in.)

Head coefficient, $\mu = (g)\ (H)\ /U_2^2$

Where

g = gravitational constant (= 32.17 lb_m-ft/lb_f-sec^2)

H = head (ft-lb_f/lb_m)

U_2 = impeller tip speed (ft/sec) = $(N)\ (\Pi)\ (d_2)\ /\ 720$

In some cases, the best way to determine the inter-relationships among the different operating points is to make a base component selection and determine if this selection can accommodate all of the specified operating points. This will help to define how the selection must be adjusted to best meet all of the required points. If there is a very wide operating range, recycle flow may be required for low flow points. For refrigeration compressors, flow range from surge to choke in each compression section is typically smaller than for other types of centrifugal compressors due to high impeller Mach number and high volume ratio effects.

For processes with a wide range of operating points, it is beneficial if the points can be prioritized or ranked based on the percentage of time the plant is operating at each condition or on their significance to overall production. This can help the designer to match the compressor staging slightly more to one side of the flow range or the other to ensure the higher priority cases are always met. If compromises are required, they can be applied to the lesser "important" operating points.

Compressor vendors have different types of stage designs from which to select. Some designs provide high head per stage, but may have lower head rise to surge. Others may have higher efficiency, but also have longer axial space requirements. Some stages may be more costly than others, and some may have wider operating ranges but lower efficiency levels. Due to variations in sectional head requirements, impellers may need to be adjusted to a slightly different base diameter or the blade tips may need to be cut back to a smaller diameter. As the various operating points are analyzed, further adjustments may have to be made to the individual stages in order to have each one operating within a normal flow range between surge and choke. All of these considerations must be taken into account to provide the best selection to meet the customer's needs.

The application engineer also needs to make note of the molecular weight of the gas being compressed, as this will affect how internal components are selected. An ethylene compressor, at a molecular weight of 28.05, can run at higher absolute speeds than a propylene/propane compressor (MW = 42.08/44.09). Volume ratios from stage to stage are much higher in propylene/propane service than for ethylene service. One needs to be more mindful of Mach number limitations for propylene/propane applications than for ethylene duty. Inherent to the method used to calculate relative inlet Mach number, higher flow coefficient impellers have higher Mach numbers than do lower flow impellers at the same gas conditions. Full or partial inducer, axial inlet type impellers will have lower Mach numbers than do non-inducer radial inlet type wheels. This is simply a function of the blade inlet diameter at either the mean or shroud location. In and of itself, high impeller inlet and tip Mach number values do not pose any serious mechanical or aerodynamic operational problems. For example, impellers in turbochargers and industrial air compressors routinely operate at higher Mach numbers than refrigeration compressors. They usually are unshrouded to handle the higher mechanical speeds and resultant stress levels. High Mach numbers do, however, limit the compressor flow range and can reduce the efficiency. Please see Figure 1. On this plot, performance curves at lower Mach numbers are represented by the "+" and "*" symbols; higher Mach numbers are represented by the "x","y" and "z" symbols. Flow range from surge to stonewall is obviously smaller at the high Mach numbers. The margin from design to maximum flow capacity is reduced because the inlet relative velocity in the impeller blade passage reaches sonic velocity at the throat sooner. Likewise, the flow margin to surge is reduced due to high positive incidence angles at the impeller blade leading edge and lower absolute flow angles in the radial diffuser. The outer set of lines at the top of the plot indicate the locations of the surge and choke flows for various impeller tip Mach numbers, while the inner set defines the nominal flow range for selecting a particular stage. This allows each stage to operate with a built-in flow margin to the predicted surge and choke flows.

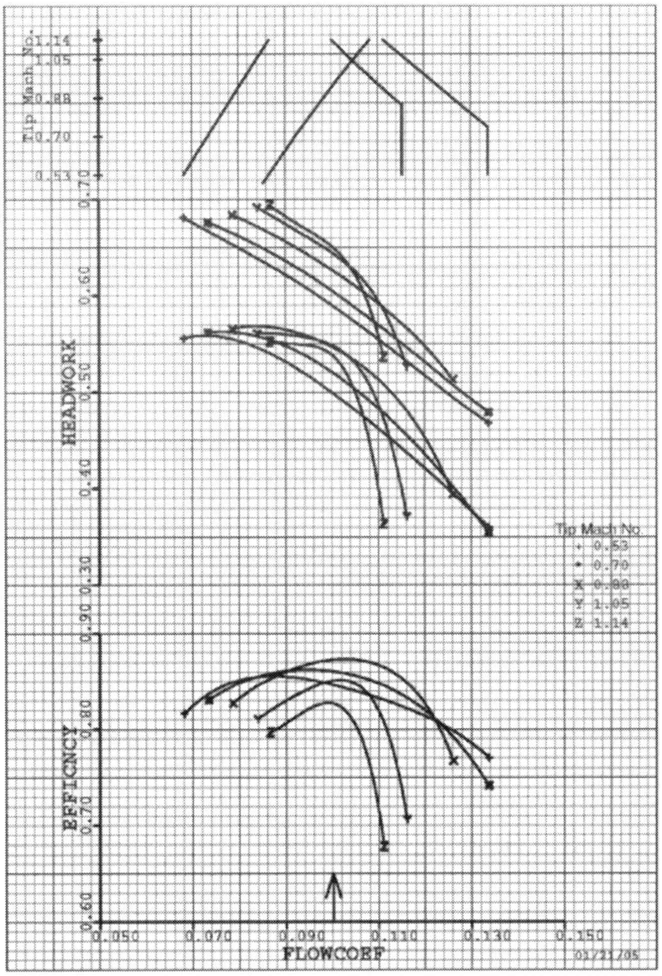

Figure 1: Typical Single Stage Performance Curve

It is important that internal compressor leakages be properly accounted for in the compressor selection process. For in-line compressors, balance piston recirculation flow must be included. The flow of "hot" gas at discharge temperature on the outboard side of the balance piston is (usually) piped back to the compressor inlet to equalize the pressures at the inlet and discharge end main casing seals. Not only does this increase the mass flow to the first impeller after the main inlet, it also raises the temperature of the gas prior to entering the eye of the first impeller. The net effect is an increase in both the head required to meet the final discharge pressure and increased power consumption. In some cases, the balance piston flow can be an appreciable percentage of the overall mass flow and can have a significant effect on the stage inlet temperature. In these instances, one could consider using a non-equalized seal arrangement (inlet end seal and discharge end seal operating at different pressure levels) and running the balance piston line back to a sideload or extraction nozzle. Although this may work to raise the overall efficiency level of the compressor and reduce the amount of power required, it also reduces the amount of thrust compensating force generated by the balance piston and adds cost/complexity to the seal oil or seal buffer system.

For smaller size compressors with buffered seals, buffer gas leakages should be factored into the compressor selection to account for the increased weight flow and any heating effect the buffer gas has at the main inlet. However, for larger refrigeration machines, buffer flow effects can be considered negligible.

Mechanical / Performance Issues

Fixed vs. Variable Speed

The type of driver affects how the compressor is designed. For variable speed applications, the best place to start is with the operating conditions identified as the guarantee point. For the initial selection, stages are selected to achieve maximum efficiency for this point and a speed is calculated to meet the sectional and overall head requirements. Operating conditions for the specified alternate points are then analyzed using the same stages selected for the guarantee point to determine if all points can be met. Several iterations may be necessary to find the stage selection that best accommodates all of the specified operating points. This may require that the guarantee point run at less than optimum efficiency. Variable speed offers greater flexibility to meet all of the specified operating points through simple speed changes. Alternate operating points can be met with lower horsepower consumption since the inefficiencies of suction or discharge throttling (required for fixed speed) are avoided.

Fixed speed refrigeration compressors are a little more challenging to select, especially when there are several operating points. The user must identify how process control is to be accomplished – suction throttling, discharge throttling, bypass. It is imperative that the selection of aero components be based upon the highest head case. Individual stages must be selected to allow the greatest flexibility to meet all of the specified operating points. Even so, there are often cases that can be met only by recycling gas to one or more sections. Recycle flow may also be required for variable speed machines, but less frequently.

Fixed speed compressors in refrigeration service can have operational complications due to the low temperatures involved. Thermal effects require careful attention during start-up, and can create limitations with respect to re-starts. Slow-rolling the compressor rotor can alleviate some of these complications, allowing the rotor to rotate with more even temperature distribution instead of localized high or low temperature spots. Advantages that fixed speed applications have are that, with no speed range, it is easier to obtain acceptable critical speed margins, and rotor stability may be slightly better since there is no overspeed margin to worry about.

Wide Speed Range

It tends to be more difficult to meet critical speed margins for compressors with wide speed margins (minimum speed of 60% or 70%) versus units with a minimum speed of 80% or 85%. The user needs to evaluate whether the compressor will ever actually run at speeds as low as 60% or 70%. The resultant stage-to-stage flow mis-match, particularly in propylene/propane service, significantly reduces the overall flow range at very low speeds.

While the user may feel it is more conservative to specify a wide speed range, the effect can sometimes be the opposite. Lowering the minimum continuous speed to a value that might not be necessary drives down the required first critical speed for the compressor, which, in turn, generally reduces the stiffness-to-damping ratio, thus reducing the mechanical stability of the compressor.

Settle Out and Start-Up / Alternate Operating Modes

The user needs to specify the settle out pressure for the compressors, as well as identify how the compressors will be started up (pressures, temperatures, gas compositions). These can affect driver sizing, especially in the case of a fixed speed application. A higher power rating is required if a driver must be oversized to accommodate start-up at full settle out pressure with no suction throttling.

Alternate operating modes, such as air dryout, must also be identified. The user must specify inlet pressure and temperature, relative humidity (for air runs) and any desired discharge targets. For refrigeration machines, it is likely that one or more of the intermediate nozzles can be blanked off for air operation. In some cases, additional flow may have to be introduced into an intermediate nozzle to avoid surge or extracted from one of the intermediate nozzles to avoid choke. This depends on the specific selection and relative design flows for each section, and varies from compressor to compressor.

Experience Requirements

Many customer specifications require that particular compressors or compressor components must have a minimum of one or two years satisfactory operating experience or require single stage testing of proposed staging. With the increasing size of ethylene plants, it is highly unlikely that every component quoted will have a wealth of operating experience. Also, with all of the variations in processes, it is improbable that the exact machine being quoted will duplicate an earlier machine. In these cases, single stage reference test data (collected during stage development) is usually offered as supporting evidence of a new design's reliability. Some turbomachinery manufacturers minimize design work and manufacturing costs by utilizing scaling of major components as opposed to custom designing each component. Scaling creates the opportunity for experience in one size compressor to be applicable to a different size compressor. Using this design technique, all aerodynamic flow paths are exactly scaled up or down using one geometric factor. This preserves all performance parameters when referenced non-dimensionally.

Performance Issues

Curve Shape Guarantees

Sometimes, special curve shape guarantees are imposed over and above the requirements of API 617. These can include: surge flow guarantees, overload flow guarantees, pressure rise or head rise to surge guarantees, a specified head level at a specified overload flow and intermediate (sideload / extraction) pressure guarantees. When such assurances are required, it affects the way the machine is selected by the compressor vendor. For instance, if there is a head or pressure rise to surge requirement, and if a target level is specified, aero components will be selected farther out toward overload to ensure that adequate margin is achieved. To make certain that such a guarantee is met, the designer will usually require *predicted* HRTS/PRTS to be some degree above the *guaranteed* level. The level of guaranteed HRTS/PRTS is subject to debate and differs for different end-users. It is important to note that the amount of achievable HRTS/PRTS depends upon the number of impellers in the compression section being considered, the volume ratio, and the curve shape(s) of the individual stages used in that section. In his discussion of this subject, Lucas (1989) suggests that pressure rise on the order of 3% - 6% is reasonable. It may be quite difficult to meet a requirement of 5%-6% HRTS/PRTS for a compression section operating at high mach number (high equivalent tip speed), while still preserving adequate flow margin to choke.

If curve shape guarantees are imposed only on one end of the curve (low flow end - surge and/or HRTS/PRTS or the high flow end – overload flow and/or specific head level at a designated flow), the selection of aero components can generally be made to meet what is required. Curve shape guarantees on both the high and low flow ends are much harder to design for and to meet. In general, adjustments to improve surge flow margin or HRTS/PRTS can reduce overload margin. Likewise, adjustments to improve overload margin will usually reduce surge flow margin or HRTS/PRTS. For this reason, it is important that the end-user determine what is most important to their process with respect to curve shape guarantees and that they only impose *those* requirements. Also, curve shape requirements can be somewhat inconsistent. For instance, a customer may concurrently require 5% HRTS and that the head at 115% flow be no lower than 85% of design head. This requires a steep head curve on the low flow end and a relatively flatter head curve on the high flow end which can be inconsistent with normal compressor curve shape, especially for high Mach number applications.

Requested vs. Actual Operation

In many cases, the end-user or contractor will apply head rise or pressure rise requirements on a machine in the quotation stages, but in actuality, run the compressor "flat-out" toward high flow to maximize ethylene production. In such cases, it would be wiser to build "extra capacity" into the machine by selecting a design point with more flow margin to choke (larger flow capacity impellers) and slightly reduce the amount of HRTS/PRTS specified. This could provide a better fit for the intended duty without sacrificing too much turndown capability.

Sectional Power Guarantees

Some customers elect to impose sectional power guarantees on multi-section compressors. Such requirements are very restrictive and usually unnecessary. While it is agreed that the overall power required to meet a specified discharge pressure given a set of inlet conditions is important, individual section powers are not as critical. Usually, the refrigeration process has some flexibility with respect to intermediate pressure (head) levels, which would allow variation in sectional power levels. Enforcing sectional power guarantees could potentially force the compressor vendor to compromise the compressor selection or to make adjustments to the machine on the test floor (schedule delays) that may have little impact on the compressor operation at site.

Induction / Extraction Design

Inlet Sidestreams

A unique feature of compressors used in refrigeration service is the use of induction or sidestream inlets. It is typical to have as many as three sidestream inlet connections, which allow gas that has boiled off from liquid refrigerant (refrigerant vapor in a flash drum), to enter the compressor at various temperature and pressure conditions. The use of sidestreams in the process has an effect similar to intercooling in other compression systems. Rather than achieving the required discharge pressure all in one step to meet the condenser specifications, sidestreams provide a means of reducing the inlet temperature to multiple sections of compression. The colder, saturated vapor from the sidestream is mixed at the same pressure with the warmer, superheated vapor from the upstream compression section. This has the effect of de-superheating the gas, resulting in a lower mixed gas temperature. The lower inlet temperatures to the subsequent compression sections and the overall process of throttling the condenser pressure down to an evaporator in incremental steps (sidestream pressures) have the twofold effect of reducing the overall compressor work (or power) and increasing the refrigeration capacity in the process. For this reason, sidestreams have been referred to as economizers since the horsepower per ton of refrigeration is reduced.

To properly receive these additional gas streams, the compressor staging upstream of each entry point must be designed to produce an internal static pressure that is at the same level as the sidestream flow. Calculation of the pressure drop from the sidestream flange to the exit of the inlet guide vanes just prior to mixing is therefore necessary to properly match these two pressures. The specific design configuration of the side entry flow path will determine how well it merges with the existing compressor through-flow. It has been found from internal CFD (computational fluid dynamics) studies and actual shop and field performance tests that the design should minimize the pressure loss due to mixing. This is accomplished by merging the two streams at similar velocities, in a near parallel direction, such that pressure losses due to turbulence and flow mixing are small. Figure 2 shows an example of the predicted velocities of the two flow streams for this type of design. Past experience has proven that designs with excellent mixing properties but high levels of turbulence can cause severe performance deficiencies. Some of the compressors designed in the 1960's and 1970's were designed so that the gas streams merged perpendicular to the upstream through-flow. As a result, it was not uncommon that some of these machines required operation in the field at near maximum continuous speed to maintain the necessary internal pressures and meet the design discharge pressure. A rerate (existing compressor casing with modified internal components) using modern sidestream diaphragms can help to restore performance by reducing mixing losses, thus improving the efficiency and lowering the operating speed.

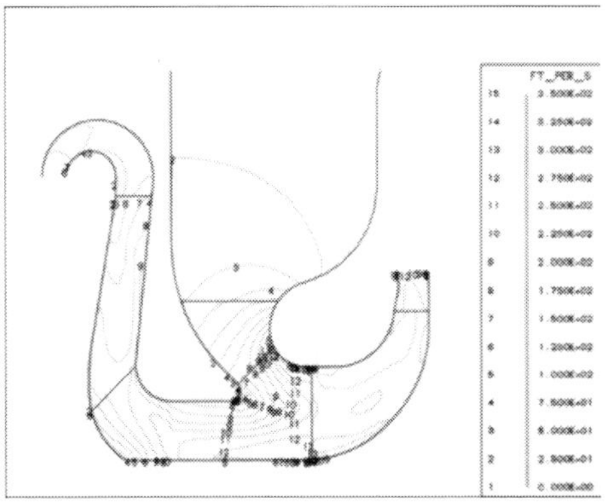

Figure 2: Average Flow Velocity Contours

Proof tests of sidestream compressors require that internal instrumentation be installed to facilitate calculation of the sectional flow, head, and efficiency. Please refer to Figure 3. Normally, static pressure taps and thermocouples are installed at the discharge of the upstream section (return channel vane exit) prior to mixing. Static pressure taps are also installed at the exit of the sidestream passage prior to mixing. Using these measurements, the total (stagnation) conditions for each stream are calculated and combined through mass averaging to determine the inlet conditions to the downstream stage. Since the sidestream temperature is lower than the upstream through-flow temperature, some radial stratification can exist when entering the downstream impeller due to the low loss mixing design. This aspect of the sidestream compressor has been predicted using CFD. Temperature mixing will therefore occur primarily in the impeller. Because some customers occasionally express concern with this issue, additional CFD analysis was performed to evaluate the influence of inlet temperature variation on the head and efficiency of the downstream impeller. Results indicated there was no negative impact on impeller performance due to radial temperature stratification at the inlet.

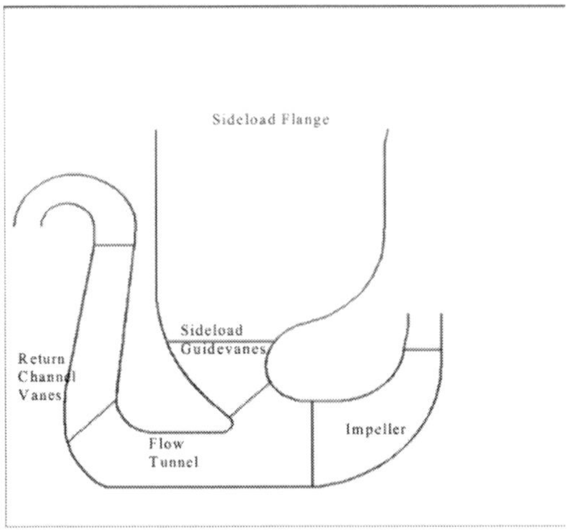

Figure 3: Typical Inlet Sidestream

The exit area of the sidestream and the upstream area of the through-flow are typically sized for one set of flow conditions. This is usually the rated or guarantee point, where often a tolerance range on the sidestream flange pressures are required to be guaranteed. This necessitates accurate predictions of the individual stage aerodynamic and thermodynamic performance. The pressure ratio between the sidestream flanges determines the level of head required at each section of compression. These flange pressures are a direct function of the specific refrigeration process employed by the customer. In order to meet the specific pressure levels at the inlet flanges, various types of individual compressor designs must be considered. Head levels for each section can be adjusted using techniques such as: impeller blade cutbacks or blade extensions, impeller scaling to larger or smaller diameters, or selection of different impeller families with varying amounts of design head (ie: blade backward lean). Also, depending on the amount of flow being added at each inlet, it is not uncommon to have high flow coefficient staging required in the first position of each section.

Extraction Design

Extracting flow from the compressor tends to be a simpler process than adding flow in with a sidestream. In this case, the designer need only be concerned with choosing a location within the internal compressor flow path to divert a portion of the gas to the extraction nozzle. Of course, it is necessary to design the upstream staging to provide the required pressure at the extraction flange for the customer's process. One extraction design method is to use a volute that is geometrically sized for the amount of flow being extracted. The volute is located at the end of the diffuser or crossover (U-bend) and connects to the extraction nozzle similar to normal discharge volute/nozzle designs. Please see Figure 4. The remaining flow continues into the return channel (de-swirl vane passage) and into the next impeller. Another approach is to simply open a passage in the internal annulus between the return channel exit and the downstream impeller, allowing the gas to flow radially out to the extraction nozzle. This design has a somewhat higher pressure loss than the volute design, but can be less expensive to manufacture. In some cases, the process requires that a particular nozzle location be used as both a sidestream and an extraction point. When this occurs, the sidestream design geometry is favored, with higher losses estimated for the extraction flow exiting from the internal annulus, through the sidestream guide vanes, and out through the circumferential plenum to the nozzle.

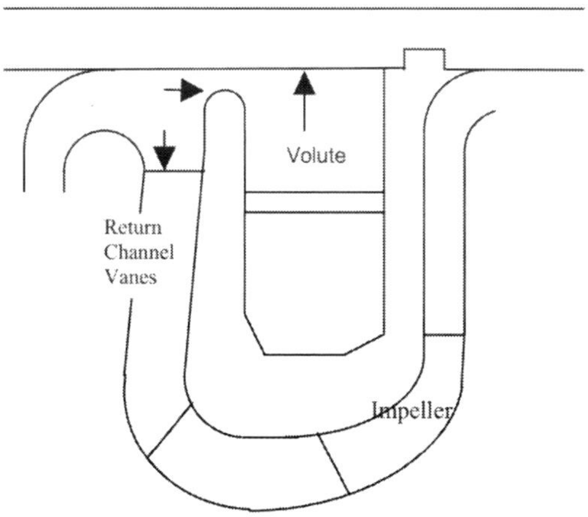

Figure 4: Volute Type Extraction

Testing of extraction compressors requires similar internal pressure and temperature measurements as the sidestream compressor at the return channel exit for evaluating section performance. Since no mixing is required, the downstream inlet volume flow is based on the previous section discharge specific volume and the net mass flow after the extraction.

Rotor Dynamics

A refrigeration compressor generally has a longer bearing span than a straight-through compressor since the introduction of a sideload flow or siphoning off of an extraction flow requires more axial space (except for very low flows) than is available in a "normal" stage space.

Rotor dynamics of turbomachinery includes two areas of focus with respect to vibration: synchronous vibration (vibration that is the same frequency as the equipment running speed) and non-synchronous vibration (higher or lower frequency than the equipment operating speed). Synchronous vibration is relatively straightforward for most compressors of typical rotor and bearing construction, and most manufacturers and users have reached concurrence regarding design guidelines and calculation methods to achieve them. Non-synchronous vibration is often a function of the stability of a compressor. Stability is generally determined by three factors: the rotor stiffness, the bearing characteristics, and destabilizing (or cross coupling) forces. Typically, there is consensus within the industry for determining rotor stiffness. There is also relatively good agreement across the industry with respect to bearing characteristics; although when it comes to stability evaluation, even *that* issue is divided into two different approaches.

One approach utilizes synchronous bearing coefficients, while the other uses frequency-dependent bearing coefficients. The most controversial subject is the method for predicting cross coupling forces.

Within the turbomachinery industry, there are a variety of methods used to estimate aerodynamic cross coupling; and, in turn, to predict stability. They vary from purely mathematical evaluations of aerodynamic forces to graphical correlations of historical experience. Examples of the former are some of the mathematical models produced at various Universities, such as Texas A&M and the University of Virginia, while examples of the latter are the Fulton Curve (Fulton, 1984) and Kirk's Criteria (Kirk, et al., 1983). In between are dimensional analysis methods based on benchmark case studies. One of the most well-known examples of this is the Wachel formula (Wachel, 1982) for determining cross-coupling forces. Many methods have their origins in the Wachel formula, as it was one of the first empirical methods to be published. Most of these methods focused on small high pressure barrel compressors as their benchmarks, since these were the types of machines which first experienced stability issues. Refrigeration compressors operate at lower pressures and gas densities, resulting in relatively lower excitation forces.

The difference in approach for all of the various OEM's, users, and consultants to calculating log decrement values often results in differing values for the same physical system. It is suggested that each method and calculation result needs to be evaluated against the guidelines that arose out of that method; and the reliability of a particular method is a function of the body of work or experience behind the method.

As an example of varying results for the same physical system, a large Propane compressor was analyzed using three different methods of stability analysis. The propane compressor has three sidestreams, with a total power of approximately 70,000 hp. The bearing span-to-main shaft diameter ratio is approximately 10.8, and the running speed is approximately 3600 rpm. The three analysis methods were the API 617 7th Edition method, an OEM proprietary method, and a third party consultant method. These results are presented below:

Table 1: Predicted Stability Comparison

Analysis Method	API 617 7th Edition	OEM Proprietary	Third Party Consultant
Log Dec Value	.08	.11	-.365
Acceptable?	No	Yes	No

In spite of two of the three methods resulting in unacceptable results, this machine was built and has run well throughout its history. Overall vibration is low at 17 microns, with a synchronous component of 5.6 microns, and very little excitation (approximately 0.4 microns) is seen at the predicted first resonant speed of approximately 28 Hz. An example of an actual spectrum plot for this compressor while running in the field with full load follows.

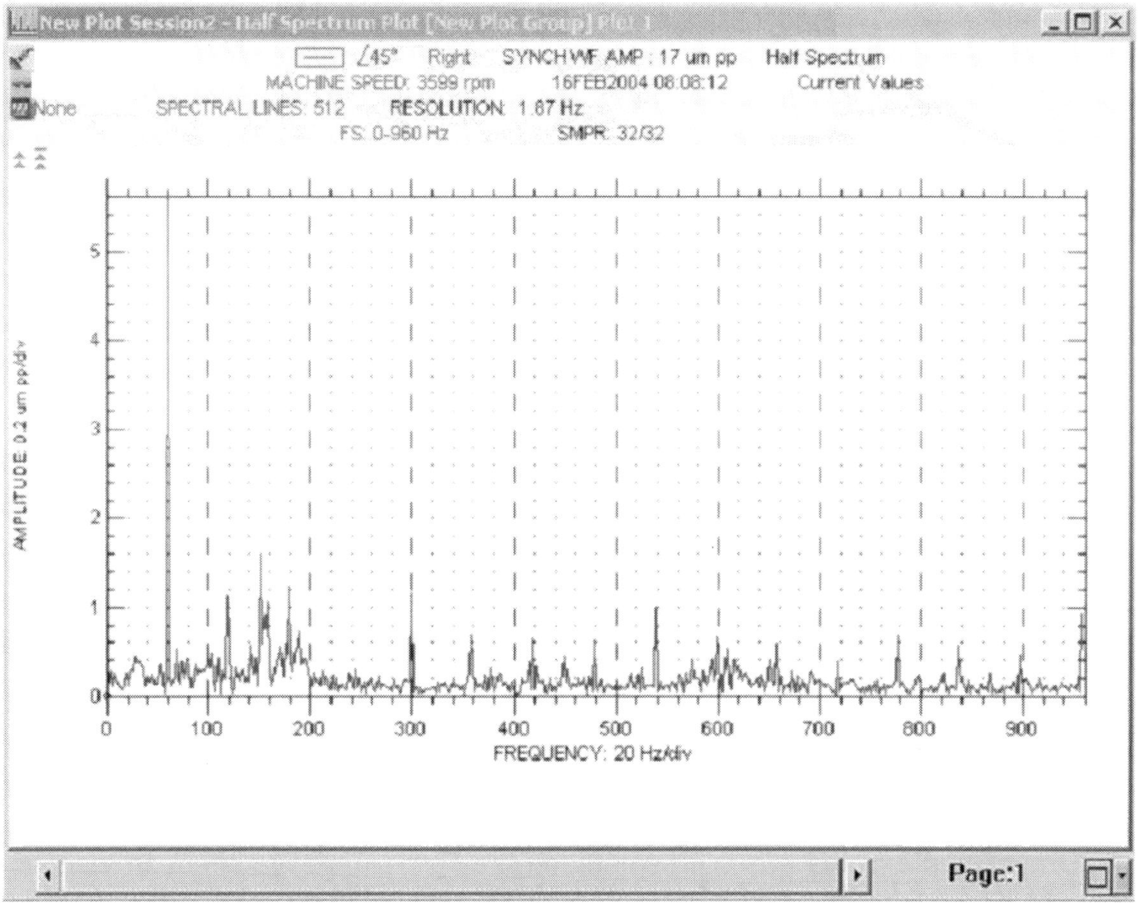

Figure 5: Operating Vibration Spectrum

As stated previously, sidestreams are very common in refrigeration compressors. Propane and propylene compressors (with relatively high molecular weight) can be relatively high in horsepower, especially in large size plants. Sidestreams and horsepower are two issues that are treated very differently in the various stability prediction methods. Sidestreams for some methods are evaluated as the beginning of a new machine, while others use averaging of conditions, or use stage-by-stage parameters. With regard to compressor power, some methods use power as a raw number in evaluating cross coupling; other methods utilize values that are mitigated by a geometry factor to establish how high in horsepower a machine is for its relative size.

Some methods have been very successful in predicting stability in the larger frames that are typical in refrigeration. Other methods realize most of their success with small, high pressure barrel compressors which were utilized as the benchmarks when these methods were developed, and have not been as successful in predicting the stability of larger compressors. False negative predictions of stability can be very damaging in the compressor selection process, as they can force a viable single compressor to instead be built as two bodies. Therefore, the track record of the manufacturer and the particular prediction method must be evaluated in their own right.

Minimum Design Temperature

Materials utilized in industrial centrifugal compressors are generally rated to a minimum temperature of -20F (-28.9C). For propylene/propane duty, minimum design temperature is usually in the range of -50F (-45.6C) to -60F (-51C); ethylene duty generally has a minimum design temperature of approximately -150F (-101C), but may range anywhere from -50F (-45.6C) to -170F (-112C).

A few things to note with regard to temperature:

1. The lower the temperature, the more specialized the material. Some low temperature materials are only available from forge shops/plate vendors in heat lot quantities. This can significantly increase material costs unless the entire heat lot can be used. Special materials require longer lead-times, which can affect delivery.

2. To determine minimum design temperature, one must also consider static (settleout) conditions. Although a settleout condition may occur at a lower temperature than during normal operation, the fact that the rotor is at rest needs to be considered.

3. Cost may be reduced by applying low temperature materials only where they are required by the duty. As temperature increases from the inlet end of the compressor to the discharge end higher temperature "standard" materials may be used.

Over-Conservatism

While conservatively designing a machine may seem to be desirable so as to avoid breaking new ground, over-conservatism has associated costs, or may increase risk in other areas. One example of this was cited earlier, regarding the negative effect on mechanical stability resulting from specifying a lower-than-necessary minimum continuous speed. Other examples are:

- Oversizing shaft ends, due to conservative material requirements or the application of high safety factors, particularly on driven-through bodies can reduce critical speeds and stability, increase coupling costs; and, in extreme cases, require larger diameter bearings and seals.

- Nozzle velocity limits can increase bearing spans by requiring larger nozzles with increased axial space, reducing critical speeds and stability.

- Consideration of low level excitation sources for torsionals, such as avoidance of torsional frequencies at multiples of running speed, can result in lateral critical speed problems.

- Overly conservative stability requirements may require artificial damping devices which could, depending on the devices used, add complexity, reduce reliability, decrease efficiency or require additional compressor bodies.

- Low thrust load or low bearing metal temperature requirements may require larger bearings at a higher cost, introduce higher mechanical losses, and adversely affect rotor dynamics.

- If the minimum temperature is too conservatively specified, operating clearances may be oversized, introducing inefficiency into normal operation. Alternative materials may unnecessarily be required for items such as seals, O-rings and sealants, which may not function as efficiently at the normal operating conditions.

- Restrictive guarantees on vibration through critical speeds can result in higher vibration in the operating speed envelope due to a shift of balance priorities.

Conclusion

Refrigeration compressors are very important to the ethylene process and their proper functioning is critical to realizing plant production goals. Variables such as multiple operating points, curve shape guarantees, intermediate pressure guarantees and special mechanical requirements have significant influences on how the compressor is designed and how individual staging is selected. It is essential that both the customer and compressor manufacturer have close communication in the design phase. This will help to ensure that the design focuses on the most important process requirements, that the various specifications do not conflict with each other, and that the purchase specifications result in equipment that is well-suited for the intended operation.

References Cited

1. Blahovec, J.F., Matthews, T. and Eads, K.S.,1998, "Guidelines For Specifying And Evaluating New And Rerated Multistage Centrifugal Compressors," *Proceedings of the 27th Turbomachinery Symposium,* Turbomachinery Laboratory, Texas A&M University, College Station, TX, pp. 215-231.

2. Lucas, C.E., November 1989, "Refrigeration System Stability Linked to Compressor and Process Characteristics", Chemical Engineering Progress, pp 37-56.

3. Fulton, J.W., "The Decision to Full Load Test a High Pressure Centrifugal Compressor in its Module Prior to Tow-out," I. Mech. E. 1984, Paper No. C45/84

4. Kirk, R.G. and Donald, G.H., "Design Criteria for Improved Stability of Centrifugal Compressors," AMD Vol. 55: Rotor Dynamical Instability, ASME, 1983.

5. Wachel, J.C., "Rotordynamic Instability Field Problems", NASA Conference Publication 2250: Rotordynamic Instability Problems in High-Performance Turbomachinery, 1982.

6. Hardin, J.R., "A New Approach to Predicting Centrifugal Compressor Sideload Pressure", IMECE 2002, Paper No. 39592.

Paper 42c

Turboexpanders in Ethylene Plants

Joseph K. Lillard

Vice President of Technology

Mafi-Trench Corporation

Prepared for Presentation at the 2005 AIChE Spring National Meeting

17th Annual Ethylene Producers' Conference

Atlanta, GA, April 10-14, 2005

Session TA010 Ethylene Plant Technology – Refrigeration Systems

April 13, 2005

Introduction

Turboexpanders have been used in industry for over 60 years to produce the low temperature refrigeration needed for the separation of gases. Ethylene processes have incorporated turboexpanders for over 40 years. The modern turboexpander is a reliable and highly efficient, integral part of the cryogenic recovery section in most large-scale ethylene plants.

Turboexpanders

A turboexpander (also called an expander) is a machine that continuously expands a fluid from a greater pressure to a lower pressure while causing the fluid to do work. When the fluid is a gas, the gas is cooled significantly while the expander produces shaft work.

Though several types of turbine design are feasible for a turboexpander, the radial inflow turbine has found the greatest acceptance in today's cryogenic refrigeration and power recovery applications. The radial inflow turbine is a compact, robust, and very efficient machine, which is readily scaled to cover a wide range of process conditions.

Figure 1.0 - EC Cross-Section

The radial inflow turbine combines two expansion stages, one stationary and one rotating. For peak efficiency, approximately half the enthalpy drop occurs across the stationary stage and half across the rotating stage. The stationary stage is an annular ring of inward directed converging nozzles, known as inlet guide vanes, or IGV's. These IGV's convert the pressure energy to kinetic energy in a nearly isentropic expansion. The IGV's direct this high velocity flow into the rotating stage, called the impeller, or "wheel". The wheel rotational speed is selected to provide an optimum match between the wheel tip velocity and the velocity of the fluid stream exiting the IGV's. After the gas enters the wheel, it is accelerated and turned from the radial inward direction to the axial direction. The discharge blading on the wheel is designed to remove energy from the swirling fluid stream and to leave only an axial velocity component. Some of the energy in the exit velocity is further recovered by an exhaust diffuser. Overall, the flange-to-flange isentropic efficiency of a modern turboexpander will usually be between 85 and 90% at the best efficiency point.

The wheel is attached to a shaft that is supported by radial and axial bearings. The power extracted from the process by the wheel is transmitted through the shaft and can be used to drive a variety of different machines. Most commonly, the power is used to drive a single stage compressor wheel integrally mounted on the end of the shaft opposite to the expander wheel. The compressor wheel can be used in the process to either increase the pressure drop available to the expander, thereby increasing the overall cryogenic refrigeration produced or to reduce the load on a separate process compressor.

Figure 2.0 - Skid Mounted EC

These units are known as Expander-Compressors (or "EC" units) and are recognized in the seventh edition of American Petroleum Institute (API) purchasing specification API 617.

Turboexpanders are available in a broad range of sizes. The smallest hydrocarbon gas processing turboexpanders use impellers as small as 45 mm (1.75 in.), running at 125,000 rpm and producing 11 kW (15 HP). The largest units use impeller diameters of 914 mm (36 in.), running at 7000 rpm and capable of producing 18,600 kW (25,000 HP).

Bearings and Seals

The modern turboexpander uses either oil bearings or magnetic bearings, as selected by the plant owner/operator. Both systems have advantages and disadvantages, but the industry trend is toward greater use of magnetic bearing systems, especially in ethylene applications.

This is mainly due to the complete elimination of lube oil, thereby preventing oil from entering into the process, which can then freeze and clog the cold box exchanger. In addition, users report lower maintenance costs for magnetic bearing units, due to having simpler auxiliary systems.

Both types of bearing systems require shaft seals between the process side (expander and compressor wheels) and the bearings. In an oil bearing unit, the seal gas prevents cold or dirty process gas from entering the bearing housing and chilling or contaminating the lube oil, while also keeping the lube oil out of the process casings. In a magnetic bearing unit, the seal gas prevents cold or dirty process gas from coming into contact with the auxiliary bearings and provides cooling for the windage losses generated by the rotor.

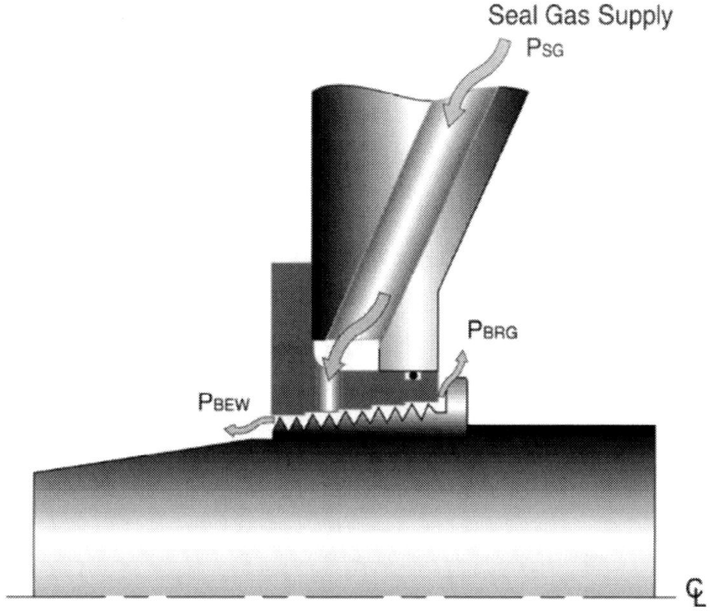

Figure 3.0 - Schematic of Shaft Seal

Process Designer Guidelines, Concerns, and Limitations

Sizing

The goal of turboexpander sizing is to offer the highest performance and reliability at the lowest cost. As a result, vendors will always offer the smallest frame size that still provides competitive turboexpander efficiency. For turboexpanders, it can be shown that the outlet volume flow is the controlling parameter for frame size. Figure 4.0 shows how wheel diameter varies with turboexpander outlet volume flow for optimum efficiency. The largest cryogenic expanders built for natural gas processing applications have outlet volumes of over 20,000 m3/hr (12,000 ACFM), and the largest built for power recovery applications have outlet volumes of over 60,000 m3/hr (35,000 ACFM).

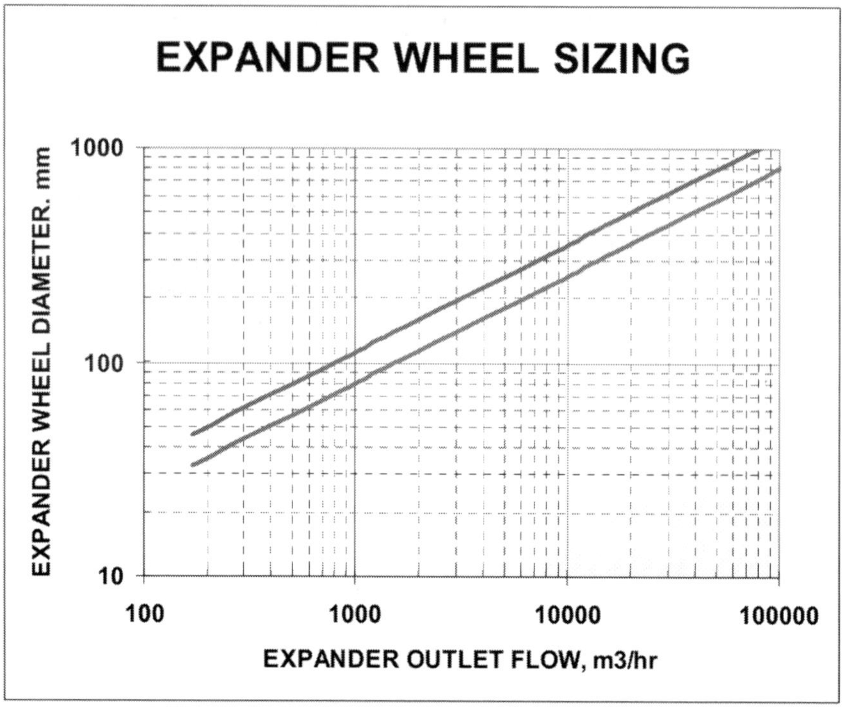

Figure 4.0 – Graph of Flow vs Expander Wheel Diameter

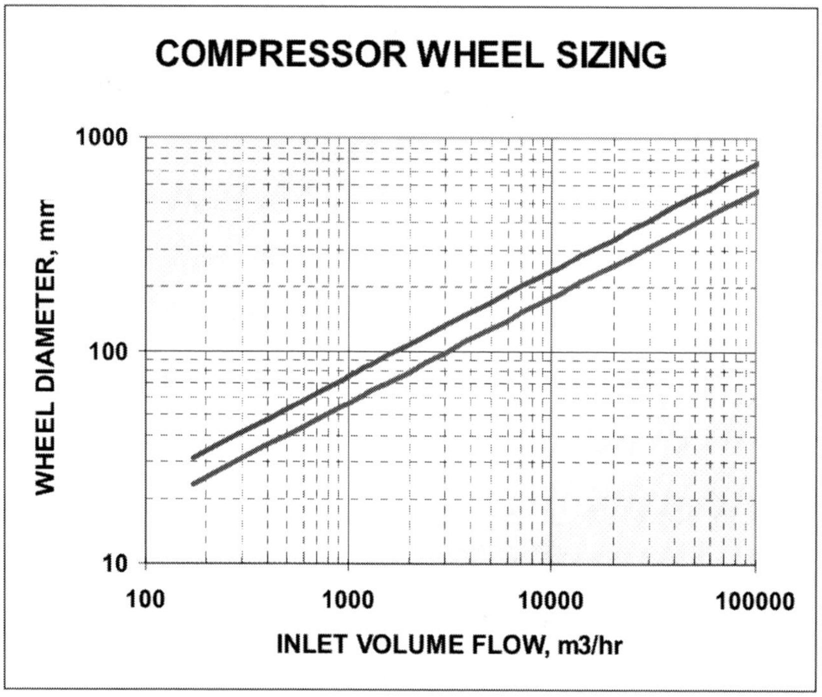

Figure 5.0 – Graph of Flow vs Compressor Wheel Diameter

The compressor impeller must be sized to take the power produced by the turboexpander and use it to compress its process stream as efficiently as possible. Figure 5.0 shows how wheel diameter varies with the compressor inlet volume flow for optimum efficiency.

Mechanical Design Considerations

Turboexpanders for ethylene plant applications will have the following characteristics within the limits of the frame size:

1. Low power
2. High speed
3. High tip speed (impeller peripheral speed)
4. High bearing journal velocity (journal peripheral speed)

Table Figure 6 - Table of Frame Size Limits

Frame Size	Approx. max Power (kW)	Flange Size	
		Exp	Comp
Frame 1	200	3"/4"	6"/6"
Frame 2	1200	4"/6"	8"/8"
Frame 2.5	2000	6"/8"	10"/10"
Frame 3	3500	8"/10"	12"/12"
Frame 3.5	6000	10"/12"	18"/18"
Frame 4	8000	12"/14"	20"/20"
Frame 5	10000	20"/24"	24"/24"
Frame 6	18000	24"/30"	36"/36"

By "low power", manufacturers mean that the unit is well below the maximum power capability of the frame size. Turboexpanders use the same rotor/bearing systems for natural gas processing applications, which typically operate at much higher gas densities and power levels.

By "high speed", manufacturers are really referring to both high tip-speed and high bearing journal velocity, compared to the design limitations of the frame size. For a given application, these limits are determined by considering several factors at once, therefore there is no absolute limit for each individual parameter.

Design considerations for high tip-speed include stress and deflection in the impeller disk and blades, and avoidance of impeller resonance in the operating speed range of the unit. With the use of Finite Element Analysis (FEA) and a design methodology based on years of successful experience, tip speeds of turboexpander-compressor impellers at maximum continuous speed have been pushed up to over 450 m/s (1500 ft/s).

High tip-speeds result in large wheel diameters for the frame size, and thus the rotordynamic design must be carefully analyzed. EC units should be designed so that the first shaft bending critical speed (forward mode) is at least 25% above the maximum continuous operating speed.

Design considerations for high bearing journal velocity depend on the type of bearing selected, and both oil bearings and magnetic bearings have their own limitations. For oil bearings, the limitations are oil film stability and adequate cooling. For magnetic bearings, the limitations are lamination stack stress and deflection, thrust disk stress and deflection, and auxiliary bearing DN limits. Cooling of the magnetic bearings must also be carefully addressed in the design.

Liquid at the Turboexpander Inlet and Outlet

Small amounts of liquid (1-2% of process flow by weight) can be tolerated at the inlet of the turboexpander, as long as the liquid droplets are atomized. Particle size is dependent on the impeller size, tip speed, and gas density. Figure 7.0 shows approximate allowable particle size as a function of wheel diameter for a typical ethylene processing application. Liquid particles of this diameter or smaller will be carried through the expansion stage by the gas and will not cause erosion of the impeller or IGV's.

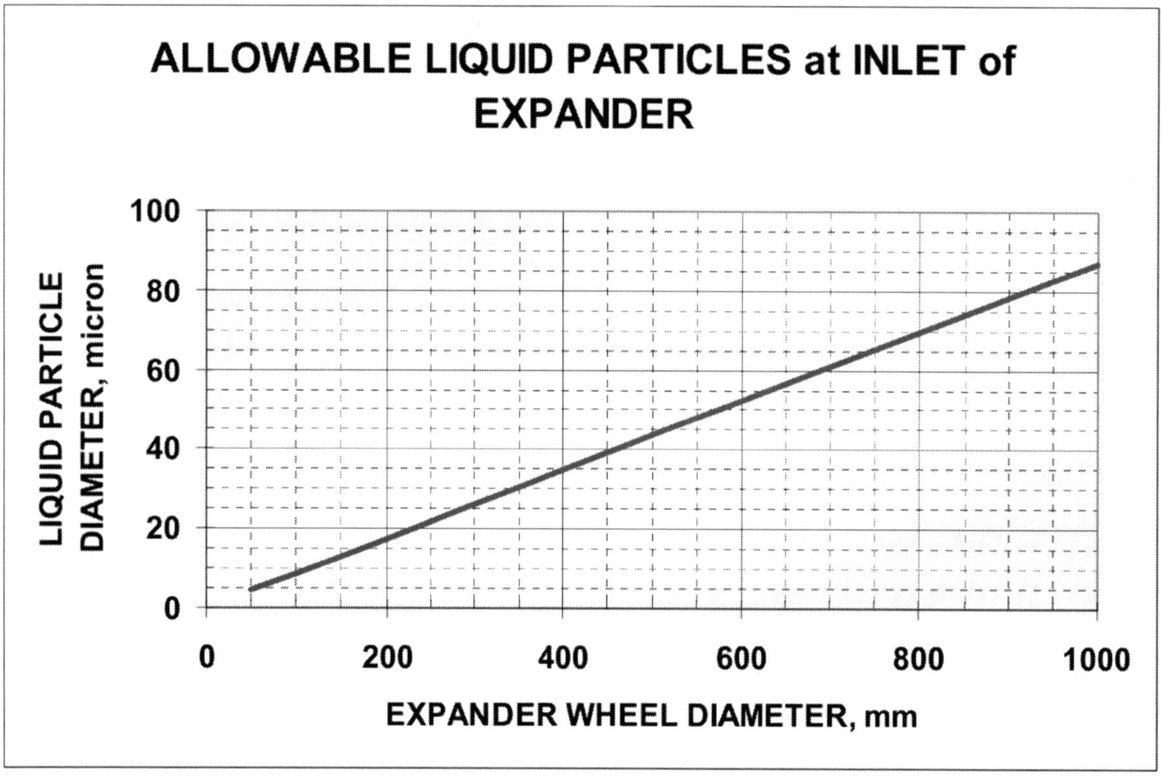

Figure 7.0 - Graph of Diameter vs Liquid Particle Size

At the discharge, very large amounts of liquid can be tolerated without loss of efficiency or reliability. Natural gas processing applications have been in operation for many years with up to 50% liquid by weight at the expander discharge. Ethylene plant applications rarely see more than 20% liquid by weight, due to the nature of the process design.

Enthalpy drop, Tip Speed, and Staging

In turboexpander aerodynamic design, it can be shown that efficiency of a turbine stage is related to the value of tip speed divided by the square-root of the isentropic enthalpy drop, a ratio commonly denoted by 'U2/Co'. This performance relationship can be plotted in a way that is more useful to the process designer [Fig. 8.0].

The plot shows the maximum attainable efficiency for a large single-stage turboexpander with optimized mechanical and aerodynamic design. Note that when high tip-speeds are possible, the maximum efficiency remains flat all the way up to 230 kJ/kg, and is still as high as 80% up to 380 kJ/kg.

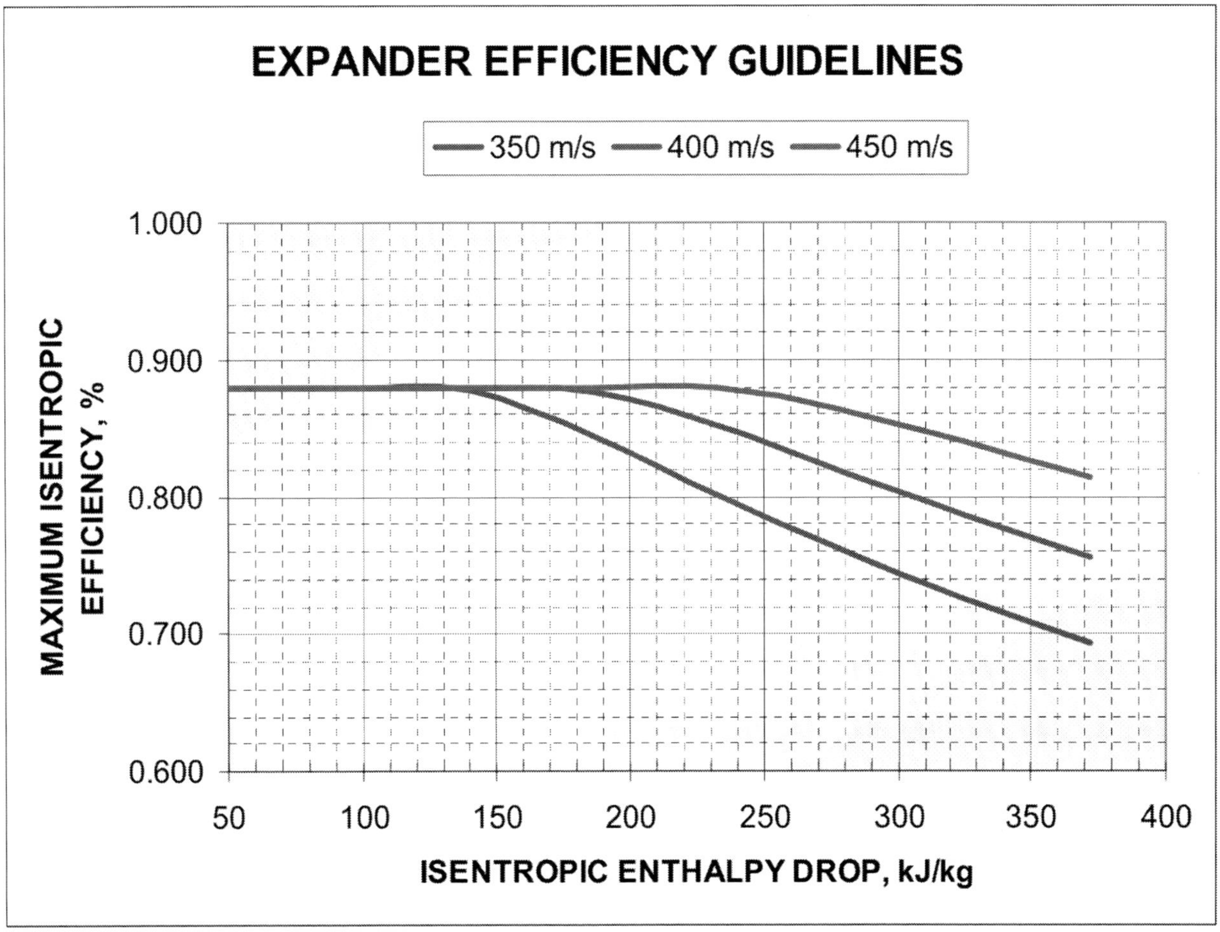

Figure 8.0 - Graph of Enthalpy vs Efficiency

One problem still remains: how can this power be used by a single compressor wheel without exceeding the mechanical design limits? For many ethylene plant processes, the compressor flowrate is the same (or nearly the same) as the expander flowrate. In this case, the compressor tip speed will need to be at least 10-20% higher than the expander tip speed, in order to provide a power balance at the chosen expander operating speed. It is often the case that the compressor wheel tip speed, rather than the expander wheel tip speed, limits the enthalpy drop (and therefore the amount of refrigeration) that can be achieved with a single stage expander wheel.

Mafi-Trench Corporation has patented an internal flow re-circulation mechanism that allows the compressor outlet flow to recycle back to the compressor inlet flow, eliminating the need for an external anti-surge valve and piping system. This "internal recycle valve" is controlled via pneumatic or electric actuator mounted external to the compressor casing. The compressor wheel is provided with a higher mass flow than the expander wheel. This reduces the specific enthalpy required to be absorbed by the compressor wheel, which can then be designed to have a mechanically acceptable tip speed. This design enables a single stage expander to have a higher enthalpy drop, at the sacrifice of some compressor pressure rise.

Pemex operates a single-stage MTC Frame EC-3.0 turboexpander-compressor in their ethylene plant in Cangrejera, Mexico, commissioned in 2003. The unit produces 1000 kW (1300 HP) of refrigeration, running at 34,000 RPM. The isentropic enthalpy drop across the single expander stage is 337 kJ/kg (145 Btu/lb), with an isentropic efficiency of 76%, at a tip speed of 374 m/s (1228 ft/s). Note that the efficiency would be up to 84% at a tip speed of 450 m/s (1476 ft/s). The single-stage compressor wheel mounted to the other end of the shaft uses the internal flow re-circulation system, and still produces a polytropic enthalpy rise of 121 kJ/kg (52 Btu/lb). The internal flow re-circulation system is controlled as both a speed limiter, and as anti-surge protection, making use of MTC's internal flow measuring device.

U.S. Patent Jan. 6, 2004 Sheet 2 of 6 US 6,672,826 B2

FIG. 2.

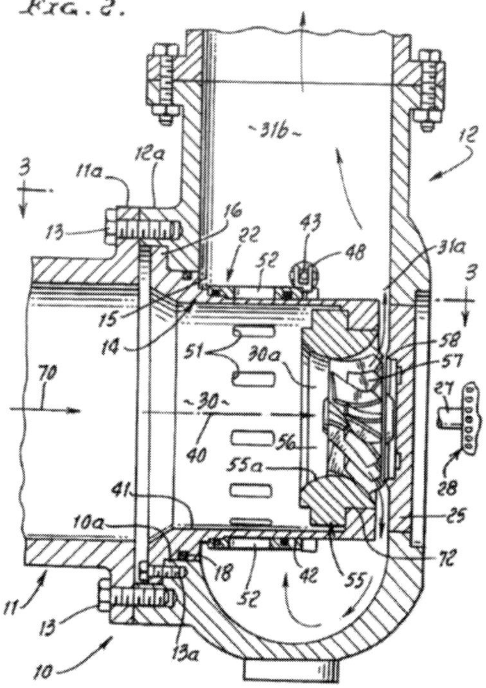

Figure 9.0 – Internal Flow Recycle Mechanism
Multiple Stage Systems

The desire for maximum efficiency in high enthalpy drop applications may result in a staged system. When economically justified by the value of the recovered product, two or more turboexpander units are arranged in series. Up to 4 (four) EC units in series have been supplied into ethylene plant cryogenic recovery sections. By staging the units in series, the expander and compressor enthalpies across each stage can be reduced so that compressor flow recycle is not required, and so that the overall efficiency can be optimized. If the expansion results in liquid formation at the discharge of the interstage units, then liquid knockout vessels should be provided, so as to avoid any significant quantity of liquid entering the downstream expander.

Figure 10.0 – Three Stage EC System

Seal Gas/Cooling Gas

As mentioned earlier, EC units require some type of warm, dry, superheated buffer gas at adequate pressure and temperature to be used for seal gas (oil or magnetic bearing units) or for cooling gas (magnetic bearing units only). All of this seal gas is recovered and returned to the process at the compressor suction pressure. As a ballpark estimate, the minimum required supply pressure is equal to the square root of the product of the turboexpander inlet and outlet pressures. For multiple stage systems, use the first stage turboexpander inlet and outlet pressures to calculate the seal gas supply pressure. Preferably, a warm process gas source can be found at suitable pressure. If not, then expander inlet gas can be heated (using a process exchanger, electric heater, or steam sparger) to the desirable supply temperature. It is important to ensure that liquids do not form at any time in the seal gas system or in the shaft seals. Liquid hydrocarbons will contaminate the lube oil in oil bearing units, and will cause erosion in the labyrinth seals in both oil bearing and magnetic bearing units.

Summary and Conclusions

Improvements in analytical tools and inventive design approaches have increased the performance, reliability, and cost-effectiveness of turboexpander-compressors. Expander efficiencies of 88% and compressor efficiencies of 85% are now achieved in larger, fully optimized units. The use of magnetic bearings can eliminate the possibility of process contamination from lube oil, and provide an auxiliary system with lower maintenance expense. Very high expander enthalpy drops can be efficiently recovered in a single-stage unit using innovative features such as the internal recycle system. Finally, large Frame turboexpander-compressor units developed for large natural gas processing plants are capable of handling the requirements of the largest ethylene plants presently under study.

References

1. Lillard, J. and Nordwall, G., "Using CFD Tools in the Design of Turboexpanders", Compressor Tech, Jan-Feb 2003, pgs, 10-16.

2. Lillard, J. and Nicol, G., "An Operating History of Turboexpanders in the Supercritical Fluid Regime", 73rd Annual Gas Processors Association Proceedings, 9 March 1994.

3. Jumonville, J., Ramsay C. and Andrews, F., "Specifying, Manufacturing, and Testing a Cryogenic Turboexpander Magnetic Bearing System", Proceedings of 20th Turbomachinery Symposium, 1991.

4. U.S. Patent No. 6,672,826 B2 issued 6 January 2004.

5. Bergmann, Dirlam, Lillard, Toogood, et al, "Mafi-Trench Corporation Turboexpander Handbook", 2003.

Paper 42d

Mixed Refrigerant Systems for Ethylene Plants

Sanjeev Kapur and Qi Ma
ABB Lummus Global
Houston, Texas
e-mail: sanjeev.kapur@us.abb.com
qi.ma@us.abb.com

Prepared for Presentation at the
American Institute of Chemical Engineers
Session TA010 – Ethylene Plant Technology – Refrigeration Systems
17th Ethylene Producers Conference
2005 Spring National Meeting
Atlanta, GA
April 2005

Session: Ethylene Plant Technology
Refrigeration Technology as Practiced in Olefins Plants

Mixed Refrigerant Systems for Ethylene Plants

Introduction

The cracked gas from a steam cracking heater is a mixture of various hydrocarbon components and hydrogen. The separation of these components involves compression and chilling to very low temperatures. The conventional ethylene flow sheet utilizes three cascade refrigeration systems to provide process cooling from ambient to -140°C. These include a propylene refrigeration system, an ethylene refrigeration system, and either a methane refrigeration system or an expander/recompressor system. Some schemes utilize a methane recycle stream to the cracked gas compressor instead of a separate methane compressor.

The cracked gas composition and the operating pressures of the processing scheme dictate the refrigeration loads at various levels. For a given processing scheme, the process refrigeration requirements are determined mainly by the cracked gas composition. Figure 1 shows the variation in cracked gas composition for various feeds.

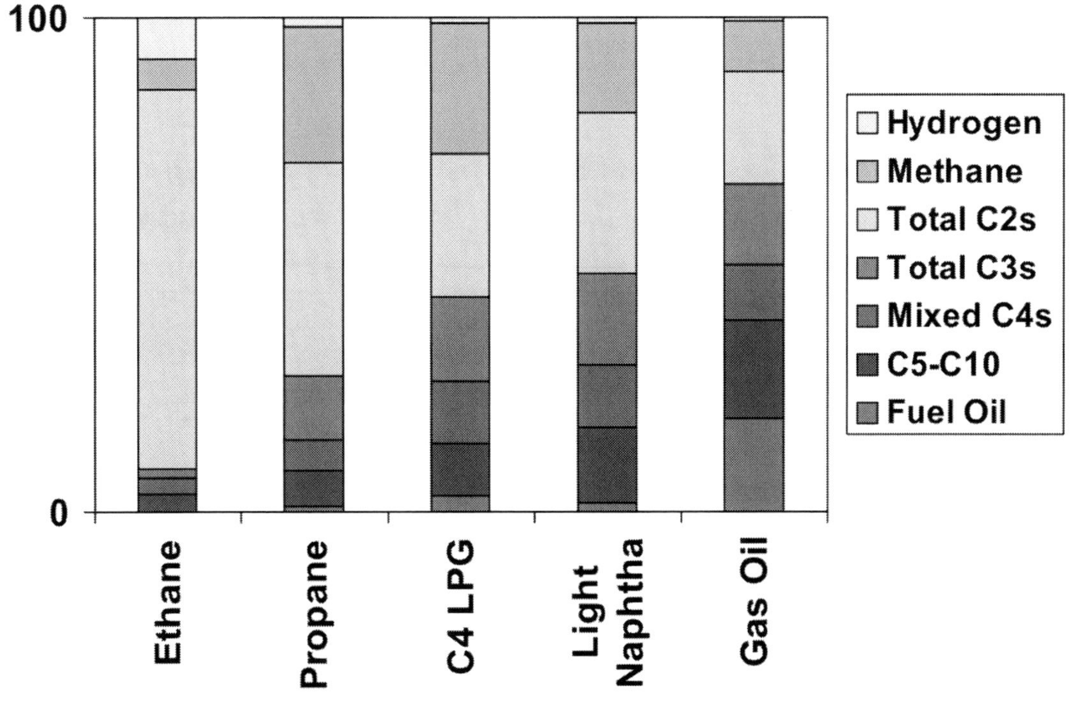

Figure 1 – Product Distribution

For a given ethylene capacity, the relative content of hydrogen, methane, propylene, and C_4s varies depending on feedstock. This dictates the refrigeration needs at different temperature levels. The cracked gas composition will also vary with changes in cracking severity for a given feedstock.

The dynamics of feedstock pricing and the competitive nature of the ethylene business require crackers to have sufficient design flexibility to accommodate the constantly changing feedstock slates and cracker severity. Therefore, the refrigeration system(s) design has to be able to adapt to these varying demands in a controlled and reliable manner. This differs from the mixed refrigeration systems for LNG and gas separation plants where the refrigeration demands are directly dependent on feed flow and composition due to the physical nature of these processes.

For a conventional ethylene plant flow sheet, approximately 20–25 percent of the ISBL equipment cost is in the compressor account. Therefore, one of the main objectives in developing the new refrigeration system was to reduce the number of rotating equipment systems. This reduction has the added benefit of increased reliability and reduced maintenance costs.

ABB Lummus Global developed the patented mixed refrigerant systems to provide refrigeration needs from ambient to -140°C. As illustrated in Figure 2, Lummus approached the combining of the refrigeration systems in two steps.

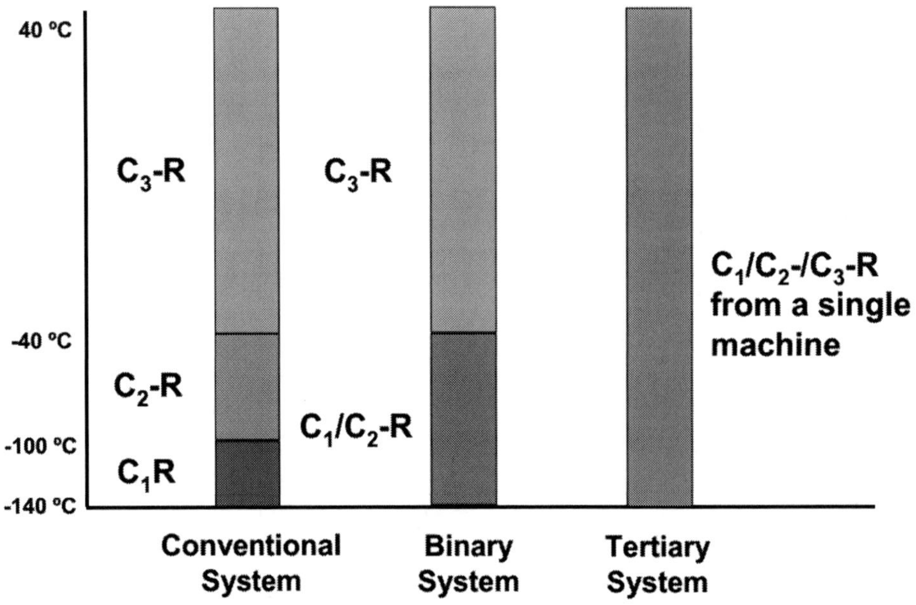

Figure 2. Tertiary Refrigeration Concept

The first step was to combine the ethylene and methane refrigeration levels, providing refrigeration between -40°C and -140°C with the binary refrigeration system. The binary refrigeration system uses a mixture of ethylene and methane. The system is successfully operating in two ethylene plants in China, with the first unit in operation since 2001. Both units operate nearly flawlessly, with system reliabilities higher than the separate ethylene and methane refrigeration systems. The development of the binary system involved extensive use of dynamic simulation modeling tools for establishing the design approach and control scheme, and to understand transient behavior during start-ups and major upsets. Figure 3 shows the results of dynamic simulation for a change in binary composition after introduction of a methane stream during start up. Figure 4 shows the inventory build-up in the accumulator during the same period.

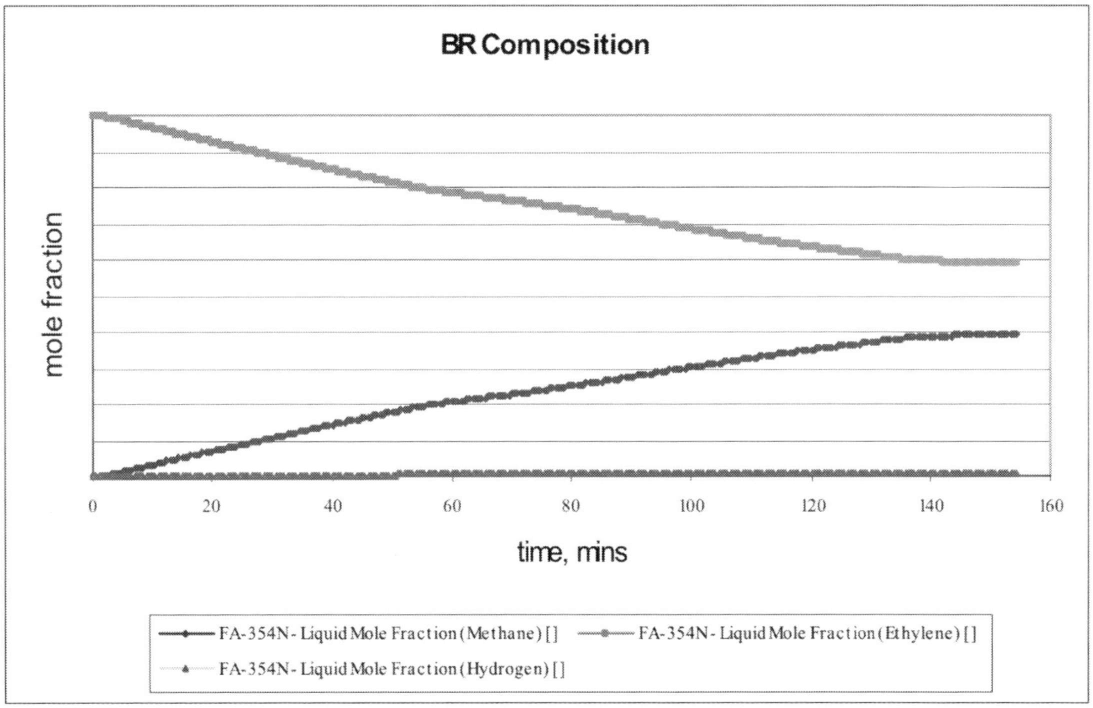

Figure 3 – Binary Composition During Start-up

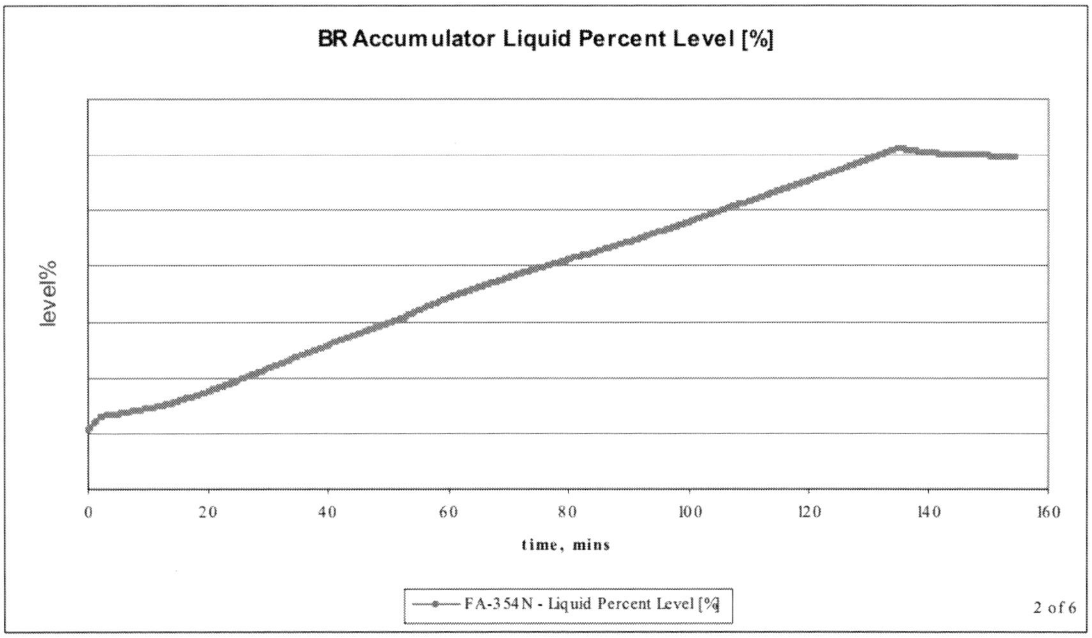

Figure 4 – Binary Refrigeration Accumulator Inventory During Start-up

With the successful commercialization of the binary refrigeration system, the next logical step was to introduce the tertiary refrigeration system. The tertiary refrigeration system utilizes a mixture of propylene, ethylene, and methane to provide refrigeration in a single system for temperatures from ambient to -140°C. Similar to the binary system, the tertiary refrigeration concept was also verified with dynamic simulations for controlled operation and system stability during major upset scenarios.

The tertiary refrigeration unit started up in October 2004 as part of an ethylene plant expansion at Qilu Petrochemicals and is shown in Figure 5. Tertiary refrigeration technology was a vital component in the client's economic justification of the expansion project.

Figure 5 – Tertiary Refrigeration Machine

In addition, five projects based on mixed refrigeration concepts are in various stages of engineering design, construction and commissioning.

Description

Like a conventional cascade system, the mixed refrigeration system is based on a mechanical refrigeration cycle. A typical example of a mixed refrigeration cycle is shown in Figure 6. In a cascade refrigeration system, warmer refrigerants condense the cooler ones. In a mixed refrigeration system, in addition to condensing using warmer refrigerant, auto-refrigeration is also used to enhance system flexibility. The use of the tertiary refrigerant provides all the refrigeration loads and temperatures required for an ethylene plant while obviating the need for two or three separate refrigeration systems.

Refrigeration cycle

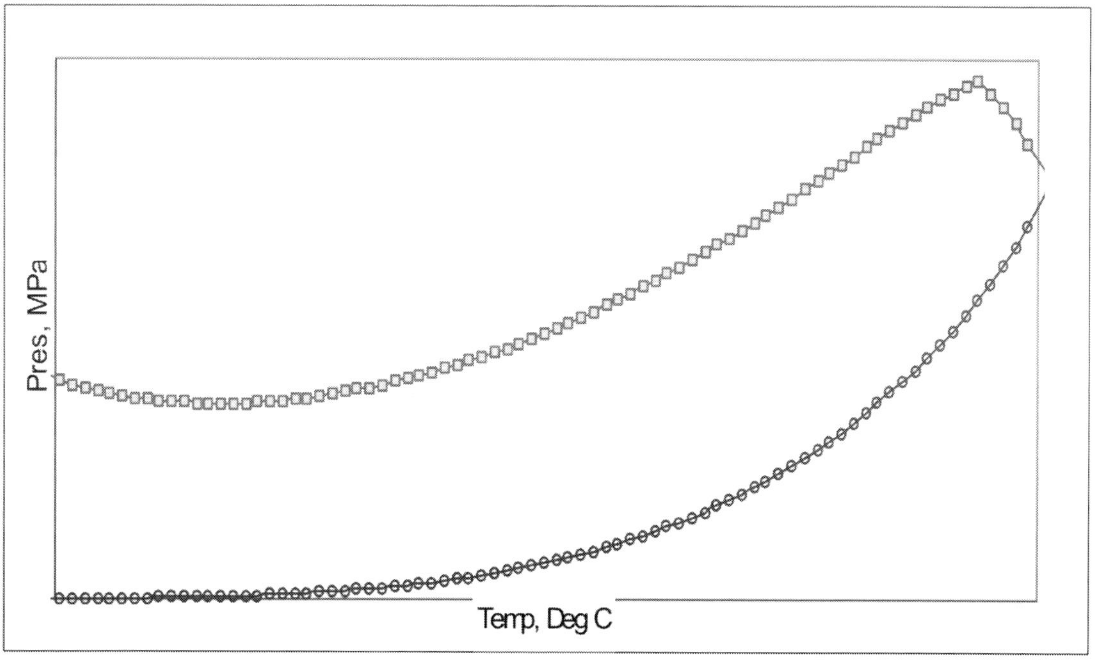

Figure 6 – Mixed Refrigerants System Example

Table 1 shows the number of equipment pieces that are eliminated. The number of rotating equipment casings is reduced from six with conventional technology to three using the tertiary refrigeration system. This reduction in rotating equipment casings lowers maintenance costs and improves reliability.

Table 1 – Equipment Piece Count Comparison

	Convention Cascade Refrigeration	**Binary Refrigeration**	**Tertiary Refrigeration**
Compressor Casings	2-3	2	2
Turbines/Drives	2-3	2	1
S&T/PF Exchangers*	29	22	17
Drums	14	12	6
TOTAL	47 – 49	38	26

* PF – Plate Fin (Brazed Aluminum), number represent services and not cold box.

The tertiary refrigeration system eliminates equipment, saves capital, improves reliability, and reduces maintenance costs, thereby improving return on investment. For a typical world scale ethylene plant, the tertiary refrigeration system can result in capital savings in excess of two percent on an installed cost basis.

Summary

The development of mixed refrigeration systems was carried out in two steps:first with binary refrigeration and then with tertiary refrigeration. For both concepts, extensive dynamic simulation studies were conducted to verify the design approach, control scheme, and system behavior during major upsets. In addition, various start-up scenarios were looked at to understand the system behavior and timing. These studies helped in the smooth start up of the three systems currently in operation. Five additional mixed refrigeration systems are in various stages of engineering and construction. This represents a major development in refrigeration systems that can meet the needs of the olefins industry in a reliable manner and result in significant savings in capital investment.

Paper 42e

LNG & Ethylene Plant Refrigeration Systems

Julio Rios
Process Manager LNG / Gas Processing
Kellogg Brown and Root

Steven Borsos
Olefins Technical Manager
Kellogg Brown and Root

Prepared for Presentation at the 2005 AIChE Spring National Meeting
17th Annual Ethylene Producers' Conference
Atlanta, GA, April 10-14, 2005

Session TA010 Ethylene Plant Technology – Refrigeration Systems
April 13, 2005

Abstract:

This presentation will compare and contrast refrigeration systems as used by current world-scale Liquified Natural Gas and Ethylene plants. This overview will cover evolution, configuration and basic design and operational issues.

The PowerPoint presentation was not available in time for publication. If you are interested in a copy of this presentation please contact:

Julio Rios

julio.rios@halliburton.com

Paper 43a

How Industry Sponsored Air Monitoring is Being Used to Drive Air Quality Improvements in the Houston Area

Walter Crow – URS Corporation
Steve Smith – Lyondell Petrochemical Company
Steve Hansen – Shell Chemical Company
G. Michael White – ExxonMobil
Katherine Lumbley – URS Corporation

Prepared for Presentation at 2005 AIChE Spring National Meeting/Atlanta, GA/April 10-14, 2005/Ethylene Producers' Conference/Environmental Session

Unpublished

ABSTRACT

The Houston metropolitan area is currently non-attainment for both the 1-hour and 8-hour National Ambient Air Quality Standards for ground level ozone. In 2003, the Houston-Galveston-Brazoria region routinely exceeded the 1-hour ozone standard approximately on 38 days and it exceeded the 8-hour ozone standard on 53 days. Recent studies have shown that emissions of light olefins from stationary point sources when combined with emissions of oxides of nitrogen can react to form transient high levels of ground level ozone. Houston has until 2007 to achieve attainment of the 1-hour standard for ozone and until 2010 to reach attainment for 8-hour ozone standard.

Currently, an industry consortium is funding the operation of an air quality-monitoring network that consists of nine air monitoring stations. The Houston Regional Monitoring (HRM) network is equipped with state-of-the-art air monitoring sensors and is being used to gather scientific data that can be used to understand those chemical and meteorological processes that control the formation of ground level ozone. The knowledge gained from the HRM network coupled with data from other air monitors in the region is being used to help develop cost effective emission reduction strategies that will enable the area to achieve attainment of the ozone standards.

Data from HRM is also being used as part of a near-real-time Environmental Monitoring and Response System (EMRS) pilot study to alert industry when elevated levels of highly reactive volatile organic compounds (HRVOCs) are detected. This information is then used to investigate those plant activities that could be contributing to the observed levels of reactive VOCs in the atmosphere. It is hoped that these data can provide insights into "best management" practices that can be voluntarily implemented to reduce the emissions of ozone precursors, thereby avoiding additional regulation.

This paper describes the HRM Network and how data from this industry-sponsored network is being used to further our current understanding of ozone formation in the Houston non-attainment area. The paper will describe initiatives that are currently underway to analyze the data and help craft new emission reduction strategies attain compliance with the National Ambient Air Quality Standard (NAAQS) for ozone in Houston.

OVERVIEW OF HRM AIR MONITORING NETWORK

Because the Houston area is classified as a severe area for attainment of the 1-hour ozone National Ambient Air Quality Standard (NAAQS) and moderate for the 8-hour ozone NAAQS, the area boasts one of the most extensive ambient air-monitoring networks in the United States. Currently there is an extensive network of approximately forty-five ground based monitoring stations that are operated by both public and private groups. This extensive network of air monitors is operated the Texas Commission on Environmental Quality (TECQ), the City of Houston, Harris County, the University of Houston, and private industry.

The Houston Regional Monitoring Network consists of 9 air monitoring sites. This network is voluntarily supported by forty-five industrial facilities that have banded together to fund the operation of the network. HRM is in its twenty-third (23rd) year of continuous operation, and represents one of the largest and longest running privately sponsored air monitoring networks in the United States. The Houston Regional Monitoring (HRM) network is equipped with state-of-the-art air monitoring sensors and is being used to gather scientific data that can be used to understand those chemical and meteorological processes that control the formation of ground level ozone. The HRM network is designed to gather representative air quality information in a 900 square-mile area that includes the Houston Ship Channel Industrial corridor. The monitoring objectives of HRM are to:

- Better understand the air quality in the Houston area including the regional monitoring of trends;
- Improve public awareness of ambient air quality by monitoring criteria pollutant and volatile organic compound (VOC) concentrations at locations around the Houston area;
- Demonstrate attainment or progress toward attainment of National Ambient Air Quality Standards; and
- Provide decision-makers with current air quality information based on sound science.

Since 1981, the participating HRM member companies have accumulated ambient air measurement data from as many as 12 monitoring stations; including measurements for all six criteria pollutants and meteorological parameters. The monitoring program currently consists of a nine station ambient air-monitoring network (see Table 1 and Figure 1). Since 1981 routine VOC measurements have been made for a broad list of over 150 VOCs. VOC samples are collected in SUMMA canisters that are analyzed using EPA Method TO-15. The VOC samples are collected over a 24-hour period once every sixth day. The VOC measurements serve two purposes. These measurements are useful in gathering long-term air quality trends that can be used to assess potential community exposure to a subset of potentially hazardous air pollutants. Secondly they are useful in assessing emission trends of ozone precursors. This body of data is recognized and accepted by TCEQ, EPA, and the City of Houston for its validity and thoroughness. In 2003, the network was further modified to include continuous measurements for a target list of approximately 50 VOC precursor compounds at three site locations. Information from the three Photochemical Assessment Monitoring Stations (PAMS) is providing valuable information to help further our understanding of the role that VOCs play in the formation of ground level ozone.

Table 1. HRM Network Configuration

Criteria Pollutants	Non Criteria Pollutants	Meteorological Data (All Sites)
Ozone (All Sites)	H_2S (Sites 1, 3, and 7)	Wind Speed
CO (All Sites)	Continuous Sulfate (Site 3)	Wind Direction
NO_x (All Sites)	NO_y (Site 3)	Wind Gust
SO_2 (Sites 1, 3, 7 and 11)	Organic and Elemental Carbon (Site 3)	Wind Vector
$PM_{2.5}$ (Site 3)	VOCs (All Sites) – Every 6th day	Temperature
PM_{10} (Sites 1, 3, and 7)	Emergency Accidental Release System	Precipitation
Lead (Site 1)	PAMS[1] (3 Sites) HRM Site 3, Lynchburg Ferry, Wallisville Rd.	Net Radiation

1. Photochemical Assessment Monitoring Stations

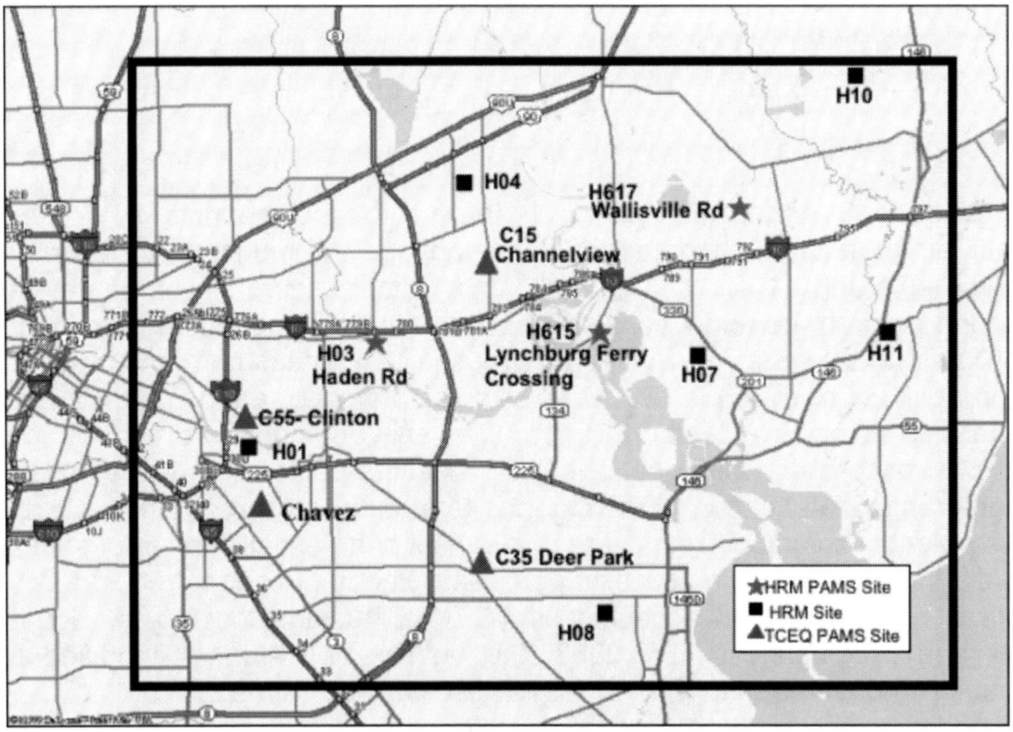

Figure 1. HRM Service Area

FACTORS WHICH INFLUENCE OZONE FORMATION IN HOUSTON

Since the mid-seventies, control strategies designed to reduce the concentration of ground level ozone in Houston have focused largely on the reduction of volatile organic compounds. More recent photochemical modeling conducted by the TCEQ in the late nineties suggested that in addition to reductions in VOCs, significant reductions in oxides of nitrogen would also be needed if the Houston-Galveston-Brazoria (HGB) area is to attain the 1-hour National Ambient Air Quality Standard for ozone by 2007. The modeling indicated that substantial reductions of stationary point source NO_x (approximately 90%) from a projected 2000 baseline would be needed.

In the summer of 2000, the TCEQ embarked on a collaborative effort to conduct a large-scale field study to gather scientific data that could be used to better understand those factors that influence ozone formation in the Houston area. In addition, the study provided needed information that could be used to ground truth regulatory photochemical modeling efforts, and to help quantify the uncertainty of the current emission inventories. The Texas Air Quality Study (TexAQS) was a unique collaboration between the TCEQ and over forty other public and private organizations and institutions. The field intensive portion of the study was performed over an approximate 2-month period in July and August of 2000. The study included the use of several instrumented aircraft, radar profilers, and several enhanced ozone chemistry monitors in addition to the existing (extensive) ground based network of long-term air monitoring stations currently in place in Houston.

Several key findings emerged from the TexAQS study. One significant finding indicated that in isolated industrial plumes in the Houston ship channel ozone forms more rapidly and efficiently than in non-industrial and urban areas of Houston. Aircraft measurements which included repeated transects of isolated industrial and urban plumes showed ozone production rates in Houston were 2 to 5 times higher than that measured in four other US cities (Kleinman et. al 2002). A further examination of the source of reactive hydrocarbons indicates that light olefins (i.e., ethylene, propylene, 1-3 butadiene, and isomers of butene) account for the majority of ozone reactivity (see Figure 2). It is not uncommon for ozone to form very rapidly on some days in the Houston ship channel at rates of up to 100 ppb/hr. The results from the TexAQS 2000 study linked the rapid rise in ozone levels to industrial sources of highly reactive VOCs. An examination of measured hydrocarbon/NO_x ratios (from aircraft measurements made in isolated industrial plumes) compared to those reported in the emissions inventory suggests that the hourly (day specific) inventories used to perform the regulatory modeling may be significantly under reported by as much as a factor of 6 to 10. These findings would prove to have significant consequences on future emission reduction strategies in the HGB area.

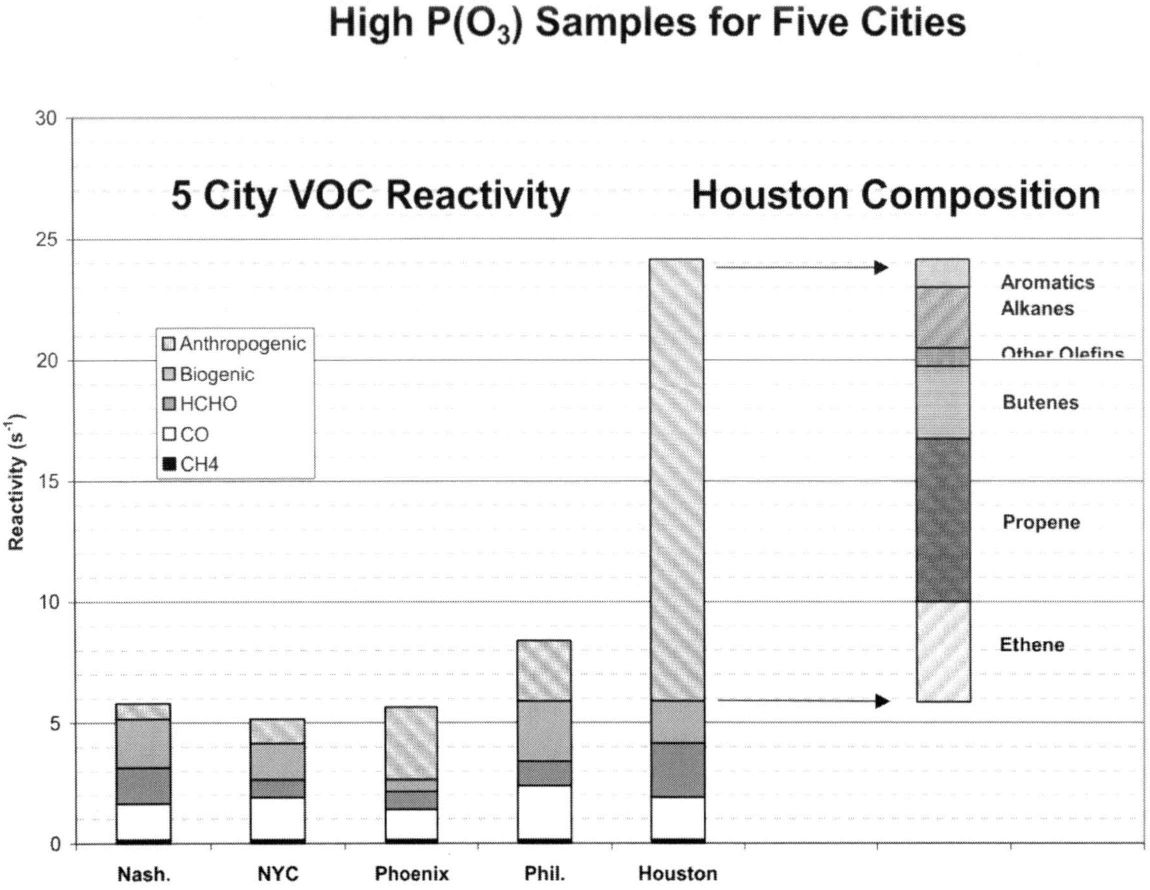

Figure 2. Ozone Reactivity for Five US Cities (Klineman et.al. 2002)

ENHANCED INDUSTRY SPONSORED MONITORING NETWORK

The results of the TexAQS 2000 study has had a profound impact on SIP (State Implementation Plan) planning and development in Houston, and has caused the TCEQ and industry to rethink its emission reduction strategies going forward. It has also brought to light the need for more air monitoring data. Based in on the results of the TexAQS 2000 study, the TCEQ led an effort to expand the network of ground-based gas chromatographs in the Houston area to gather near real-time measurements of ozone precursors. This network of seven enhanced industry sponsored monitors (EISM) is being used to gather information to (1) better understand the role these compounds play in ozone formation, (2) to help understand the accuracy of the emission inventories, and (3) use the information to provide feedback to industry to help understand and control short-term (periodic) emission events.

In 2003, HRM working in cooperation with the TCEQ expanded its base-monitoring program to include Photochemical Assessment Monitoring Stations (PAMS) at three locations in the Houston Industrial area. The PAMS network is designed to determine the ambient air concentration of ozone, NO_x, VOCs and HRVOCs, and meteorological parameters that have potential contribution to ozone formation. The PAMS monitoring stations provide near real-time ozone, NO_x and meteorological data, and continuous measurement for 54 common VOCs (see Table 2). HRM shares the hourly VOC (auto-GC) data, and 5-minute average O_3, NO_x and met data from the three PAMS Sites with the TCEQ.

The PAMS network data has been used to develop a speciated ambient air database, characteristic of source emission impacts, and representative of ambient VOC profiles. Long-term PAMS data is used to assess ambient trends for speciated VOC, NO_x and toxic air pollutants, and track multiple statistical indicators during ozone events. The exhaustive database of hourly VOC measurements is used to assess VOC reactivity trends over time, and to determine the possible future need of extending the HRVOC rule requirements to other VOC compounds or classes. Below is a table of VOCs measured in the PAMS network:

Table 2. PAMS Target List

ETHANE	2-METHYLHEXANE
ETHYLENE	2,3-DIMETHYLPENTANE
PROPANE	3-METHYLHEXANE
PROPYLENE	2,2,4-TRIMETHYLPENTA
ISO-BUTANE	N-HEPTANE
N-BUTANE	METHYLCYCLOHEXANE
ACETYLENE	2,3,4-TRIMETHYLPENTANE
TRANS-2-BUTENE	TOLUENE
1-BUTENE	2-METHYLHEPTANE
CIS-2-BUTENE	3-METHYLHEPTANE
CYCLOPENTANE	N-OCTANE
ISO-PENTANE	ETHYLBENZENE
N-PENTANE	M&P-XYLENE
1,3-BUTADIENE	STYRENE
TRANS-2-PENTENE	O-XYLENE
1-PENTENE	N-NONANE
CIS-2-PENTENE	ISO-PROPYLBENZENE
2,2-DIMETHYLBUTANE	N-PROPYLBENZENE
2,3-DIMETHYLBUTANE	M-ETHYLTOLUENE
2-METHYLPENTANE	P-ETHYLTOLUENE
3-METHYLPENTANE	1,3,5-TRI-M-BENZENE
ISOPRENE	O-ETHYLTOLUENE
N-HEXANE	1,2,4-TRI-M-BENZENE
METHYLCYCLOPENTANE	N-DECANE
2,4-DIMETHYLPENTANE	1,2,3-TRI-M-BENZENE
BENZENE	M-DIETHYLBENZENE
CYCLOHEXANE	P-DIETHYLBENZENE
	N-UNDECANE

OZONE EXCEEDANCE DAY ANALYSIS

The knowledge gained from the vast body of data collected from the HRM network coupled with data from other air monitors in the region is being used to better understand ozone formation in the Houston area. During the Houston ozone season, which typically lasts from March through September, HRM creates Ozone Fact Sheets for each of the area's ozone exceedance events. An analysis of each ozone episode (in Houston) is made to assist HRM members in identifying conditions under which ozone forms in Houston, particularly along the industrial Ship Channel. This analysis is reported usually within 24-hours of an exceedance of an NAAQS violation. Each analysis includes a fact sheet summary of the location of the monitored exceedances of the ozone standards; the maximum values reported, the duration of monitored exceedances; prevailing meteorological conditions, including wind speed, wind direction, temperature, sun and cloud conditions; and ozone data statistics for year-to-date episodes. In addition to the monitoring data, the fact sheet includes an air parcel trajectory analysis; analysis of unusual contributing events such as agricultural fires, upsets, and natural events; PAMS reactivity analysis; and an analysis of the probable causes of the ozone exceedance episode.

An example of a typical ozone exceedance day is provided for illustrative purposes. This example includes information for ozone exceedances that occurred on Saturday, November 6, 2004. This day was characterized by light winds from 1 to 4 mph in the early morning shifting to the southwest in the late morning to early afternoon. The skies were clear and the temperatures for the day peaked in the mid-seventies. On this particular day, five monitors (all in the HRM network) exceeded the 1-hour NAAQS for ozone, and four of the five monitors exceeded the 8-hour ozone NAAQS. Each of the five sites was located in narrow corridor at the eastern end of the Houston ship channel (see Figure 3). All 1-hour NAAQS exceedances occurred over a 3-hour span from 11 AM to 2 PM, and ranged from a high of 187 ppb at HRM Site 7 (C-607) to a low of 138 ppb at the Wallisville Rd. site (C-617).

A back trajectory plot using 5-minute average surface wind data is provided to help visualize the track of individual air parcels back in time to midnight of the exceedance day. As can be seen in this plot the air parcels for each of the five sites where the ozone standards were exceeded each passed in close proximity to stationary point sources that were rich in reactive volatile organic compounds. Figure 4 shows a time series plot of 5-minute ozone concentrations for each of the 5 monitors and hourly MIR (Maximum Incremental Reactivity) plots computed from auto-gas chromatographic data obtained at the Lynchburg Ferry. This plot and other similar plots developed from the other PAMS monitors in the ship channel on this day, provide strong evidence that emissions of reactive volatile organic compounds from stationary point sources in the Houston ship channel most likely contributed to the ozone exceedances measured at each of the five monitors. Because of intense stagnation on this day there is no evidence to suggest that there was significant mixing with the Houston urban plume.

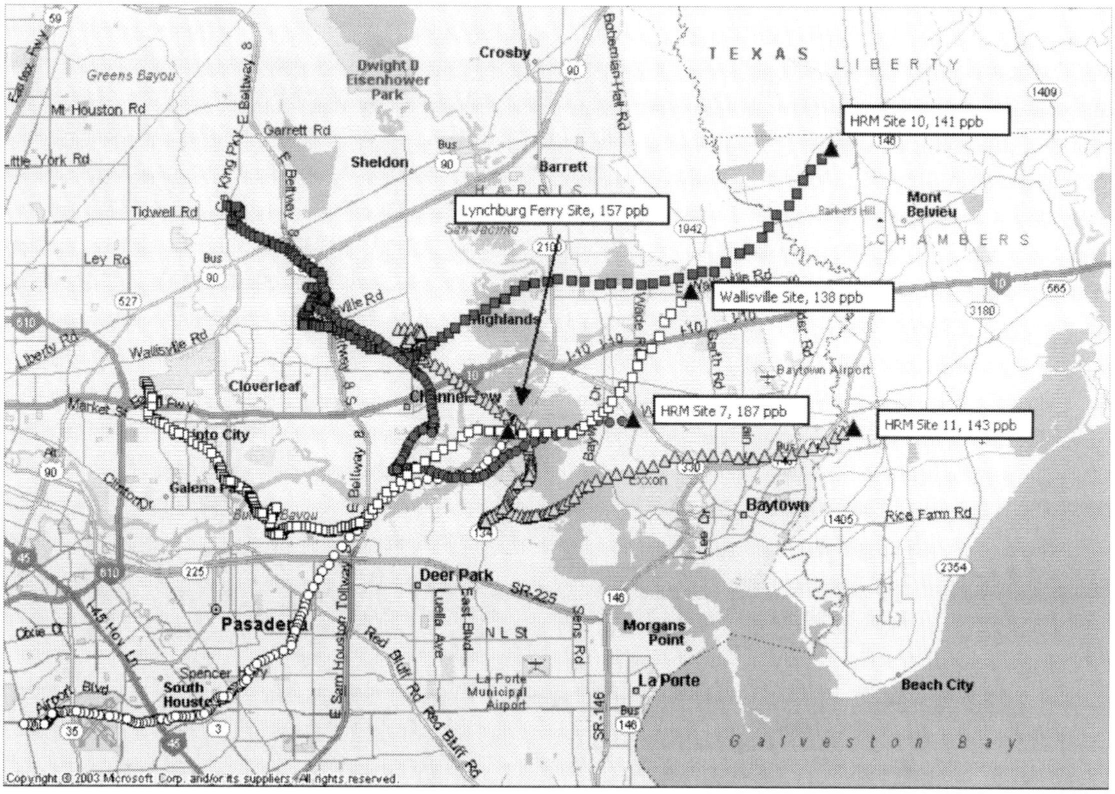

Figure 3. November 6, 2004 Back Trajectory

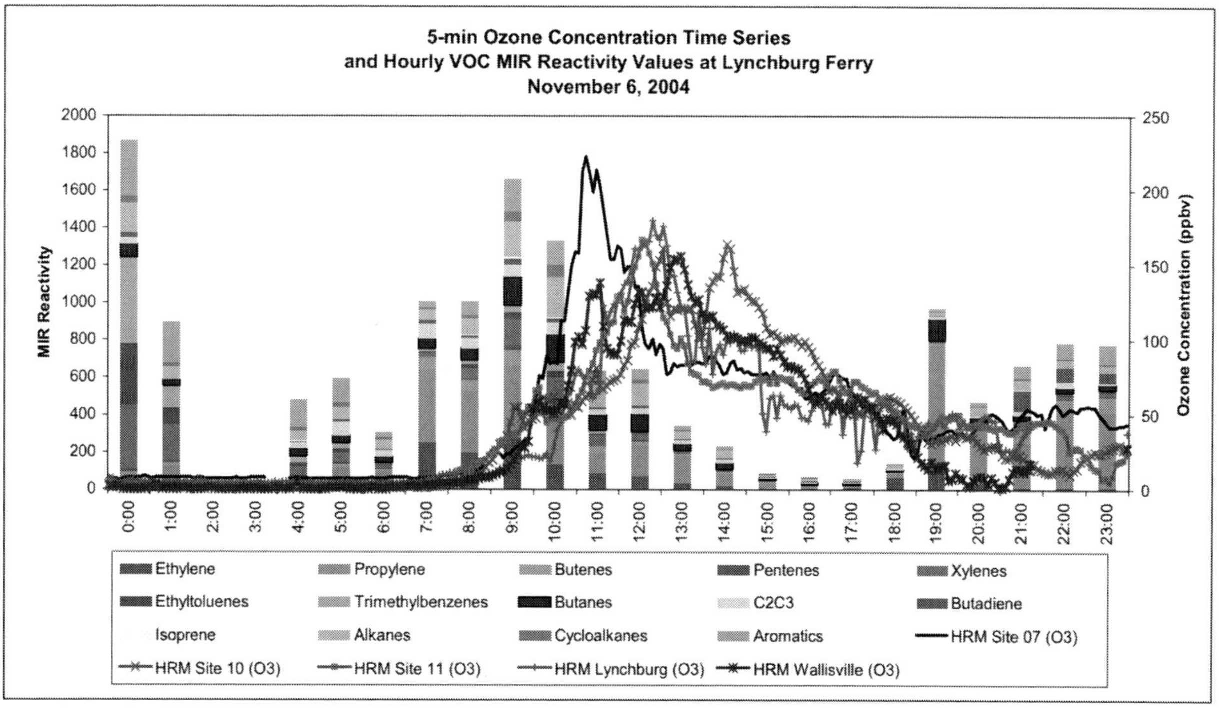

Figure 4. November 6, 2004 Ozone Reactivity Plot (Lynchburg Ferry PAMS Monitor)

ENVIRONMENTAL MONITORING AND RESPONSE SYSTEM

Data from HRM PAMS GC sites is also being used as part of a near-real-time pilot study designed to alert industry when elevated levels of HRVOCs are detected. During this 6-month collaboration with the TCEQ, the Environmental Monitoring and Response System (EMRS) pilot program notifies participating companies located along the industrial Ship Channel if elevated ambient HRVOC levels are detected in the vicinity of the company. This information is then used to investigate those plant activities that could be contributing to the observed levels of reactive VOCs in the atmosphere. Figure 5 illustrates the areas covered by PAMS GC sites during the EMRS pilot program.

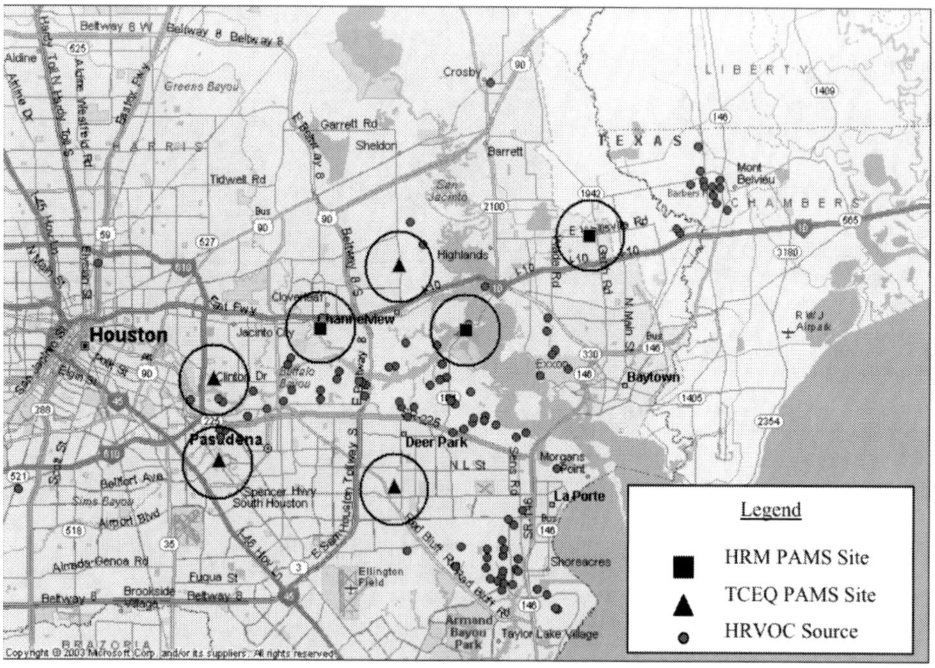

Figure 5. EMRS Coverage Area

During the EMRS Pilot over sixty industrial facilities participated in the program. The Houston Industry and the TCEQ have shared information and have communicated well during the pilot program. The companies that receive an alert voluntarily submit an electronic web based form detailing the activities of the plant during the day of the alert (i.e., any scheduled start-up, shutdown, maintenance, loading/unloading operations that involved the triggered HRVOC). Once the form is submitted, it is logged into relational database, and a duplicate electronic form is also submitted to the TCEQ. It is hoped that these data can provide insights into "best management" practices that can be voluntarily implemented to reduce the emissions of ozone precursors, thereby avoiding additional regulation.

OBSERVATIONS AND CONCLUSIONS

Houston industry has been actively engaged in funding programs to better understand the air quality issues in the Houston area. This fact is best illustrated by support of the Houston Regional Monitoring Network. HRM continues to fund and operate a network of nine air quality monitors in heart of the Houston ship channel. HRM is voluntarily program that is supported by forty-five industrial facilities located in the Houston ship channel. The program, which is in its 23rd year of operation, provides valuable data and information to better understand air quality issues of the region.

Information from HRM and other regional monitors have been instrumental in helping to understand those factors that influence ozone formation in the region. Recent data from the program indicate that light olefins may play a more significant role in ozone formation than was previously understood. Recently the HRM program has undergone significant upgrades to include the continuous speciation of VOCs that participate in ozone formation. These upgrades, which were made as part of Enhanced Industry Monitoring Initiative, are providing valuable information to better understand:

- Those factors which influence rapid ozone formation in the ship channel;
- Provide valuable data to help develop cost effective emission reduction strategies to achieve and maintain of the 1-hour and 8-hour NAAQS for ozone;
- The role that reactive VOCs play in ozone formation in the Houston area; and
- The uncertainty of emission inventories needed to develop effective emission reduction strategies.

In addition, the automated PAMS monitors operated by HRM were used as part of a year-long pilot study to evaluate the use of real-time measurement data to provide the core of an Environmental Monitoring and Response System. It is hoped that information from this program will provide near-real time alerts to industry when ambient concentrations exceed pre-established norms. This will prompt an investigation to determine what activities may be contributing to elevated concentrations of certain compounds. It is hoped that this knowledge will result in "best management" practices that will lower the emissions of ozone precursors and will result in significant air quality improvements.

REFERENCES

Kleinman, L. I., Daum, D. Imre, Lee, Y, -N, Nunnermacker, L. J., Springston, S.R, Winstein-Lloyd, J., Rudolph, J.; Ozone Production Rate and Hydrocarbon Reactivity in 5 Urban Areas: A Cause of High Ozone Concentrations in Houston; Geophysical Research Letters, Vol. 29, No. 0, 1029/2001GL014569, 2002.

Paper 43b

Title V Reporting and Investigation Process for Texas Operating Sites

Stephanie King

Environmental Engineer

Lyondell Chemical Company

LaPorte Plant

LaPorte, Texas

Prepared for Presentation at 2005 AIChE Spring National Meeting/Atlanta, GA

April 10 – 14, 2005/ Ethylene Producers' Conference/ Environmental Session

UNPUBLISHED

The Title V Process for most complex chemical sites became effective with the issuance of site operating permits in Texas in 2001. Other less complicated sites had effective General Operating permits in the late 1990s. This paper will discuss the general structure of Texas site operating permits under the Title V Program, auditing and reporting against the permit, types of reports required by the permit, enforcement policies and procedures at the Texas Commission on Environmental Quality (TCEQ), and agency investigations following Title V Report submittal.

Content of the Texas Title V Permit

In order for sites to understand their reporting obligations under the Title V Permit, the site needs to understand the content and structure of the Title V Permit. In Texas, the Title V Permit consists of several distinct types of information, each of which must be evaluated for reporting purposes.

The first portion of a site's Title V Permit consists of "General Terms and Conditions". This section of the permit outlines the permittee's responsibilities in terms of the Chapter 122 requirements in the Texas rules, which are the Texas Title V Program Requirements. If a permittee failed to appropriately certify compliance, this section of the permit contains regulatory citations referencing the Chapter 122 Certification requirements.

The second section is called, "Special Terms and Conditions." These conditions are also called "Site Wide" conditions. These regulatory applicabilities would be appropriately applied to activities that occur over multiple units, or that occur outside of units in the facility. An example would be gasoline and diesel fueling stations that service an entire site complex. The regulatory terms and conditions governing this activity would be codified in the Special Terms and Conditions section of the Title V Permit. Another type of regulatory citation found in this section of the permit are high level terms and conditions. Some periodic monitoring, such as annual visible emissions observations, may also be found in this section of the permit. Permittees should review this section of the permit and familiarize themselves with the types of equipment, which might carry these regulatory requirements. The permit will not specify the individual sources to which these rules apply, so the permittee must have some understanding of the types of equipment to which these terms and conditions might apply.

The next section of the permit is the Unit Summary. This is the section of the permit where permittees may find if specific equipment has been addressed by the subsequent sections of the operating permit. The Unit Summary will also show what regulatory applicabilities the equipment has so that the section of the operating permit that refers to the equipment is more readily identified.

The "Applicable Requirements Summary Table" is the next section of the Title V Permit. This is the most detailed section of the permit where each applicable regulatory citation is listed for each individual or grouped type of equipment. Regulations are broken out into "Standards," "Monitoring and Testing," "Recordkeeping," and "Reporting" requirements. This distinction is important in the reporting of deviations on the Texas Deviation Report Form. In theory, the owner/operator should be able to determine all of the source's reporting obligations by reading the Reporting column of this section of the permit. But permittees should not assume this section is the only section of the operating permit where requirements will be listed. The regulatory requirements identified in previous sections of the permit are deviation reportable as well. Each regulatory listing carries a brief summary of the requirement to give the reader some information about the regulation that is being listed. The majority of deviation information will be found in this section of the permit.

Periodic Monitoring is issued by the state when a rule or requirement did not contain adequate monitoring to demonstrate compliance. Periodic Monitoring Summaries follow the Applicable Requirements Summary Table. Permittees should carefully review the contents of the Periodic Monitoring Summaries.

These requirements are drafted by the Title V Permit engineer during the application phase of the permit, and vary somewhat from permit writer to permit writer. The differences in language in the Periodic Monitoring summaries may lead to more deviations site-to-site and permit-to-permit. For instance, one permit's Periodic Monitoring Summary may require a documented inspection of a cold solvent degreaser for free board ratio compliance in addition to documenting an inspection of operating requirements such as the lid being closed when not in use. A second permit at the same site may only require inspection against the requirement to have the lid closed. Both permits are responsible for insuring compliance with the underlying cold solvent degreaser free board ratio requirement, but one permit requires documentation where the other permit did not. Differences within a site between Periodic Monitoring Requirements should be carefully considered when determining if Title V deviations are occurring.

Following the Periodic Monitoring Summaries, the operating permit has a listing of agreed upon negative applicabilities. This set of information was presented by the permittee during the application process, and with supporting information provided to the state, received a listing of the requirement as negatively applicable. Permittees may find this section of the permit useful in determining why a particular issue is not being deviation reported on. If the requirement does not apply for a specific reason, a brief description of that reason is listed in this section of the permit.

The last portion of the Texas Operating Permit is the authorizations section. Each unit in the permit will be listed with its specific authorization. There are appropriate ways in which a unit may not be individually listed in this section of the permit. One way is by permit by rule (PBR) authorization. For any source, which is listed on the maximum allowable emission rate table (MAERT) of the air permit, a specific listing should occur in this section of the operating permit. In addition to understanding a source's authorization from this section of the operating permit, one of the special terms and conditions of the operating permit is a requirement to abide by all Chapter 116 requirements. This means that all MAERT limits and all special conditions of the permit are conditions of the Federal Operating Permit. Sites should understand all special conditions and all MAERT limits of authorized sources in order to understand the operating permit requirements.

Auditing and Reporting Against the Permit

With a complete understanding of the permit's structure and requirements, sites have an obligation to perform reasonable inquiry as to the site's compliance with the Title V requirements. Sites have an obligation and responsibility to know when they have deviated from their Title V Permit, and to report those instances of deviation. In Texas, two semiannual deviation reports are submitted during the year and an annual compliance certification is submitted at the end of the year. Permittees must consider credible evidence and with reasonable inquiry evaluate the requirements of the Title V Permit for reporting. Credible evidence is best described in the EPA's Credible Evidence rule. In its most basic terms, it means that all data, whether gathered for a regulatory purpose or not, has to be considered when reviewing the compliance status of a source. Reasonable inquiry is another concept that has no direct definition. In practice, it means that data reasonable to include in the review to determine compliance had to be considered.

There are as many ways to conduct reasonable inquiry as there are sites and permits. Some sites use auditors who check the required data on an auditing schedule. Other sites have elaborate database systems with intricate e-mail networks where site personnel confirm that they have completed their tasks either by answering an e-mail or by taking data into a PDA-type device and docking and downloading the data. Other sites use a cascading system of certifications starting at lower levels of the organization and cascading up to the responsible official where each level of the organization certifies to the compliance status of the operating unit. However sites choose to accomplish reasonable inquiry and certification of compliance, the site should be conducting inquiry into the compliance status of the regulatory obligations of the Title V Permit, auditing the facility against the requirements, and reporting instances where the compliance obligations were not met.

Sites will be constructing tools and systems to manage the rule requirements, tasks, and data. Sites will be training personnel on the environmental tools and systems used to implement the Title V reporting process. The tools and systems created will help identify Title V deviations and help manage the new reporting requirements. Title V tools vary widely from site to site. But the underlying principle is that some system will be formalized which will assist in the management of Title V data and Title V tasks. Sites will use these tools to help identify Title V deviations and manage reporting data.

The next hurdle sites have to clear is how to house the deviation information and how to communicate the information to effected personnel. Again, this varies site to site even within the same organization. Over time, sites will find that keeping a real-time system to be much easier than performing all of the work at the end of the reporting period. As issues are found, they can be added to the system and details documented for reporting at a later date. Additionally, keeping real-time deviation information allows for trending to occur so that deviation information can be relayed to the operations, maintenance, and technical groups allowing those groups to address their issues without waiting until the end of the reporting period. For instance, if failing to hang a leak identification tag was occurring from a particular group assigned to hang leak identification tags, and if evaluating the real-time deviation data showed the trend occurring with this one group, training could be conducted before the end of the period and further deviations prevented.

Once deviation information is assembled, it is reviewed with the responsible official so that he or she can sign the periodic reports. Obtaining management support to correct deviations prevents deviations from reoccurring or persisting. Also, in today's facilities, there are many competing priorities. When Title V deviations are made into a "scorecard" item or into a "metric" for the facility, then progress can be tracked and deviations eliminated.

Types of Reports Required by the Title V Permit

In addition to the reporting required by the underlying rules and regulations, two new types of reports are required by the Title V Permit. Those are the semiannual deviation reports and the annual compliance certification. The two reports differ in intent. The semiannual deviation report identifies terms and conditions that the owner/operator may not have complied with during the reporting period. The annual compliance certification certifies compliance with all terms and conditions of the operating permit with exception of the deviations identified in the semiannual deviation reports. People sometimes confuse the annual compliance certification with deviation reports and wonder why no deviations are reported in the annual report. The reason is that the two reports are different. The annual compliance certification certifies compliance, lists reports submitted during the period, summarizes deviation information from the two semiannual reports, and identifies which of multiple monitoring options the permittee is certifying to. Permittees should understand the annual report even when constructing their first semiannual deviation report because the annual report certifies compliance with all permit terms and conditions with the exception of any terms or conditions identified as deviations. Doing a thorough deviation review in each period makes the annual compliance certification less time consuming at the end of the annual reporting period.

Understanding the State of Texas' Enforcement Policies and Procedures

Texas Operating sites submitting Title V Deviation Reports and Annual Compliance Certifications should understand the enforcement processes and procedures with which the report will be reviewed. Sites should familiarize themselves with two TCEQ policies and one TCEQ rule prior to the agency investigation: the Enforcement Initiation Criteria (EIC) (policy); the Penalty Policy; and the Compliance History Rule. The EIC is a policy, which tells the Investigator the necessary category the violation should be assessed. If the EIC requires a specific action from the agency for sites, which have violated a rule in a particular way, the possibility that the site will be able to negotiate any other action from the agency is unlikely.

In the EIC, Category A violations are issued a Notice of Enforcement and will be assessed a fine through the enforcement process. Category B violations typically are issued a Notice of Violation (NOV) which may or may not require a response from the permittee to resolve. Repeat Category B violations in subsequent periods on the same issue may raise the issue to a Category A in future investigations. Category C violations are violations which do not have emissions associated with the violation (such as a recordkeeping violation) and are typically only issued a NOV . The A, B and C categories are specific to the state's EIC policy, but apply to all kinds of violations, not just Title V deviations.

Another policy that permittees should be aware of is the state's Compliance History Rule. In the Compliance History Rule, sites are evaluated based on their environmental performance and a compliance history rating computed based on a formula which takes into account the site's complexity and investigation history. The site's compliance history rating is posted on the TCEQ's website for review by anyone interested in the rating. The rating is also utilized in assessing penalties in the penalty policy phase of an Enforcement process and in assessing a site's ability to obtain permits and authorizations. Two negative ratings in this rule are "poor performer" and "repeat violator." Poor performers and repeat violators may not have the ability to obtain required authorizations such as air permits, and may have to have agency supervision for their compliance processes. A site can become a repeat violator by having four major violations in a five-year period. Sites submitting Title V Deviation Reports can expect multiple violations at the end of the reporting period and investigation if multiple Title V deviations are reported. If any four violations are assessed as "major" violations, the site will be classified as a repeat violator. In the Compliance History Rule, the TCEQ lists types of violations that should be appropriately classified as "major." Basically, those boil down to violations that demonstrate a significant disregard for the state's environmental laws and regulations. However, in practice, sites have had mistaken classifications where issues were inadvertently classified as major, but in all likelihood, should not have been. When that happens, and when sites disagree with the rating, they should be immediately discussing the issue with regional personnel trying to understand why the issue was classified as a major violation and trying to get any errors resolved. In the Compliance History Rule, issues can be classified as major, moderate or minor where each term is defined in the Compliance History Rule.

So, how should sites know what the compliance history rating of a particular violation is? They have to ask the regional investigator that issued the violation. In Texas, the major, moderate or minor classifications will not be listed in the written communication on the issue from the state. Operators in Texas will have to ask enough questions from their regional investigators to understand the rating that the agency thinks is appropriate for the issue being discussed. In the state's enforcement policies, it is not required that the state notify the operating site that they have received a classification of "repeat violator." In the worst case scenario, operating sites could find out that they have received the repeat violator status when their application for an air permit authorization is denied, and not before. It is recommended in the state of Texas that operating sites request a detailed Compliance History Report at least annually and review the report for accuracy. If there were any discrepancies, they should be resolved with the regional inspectors before a significant issue (such as the denial of an air permit authorization) occurs.

The other type of classification to note is the poor performer classification. A site may become a poor performer when their compliance history rating reaches a particular level. Becoming a poor performer typically happens over time and is a more gradual process than the process to become a repeat violator, which could happen in literally a matter of days or weeks if a series of major violations were assessed. Again, annually reviewing the site's compliance history rating and evaluating all violations assessed at the site will help prevent accidental negative classifications that neither the agency nor the site intended to have applied to the site.

Another Enforcement policy that sites should be familiar with in Texas is the Penalty Policy. Penalties are additive over a five year period of time and are tied of the major/minor/moderate classifications previously discussed. Multiple major violations are assessed additive 25 percent penalties for each major violation. Four major violations will cause a site to be a repeat violator and have a 100 percent penalty increase for the issues that are fined. Some issues have a 30 percent penalty increase and are required to be issued as findings orders. Texas has a guidance document on the appropriate use of findings orders, which tells the regulated community what findings orders are, and when they are required to be used. Sites who find out that their violation is being classified as a findings order will need to review the State's policy regarding that type of Enforcement order and understand what it means functionally to the site.

Agency Investigation of Title V Deviation Reports

Once both semiannual Title V Deviation Reports and the Annual Compliance Certification have been submitted, in Texas, an agency investigation will occur. An investigator from the TCEQ will review deviation information submitted to gather information necessary to classify the issue within the state's policies. Typically, an office review of the reports occurs prior to contact from the agency to the site. All sites submitting deviation reports and compliance certifications should receive an investigation annually. The timing of the investigation, however, varies from site to site and depends on TCEQ regional schedules. Sites may receive notice of an impending investigation days after submitting a compliance certification while other sites may not receive an investigation for several weeks or months after submittal. The TCEQ has committed that all sites will receive an investigation following submittal of the annual compliance certification and sites will be investigated at least annually.

Texas Operating sites may save time and effort in the Title V investigation process by providing enough data in the Title V Deviation Report for the investigator to classify the issue in the state's policies. For instance, if 5 percent monitoring system downtime is classified as an enforcement issue in the EIC, then operating sites reporting monitoring system downtime might note that less than 5 percent downtime occurred during the reporting period in the Title V deviation report. By doing so, sites can prevent having to go back and provide that data on issues that may have occurred a full year prior to the investigation. If sites provide the data at the time of the deviation report, it can actually make the investigation process go more smoothly and will be less work for the site at the time of the investigation. If sites understand the EIC policy and evaluate their Title V deviations against those criteria, then at the time of the investigation, there should be few surprises for the TCEQ or for the operating site.

Some sites question whether the submittal of deviation information in a deviation report will cause the site to receive a notice of violation or a notice of enforcement. The state's perspective is that if a site has notified the regulatory agency of a violation of a state or federal rule, then the issue is a violation and the site will receive a violation letter and enforcement if appropriate. The process by which the decision is made to determine if a NOV or a NOE will be issued involves evaluating the deviation against the state's Enforcement Policies such as the EIC.

The process of the site investigation varies by investigator and complexity of site. Some investigators will contact a complex site and arrange a site visit to review Title V information. The scope of the Title V investigation is not limited to the information provided on the deviation report. The investigator will ask to see records, which go beyond the deviation report and verify that the site is indeed in compliance with common types of deviations not reported on the deviation report. The investigator will review all Title V deviations during the investigation. Typical questions target gathering the required data necessary to classify the issue within the state's enforcement policies such as the EIC, discussed above. Sometimes, the investigator will leave a list of required documentation with the site, which has to be provided to the agency. A typical Title V investigation lasts approximately a day with follow up taking from a few days to a few weeks depending on the magnitude of additional data to be provided.

Managing the Issues

The site has researched the Title V Permit, understood it's contents, trained the plant, audited the site's compliance, applied reasonable inquiry, considered credible evidence, produced a deviation report, produced an annual compliance certification, worked with site personnel to eliminate future deviations, and assisted the TCEQ in an investigation of the Title V issues. So, what is there left to do? Going forward, sites must stay engaged in the enforcement process that follows submittal of a Title V Report. Title V deviations and subsequent investigations lead to enforcement actions, whether that is an NOV or an NOE. And sites must understand the process that is occurring.

Going forward, site environmental professionals will be helping to provide the data in the form and format that the agency needs to classify the issue in the existing enforcement policies. Sites should be communicating with the TCEQ each step of the way by providing data necessary for the state to do their job. But in addition to providing data, they should be asking the state what each step of the process means so that site personnel understand what the expected outcome of the issue will be. Site personnel need to understand the enforcement policies and question their application where appropriate. With a detailed compliance history report from the TCEQ printed annually, sites know their current status. With knowledge of the enforcement policies and procedures, they can predict the impact of current enforcement issues and communicate to the site on this issue.

Conclusion

Sites with Texas Title V Permits must understand the structure and content of the Title V permits which cover operations at their site. They must audit and report against the permit's terms and conditions. Systems must be created to manage compliance tasks and data, and training must reiterate site personnel roles in environmental compliance. Sites should understand the enforcement process that follows Title V deviation report submittal and should be prepared for the resources that it requires. Texas Operating sites submitting Title V Deviation Reports and Annual Compliance Certifications should be focused on deviation elimination at the site as a first priority. Until deviations are eliminated, sites should understand the state's processes that govern the review of the reports and the processes that follow.

Paper 43c

Experiences in the Installation of Hydrocarbon Monitoring Systems for Flares and Cooling Towers

Randolph H. Smith
Principal Engineer – Hydrocarbons Production Engineering
Equistar Chemicals, a Lyondell Company
P.O. Box 777
Channelview, Texas 77512

Prepared for Presentation at 2005 AIChE Spring National Meeting/Atlanta, GA/April 10-14, 2005/Ethylene Producers' Conference/Environmental Session

Unpublished

Abstract

As a part of the Texas State Implementation Plan (SIP) for the Ozone Non-attainment Area including the Houston-Galveston Region, the Texas Commission on Environmental Quality (TCEQ) promulgated new regulations that require the continuous measurement of ethylene, propylene, 1,3-butadiene and isomers of butane [highly reactive volatile organic compounds (HRVOC)], from flares and cooling towers. While analytical methods exist for measuring these compounds in various process applications, they typically do not involve applications where the concentration ranges between the ppm levels up through significant weight percent levels as might be found in a flare sample. Likewise, the measurement of these compounds in a cooling water system presents its own unique problems. The regulation requires use of the "El Paso Stripper" type air stripping device or to demonstrate by actual testing an equivalent technology. When these requirements are coupled with a deadline of having such a system operational by the end of December 2005, the challenges can be significant.

This paper describes some of the experiences with the installation of these systems at the Lyondell Chemical Company, Channelview Chemical Complex. This facility is installing six flare monitoring systems and eight cooling water monitoring systems to meet these requirements and adding six new analyzer shelters with all of the associated infrastructure to support them. Since the required sampling locations in most cases are not close enough to be able to use existing process analyzer shelters, the logistics of the installations were a significant part of the project. Overall the Lyondell facilities are installing 43 monitoring systems in 33 shelters located at 10 different facilities in the Houston-Galveston area.

The methodology associated with these systems can best be described as equivalent to compliance stack sampling with the added requirements that it be done continuously (i.e., have a 95+ percent on-stream time and an analyzer cycle time of not more than 15 minutes). Sample location issues created challenges not envisioned when the rules were developed. For example, flares located hundreds of feet from adjacent operations and underground cooling water lines.

Background

In 2002, the Texas Commission on Environmental Quality (TCEQ) published new regulations relating to what they termed "highly reactive volatile organic compounds (HRVOCs)." These compounds include ethylene, propylene, 1,3-butadiene, and the isomers of butene. These new regulations were developed as part of the Texas State Implementation Plan (SIP) for the Ozone Non-attainment Area which includes the Houston-Galveston Region. The regulations were developed to reduce ozone events as required by the Clean Air Act. These rules followed earlier changes for controlling industrial NOx emissions and other fired sources.

The Houston-Galveston area experiences two distinct patterns of ozone (see Figure 1). The first is the typical pattern caused by routine, continuous emissions and is characterized by small increases and decreases in observed ozone values (the same as are seen in other cities). The second is characterized by one-hour spikes in the ozone concentration that are more than 40 ppb higher than the previous hour.

Figure 1 - Two observed ozone patterns in Houston

Typical Ozone (Part 1):

- Daily smooth small increases and decreases in observed ozone values (same as other cities).
- Typical ozone - caused by routine, continuous emissions.

Spike Ozone (Part 2):

- One-hour "spikes" that are more than 40 ppb higher than the previous hour.

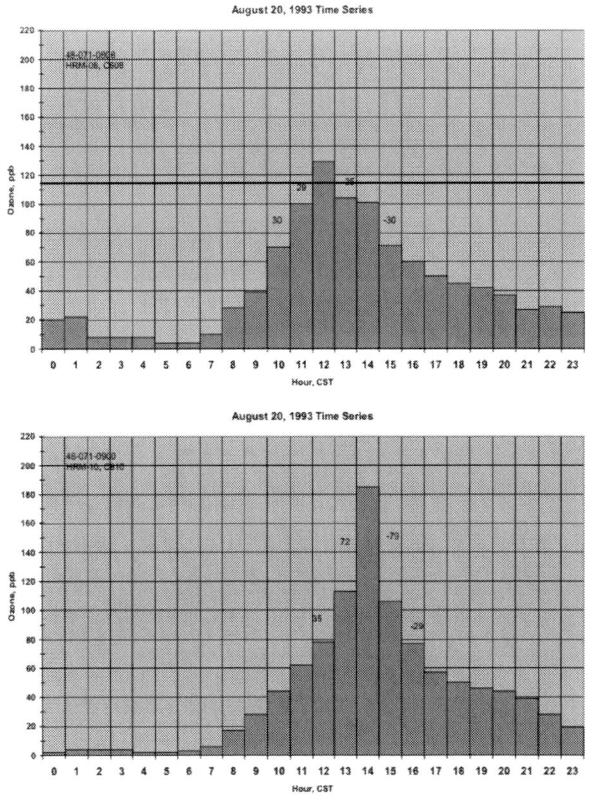

The selection of these HRVOC compounds was based on the results of a study (TEXAQS2000) in which unique short-term ozone formation was found to be correlated to concentrations of light olefin compounds in the atmosphere. The study involved ambient sampling using airborne instruments for ozone and various hydrocarbons (see Figure 2).

Figure 2 – Finding the Ozone Plume (Airborne Flight Path)

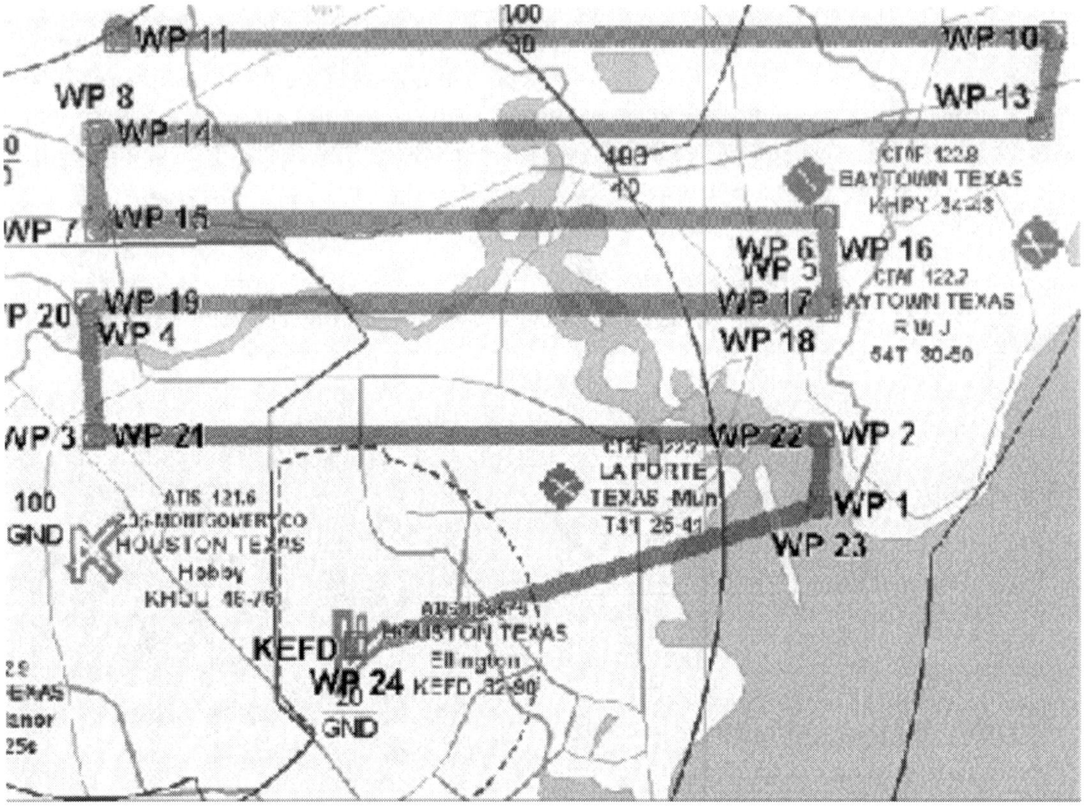

Sources of HRVOCs associated with routine industrial emissions (see Figure 3) are predominately from fugitive sources (48 percent). Flares are a close second (30 percent), followed by vents, cooling towers, and miscellaneous sources (about seven to eight percent each).

Figure 3 - HRVOC Emissions from Point Sources - Routine

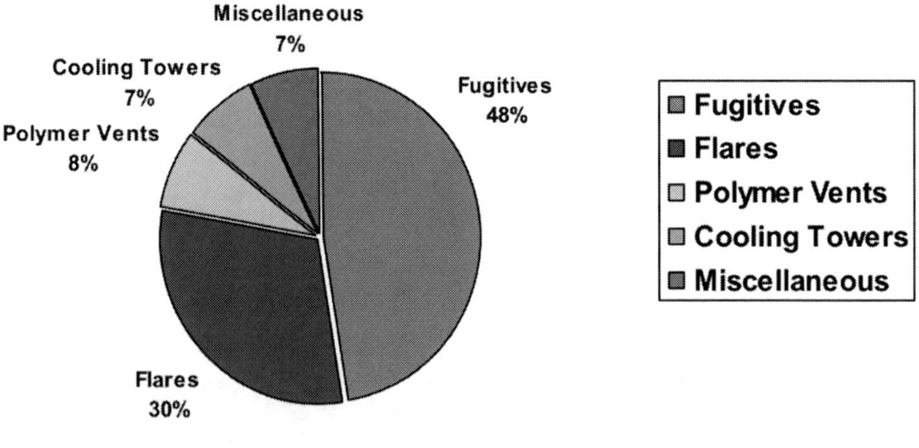

When upset or sudden release emissions were evaluated (see Figure 4), fugitive sources accounted for only 17 percent while flares increased to 73 percent and cooling towers to 10 percent. Vent and miscellaneous sources decreased to essentially insignificant levels for these events.

Figure 4 - HRVOC Emissions from Point Source - Upsets

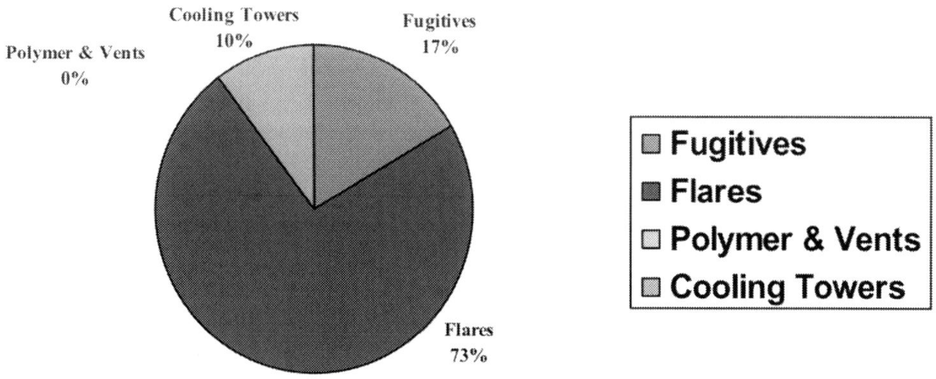

<u>Requirements</u>

The HRVOC rules increased controls on the emission sources for these compounds (such as fugitive emissions from valves, flanges, pumps, etc. and vent sources). Additionally, the TCEQ issued requirements to have continuous monitoring systems on flares and cooling water systems that potentially contain these compounds.

Basic requirements for flares are:

1. Continuous flow and velocity measurement
2. Measurement of total VOCs and speciated HRVOCs
3. Determination of net (lower) heating value
4. If the analyzer is non-functional for more than eight hours – grab sample daily

Basic requirements for cooling water systems are:

1. Continuous flow measurement
2. Measurement of total strippable VOC
3. Monthly speciation of HRVOC's (daily if total strippable VOC is > 50 ppb)

The methodology associated with continuous monitoring systems can best be described as equivalent to compliance stack sampling with the added requirements that it be done continuously (i.e. have a 95+ percent on-stream time and an analyzer cycle time of not more than 15 minutes).

Analytical methods exist for measuring these compounds in various process applications. Typically, they do not involve applications where the component concentration range varies between ppm levels through significant weight percent levels. Flare sampling is an example. Likewise, the measurement of HRVOC compounds in a cooling water system presents its own unique problems. The most significant is the TCEQ's requirement to use the "El Paso Stripper" type air stripping device or demonstrate equivalent technology by actual testing. This part of the sample preparation system ahead of the analyzer involves managing very low volume but precise flow rates in precise, defined ratios to each other.

Even a requirement such as the determination of the net (lower) heating value of the flare gases is not without issues. A decision was made to calculate the heating value from the analyzer information rather than install calorimeters.

Experiences and Issues

In the Channelview Chemical Complex, there are six flare monitoring systems and eight cooling water monitoring systems that have been installed to comply with these requirements. These monitoring systems are housed in six new analyzer shelter buildings, along with the associated support infrastructure. None of these new sample locations were located close enough to utilize existing process analyzer shelters. (Figure 5 shows an example of one of the cooling water analyzer shelters). Networking challenges were a significant part of the project as the new equipment had to interface with a variety of process data systems. Overall Lyondell is installing 43 monitoring systems in 33 shelters at 10 different Houston-Galveston facilities.

Figure 5 – Analyzer Shelter for Cooling Tower (with six analyzer systems)

Sampling all the various flare sources was the first step to determine the range of components and concentrations. This data and expertise from the corporate analytical laboratory group defined the analyzer system requirements. After evaluating this information, a prioritized installation plan was developed based upon the number of systems for a given location, plus the potential complexity of those systems. The Channelview Chemical Complex, which has the largest number of monitoring systems, was selected to receive the first installations and to serve as the test-bed for the entire process. A decision was also made to standardize flowmeters, analyzer systems and associated shelters where possible. This simplified construction, installation and start-up.

It also allowed learnings from the problems that were discovered in the initial installations to be used to correct them before they occurred in the other locations. Additionally, the project focused on incorporating reliability into both the systems themselves as well as their installation and startup. The selection of the analyzers and flowmeters was based on state of the art analytical or measurement techniques, reliability and maintainability.

The project team included in addition to the normal technical and project personnel, plant representatives from each location from operations, technical, environmental, and reliability along with a dedicated senior analyzer technician who followed the engineering, construction and startup of all the systems. Plant analyzer technicians participated in the factory acceptance testing for their own new analyzers as well as receiving specific training on the analyzers and the new applications being installed. This approach has made the startup of these systems as smooth as possible since the people that will be using and working with the systems have been involved throughout the development and implementation.

Although the regulations only required that the defined HRVOCs in the flare stream be monitored, it was decided to also monitor the other compounds that might be present in the same stream (i.e. to monitor C4 and lighter hydrocarbons plus the fixed gases including hydrogen). This additional monitoring provides expanded information about the flare stream as well as addresses the determination of the net (lower) heating value. Standardized calculations were developed for the heating value, the quantities of HRVOCs being emitted and the flare tip velocities. The effect on flare destruction efficiency as a result of both heating value and flare tip velocity was built into the calculations as specified in the regulations.

A decision was made regarding the cooling water analyzer systems that went beyond the minimum required by the regulations. Rather than just installing a continuous total strippable hydrocarbon detector as required and doing the monthly analysis to speciate the HRVOCs (or daily if the total strippable hydrocarbon concentration determined in the water was greater than 50 ppb), it was upgraded to the same speciated HRVOC analyzer installed on the flare systems. This would also provide more information about the other VOCs potentially present in the cooling water system and would aid in troubleshooting that operation.

In some instances the flowmeter installations were simple and straight forward. However, the flare lines involved in the Channelview Complex ranged in size from six inches to 60 inches in diameter and therefore complicated the process. The ultrasonic flowmeters were selected as many of the flare systems function as both a continuous flare at low flow rates and also as an emergency flare at high flow rates. The flowmeter must be to be able to handle accurately the entire range of potential operations. In some cooling water locations, installing flowmeters on multiple lines that are underground until just a few feet before they merge into a header system was necessary, as it was not possible to use of a single common monitoring point. It was decided to utilize the option the TCEQ offered of using the manufacturer's pump curves and the supply header pressures for those locations where a single return header existed. In the cases where flowmeters were used, an externally mounted ultrasonic type meter was installed to avoid potential issues with installing and maintaining internal meters in systems that are not routinely or frequently shutdown. It was possible with these systems to optimize the locations to obtain the best flow measurements in terms of signal to noise and signal strength.

The use of the "El Paso Stripper" device, which is basically a packed column used to air strip potentially present hydrocarbons from the water, presented several issues for the cooling water monitoring systems (See Figure 6).

Figure 6 – El Paso Stripper Schematic

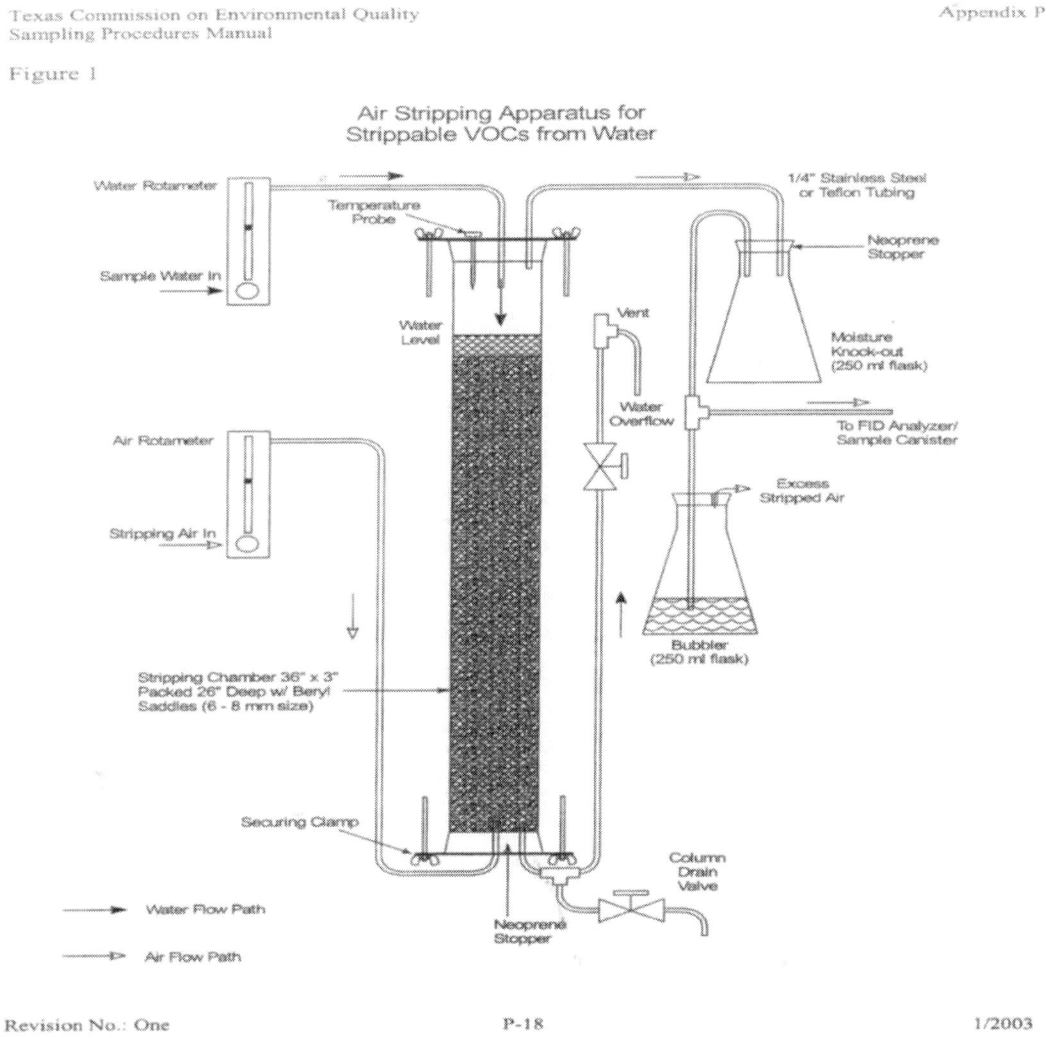

As originally described by the TCEQ in their guidance document for determining emissions from cooling water systems, this device used rotometers to manually set the required flows for both the water stream and the stripping air flow. Manual rotometers are susceptible to inconsistent flow rate if they become plugged. When this guidance was originally developed, the system was designed to be used to collect discrete samples for analysis. The use of manual rotometers was addressed by checking the flows before and after each sampling event as prescribed in the guidance document. However, for a continuously operated system, such manual adjustments would not be appropriate. Cooling water systems have a tendency to scrub dirt and dust out of the air. Therefore, the issues of plugging occurring in the flowmeters, flow controllers and valves that were installed to supply a constant defined flow rate of water to the stripping device had to be addressed. Initially it was understood that no filters were allowed in the system and this caused some serious concern as to meeting the required 95+ percent on-stream requirement as well as the obvious maintenance workload that would be required to service the system. It was later determined (after discussions with the TCEQ staff) that a solids removal system could be used if it did not also remove the hydrocarbons being monitored. The right size and type of system, to remove the solids while also allowing the maximum time between required maintenance activities, had to be found to correct this problem.

The source of water used in the cooling water system, does at times, contain small and varying amounts of a chlorine containing compound, which not only interfered with the analysis, but also potentially would degrade the separation columns. The first cooling water analyzer systems were started-up during a period when the concentrations were at a higher level and therefore the issue was noted and could be addressed. With a revision to the separation columns and operational programming in the analyzer, both the potential interference issue and the column degradation issue were resolved.

The use of the analytical laboratory services provided a quick means of determining how accurate the flare analyzer systems were during the initial testing phase of their installation. Grab samples were collected from the manual sample points for comparison with the analyzer data. While the specific analytical equipment used by the laboratory and by the field analyzers is not identical, the results compared very closely. This comparison was used to assist plant operations in determining when the new systems were ready to be placed into normal operation. The comparisons also helped to troubleshoot any of the typical analyzer start-up problems (i.e. leaking sample systems, plugged filters, etc.). The first few flare systems that were installed had analyzers for several different flares in the same analyzer shelters. This provided a means of troubleshooting problems by using both analyzers to sample the same source. Problems with an individual analyzer could be quickly determined as well as isolated from problems with the manual sampling process. A commercial laboratory that specializes in ultra low level air analysis was used to provide a similar service for the testing of the cooling water analysis systems. However, the turnaround time for getting these results made this a much slower and more difficult process. Since these analyses were at low ppm or less concentrations even tiny errors with the manual samples could result in large differences in the results when compared to the continuous analyzers. In the future, when there are analyzer system outages, it will be required that grab samples must be collected to demonstrate compliance. The procedures for collecting those samples, the process for shipping them to the analytical laboratory and the method for integrating that data into the official data collection system were also developed. The TCEQ guidelines specify how these samples must be collected and analyzed. A standardized protocol was developed to address this process.

As the new systems were brought online, the collection of data in the plant data historians began and it became possible for operations personnel to analyze the data. For the first time is was possible to monitor in real-time compliance with the various permit limits for both flare and cooling water systems. The information from the new systems was processed in the data historian system to generate the flare gas heating values as well as the HRVOC and total VOC mass emission rates for both the flare and cooling water systems. This was then fed back into the process computer systems to allow the operations technicians to monitor the systems. A simple generic spreadsheet analysis tool was also developed to allow personnel to look at both discrete values as well as trends. Before the systems became operational, the operations personnel were trained on the new monitoring systems and the required responses to potential results that might be seen. Additionally, routine data analysis meetings were begun with operations and engineering staff to review and discuss the data and the underlying events that had occurred. These meetings provided a starting point of additional efforts to better understand what was being seen and develop processes and procedures to reduce emissions or better optimize the associated processes.

Benefits Gained

While there have been a number of issues that were discovered and had to be addressed during the initial installation and start-up of these systems, there have been some definite benefits noted. In several instances, plant operations have been able to troubleshoot minor process upsets such as thermal pressure relief valves leaking into the flare system or minor leakage from pressure control valves leaking into the flare that otherwise might have gone undetected for longer periods of time. Continuous measurements of flare composition and flow have also shown that previous estimates of potential emissions associated with clearing equipment for maintenance activities may have been more conservative (i.e. larger) than they actually were. Plus, with this new tool, we can better evaluate alternative methods for clearing equipment that not only result in lower emissions but also improve recovery of products during those operations. Additionally, work is ongoing to optimize the amount of natural gas used to maintain the required minimum heating value of the flare gas which can result in a significant cost savings. While field testing some of the first cooling tower analyzers, it was possible to detect and eliminate a cooling water exchanger leak before it reached the level at which we previously would have been able to detect it and well before it became an event that exceeded the corresponding permit limit.

Paper 43d

Impact of Ethylene Emissions on Commercial Crops

Kenneth R. Foster*, Thomas R. Jack, Jim Dixon***, Xiaomei Li****, and Daniel J. Archambault*******

*AMEC Earth and Environmental, A division of AMEC Americas Ltd., Calgary, Alberta, Canada
**NOVA Chemicals Research & Technology Centre, Calgary, Alberta, Canada
***NOVA Chemicals Corporation, Joffre, Alberta, Canada
****Alberta Research Council Inc., Edmonton, Alberta, Canada
*****Alberta Research Council Inc., Vegreville, Alberta, Canada

Prepared for Presentation at 2005 AIChE Spring National Meeting/Atlanta, GA/April 10 - 14, 2005/Ethylene Producers' Conference/Environmental Session

Unpublished

Introduction

As a plant hormone, ethylene regulates plant growth, development, and reproduction at very low concentrations. Vegetation responses to ethylene vary widely, being dependent not only on the concentration and duration of ethylene exposure, but on the environmental conditions during which exposure occurred, and responses vary by species and growth stage. The discovery of the effects of ethylene on plants dates back to the 1700's, and there has been a large scientific effort dedicated to understanding the role that ethylene plays in regulating natural plant growth. Because of this effort, a large body of scientific information exists on the production of ethylene by plants, and on their responses to this naturally produced ethylene.

However, a lesser effort has been dedicated to the investigation of vegetation responses to ethylene produced and released from non-natural sources, which include petrochemical facilities. With the expansion of the petrochemical industry in Alberta, Canada, through the 1990's, the issue of ethylene releases and regulatory control of such releases became relevant. It's also worth noting that combustion of fossil fuels also results in ethylene production, and the ever increasing numbers of vehicles became an ethylene source requiring consideration.

In the absence of a substantive scientific database on vegetation responses to ethylene released from industrial complexes, derivation of regulatory ambient air quality guideline limits was difficult. Sufficient data were available to suggest ethylene concentration limits for 1 and 30 day periods and limits based on these data were established and presented in an Interim Guideline for Ethylene (Alberta Environmental Protection, 1997). These limits became the basis for regulation of ethylene emissions and effects in Alberta.

It was recognized by industry, government, and stakeholders (e.g. residents near petrochemical facilities) that while the interim guidelines were a reflection of the scientific understanding of ethylene effects, additional scientific work was required to validate or adjust these limits. In addition, industry was interested in determining the levels of ethylene that would pose a risk to field crops, as such effects would pose risks (economic, public perception, regulatory) to the operation of petrochemical facilities.

Upon agreement to these objectives, the multi-stakeholder Alberta Crop-Ethylene Research Program was initiated. The two stated objectives of the research program were:

- To provide scientific information to Alberta Environment for further development of ambient air quality guidelines for ethylene; and
- To provide the petrochemical industry and communities adjacent to petrochemical facilities with scientific information for site-specific risk assessments.

Achievement of these objectives would ensure that vegetation in Alberta was protected from the harmful effects of ethylene emissions, while also providing assurrance to the petrochemical industry that their developments could proceed within known and established regulatory limits based upon data relevant to the Alberta petrochemical industry.

Team Structures

It was determined that to succeed, the research program must meet three criteria. First, clearly defined research objectives would be stated, and the overall management (budget, schedule) of the program would be conducted within the bounds of the stated objectives. Ensuring that this criterion was met was the responsibility of the Management Team, co-chaired by industry and government, and composed of representatives of the various petrochemical facilities in Alberta and government agencies responsible for guideline development. Funding was provided by industry and government. The agreed-upon objectives are stated above.

The second criterion was that the project would be developed and executed in a scientific manner – this responsibility was assigned to a Technical Team. Experimental approach, design, execution and interpretation of results fell within the mandate of this team. Team membership included specialists knowledgeable in the various aspects of vegetation responses to ethylene, ethylene production, agricultural practices, and research techniques. Included on this team was a neighbouring farmer of one of the petrochemical facilities. This person was very knowledgeable in current agricultural practices, in the issue of ethylene emissions and their potential impacts to crops, and in the regulatory process used to evaluate emissions and to set limits. Participation by this person ensured that community concerns would be evaluated by the Technical Team, and that appropriate experiments would be conducted to address these concerns. This team was also the primary link to the consultant team, the Alberta Research Council, which performed the actual experimental programs.

The third criterion was scientific credibility. This was met through the establishment of a panel of four experts in the field of vegetation effects due to ethylene exposure. Three members of this Expert Review Panel were drawn from North American universities, each of whom have operated a long term plant physiology research program relating to ethylene, including field programs relating to the petrochemical industry. The fourth panel member was an independent consultant who had extensive experience in the area of researching and managing air pollution effects on plants, including those related to ethylene exposure. This independent panel reviewed the experimental approach and design, project reports, interpretation of data, and conclusions, and provided independent reviews to the Technical and Management Teams. Scientific issues were resolved between the Technical Team and Expert Review Panel members; such issues were resolved through adjustment of experimental protocols, addition of experiments, and revision of interpretations and conclusions drawn from the data.

Issue Identification

The absence of direct, Alberta-relevant data on vegetation responses to ethylene emissions was the primary issue that led to the initiation of the Alberta Ethylene-Crop Research Project.

In order to succeed, the research project needed to provide data and information not only relevant to Alberta regulators (guideline verification or adjustment), but also to the industry members and to their neighbours. Specific issues that required resolution at initiation of the project were:

- what would constitute an effect of ethylene exposure sufficient to be considered in guideline validation or adjustment?, and
- what species, and cultivars of these species, would be appropriate to include in the study, and how would these be selected?

The majority of petrochemical facilities in Alberta are located in rural, agricultural settings. Although ethylene can affect crops at any stage of development, most issues arising from community members related to concerns regarding changes in plant growth that would have economic implications. As cereal and oilseed crops predominate in these areas, the primary concern was one of ethylene effects on seed yield. Therefore, it was unanimously agreed that the research study would focus on yield responses, although responses in terms of vegetative growth would also be monitored and reported. In all experiments, three plants per pot were grown, treated and harvested. Yields are reported on the basis of grams of seed per pot.

To facilitate communication on a wide range of issues, each of the petrochemical facilities in Alberta had established a community advisory panel, comprising local community residents and stakeholders. This afforded an opportunity to communicate the objectives, progress and results of the research project to interested and affected individuals, and to solicit their input and feedback. To guide the selection of specific crop types, and cultivars within these types, a survey form was developed and distributed through the community advisory panels. This survey requested information on crop types and cultivars currently being grown in the vicinity of petrochemical facilities, and asked for recommendations as to the crops and cultivars that should be included in the study. On the basis of these responses, four crop types (and cultivars within these types) were selected by the Technical Team, each of which would be included in the screening phase of the project.

The results of the survey and the Technical Team's selection of crops and cultivars was confirmed with community residents at an Open House held at the Alberta Research Council facility. Representatives from the government, industry, and community advisory panels reviewed the selections, and toured the research facility. At the open house, the owners of a greenhouse in which tree seedlings grown and sold for reforestation in northern Alberta and Saskatchewan voiced concerns regarding potential ethylene effects on their operation. This concern resulted in the addition of a series of experiments directed towards evaluating potential effects on spruce and pine seedlings that could negatively impact the economic value of the seedlings.

Experiments & Results

Plant Growth and Treatment Systems

Six large growth chambers were retrofitted and coupled to a computerized ethylene delivery and concentration monitoring system to provide the ability to expose plants to ethylene at specified concentrations, for defined periods of time. An automatic fertilizer and water delivery system was added to minimize the need to open the chamber doors. The growth chambers were intensively evaluated and upgraded to ensure that variability among them (light intensity, temperature, humidity) was minimized to the extent possible, thus maximizing the potential to detect smaller ethylene effects.

To expedite the program, and to maximize the use of these chambers, plants were germinated and grown in a greenhouse, moved to the chambers for an acclimation period before ethylene exposure, exposed to ethylene at defined concentrations and durations in the chambers, then moved back to the greenhouse until plants had reached maturity and were harvested. Care was taken to treat all plants the same way, such that any effects of transport between chambers and greenhouse, or arising from the change in environmental conditions between the two systems, were equivalent between the control plants exposed to background ethylene levels and those subjected to higher concentrations or durations of ethylene exposure.

While this approach allowed the ability to maximize the use of the chambers and to conduct several experiments at one time, the approach and long growth period to maturity of the crops together limited the ability of the Technical Team to respond to unexpected results. Thus, while the experiments below are discussed sequentially, experiments were run concurrently. Results from earlier experiments were sometimes not available until well after the treatments had been applied in later experiments. In spite of the delay in acquiring results upon which to build and design further trials, the Technical Team continually received and interpreted new data, and designed further experiments to probe the responses of field crops to ethylene, always with a view to examining responses to exposures that would represent exposures that might occur in the vicinity of a petrochemical facility, and that would also assist in guideline development. Any result that was of interest, but not in this specific context, was noted but not pursued. In this manner, the research project operated differently than one that would be conducted within an academic research context.

An aphid infestation in the greenhouses occurred during the course of this two-year research project, becoming more prevalent towards the end of the research program. As required, plants were fumigated in the greenhouse with nicotine for an overnight period. Experiments later in the program required more frequent nicotine fumigations. All responses of ethylene-treated plants were evaluated against the control plants, thus ensuring that effects of nicotine did not confound the interpretation of the results. Nicotine fumigation had no visible effect on the plants.

Crop and Cultivar Screening

An experimental approach was derived that would permit the team to first evaluate crop responses to ethylene at a broad scale, and on the basis of these results, refine the experiments to focus on ethylene effects that could potentially result in effects that could impact economic returns.

Based upon the results of the community surveys, two cultivars of each of barley, wheat and canola, and one each of oats and field peas were selected for initial screenings for responses to ethylene. The objective was to select three crops representing the most sensitive to ethylene exposure.

Evaluation of yield responses required growing the plants to full maturity, a four to five month period. Limitations on growth chamber space required that an alternate approach that provided a larger space for testing be used; greenhouses provided that opportunity. Ethylene treatments in a greenhouse were impossible, so spray treatments with the ethylene-releasing agricultural product Ethephon (2-chloroethylphosphonic acid) were used. Each of the selected cultivars was sprayed with Ethephon at selected growth stages, grown to maturity in the greenhouse, and yields were measured. All plants responded to Ethephon treatment in a manner consistent with that expected for a large ethylene exposure. On the basis of the responses to Ethephon, the following crops (cultivars) treated at their most sensitive growth stages, were carried forward into the next stage of the study:

- barley (cv. *Harrington*), at the spike (seed head) emerging stage;
- canola (cv. *Quantum*), at the many flowers open stage; and
- field pea (cv. *Carrera*), at the flat pod stage.

Short-Term Exposure

Responses of the selected crop cultivars to single exposures (at the selected growth stages) of relatively high ethylene concentrations and for short-term periods were evaluated. As exposures under field conditions tend to be for relatively short periods of time, exposure durations of 1.5, 3, 6 and 12 hours were selected. Concentrations of ethylene used were 10 ppb (control; approximate background concentration), 75, 150, 300, 600 and 1,200 ppb. No statistically significant effects of ethylene exposure, including 12 hours of exposure to 1,200 ppb ethylene, were observed.

It was concluded that short term exposures to relatively high levels of ethylene either induced no effect on these crops (selected for their sensitivity to ethylene), or that any induced effects were transitory and reversible.

Long-Term Exposure

Longer exposures to ethylene, at lower concentrations, may also affect plant growth, specifically, yield. The three selected cultivars were exposed to 10 ppb (control;

approximate background concentration), 50, 100, 150, 250 and 400 ppb ethylene, for the duration of their defined sensitive growth stage (barley, 26 days; canola, 16 days; field pea, 31 days). Because effects were observed in barley exposed to 50 ppb, a follow up experiment in which a 30 ppb treatment was substituted for the 400 ppb treatment, and all treatment periods were shortened to 14 days, was completed. In this latter experiment, an aphid infestation occurred requiring the treatment of plants with nicotine for an overnight period. Although yields were reduced in all plants in this experiment (compared to the previous experiment), the relative responses of the barley subjected to different ethylene concentrations (i.e. compared with control) between the two experiments were similar.

Barley yields were reduced (relative to control) by nearly 60% in the 26-day, 50 ppb treatment, and to nearly zero following a 26-day exposure to 250 ppb and 400 ppb (Figure 1). Yield reductions were also observed after a 14-day exposure to 30 ppb (by 50%); both effects were statistically significant. Decreased yields were generally coupled with increases in biomass of below-ground and above-ground plant parts (excluding seeds), indicating that the plants were responding to the ethylene by redirecting resources away from seeds and into other organs (roots, stems). This effect was consistent with that observed in other research programs (Foster et al., 1991; Reid and Watson, 1981).

Similar to barley, the response of canola (cv. *Quantum*) to long-term ethylene exposure indicated a redistribution of resources, away from seed production to other plant parts. While the size of individual seeds was not affected, the number of seeds produced decreased markedly at the highest ethylene concentration used (400 ppb, 31-day exposure) resulting in an overall yield decrease of 77% (Figure 1). Yield decreases were apparent following all treatments, including the 31-day exposure to 50 ppb (20% decrease).The quality of the seeds produced was not affected (protein, calcium, phosphorus and oil contents; data not shown).

Field pea (cv. *Carrera*) yield responses to ethylene (16-day exposures) were not apparent at concentrations up to 100 ppb (Figure 1). Two measures of yield were taken – harvestable yield and actual yield. Harvestable yield was defined as the yield that would be realized if the plants were harvested when the control matured, which would represent the yield in a field setting as any delays in maturation would put the crop at risk of frost damage and further yield impacts. The actual yield was measured when the plants reached maturity, regardless of the delay incurred in reaching maturity.

The 16-day, 150 ppb exposure resulted in a harvestable yield decrease of about 25% (Figure 1). The effect was more pronounced following exposure to 250 and 400 ppb, both of which reduced harvestable yields by about 70%. As noted in the barley and canola experiments, field pea responded to ethylene treatment by diverting resources to other plant parts, including the production of additional seed pods. Providing extra time for these pods to mature resulted in a slight increase in actual yields (relative to control) in the 250 and 400 ppb, 16-day ethylene treatments (Figure 1).

Long-term ethylene exposures, at their most sensitive growth stages to ethylene, resulted in yield decreases in the three plant types selected for their sensitivity to ethylene. Concentrations as low as 30 ppb, provided for a duration of 14 days, resulted in significant (statistically) reductions in yield. Barley (cv. *Harrington*) was found to be more sensitive to ethylene than the other two species/cultivars. While yield was negatively affected in both canola and field pea, the effects were at higher concentrations, provided over a longer period of time.

Exposure Duration

The results of the short and long-term series of experiments established bounds on concentrations and exposure durations within which effects of ethylene on the yield of agricultural crops occur. A concentration greater than 1,200 ppb and/or an exposure duration of longer than 12 hours is necessary to induce a yield effect, while a concentration of 30 ppb for 14 days significantly reduced barley seed yield. Since crops near an ethylene release source would be exposed to low concentrations more frequently than to higher levels, it was decided to examine the duration of crop exposure to a low level of ethylene, and to identify if possible a minimum exposure duration necessary to induce a negative yield effect.

Barley (cv. *Harrington*) was exposed to 50 ppb for 0, 3, 6, 12, 18 or 24 days, centered on the sensitive growth stage. Field pea (cv. Carrera) was exposed to 50 ppb for 0, 12, 16, 20, 24, and 28 days. Plants treated for periods shorter than the full 24 or 28-day experimental period remained in the growth chambers for the full period to eliminate growth chamber effects. During this time, an ethylene level of 10 ppb was established to mimic natural levels. Before and after the completion of the exposures, plants were maintained in a greenhouse.

Barley responded to a 50 ppb ethylene treatment as short as 3 days, a treatment that reduced yield by about 40% (Figure 2). Yields decreased to a greater extent as the exposure duration increased. Field pea did not respond to any of the exposures in terms of harvestable yields, nor were any other attributes of plant growth and development affected.

Exposures of barley to 50 ppb for a period of 3 days (during the sensitive period of growth) established that barley was the most sensitive of the three crops evaluated to this point in the program. Barley is a common crop in Alberta, grown over a wide area every year. Because of its observed sensitivity to ethylene, barley was selected as the single subject for further study. Removal of the other species from the study provided greater growth chamber and greenhouse availability for barley trials.

The results of this trial narrowed the bounds within which an ethylene effect on crop yield would be induced. A concentration greater than 1,200 ppb and/or an exposure duration of longer than 12 hours is necessary to induce an effect, while a concentration of 50 ppb for 3 days significantly reduced barley seed yield.

Intermittent Exposures

In a rural setting that includes a potential ethylene release source (e.g. petrochemical facility), exposures to ethylene will generally be intermittent through the growing season, with each individual exposure being of a short-term nature measured in hours, separated by relatively long period (days) during which ethylene levels are at natural background concentrations. Given that a constant concentration (30 ppb) applied continuously for a period of 3 days had been demonstrated to cause a negative yield effect, effort was put into dissecting the type of exposure required within a 3-day period that would result in a reduction in yield.

It is known that plant responses to ethylene may vary with the time of day in which the exposure occurs (Tonneijck et al., 1999, 2000). To evaluate if this was the situation with barley, a protocol including a relatively high level of ethylene provided to the plants during one 6-hour period for each of 30 days was derived. Plants were exposed to 200 ppb ethylene:

- between 0400 and 1000 hrs, which spanned the dark to light transition period in which many plants are responsive to gaseous exposures, possibly including ethylene;
- 1000 to 1600 hrs, full light period;
- 1600 to 2200 hrs, full light period; or
- 2200 to 0400 hrs, full dark period.

Use of the 200 ppb ethylene concentration, applied during one of the periods above on each of 30 consecutive days ensured that a plant yield response, if any, would be detected, and that any differential sensitivity to ethylene during the day would be identified.

Results of this study indicated that barley is more sensitive to an ethylene exposure occurring during the 1000 to 1600 hour period, which was established as a full light period (4 hours after lights on, 6 hours before lights off). This treatment resulted in a reduction in yield by about 45% (Table 1). The other three treatments did not differ from the control (10 ppb for the duration of the 30-day treatment period). While of interest, use of these data in a guideline or facility management context would be very difficult. Nevertheless, the study demonstrated that plant sensitivity varied according to growth stage and time of day exposure timing aspects. Exposures in subsequent experiments were timed to coincide with the period of day when barley was most sensitive to ethylene.

In a parallel experiment, barley was exposed to 250 or 500 ppb from 1000 to 1600 hrs up to five times over the sensitive growth stage (25 days) at intervals of 0, 1, 3, and 5 days, or three times with an interval of 11 days between exposures. Exposures were centred on the sensitive growth stage (spike emerging). The objective of this experiment was the identification of a critical interval length that would allow for plant recovery, thus avoiding a permanent (i.e. yield) effect. This experiment was repeated times to overcome data deficiencies resulting from an increasingly prevalent aphid infection, and to provide an opportunity to fine tune treatments.

Yields were reduced in most treatments, with shorter intervals generally resulting in greater yield impacts (Table 2). Yield decreases of 50% to 80% were observed in the 0-day interval exposure regime, while yield decreases of 30% to 45% occurred in the plants subjected to the 5-day interval regime. However, yield in barley exposed to 250 ppb ethylene at 1-day intervals was higher than in the control (experiment #1), and the pattern was reversed in the 500 ppb treatments in experiment #2.

Standardizing exposures for the purposes of comparison of results among experiments was attempted using the simple dose response equation:

$$\text{concentration (ppb)} \times \text{exposure duration (days)} = \text{dose (ppb days)}$$

There is no evidence indicating that this equation properly represents the response of barley (or any other crop) to ethylene, however, insufficient data are available upon which to construct an alternate, defensible dose-response equation. Use of this simple dose-response equation indicated that the same dose applied intermittently results in a lesser yield effect than one applied over a continuous period of time. This result suggests that plants exposed to periods of background ethylene levels between exposures to higher levels can reverse the impacts induced by an exposure to elevated levels of ethylene if the interval between exposures exceeds 5 days. Given the use of 250 ppb to establish this effect, it is possible that plants exposed to concentrations lower than 250 ppb would require a shorter interval for recovery. The response of barley to 250 ppb at 1-day intervals, and to 500 ppb at 1 and 3 day intervals, showed the opposite trend. These results were not sufficiently definitive to draw any conclusions regarding the reversibility of ethylene-induced effects on seed yield in barley.

A final test of barley response to ethylene was conducted with the objective of determining if a pattern of ethylene exposure derived to mimic one that could occur in the field would decrease yield. This "demonstration exposure pattern" (Figure 3) was derived on the basis of monitoring data acquired from the ambient ethylene continuous monitoring program operated at the NOVA Chemicals Joffre, Alberta facility. The monitoring data were selected and transformed into the experimental "demonstration exposure pattern" as follows:

- the month on record within which the highest average ethylene concentration was recorded (September, 1995) was selected;
- 1-hour periods in which concentrations of 5 ppb or less occurred were set to 10 ppb (the background concentration used in all other experiments) for treatment purposes;
- from the time at which ethylene levels exceeded 5 ppb to the time that levels returned to 5 ppb or less, the highest 1-hour concentration reported in that period was applied for the duration of that period, rounded to the nearest 10 ppb.

The pattern derived was somewhat conservative in that barley plants were exposed to higher ethylene levels for longer periods than would have occurred in September 1995. Consistent with earlier experiments, the demonstration exposure pattern was applied for a 30-day period, within which the spike emerging stage (most sensitive stage) of plant growth occurred.

Barley plants treated with ethylene in this pattern increased above- and below-ground biomass production, indicating that a response to ethylene likely occurred. However, no changes in yield were observed. Given this result, it was accepted that no effects on crop yield would have been expected from increases in ambient ethylene levels from historic emissions, or resulting from current emissions.

Conifer Seedlings

At the open house held early in the research program, the operators of a greenhouse near a Joffre, Alberta petrochemical facility voiced concern about potential effects of ethylene on conifer seedlings produced in the greenhouse for use in reforestation programs in northern Alberta and Saskatchewan. This concern was valid, and consistent with the direction adopted by the Management Team a series of experiments were designed to evaluate potential negative economic effects resulting from ethylene to conifer seedlings. Negative economic impacts were defined on the basis of marketability criteria stated in the contract between the greenhouse and the purchasers of the seedlings:

- reduced germination;
- reduced seedling vigour, assessed using established greenhouse criteria; and
- reduced growth resulting in seedling rejection according to criteria (height).

The effects of ethylene on white spruce (*Picea glauca*) and lodgepole pine (*Pinus contorta*) germination and seedling growth (approximately 3-month-old seedlings) were examined by exposing the germinating seeds or seedlings in the rapid growth phase to the short-term ethylene exposure treatment series. Insufficient greenhouse space required that the seedlings be produced in the Joffre greenhouse, and be transported to the ARC for treatment. Because mechanical stimuli are known to cause plants to produce ethylene (reviewed in Abeles et al., 1992), ethylene generation during transport could result in an ethylene effect independent of the ethylene treatments applied in the growth chambers. For this reason, half of the plants were returned to the Joffre greenhouse after treatment, and half remained at the ARC greenhouse until final measurements.

No effect of ethylene on germination, vigour, or rapid seedling growth up to 1,200 ppb for 12 hours was observed in either tree species. Because the effects of concern were related to short term exposures in young plants prior to sale for use in reforestation, it was concluded that no economic impacts are likely in this scenario and that further trials with longer exposures were not necessary.

Interpretations & Conclusions

Regulatory use of Data and Findings

A primary objective of the research program was to generate data relevant to the validation or revision of Alberta Environment's draft interim ambient air quality guidelines for ethylene. The experiments were conducted in a fashion that would provide data on responses of a sensitive cultivar of a sensitive crop species, exposed to ethylene at a sensitive growth stage; guideline levels predicated on these data would therefore be protective of the majority of other Alberta crops and cultivars, as their sensitivity would be less than that shown by barley, cv. *Harrington*.

The data were evaluated by the provincial environment ministry (Alberta Environment), and guideline levels were proposed that incorporated the findings of these studies. The process which led to the establishment of the final ethylene guideline levels and averaging periods is discussed in a companion paper in this volume (Dixon et al., 2005).

The second, primary objective was to provide data and information to the petrochemical industry to support site specific risk assessments relating to risks and effects arising from ethylene releases from their facilities. The absence of a negative economic effect from the demonstration pattern experiment provided confidence to all parties (regulators, industry, farm community) that ethylene releases were not having a negative effect on the commercial viability of neighbouring farms. Similarly, the absence of a negative economic effect in the tree seedling experiment indicated that ethylene releases were not affecting the viability of the neighbouring greenhouse's tree seedling operation. In addition, these results assist industry in the interpretation of ambient monitoring results, insofar as being able to identify occasions when ambient ethylene concentrations may be reaching levels that could cause a negative crop response, and to direct effort to determining and resolving the sources of ethylene emissions causing the elevated ambient levels.

Publication of the results of the Alberta Ethylene-Crop Research Program is in process (Archambault et al., submitted), a final step in the achievement of scientific credibility for the program.

Photo Guide to Ethylene Effects

Throughout the project, a photo record of plant responses to ethylene exposure (Ethephon and gaseous ethylene) was generated. Photos demonstrating the various symptoms of ethylene exposure, including effects on height growth, tiller and branch production, extra pod production in pea, and effects on flowering structures, were selected and included in the *Photo Guide to Ethylene Effects.* This document has been provided to each of the participating industries, for use in the field if and when an evaluation of potential ethylene effects is required at a field scale.

Acknowledgements

Financial support for the Alberta Ethylene-Crop Research Project was provided by Alberta Environment, NOVA Chemicals Corporation, Union Carbide Canada Inc., Dow Chemical Canada Inc., BP Canada Chemical Company, Shell Chemicals Canada Ltd., Celanese, AT Plastics, and the Alberta Research Council Inc.

Literature Cited

Abeles, F.B., Morgan, P.W. 1992. *Ethylene in Plant Biology*. 2nd Edition. Academic Press, Inc. New York. 414 pp.

Archambault, D.J., Li, X., Foster, K.R., Jack, T.R. (submitted) A screening test for the determination of ethylene sensitivity. Submitted to: Environmental Monitoring and Assessment (2005).

Archambault, D.J., Li, X., Foster, K.R., Jack, T.R. (submitted) Response of barley, field pea and canola to ethylene exposure. I. Short-term exposures. Submitted to: Environmental Monitoring and Assessment (2005).

Archambault, D.J., Li, X., Foster, K.R., Jack, T.R. (submitted) Response of barley, field pea and canola to ethylene exposure. II. Long-term exposures. Submitted to: Environmental Monitoring and Assessment (2005).

Dixon, J. et al. 2005 – Developing and Managing a New Air Quality Objective for Ethylene Emissions in Alberta, Canada: A Multi-Stakeholder Approach [confirm and complete this citation when paper complete]

Foster, K.R., Reid, D.M, Taylor, J.S. 1991 Tillering and yield responses to ethephon in three barley cultivars Crop Sci. 31:130-134.

Reid, D.M., Watson, K. 1981. Low levels of ethylene as a phytotoxicant. Alberta Environment Research Trust, Alberta Environmental Protection, AB Canada. pp. 49.

Tonneijck, A.E.G., ten Berge, W.F., Jansen, B.P., Bakker, C. 1999. Epinastic response of potato to atmospheric ethylene near polyethylene manufacturing plants. Chemosphere 39:1617-1628.

Tonneijck, A.E.G., Jansen., B.P., Bakker, C. 2000. Assessing the effects of atmospheric ethylene on epinasty and tuber yield of potato (*Solanum tuberosum* L.) near polyethylene manufacturing plants. Environ. Monit. Assess. 60:57-69.

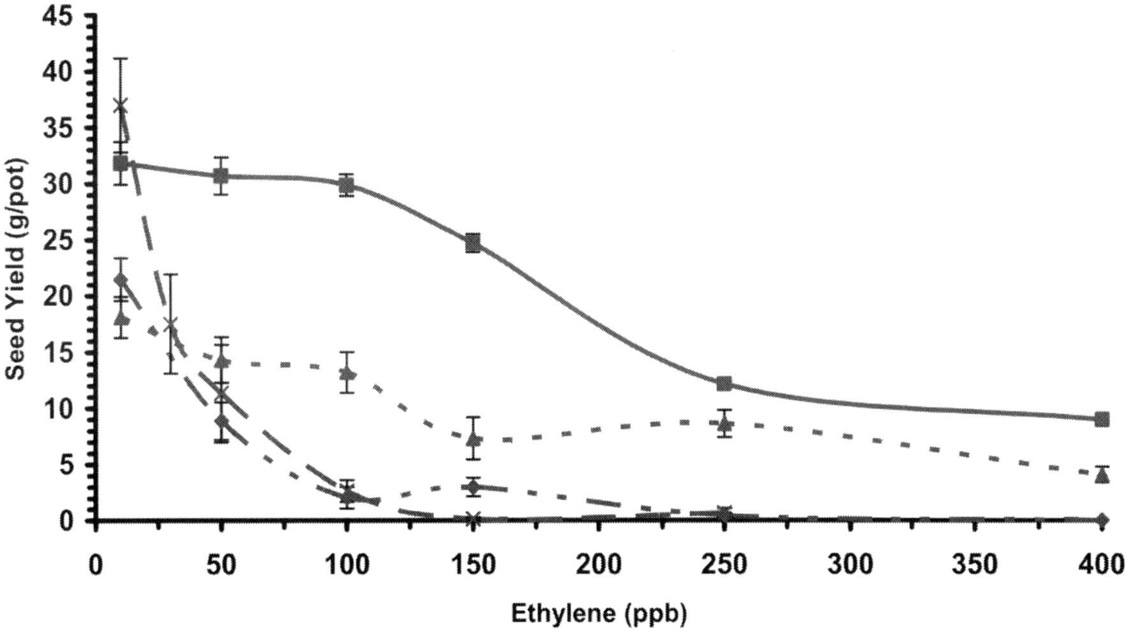

Figure 1. Yield responses of barley (▲ ; first experiment), barley (x ; second experiment), field pea (■) and canola (◆) to continuous ethylene exposures of varying concentrations (mean ± SE). Barley yields in the second experiment have been multiplied by 10 to show on the same scale as the other results. Yield is the total of three plants per pot.

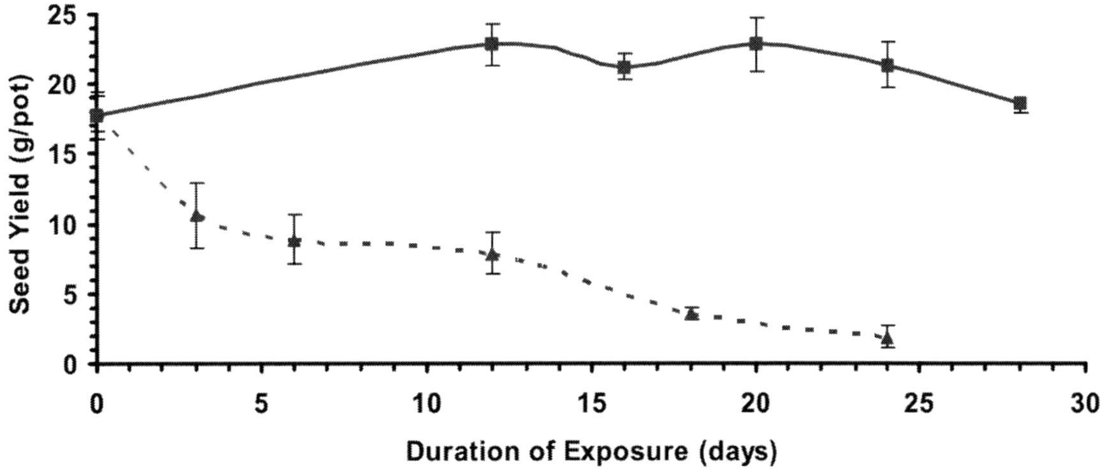

Figure 2. Yield responses of barley (▲), and field pea (■) exposed to 50 ppb ethylene for varying durations (mean ± SE). Yield is the total of three plants per pot.

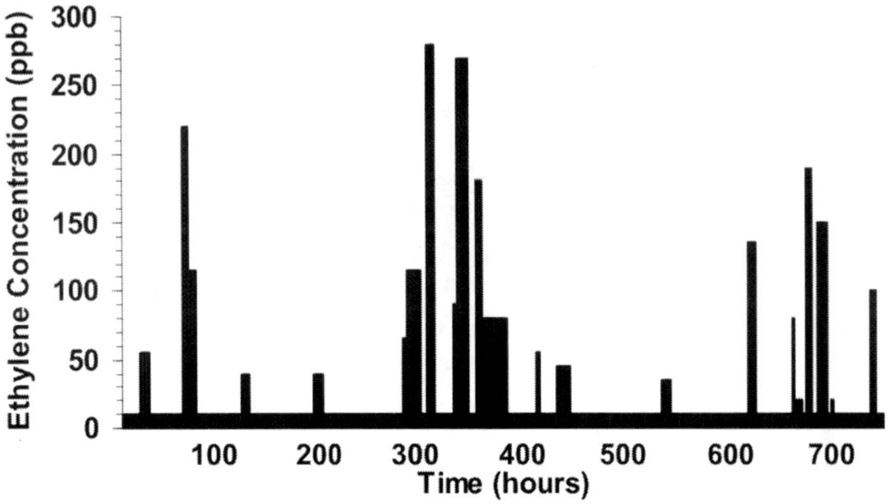

Figure 3. "Demonstration Pattern" ethylene exposure regime based upon September 1995 ambient ethylene monitoring data at the NOVA Chemicals Corporation ethylene production facility in Joffre, Alberta.

Table 1. Barley yield following exposure to 200 ppb ethylene at various times of day, for 30 days. Yields followed by different letters indicate statistically significant differences ($p<0.05$) among yields.

Ethylene Treatment	Barley Yield (g/pot)	% of Control
10 ppb (Control)	7.8 a	100
200 ppb, 0400 hr to 1000 hr	7.5 a	96
200 ppb, 1000 hr to 1600 hr	3.8 b	49
200 ppb, 1600 hr to 2200 hr	9.9 a	127
200 ppb, 2200 hr to 0400 hr	8.5 a	109

Table 2. Barley yield following treatments of 250 or 500 ppb ethylene at intervals of 0, 1, 3, or 5 days (5 treatments) or 11 days (3 treatments) at the sensitive time of day, centred on the sensitive growth stage,. Yields followed by different letters indicate statistically significant differences ($p<0.05$) among yields.

Ethylene Concentration (ppb)	Length of Interval Between Ethylene Treatments	Barley Yield (g/pot)	% of Control
Experiment #1			
10 (Control)	N/A	34 b	100
250	0	17 c	50
250	1	42 a	124
250	3	22 c	65
250	5	25 c	74
250	11	22 c	65
Experiment #2			
10 (Control)	N/A	10.8 a	100
250	0	2.5 cd	23
250	1	4.5 bcd	42
250	3	6.0 bc	56
500	1	7.9 ab	73
500	3	0.4 d	4

43e Developing and Managing a New Ambient Air Quality Objective for Ethylene in Alberta, Canada: A Multi-Stakeholder Approach

Jim Dixon, Kenneth Foster, Lois Cramer, Mike Kennedy

Ethylene is a naturally produced biologically active compound that can influence the development of vegetation at very low levels. The advent of a major petrochemical industry expansion in Alberta prompted the development of regulatory criteria for ethylene emissions. This paper reviews the nature of ethylene emissions from petrochemical facilities, existing regulatory guidelines, and the multi-stakeholder processes that were used to develop and implement a new scientifically based provincial air quality objective. The implications of this new ambient air quality objective will be reviewed in terms of the nature of petrochemical operations, their design and management.

A Low NO_x Burner Developed for ExxonMobil Ethylene Furnaces

George Stephens
Olefins Technology
ExxonMobil Chemical
Baytown, TX.

and

David Spicer
Olefins Technology
ExxonMobil Chemical
Baytown, TX

Presented at the
American Institute of Chemical Engineers
2005 Spring National Meeting
17th Annual Ethylene Producers Conference
Atlanta, GA

A Low NO_x Burner Developed for ExxonMobil Ethylene Furnaces

AIChE - Ethylene Producers Conference
Atlanta - April 2005

Presented by: George Stephens
Co-Author: David B. Spicer

Overview

- ExxonMobil Burner and Ethylene Furnace Primer
- Brief Development Summary
- Results - NO_x and Stability Test Results
- Conclusions

ExxonMobil Chemical

EMCC Staged Air FGR Burner Primer (Previous Generation)

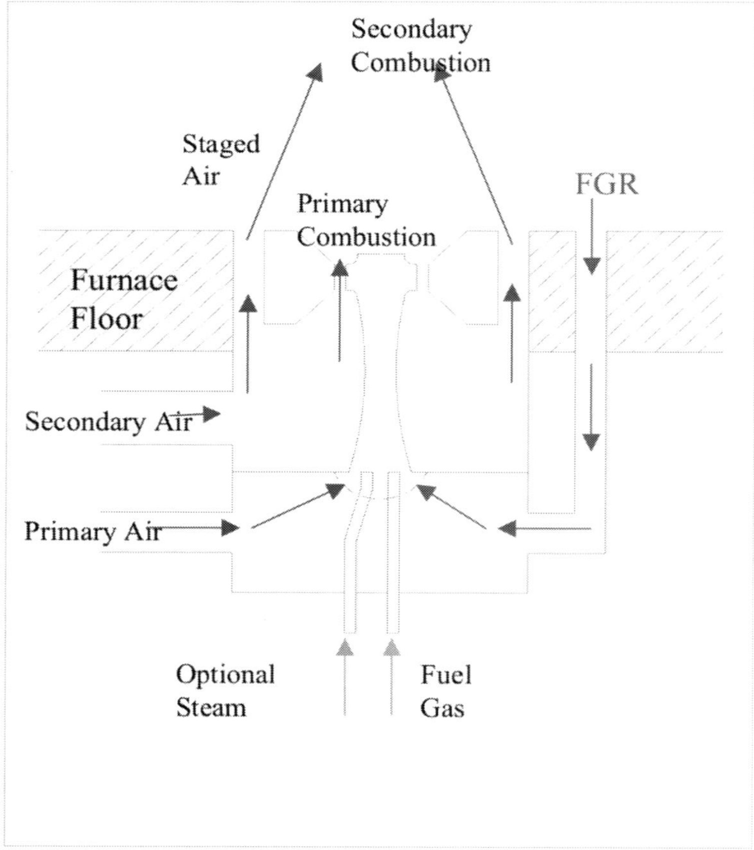

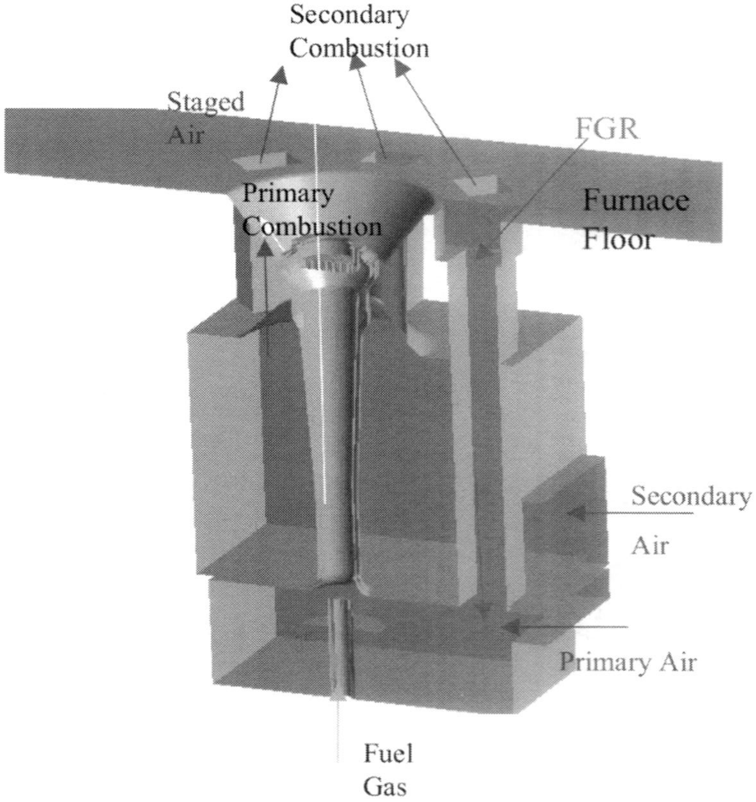

Ref: US Patent 5,092,761
US Patent 4,629,413

Features of the New Generation EMCC Burner

- Staged-air FGR premix burner with steam injection capability.
  - Development of previous generation burner, not a new concept.
- Designed for significantly higher FGR rates.
- Tighter control of fuel-air ratio in the primary combustion zone
  - Achieved by burner geometry
- Better mixing of streams internal to burner.
- Burner design reduces the interaction of high temperature zones with areas of high oxygen.
- New burner and its features are subject of several patent applications.

ExxonMobil Chemical

Benefits of the New EMCC Burner (Ex-NO_x)

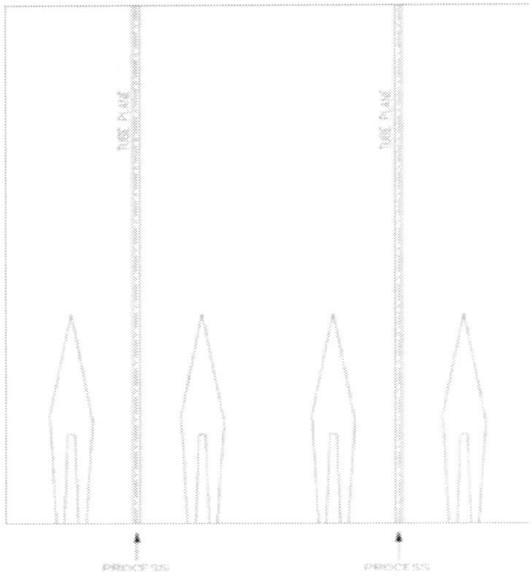

Works well with the 2-tube plane design. Accommodates the 2-tube plane furnace design.

Highly resistant to flame roll-over and outstanding flame stability. Tall, high velocity flame is extremely stiff and highly resistant to flame rollover.

No need for wall burners - Helps overall energy efficiency.

+ Heat flux profile integrates well with the pyrolysis process and tube metal temperatures are minimized.
+ No wall burners make burner balancing and O_2 minimization very simple which helps energy efficiency. This burner is capable of operating in a commercial furnace with 1% O_2 and 0 ppm CO.

Generally not susceptible to fouling. Routine cleaning is not required.

+ Single, large-diameter fuel gas per burner is well isolated from the radiant zone.
+ Negates the need for fuel gas filters/ coalescers and stainless fuel piping.

ExxonMobil Chemical

Ref: US Patent 4,499,055

ExxonMobil Chemical

Brief Development Summary

CFD Model

Benefits:

- Detailed insight into flow patterns
- Regions of high O_2, high Temperature identified.
- Qualitative assessment of NO_x
- Able to test large geometry changes relatively quickly
- Flame-to-Flame interaction assessment.

Shortcomings:

- Large models, difficult convergence, relatively long run times.
- Not able to provide absolute NO_x predictions.
- No indication of flame stability
- *Need to supplement with combustion specialist experience, knowledge of real-world burner behavior and hand calculations.* Should never use CFD as the sole tool for burner development or evaluation.

Test Furnace:

Adjust geometry for optimum NO_x and stability.

Commercial Test

Full installation in a commercial ethylene furnace.

ExxonMobil Chemical

NO_x Results

Success!

- **Low NO_x.** HGA NO_x target of 0.035 lb/MBtu HHV reached for both ambient and 350F air preheat cases. NO_x as low as 0.025 lb/MBtu HHV is achievable in a commercial ethylene furnace (ambient air with steam injection).
- **Outstanding Flame Stability.** The new burner was stability tested at extreme conditions Burner is the most stable ethylene furnace burner used to date by EMCC.
- **deNO_x steam is extremely effective in the EMCC burner.** Steam is pre-mixed with the fuel, (without tying into the fuel line) and is much more effective than injecting elsewhere.
- **Because of the effectiveness of deNO_x steam, the economics of steam injection are very favorable when compared to other NO_x reduction technologies (such as SCR, External FGR).** Only low pressure "waste steam" is needed.
- **H2 had no measurable impact on NO_x** (tested to about 50-60 vol% H2). An unusual, but beneficial characteristic for a steamcracker burner.
- **Low retrofit cost.** Burner footprint is similar to or exactly the same as previous generation EMCC burners.
- **Reproducible results**: All 3 commercial installations to date have provided very similar performance.
- **Cold furnace box flame stability:** Staged-Air design makes burner very stable during cold start-up.

ExxonMobil Chemical

NO_x Results

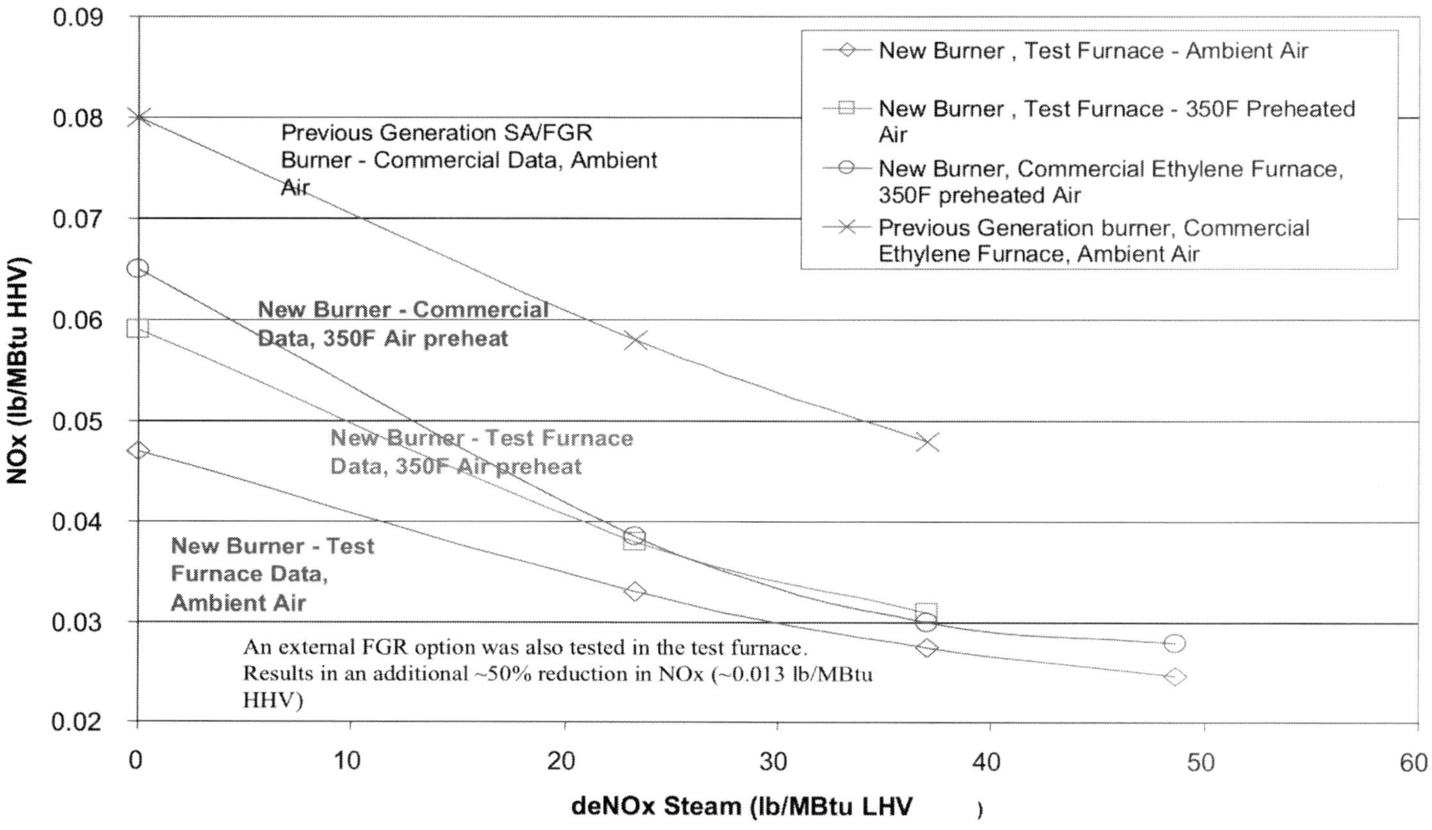

ExxonMobil Chemical

Stability Testing

Burner is stable at extreme conditions. (These tests were performed under close scrutiny to understand how this new burner behaves at limiting conditions.)

50 psig, 100% natural gas light off, high draft, with and without pilots.

- **Extremely high steam rates injected at minimum firing while maintaining good flame stability. Tested at 100 lb stm/MBtu LHV with good stability.**
- **Rapidly ramped up fuel pressure from normal (~35 psig) to maximum (60 psig) on natural gas fuel, in a cold start-up condition box with high draft (0.8") without problems.**
- **High Turndown. Normal fuel pressure about 35 psig. Turndown generally good to less than 3 psig. (However steam injection is generally stopped around 10 psig fuel pressure.)**

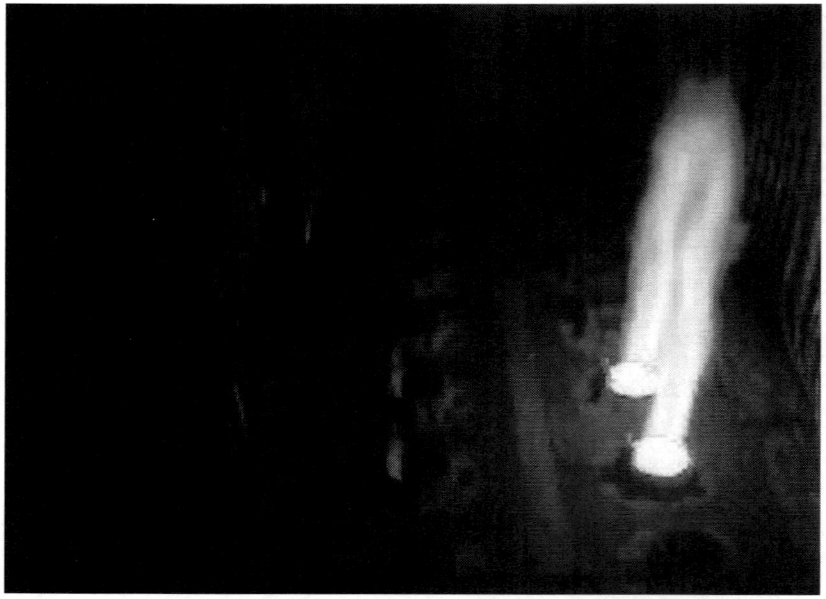

Commercial Cold Burner light-off at 50 psig (movie file)

ExxonMobil Chemical

Conclusions

New burner is a success!

- Stable - The most stable burner used in EMCC ethylene furnaces...with much lower NO_x.
- Meets the Houston NO_x targets.
- Low maintenance. Burner does not require routine maintenance.
- Flame shape and heat flux profile complement the EMCC ethylene furnace coil technology. No noticeable change in runlengths
- Successfully installed and tested on multiple air preheat furnaces to date. Ambient air commercial S/U expected in 1Q 2005.
- Further NO_x reductions possible by reducing O_2 since we are not CO constrained. Demonstrated 1% O_2 without CO. All test results are at 2% O_2.

ExxonMobil Chemical

Paper 44a

BRAZED ALUMINUM PLATE FIN HEAT EXCHANGERS – CONSTRUCTION, USES, AND ADVANTAGES IN CRYOGENIC REFRIGERATION SYSTEMS

Dan Markussen

Principal Sales Engineer

Chart Industries

Larry Lewis

CEO

SME Associates, Inc.

Prepared for Presentation at the 2005 AIChE Spring National Meeting

17th Annual Ethylene Producers' Conference

Atlanta, GA, April 10-14, 2005

Session TA011 Ethylene Plant Technology – Refrigeration Systems

April 13, 2005

BRAZED ALUMINUM PLATE FIN HEAT EXCHANGERS – CONSTRUCTION, USES, AND ADVANTAGES IN CRYOGENIC REFRIGERATION SYSTEMS

Dan Markussen, Chart Industries and Larry Lewis, SME Associates

Abstract

This paper will provide an overview of the components of a brazed aluminum heat exchanger (BAHX) and how Chart Industries designs, manufacturers, and tests them. An overview is provided of the range of uses for BAHX in cryogenic process refrigeration systems with an emphasis on debottlenecking ethylene production facilities by optimizing their refrigeration systems.

Discussion is presented on the uses of BAHX and the advantages that they provide the designer of refrigeration systems and the owners of cryogenic refining facilities.

Introduction

The continued emphasis on increasing plant efficiency driven by the increased costs of energy and feedstocks have led plant designers and owners to demand more from their heat exchangers. The required gains in efficiency and throughput targeted by a debottlenecking project can be achieved through the use of the highly efficient and flexible design of brazed aluminum heat exchangers, (BAHX.)

Over the past several decades, the use of BAHX has become quite prevalent in cryogenic refrigeration systems. Applications such as petrochemical processing, industrial gas processing, production of LNG and other hydrocarbon processing covers the general spectrum in which BAHX are commonly used.

Heat Exchanger Design and Construction

Today's heat exchangers for typical industrial processing applications generally fall under two broad categories:

- Tubular type Heat Exchangers
- Plate Type Heat Exchangers Brazed Aluminum Heat Exchangers (Compact Heat Exchangers)

Shell-and-tube tubular type heat exchangers remain the most widely used units for heat transfer in process applications. For this type of heat exchanger, tubes are mechanically attached to tube sheets, which are enclosed inside a shell with ports for inlet and outlet fluid or gas. Shell-and-tube exchangers are versatile and robust, servicing applications requiring processing of dirty, corrosive fluids, where thermal performance and compactness are not important considerations.

Historically, shell-and-tube exchangers have proven to be cost effective when the application contains warm, corrosive or dirty process fluids. While this arrangement offered serviceable performance over time, issues of durability, limited application range and high horsepower requirements tended to increase operating costs and limit system reliability.

Brazed aluminum plate fin heat exchangers, a specialty subset of Compact Plate Heat Exchangers, have become the most widely used heat exchanger in the cryogenic industry.

They offer the benefits of a compact design that is typically one-fifth the size of comparable carbon or stainless steel shell and tube heat exchanger and will weigh approximately 10 percent of the weight. BAHX smaller size is attributable to the high density of the heat transfer fins and the thermal conductivity of the aluminum material. Typical fin heights range from 0.200 to 0.380" and vary from 8 fins per inch to 25 fins per inch. Most of the surface is secondary surface with the parting sheet being the primary surface. The thermal conductivity of aluminum is very high resulting in fin efficiencies that are typically more than 80 percent. The heat transfer fin configuration is designed to create an optimum balance between pressure drop and heat transfer of the stream. Other benefits inherent in the aluminum plate fin design include close temperature approaches and a unique ability to configure multiple streams. The multiple streams can be different types of fluids, such as gas-to-gas, gas-to-liquid, two phase or any combination, with each stream balanced and optimized both thermally and hydraulically. Whereas a typical shell and tube heat exchanger is limited to two streams, brazed aluminum heat exchangers can be designed and configured in parallel or in series to create multi stream capabilities with up to 15 different streams.

Figure 1 shows the major components of a typical brazed aluminum fin heat exchanger. Brazed aluminum heat exchangers are built by stacking layers of corrugated fins separated by parting sheets and sealed along the edges with side bars.

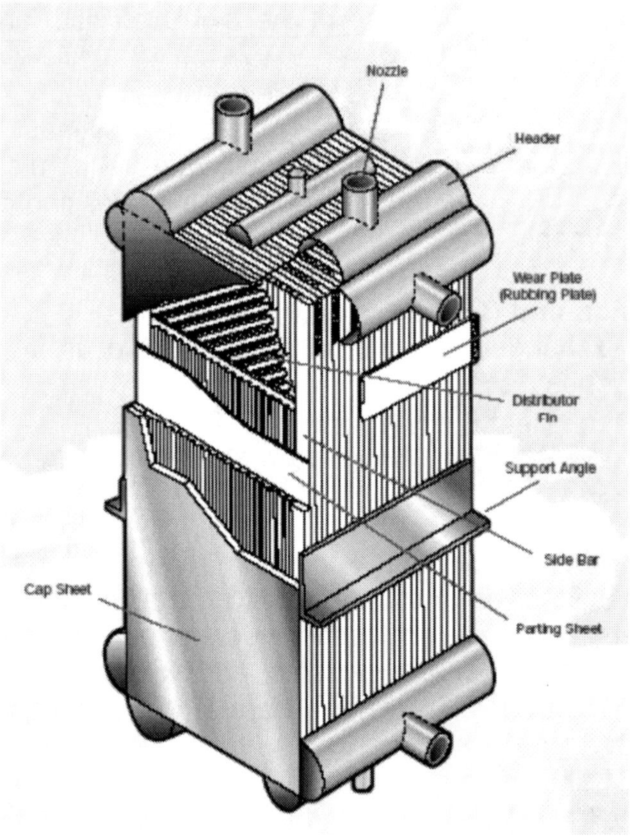

Figure 1

The matrix assembly is brazed in a large vacuum furnace to form an integral, rigid heat exchanger block. Exchanger block sizes can be in excess of 4'W x 5'D x 25'H. Headers and supports are welded onto the brazed matrix to complete the unit.

A wide range of fin patterns [plain, perforated, herringbone and serrated] accommodate different thermal and hydraulic process requirements enabling the exchanger to be custom designed for an infinite number of processes. *(See Figure 2)*

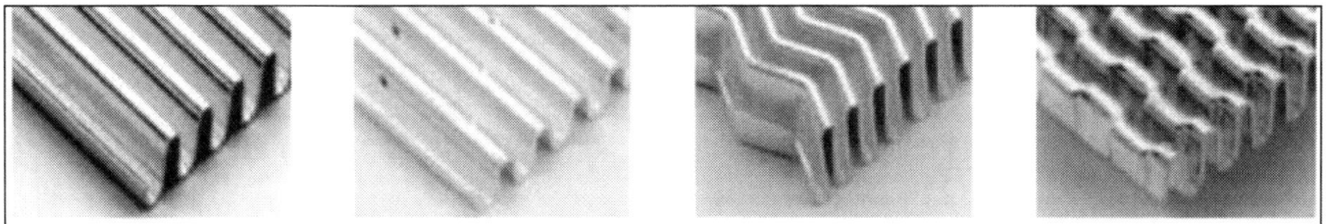

PLAIN • A sheet of metal with corrugated fins at right angles to the plates.

PERFORATED • A plain fin constructed from perforated material.

HERRINGBONE • Made by displacing the fins sideways at regular intervals to produce a zig-zag effect.

SERRATED • Made by simultaneously folding and cutting alternative sections of fins. These fins are also known as the lanced or multi-entry pattern.

Figure 2

Manufacturing

Brazed aluminum heat exchangers are manufactured as an all brazed and welded pressure vessel with no mechanical joints. The heat exchanger consists of a core block constructed of alternating layers of corrugated sheets, (fins), and flat parting sheets. Each layer is bound by bars and provided with inlet and outlet distributors. The core matrix is produced by vacuum brazing; that is joining, the fins, bars and parting sheets at high temperature in a clean, vacuum environment. Brazed aluminum exchangers, due to their inherently high surface area compactness, possess superior heat transfer capabilities and can be cost effective for non-corrosive gases and liquids as compared with traditional shell-and-tube exchangers. Historically, this type of heat exchanger has been used in cryogenic applications where small approach temperatures are important.

Codes

The design, construction, manufacturing and testing of brazed aluminum plate-fin heat exchangers are governed by recognized international codes that apply to pressure vessels. ASME Section VIII, Division I code stamp, PED, or other recognized international codes are normally applied. Associated piping is normally designed and manufactured in accordance with the ASME B31.3 Piping Code.

Supplier Considerations

Consideration should be given only to exchangers that meet one of the recognized international design codes, and adhere to strict quality control procedures, including:

- Hydrostatic and pneumatic proof testing
- Performance flow testing
- External and vacuum helium leak detection
- Dye penetrant and radiographic testing

Ideally, a total commitment to quality and reliability should be manifested through full accreditation to EN ISO9001.

Finally, customers and/or users of brazed aluminum exchangers should only consider suppliers that conform to the standards of ALPEMA (Brazed Aluminum Plate-Fin Heat Exchanger Manufacturers' Association). ALPEMA standards provide consistent quality and conformance guidelines for the design and manufacture of brazed aluminum exchangers. More information regarding ALPEMA may be obtained from www.alpema.org

Applications

In addition to design, construction, quality and performance factors, application matching heat exchanger performance lies in the ability to match the application to the appropriate technology as specifically as possible. *(see Figure 3)* Correct decisions will help to ensure low-cost, effective operation over the long term.

As stated earlier, shell-and-tube heat exchangers are versatile, all-around products intended for warm, corrosive or dirty applications where thermal performance is not highly valued. Conversely, brazed aluminum heat exchangers are ideal for cold/cryogenic; clean process streams where thermal performance, close temperature approaches, compactness and low weight are important plant considerations. Key applications include industrial gas production, natural gas processing, refinery and petrochemical processing, hydrogen and helium liquefaction and recovery, and offshore platform or FPSO processing.

A variation of the brazed aluminum fin (BAHX) heat exchanger is the Core-In-Kettle® Heat Exchanger. This approach consists of an aluminum plate fin heat exchanger assembled inside of a vessel. The aluminum plate fin heat exchanger assembly replaces the traditional tube bundle found in shell and tube kettle exchangers and other heat exchanger designs.

Core-in-Kettle designs are limited to clean fluids typical of hydrocarbon processing, natural gas processing and liquefaction. For petrochemical applications that produce ethylene, Core-in-Kettle, shell-and-tube and run back condensers are all suitable options.

Oftentimes, picking the right heat exchanger for the job is largely a process-driven task. In a refinement scenario, shell-and-tube exchangers are a natural fit for plant locations where warm conditions exist at the front end. As the product becomes more refined and the plant becomes colder, brazed aluminum exchangers then become an integral part of the heat transfer processing chain.

Applications	Products & Fluids	Typical Temperature Range	Typical Pressure Range
Industrial Gas Production - Air Separation - Liquefaction	Oxygen Nitrogen Argon Rare Gases Carbon Dioxide	+65 to -200°C +149 to -328°F	1 to 60 barg 14 to 870 psig
Natural Gas Processing (NGP) - Ethane Recovery - Nitrogen Rejection Unit (NRU) - Liquefied Petroleum Gas (LPG) - Helium Recovery	Methane Ethane Propane Butane Pentane Nitrogen Helium Hydrogen Hexane Carbon Dioxide	+100 to -130°C +212 to -202°F	15 to 100 barg 217 to 1450 psig
Liquefied Natural Gas (LNG) - Base Load - Peakshaver	Liquefied Natural Gas Multi-component Refrigerants	+65 to -200°C +149 to -328°F	5 to 75 barg 72 to 1087 psig
Refinery and Petrochemical Processing - Ethylene - MTBE - Ammonia - Refinery Off-Gas Purification - Hydrogen Recovery	Ethylene Propylene Ethane Propane MTBE Ammonia Carbon Monoxide Hydrogen	+120 to -200°C +248 to -328°F	1 to 100 barg 14 to 1450 psig
Refrigeration Systems - Cascade Cooling - Liquefaction	Helium Freon Propane Ethylene Propylene Nitrogen Hydrogen Multi-component Refrigerants	+100 to -269°C + 212 to -452°F	15 to 45 barg 217 to 650 psig

Figure 3

Cost is another consideration. While brazed aluminum offers many benefits over its tubular counterparts, a smaller or more fundamental application may benefit from an equally utilitarian shell-and-tube heat exchanger or exchanger system. Moreover, as the availability and affordability of energy used to operate plants varies globally, the energy-efficient benefits of brazed aluminum heat exchangers may not dictate its use in certain applications.

Mercury Contamination

Longtime fears about mercury's corrosive effects on aluminum have fueled some reluctance to embrace brazed aluminum technology. In reality, mercury will only react with certain alloys of aluminum and then only when liquid mercury is allowed to exist in contact with the heat exchanger and there is water present.

Mercury removal systems upstream of the exchanger are commonly installed so that brazed aluminum plate-fin heat exchangers may be used with streams containing mercury. Operators may also use special shutdown procedures to restrict moisture and maintain temperatures below 100° C. In addition to mercury removal systems, equipment from Chart Industries can be designed to be mercury tolerant. Chart's proprietary mercury-tolerant designs allow corrosive mercury to be present in the aluminum exchanger without causing the exchanger to fail.

Benefits of The Core-In Kettle Heat Exchanger Technology

Core-in-Kettle designs can possess up to ten times more heat transfer surface than traditional tube designs, providing numerous operational and performance advantages:

Close temperature approaches can be accommodated generally to 1 degree C with resultant savings in power and operating costs.

The Core-in-Kettle's smaller vessel design creates cost savings though lower liquid inventory, reduction in insulation, smaller foot-print and associated space costs, And, savings in foundation and support costs as Core-in-Kettle designs are typically one-fifth the weight of tube and shell designs.

Brazed aluminum construction eliminates mechanical joints and large diameter flanges that could leak. *(See Figure 4)*

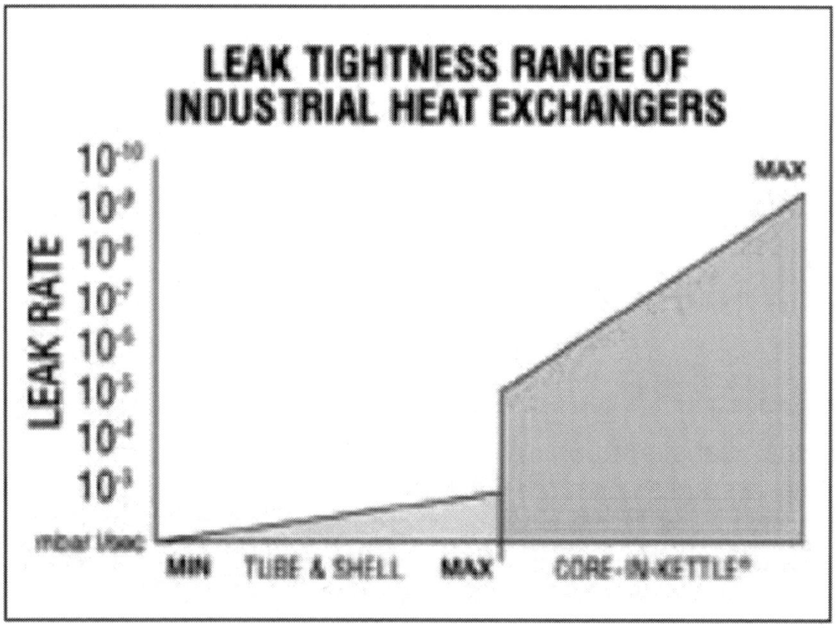

Figure 4

Core-In Kettle Applications

Core-in-Kettle heat exchangers are - ideal - for lowering initial and operating costs. In a Core-in Kettle exchanger, a brazed aluminum core is immersed in a bath of refrigerant and extremely high UA/in3 are achieved due to the refrigerant path open flow area being quite large. The reduced approach temperatures lower refrigerant compression horsepower requirements and save money. The most common applications include C2 splitter condensers, C3 splitter heat pumps, C4 splitter heat pumps, and refrigeration condensers and evaporators. *(See Figure 5)*

APPLICATIONS:		
Petrochemical	Ethylene	C2, C3, C4 splitters
	Propylene	Feed chilling train
	Ethane	Heat pump distillation columns
	Butane	Columns
	Propane	Cascade refrigeration
Natural Gas Processing	Expanding plant	Propane chillers
	LP-Gas	Cascade refrigeration
	NGL	Liquid fractionation
Gas Liquefaction	LNG	
	Nitrogen	Cascade refrigeration
	Helium	Feed chilling train
	Hydrogen	

Figure 5

Run Back Condensers

Run Back Condensers are high efficiency exchangers for the partial condensation of a range of fluids. Their sophisticated design accommodates heat and mass transfer effects while avoiding flooding. A special fin design is the heart of the refluxing stream delivering the optimum combination of free-flow area and hydraulic mean diameter. *(See Figure 6)*

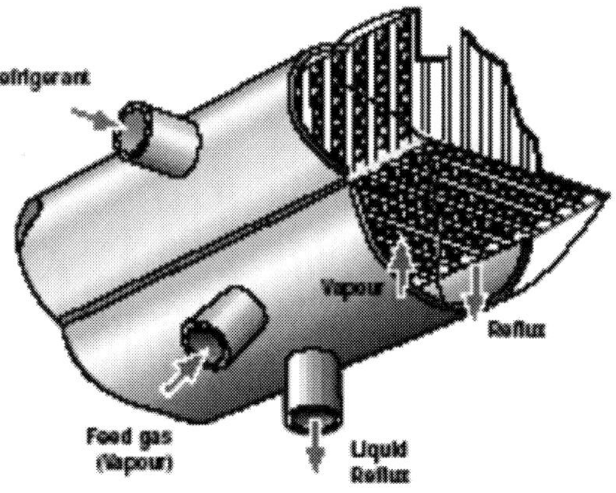

Figure 6

Runback condensers are used extensively in ethylene recovery. Other applications include Helium recovery from natural gas, hydrogen purification, condensing argon, CO_2 purification, and ammonia purge gas separation.

Benefits of Run Back Condensers include liquid runback to the tower which eliminates pumps, the elimination of several tower trays, and by shifting the refrigeration requirement for the tower to a warmer level – reducing tower cost and improving efficiency.

Cold Boxes

A cold box - in terms of BAHXs - is a steel structure enclosed, (*See Figure 7*), with thin-gauged steel panel walls. A cold box can typically contain and support one or more BAHXs, process vessels, knock out drums, and associated interconnecting piping. Cold boxes can employ the use of specially designed control valves that allow access to the valve stems and the valve internals from the outside of the box. The piping penetrations through the box are typically free floating and insulated - pipe anchor points remain at the heat exchangers and vessels. Liquid level control lines, flow meters, and thermocouples with explosion proof junction boxes can also be included in the design and scope of supply in order to assist plant operators with monitoring plant performance. Cold boxes are typically filled with perlite insulation and provided with nitrogen purging to keep the interior components cold and safe.

The compactness and ease of insulation of cryogenic process equipment makes cold boxes ideal for use in ethylene plants. Typically used in the coldest part of an ethylene plant, a cold box is usually supplied to perform the hydrogen purification requirements. Hydrogen purification typically is performed at temperatures well below -200°F and requires the use of two or more brazed aluminum exchangers, two process vessels, knock out drums, and cryogenic control valves. Although it is possible to mechanically insulate these components and its piping system, it is usually much more practical and economical to package the hydrogen purification system in a cold box.

Figure 7

Example of Heat Exchangers and Ethylene Extraction

In ethylene plants, brazed aluminum exchangers offer the advantage of pinch technology optimizing the refrigeration systems by tight approach temperatures and reducing refrigeration horsepower. Since brazed aluminum exchangers can accommodate as many as 15 streams, complex refrigeration schemes can be accomplished in services such as the ethylene recovery/purification train and the H2 purification unit.

As an example of the advantages of tighter approach temperature on refrigeration horsepower, we offer an example of a 12 MM BTU/hr Ethylene condenser (load) operating at –146.3 F @ 19.5 psia. *Figure 8* shows a cascade refrigeration system utilizing tubular exchangers with 10 F approach temperatures. A simple cycle with 2 propylene refrigeration loops and one ethylene refrigeration loop was selected for the example. The tubular exchanger example requires 9,730 HP, excluding utility HP. This equates to a Coefficient of Performance (COP) of 0.484 BTU refrigeration/BTU power.

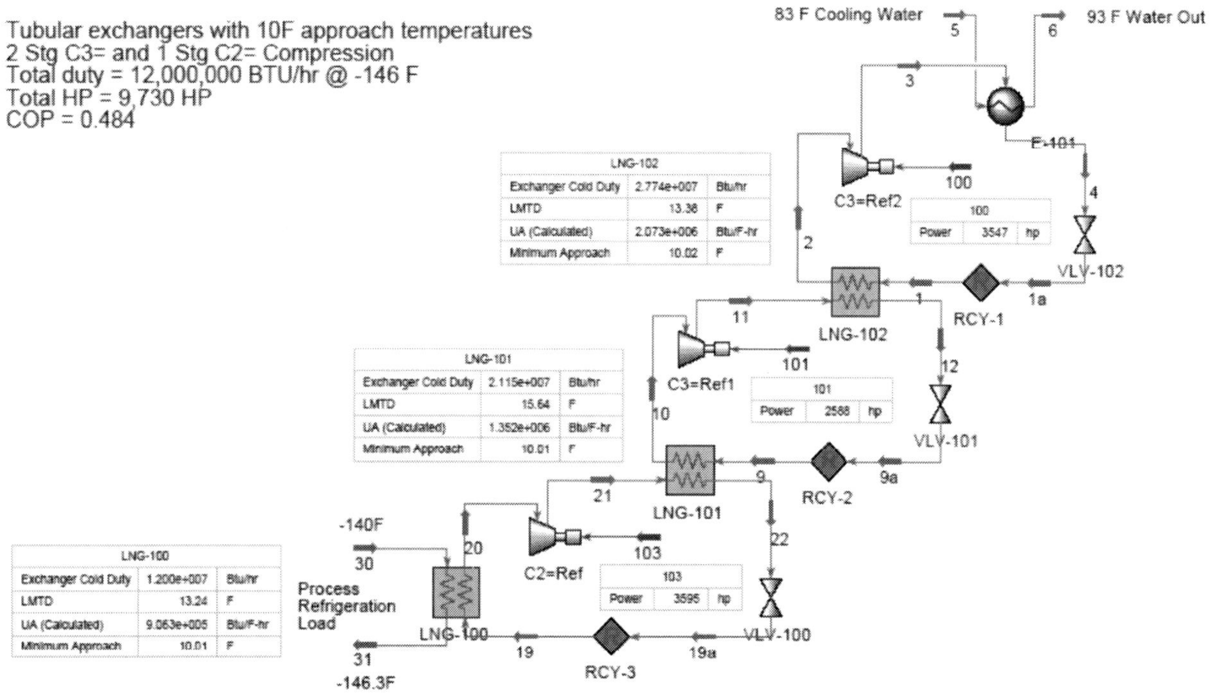

Figure 8

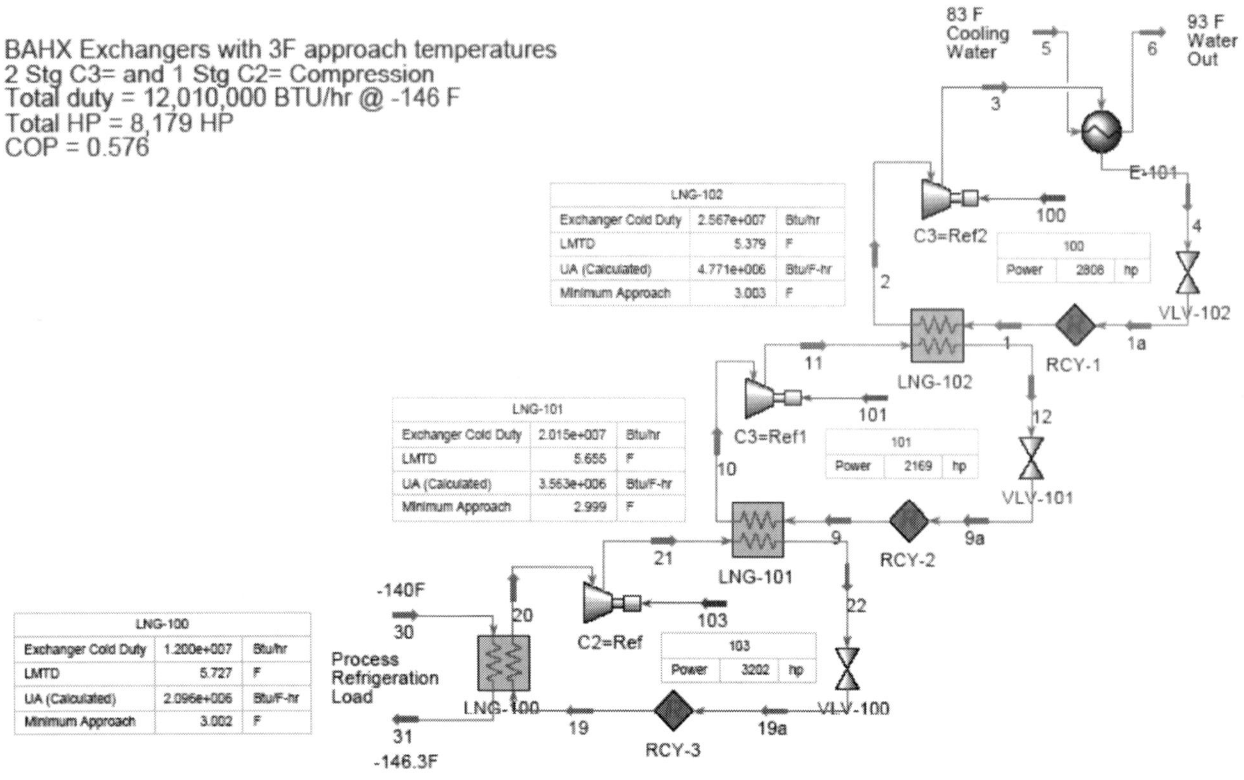

Figure 9

The same 12 MM BTU/hr Ethylene condenser load and cascade refrigeration system as shown in *Figure 9* was then simulated with BAHX exchangers achieving a 3 F approach temperature. The BAHX exchangers required only 8179 HP with a COP of 0.567. Reducing the approach temperatures from 10 F to 3 F resulted in a 16% compression HP Reduction. Composite cooling curves of both systems are shown in *Figure 10* and *Figure 11*).

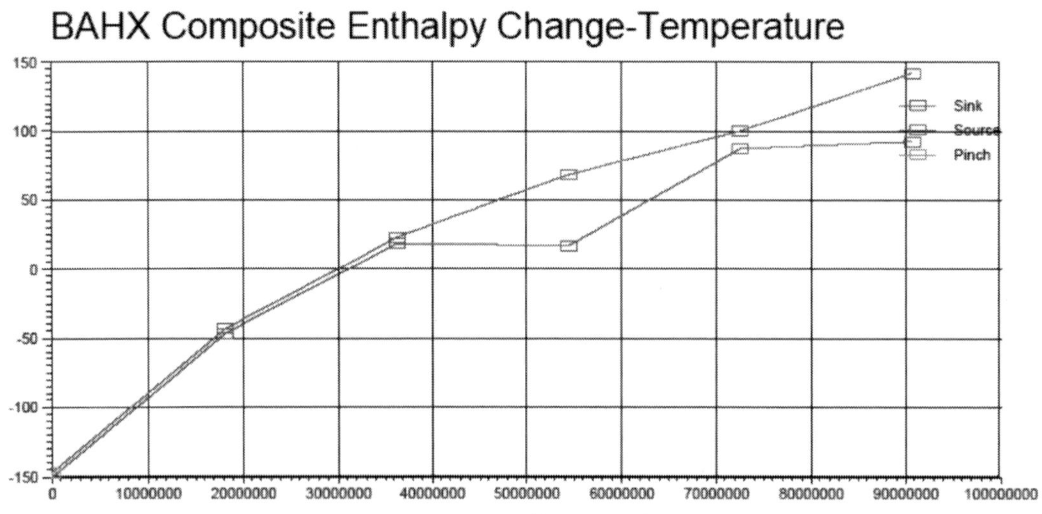

Figure 10

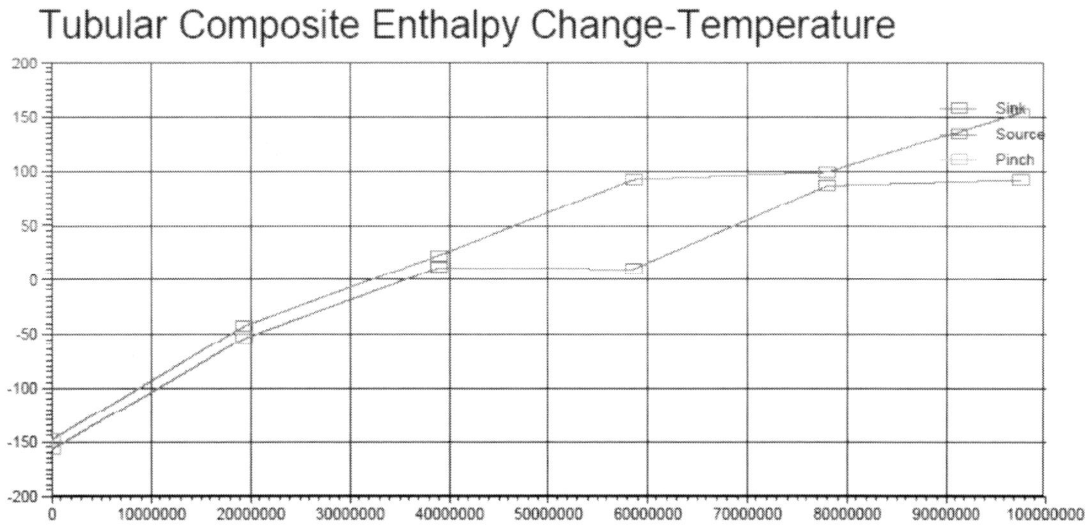

Figure 11

Further power reductions can be achieved by economizing the refrigeration loop and by increasing the number of loops.

Conclusion

By reducing approach temperatures and by taking advantage of complex refrigeration circuitry available with brazed aluminum and other compact exchangers, plants such as ethylene purification units can reduce refrigeration horsepower or debottleneck existing systems without adding compression.

Sidebar

FAQs

Although the use of brazed aluminum heat exchangers is fairly wide spread and well accepted as practical use in cryogenic processing industries, there remain many questions that are often unknown to the casual user of BAHX due to their special design and manufacturing process. The following issues are just a few that perhaps this paper has not covered or needs to be reiterated due to them often being addressed in industry.

1. **How are BAHXs tested?**
   a. A BAHX is first structurally tested. For hydrostatic test methods, each stream is pressurized, with the other streams at zero pressure, to 1.3 times the design pressure per the requirements of ASME. The unit remains pressurized for 5-10 minutes. Pneumatic test methods can be used with the pressure level being adjusted to 1.1-time design pressure.
   b. Leak testing is performed after the structural test has been successfully performed. Internal leak testing is performed by pressurizing each stream individually and monitoring small nozzle valves on the unpressurized streams using soap film solution. External leak testing is performed using soap solution on the exchanger external joints.
   c. Helium vacuum leak testing may also be used to validate the leak tightness of an exchanger, usually at an extra expense.

2. **Are the exchangers repairable if a leak is detected after the exchanger has been put in service?**
   a. Exchangers are repairable - to what extent depends on the type of leak and where it is located. Leaks develop from a number of sources - freeze failure, corrosion, and thermal fatigue are the most common. When a leak has been detected, a manufacturer of BAHX or qualified field repair company specializing in BAHX repairs should be called to consult and possible perform the repair. Repairs can include welding shut the leak or plugging of the passage that is leaking.

3. **What are the temperature and pressure limits of BAHX?**
   a. Temperature and pressure limits vary from supplier to supplier, however generally most of the world leaders in the production of BAHX, members of ALPEMA (Aluminum Plate fin heat Exchanger Manufacturer's Association) are capable of pressures from full vacuum to 1400 psi. Most manufacturers are capable of exceeding this pressure limit but the size of the exchanger and to what extent varies greatly between the different suppliers. Charts capability exceeds 2000 psig.
   b. Temperature limits are dictated by ASME code and range from -450°F to 400°F. However, practically speaking, it is not desirable to exceed 150°F, especially at elevated pressure (in excess of 200 psi).

4. **What aluminum alloys are used and what determines their selection?**
   a. Economics, availability, and code limitations dictate the type of alloy used. Generally speaking, 3003 alloy is used in the core block (consisting of parting

sheets, side/end bars, and fins) and 5083 alloy is used in the header and nozzle components. 6061 alloy is usually the alloy of choice for flanged connections.

5. **Type of stainless steel to aluminum coupling preferred?**
    a. For mechanical couplings, a RFWN flange is often used. For non-mechanical connections there are a variety of suppliers of stainless to aluminum transition joints. It is best to consult with the BAHX supplier as to what supplier is recommended. Usually the BAHX supplier has experience with each of the suppliers and is capable of making a qualified recommendation as to which supplier(s) are adequate.

6. **What process information is required for specification?**
    a. Contact a BAHX supplier, they can provide a blank data sheet to be filled in. Stream inlet and outlet temperatures and pressures are needed, as are either the physical properties at the inlet/outlet or the stream chemical composition. Chart prefers to receive a customer's HYSYS file for the specified exchanger since Chart has a program which converts the HYSYS output to Chart's design program input.

7. **Installation concerns? Piping stress on nozzles?**
    a. Consult with the supplier's installation and operating manual. Usually the supplier will have specified methods of lifting and installing the exchanger. The ALPEMA suppliers will perform thermal expansion/contraction mechanical calculations in order to provide adequate piping strength.

8. **Commissioning requirements, start up and cool down rates**
    a. There are various considerations required for start-up and shutdown which are covered in the ALPEMA standards. Typically these considerations have to do with limiting thermal stresses imposed on the heat exchanger. Visit Chart's web site for a PDF of the latest version of the ALPEMA standards (www.chart-ind.com).

9. **What are the storage conditions?**
    a. See the ALPEMA standards - www.chart-ind.com. General practice is to keep an exchanger to be stored for any length of time in a clean, dry environment away from equipment that could potentially damage the exchanger. BAHX's can be shipped with a nitrogen blanket and gages with valves to monitor the pressure. Damage in storage or shipping can be detected if pressure has been reduced or eliminated from the nitrogen blanket

10. **When is upstream filtration required? Can blockages be removed?**
    a. Due to the nature of the internal construction of BAHXs it is recommended that filtration almost always be employed. Restrictions can be removed but can be difficult at times to eliminate. Consult with the ALPEMA specifications for the options of eliminating blockages (back puffing and solvent washing are most typical).

11. **What fin types are available? What are their selection criteria? How are new fin designs developed and tested?**
    a. Fin types come in a wide variety from most of the ALPEMA suppliers. They range in height, thickness, configuration (straight, perforated, herringbone, and serrated), and fin count (fins per inch).
    b. Fins are selected on the basis of required design pressure, thermal performance, and allowable pressure drop. The optimum fin is one that does all of these things at the lowest cost.

12. **Corrosion and mercury features**
    a. Chart is the developer and world leader in providing exchangers that are suitable to perform in a mercury environment. Due to the variety of conditions for the exchanger to operate in with mercury present and the proprietary nature of Chart's design, Chart recommends that a potential customer speak to us about their individual needs. Please note, Chart has/is presenting a paper about this subject at this conference.

About Chart Industries

Chart Industries, Inc. is a leading global supplier of standard and custom engineered products and systems serving a wide variety of low temperature and cryogenic applications. Headquartered in Cleveland, Ohio, Chart has domestic operations located in seven states and an international presence in Australia, China, Czech Republic, Germany and the United Kingdom.

For more information, contact Chart Industries Energy & Chemicals Group at 888-794-7608 or www.chart-ind.com

Paper 44b

Storage of Liquefied Ethylene

John Blanchard

Principal Engineer:
Low Temperature & Cryogenic Storage

CB&I
Plainfield, IL

Prepared for Presentation at the 2005 AIChE Spring National Meeting
17th Annual Ethylene Producers' Conference
Atlanta, GA, April 10-14, 2005
Session TA011 – Ethylene Plant Technology – Refrigeration Systems

April 13, 2005

Storage of Liquefied Ethylene

John Blanchard
Principal Engineer: Low Temperature & Cryogenic Storage

Abstract:

Economic storage of ethylene requires liquefaction, which can reduce the stored volume by almost 500 times. Storage of a small liquid volume can be handled with a partially refrigerated pressurized sphere. However, larger volumes of liquid are more economically stored in flat bottom fully refrigerated storage tanks. This paper summarizes current storage concepts, design standards, materials of construction, containment options, solutions for external hazards, safety, maintenance, and construction considerations.

Introduction:

Selection of the storage concept which best fits a particular plant, requires consideration of the plant process needs, downstream use, storage material selection, and internal and external risks associated with the concept chosen. Many storage concepts are defined by international standards and variations of these concepts have been applied to meet the specific needs of projects around the world.

Partially refrigerated versus fully refrigerated

Storage and shipment of ethylene, like other gasses is economical only as a liquid. In liquid form, the volume of ethylene is reduced by almost 500 times from gas at ambient temperatures. Although gasses such as propane can be liquefied by pressure alone at ambient temperatures, due to ethylene's critical temperature of 48.5°F storage of ethylene as a liquid requires a combination of lower temperature and pressure. Combinations such as 250 psig and -25°F or 300 psig and -12°F could be used. This is referred to as partially refrigerated storage. Storage at ambient pressure and –154°F (ethylene's boiling point at 1 atmosphere) is the combination usually applied. Storage at very low pressures is referred to as fully refrigerated. The low pressure is maintained by managing the boil-off, which is typically about 0.04% of the total volume per day.

Before selection of a storage concept can be determined, material selection must be considered. Storage of liquid ethylene under pressure must consider the possibility of loss of part or all of that pressure. If pressure is lost, the liquefied gas will auto-refrigerate to maintain equilibrium. Thus the materials selected must consider this possibility. The result is a design metal temperature below the range for carbon steels; low alloy steels must then be selected. The steel selection will be the same for partial and fully refrigerated storage. Depending on the current steel market, 5% and 9% Ni are the most common materials used. Since ethylene is commonly transported fully refrigerated, there are few if any advantages to partially refrigerated storage.

Storage concepts:

Most fully refrigerated ethylene storage is designed to API 620 Appendix Q and API 2508, "Design & Construction of Ethane and Ethylene Installations" references this design standard. API 620 Appendix Q requires a primary liquid container and a separate gas vapor container separated by insulation. The outer, gas vapor, container is not designed for the liquid temperature. Leakage of liquid must be contained by a separate system, such as an earthen dike. Several other concepts are commonly used for storage of other refrigerated gasses such as methane and propane, and could be used for ethylene storage. Standard BS7777 and its soon to be released European replacement EN14620 define storage concepts that consider secondary containment as part of the storage tank. These primarily include "double" and "full containment" concepts.

Double containment places the secondary containment system close to the primary container. Secondary liquid containment is maintained, but not vapor. However, the smaller surface area of the liquid, as compared to an earthen dike, reduces the rate of vapor generation.

Full containment provides vapor containment and liquid containment. This concept is being applied to most new LNG storage facilities worldwide.

There are many variations of these concepts including combinations of steel and concrete. The actual concept chosen depends on the risks associated with each site. Concept sketches for full, double, and single containment are provided at the end of this paper.

Plant Risk (Hazard) Assessment:

When considering the storage of liquid ethylene, a company should evaluate the potential risks and how the risks can be mitigated. Storage of methane (LNG) is highly regulated, including calculating thermal and vapor exclusion distances for the protection of surrounding property and life. Ethylene has a much larger air to gas flammability range, (2.7 to 36% as compared to 5 to 15% for methane), and is denser than methane and air at ambient temperatures, (0.0725pcf as compared to 0.0415pcf for methane and 0.0724pcf for nitrogen). Cold ethylene gas is considerably more dense at its boiling point, (0.112pcf) which adds to the hazard level. Thus, volume for volume, the storage concept selected for ethylene is at least as critical as for LNG storage. Yet, most ethylene storage built worldwide relies on, and API2508 specifies, secondary containment similar to that used for liquid petroleum products: earthen dikes.

BS7777 and EN14620 provide guidance for evaluating the risks associated with liquefied gas storage. Although the risk assessment is the responsibility of the owner, the plant designer takes an important role in that assessment, defining the potential hazards and associated risks. Hazards associated with ethylene storage can be divided into internal and external events. Identifying these hazards allows the selection of the storage concept that will reduce risks to acceptable levels for the plant.

Internal events include:

- Mechanical failure resulting from thermal shock, corrosion, frost heave of the foundation, and leakage.
- Equipment failure resulting from relief valve malfunction, liquid level gauging, and pressure gauging.
- Operational and maintenance errors including overfilling and overpressure
- Roll-over (not normally a concern with pure products such as ethylene)
- Operational philosophy, practice, and upset events.

External events include:

- Natural and environmental events such as earthquakes, wind, flood, lightning and snow.
- Infrastructure events such as an aircraft crash and impact from adjacent facility accidents including fire, explosion, and projectiles.
- Site layout events including adjacent tank fire, vent fire, vapor cloud ignition and resulting pressure wave, construction, and traffic.
- Operational philosophy, practice, and plant upset events.

Potential consequences of these events include leakage of liquid and/or gas further resulting in an ethylene vapor cloud, explosion, and fire. Each of these could lead to further events escalating the risk. Hazards for people, both on-site and off-site, the environment, other parts of the plant, and adjacent facilities must be evaluated.

Selection of Storage Concept:

Considering the potential risk of an uncontrolled vapor release may lead to the selection of a full containment type tank with a concrete outer tank. This concept is costly and requires a long construction schedule. However, the selection of full containment for storage does not require that a concrete outer tank be used. The outer tank of the commonly applied double wall steel tank concept can be made of the same steel selected for the primary liquid container. This solution, which meets the full containment concept, has been applied at more than one plant and has been designed for both projectile impact and a vapor cloud ignition pressure wave.

Using the single, double, and full containment concepts defined in BS7777 and EN14620, the following generalizations relate scenarios where each tank concept may be appropriate:

Single containment tank:

Site is large and away from a congested area. External threats are either small or the resulting risk is minimal. Damage to plant from potential vapor release is considered acceptable.

Double containment:

Site is small and uncontrolled release of vapor resulting from internal threats is acceptable. Design of the outer container for external events protects the primary container.

Full containment:

Site is small, uncontrolled release of vapor is not permitted, and tank burnout risk must be at the lowest level. Design of the outer container for external events protects the primary container.

Summary

Storage of ethylene as a liquid in fully refrigerated flat bottom tanks at low pressures is the most economic storage solution. Historically, storage of ethylene has applied the single containment concept found in API 620 appendix Q. However, consideration of risks associated with ethylene storage could result in selection of a different storage concept, such as full containment. This paper presents information to make plant owners and designers aware of well defined economical solutions, which can reduce risk to people and property in and around ethylene plants.

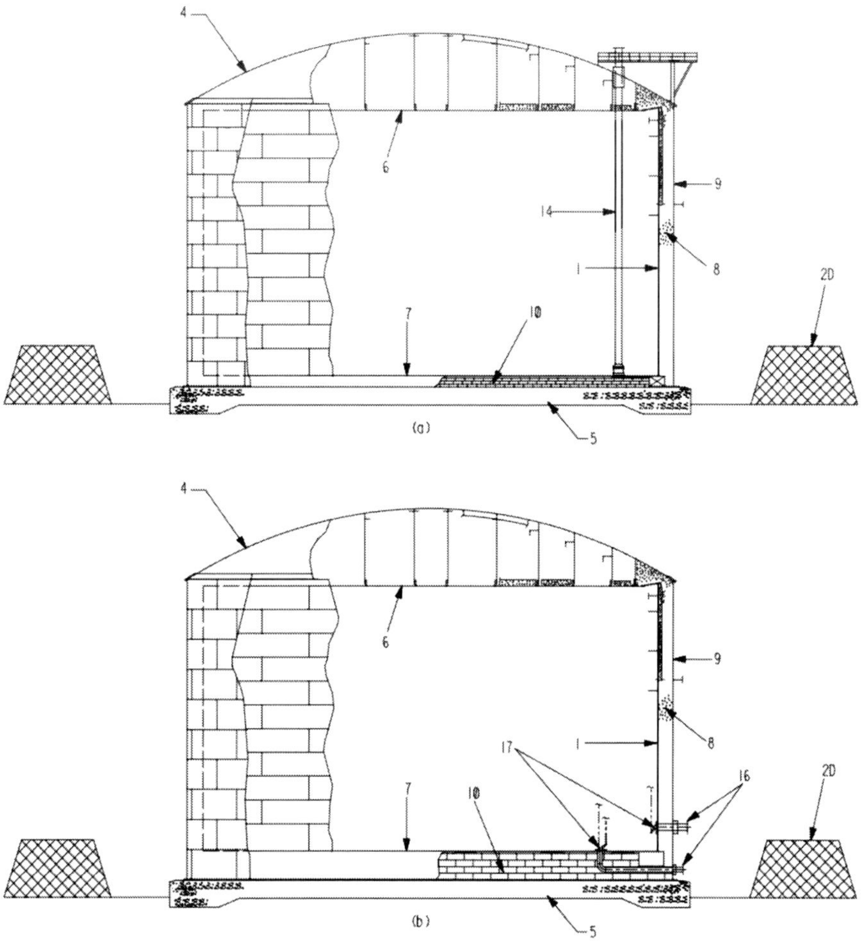

KEY

1 PRIMARY CONTAINER (STEEL)
2D BUND WALL (SECONDARY CONTAINER)
4 ROOF (STEEL)
5 FOUNDATION (REINFORCED CONCRETE)
6 SUSPENDED DECK (INSULATED)
7 PRIMARY CONTAINER BOTTOM
8 INSULATION
9 OUTER SHELL (INSULATION/VAPOR CONTAINER)
10 BOTTOM INSULATION
14 PUMP COLUMN
16 BOTTOM OUTLET/SHELL NOZZLE
17 IN-TANK SHUT OFF VALVE

EXAMPLE OF SINGLE CONTAINMENT TANK

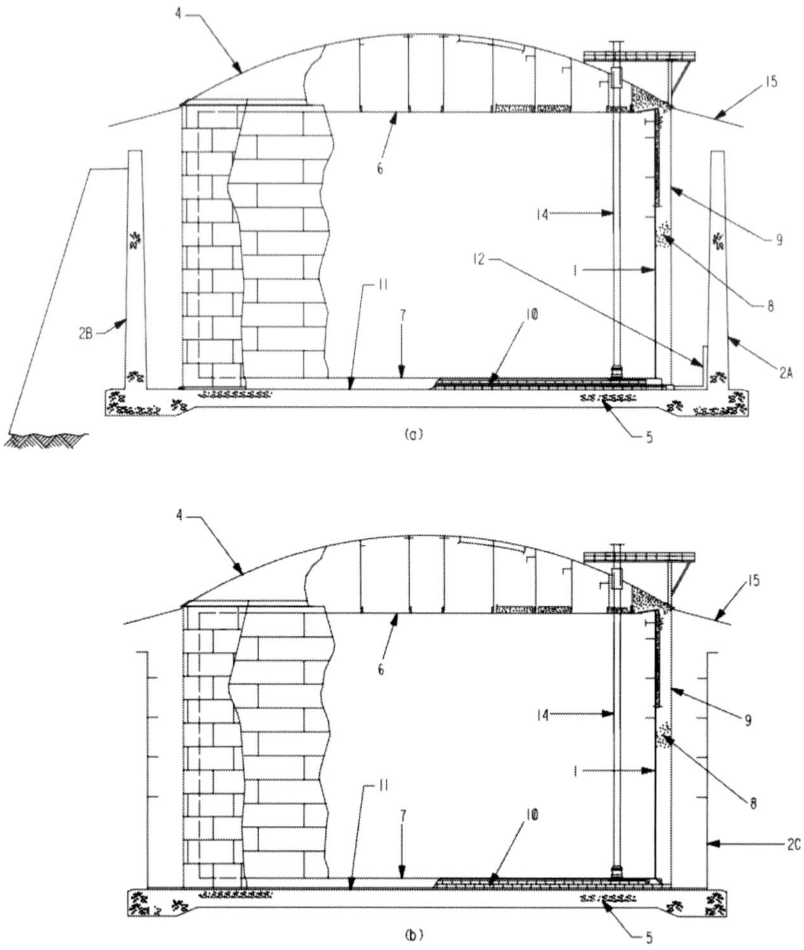

KEY

1 PRIMARY CONTAINER (STEEL)
2A SECONDARY CONTAINER (PRESTRESSED CONCRETE)
2B SECONDARY CONTAINER (REINFORCED CONCRETE W/EARTH BERM)
2C SECONDARY LIQUID CONTAINER (STEEL)
4 ROOF (STEEL)
5 FOUNDATION (REINFORCED CONCRETE)
6 SUSPENDED DECK (INSULATED)
7 PRIMARY CONTAINER BOTTOM
8 INSULATION
9 OUTER SHELL (INSULATION /VAPOR CONTAINER)
10 BOTTOM INSULATION
11 SECONDARY CONTAINER BOTTOM
12 THERMAL CORNER PROTECTION
14 PUMP COLUMN
15 COVER (RAIN SHIELD)

EXAMPLE OF DOUBLE CONTAINMENT TANK

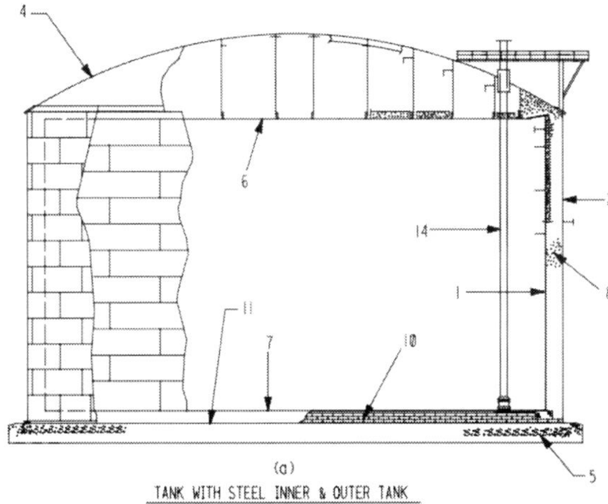

(a)

TANK WITH STEEL INNER & OUTER TANK

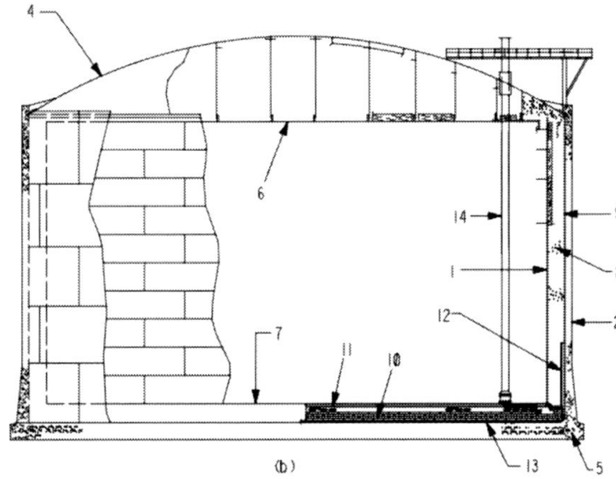

(b)

TANK WITH STEEL INNER TANK, CONCRETE OUTER WALL AND STEEL ROOF

KEY

1 PRIMARY CONTAINER (STEEL)
2A SECONDARY CONTAINER (PRE STRESSED CONCRETE)
2C SECONDARY CONTAINER (STEEL)
4 ROOF (STEEL)
5 FOUNDATION (REINFORCED CONCRETE)
6 SUSPENDED DECK (INSULATED)
7 PRIMARY CONTAINER BOTTOM
8 INSULATION
9 WALL LINER
10 BOTTOM INSULATION
11 SECONDARY CONTAINER BOTTOM
12 THERMAL CORNER PROTECTION
13 VAPOR BARRIER
14 PUMP COLUMN

EXAMPLE OF FULL CONTAINMENT TANK

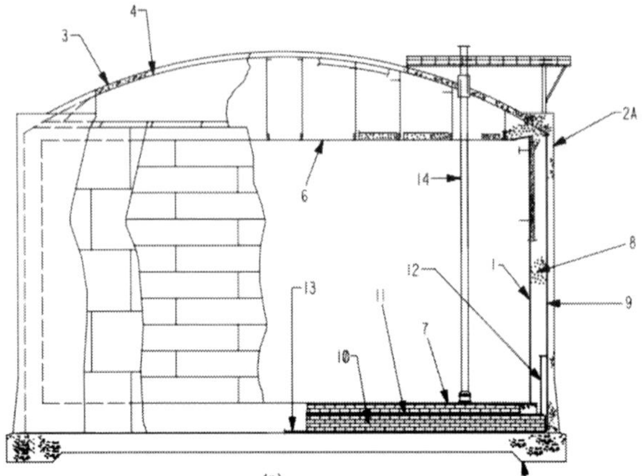

(a)

TANK WITH STEEL INNER TANK, CONCRETE ROOF & OUTER TANK

KEY

1 PRIMARY CONTAINER (STEEL)
2A SECONDARY CONTAINER (PRE STRESSED CONCRETE)
3 ROOF (REINFORCED CONCRETE)
4 ROOF LINER (STEEL)
5 FOUNDATION (REINFORCED CONCRETE)
6 SUSPENDED DECK (INSULATED)
7 PRIMARY CONTAINER BOTTOM
8 INSULATION
9 WALL LINER
10 BOTTOM INSULATION
11 SECONDARY CONTAINER BOTTOM
12 THERMAL CORNER PROTECTION
13 VAPOR BARRIER
14 PUMP COLUMN

EXAMPLE OF FULL CONTAINMENT TANKS

Paper 44c

Mega-Cracker Cold Section Energy Integration

David Gent
NOVA Chemicals Corporation
P.O.Box 5006
Red Deer, Alberta, Canada T4N 6A1

Prepared for Presentation at the 2005 Spring National Meeting, Atlanta GA, Apr 10-14th
17th Annual Ethylene Producers' Conference
Session TA011 – Ethylene Plant Technology

Mega-Cracker Cold Section Energy Integration

By:

David Gent (NOVA Chemicals)

Abstract

Modern energy efficient mega-crackers can provide some interesting challenges and opportunities with respect to energy system integration as well as to the type and size of equipment employed.

NOVA Chemicals operates one of the largest ethane crackers in the world at it's petrochemical complex in Joffre, Alberta, Canada. This paper will describe an owner's view of some of the energy integration design, operability and maintainability issues that were associated with the scale of this plant.

Rationales for the configuration of the key energy systems in the cold section of the plant will be reviewed. These will include consideration of the need for process intensification and its impact on the type of equipment that was employed. Some actual commissioning and operating experiences will also be shared.

Finally, some thoughts and conclusions about the success of the various approaches to the mega-cracker cold section energy design will be offered.

INTRODUCTION

The Ethylene 3 plant (E3) is located at the NOVA Chemicals operating site in Joffre, Alberta, Canada. This facility was designed to produce 1,270 kilotonnes per annum of ethylene from an ethane feedstock. At the time of its startup in August 2000 it was the largest capacity single-train ethylene plant in the world.

The construction of E3 appears to have heralded the era of a new generation of of so-called 'mega-crackers'. Several plants of similar and larger size are now on the drawing board or in construction.

The design of such large plants poses some interesting questions with respect to the appropriate type of process schemes and equipment that should be used in order to achieve a cost-effective result. The following paper describes some of the issues in the design and operation of the cryogenic section of the mega-plant.

MEGA-PLANT ISSUES

One of the rationales for building a mega-cracker is that there are continuing economies of scale to the capital and fixed operating costs of the plant as the size is increased. Realization of these benefits requires some focus on issues such as maintaining a single-train design concept so that parallel systems and additional equipment count are avoided.

A mega-cracker will usually be supporting the operation of several derivative units, so reliability of the design is another important factor. When the plant trips or is in an upset condition there is a potential for very large volumes of material to flow to the flare. The impacts of this to the surrounding community are another reason to place a great deal of emphasis on reliable operation.

As the scale of the plant is increased there is a tendency to drive equipment type and sizes toward first-of-a-kind applications (the 'edge of the envelope'). While some of this may be unavoidable it is desirable to use proven equipment in proven sizes as much as possible. Some design approaches can lend themselves to this goal more than others.

One means of reducing the impact of capacity increases on equipment footprint is to utilize process integration and intensification techniques. This can tend to make operation and maintenance of the plant more complex. Hence it is desirable to keep these integrations as simple as possible.

Plot space may be a significant consideration depending upon the location and nature of the site. The plot layout and spacing has to consider the potential impact of vapour cloud explosion and fire. Maintenance and construction access are also important considerations. Designs that reduce equipment piece count and physical size can be helpful with respect to these issues.

Constructability is another important issue for mega-cracker design. Equipment designs that lend themselves to shop fabrication can be superior because there is typically improved quality control when compared with field erection and less field man-hours are required. Similarly, designs that allow a significant amount of pre-fabricated modularization to occur can be very beneficial to field construction cost.

DESIGN OPPORTUNITIES

Plot Plan

There were two existing ethylene plants (E1 and E2) at the Joffre site. These plants were designed for 544 and 680 kta ethylene production respectively. A plot area had been earmarked for a third plant that was originally conceived to be of similar capacity to the first two. When the E3 project was proposed at a capacity that was greater than that of the previous two plants combined, there was an immediate concern about the ability to fit the plant into the available plot while still meeting the site's design criteria.

As can be seen in the photograph below, the final E3 footprint is fairly similar to the other two plants, yet the capacity is much greater.

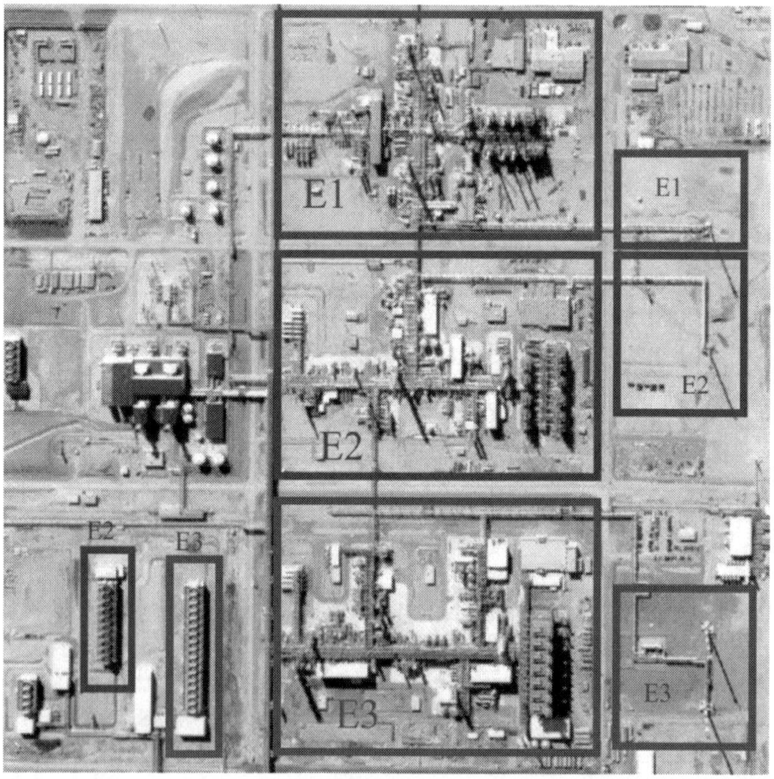

Understandably, building a plant of much larger capacity required that a number of things had to be bigger. However, there were also opportunities in many parts of the process to minimize size increases and to reduce the number of equipment items that had to be on the plot. This involved giving adequate consideration to the impact of the different choices in process schemes and operating conditions as well as to the type of equipment that would be used. One area where the opportunities were particularly significant was in the cryogenic processing sections of the plant and in their associated refrigeration systems.

The cold section of the E3 plant is highlighted in this aerial view (Figure 1).

Figure 1: E3 Cold Section

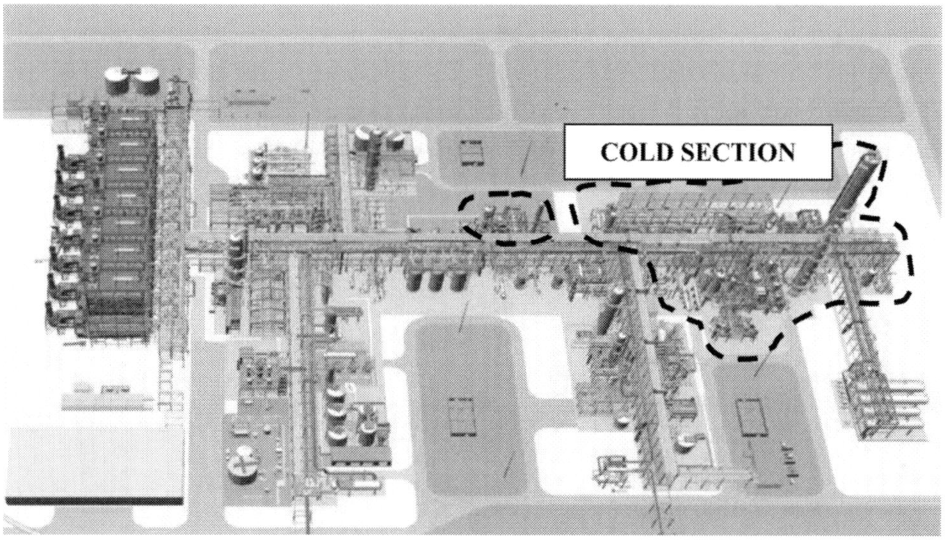

Much of the equipment in this part of the unit is typically the same size and in some cases smaller than the equivalent equipment in the other two plants. A large part of the reason for this is the progress that has been made in the energy integration and equipment efficiency from the older plants of the 1970's and 1980's to the newest designs. One of the results of these energy reductions is that more production capacity can be achieved with the same scale of energy equipment. The difference is illustrated by the fact that the specific compressor power (power per unit of product) for the E3 plant is 25% less than that required for the E1 plant.

In addition to this, the use of more compact equipment and other process intensification techniques has led to further potential for reductions in plot space requirements.

C2 Fractionator Impact on Refrigeration Load

One issue that was evident from a comparison of the refrigeration compressor loads in the existing plants and those for a new design was that the scale of the new plant could challenge the size of available compressor equipment. It was also evident that different process configurations could result in a different split of relative refrigeration compressor loads.

By far the largest single refrigeration load originates from C2 fractionation. There are two predominant approaches to providing the energy supply for this fractionation. One involves a closed-cycle heat pump that uses propylene as the intermediate refrigerant fluid. The propylene is integrated within the overall propylene refrigeration system. The C2 fractionator operates at relatively high pressures in the range of 1500-2000 kPag. The second approach is a direct heat pump that uses product ethylene as the working fluid. This is typically integrated into the overall ethylene refrigeration system but it can also be on a standalone basis. The direct heat-pumped C2 fractionator operates at low to medium pressures in the range of 400 to 700 kPag.

The propylene refrigerated high pressure C2 fractionator has been used quite prevalently in plant designs that are based on a front-end demethanizer scheme. All of NOVA's other ethylene plants have this design and it has a good operational record. Hence there were reasons to consider this type of design for a new plant. An example of the type of scheme is illustrated Figure 2.

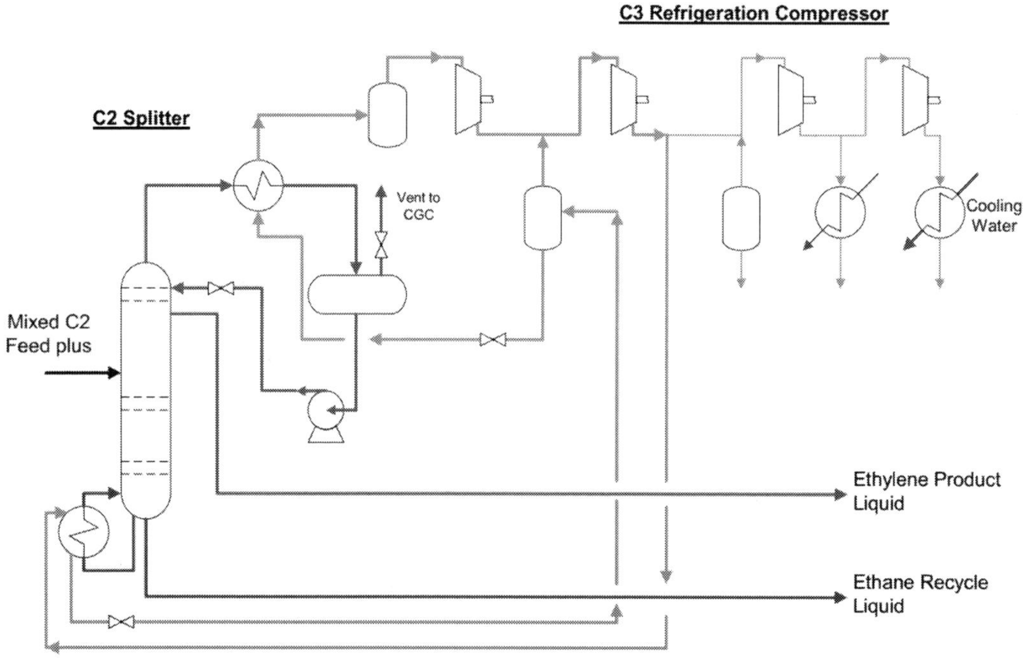

Figure 2: Indirect Heat-Pumped C2 Fractionator

One of the issues with the high-pressure design is that it requires a higher reflux ratio than those at lower pressure. This in turn translates to higher propylene refrigerant flows. Once a plant of almost double the ethylene capacity is considered the propylene flows become substantial. This imposes a very large suction volume on the propylene refrigeration compressor. In order to keep the volumetric flow within reason the compressor suction pressure needs to be increased which, in turn, typically leads to a higher operating pressure for the C2 fractionator. The alternative would be to make a significant increase to the fractionator condenser heat exchange surface. A higher tower pressure makes separation more difficult and this results in a demand for more reflux and a potential increase in tower diameter. This increases the propylene refrigeration load, which leads to an increase in compressor suction flow. This 'spiral' of increases can be mitigated by increasing the number of separation stages within the C2 fractionator resulting in a taller tower with more trays. Tower diameter increases can be mitigated by the use of high capacity trays in the tower design.

The direct heat pumped C2 fractionator has not been used as commonly in the past. However, it appears to have gained some favour in newer plants. Some of the things that may have driven this are newer heat integration schemes, improvements in front-end acetylene reactor catalyst and the adoption of dry-gas seal compressor technology. Prior to the last item there was a concern about contamination of the ethylene product with compressor seal oil. An example of this type of scheme is illustrated Figure 3.

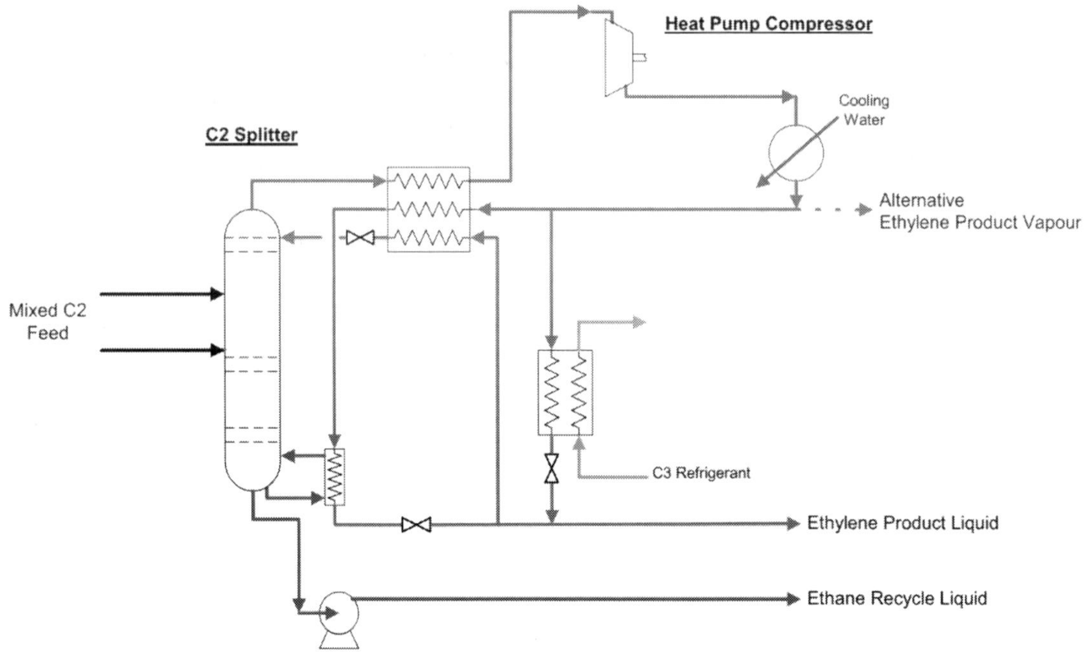

Figure 3: Direct Heat-Pumped C2 Fractionator

The use of a heat pump compressor essentially removes most, if not all, of the C2 fractionator refrigeration requirement from the propylene refrigeration compressor. If this load is incorporated into the ethylene refrigeration system the overall result is two refrigeration compressors (C2R/HP and C3R) that are of relatively equal size. This contrasts with the configuration for the propylene-refrigerated high pressure C2 fractionator system where the propylene compressor is considerably larger than the ethylene compressor. Operation of the C2 fractionator at the lower pressure results in a lower reflux requirement and a shorter tower because less separation stages are required. This mostly offsets the tendency for relative tower diameter increase due to the vapour density decrease at the lower pressure.

While the foregoing comparisons may be well known and understood, more significance may apply to some of the differences when a very large plant capacity is being considered. For example, if a taller tower is to be used to offset the impact of the high-pressure design on the propylene compressor, then the C2 fractionator may have to be fabricated in multiple sections in the field. In the case of E3, a medium-pressure C2 tower was fabricated, transported and erected as a single piece (Figure 4). This also allowed quality fabrication practices such as shop installation of the internal reboiler and the distillation trays.

It should also be considered that at some point the practical frame size limit might be reached for the propylene refrigeration compressor. Although larger frames may be available, this could cause a step change in cost. Similarly, turbine power may also become an issue.

Figure 4: C2 Fractionator Erection

Energy Integration Analysis

It is useful to have some way of benchmarking and comparing the energy use of certain flow schemes such as the heat pumps together with other process flowsheet issues and with the integration of energy in general. The approach to the E3 design involved setting some definite targets for energy performance in the cold section of the plant in terms of total refrigeration power requirements. These targets were derived from industry surveys and some knowledge of claimed state-of-the-art designs.

NOVA has found that the use of pinch technology analysis techniques can be a good way of understanding the relative benefits of various types of energy integration. These techniques can be used to define what the minimum energy use might be for a given flow scheme and thus can provide another benchmark target for the process design.

Pinch typically analyzes the heat integration opportunities for a given flowsheet. In the case of the E3 plant several flow schemes were considered and the pinch techniques were used to analyze the energy flows within the most promising ones. In some cases this analysis led to suggestions for modifications to the process scheme or to the operating conditions such as pressure.

One analysis technique that was used was the concept of distillation column targeting. When this was applied to the C2 fractionator heat pump system the effectiveness of a potential side reboiler could be clearly illustrated in terms of its impact on column exergy losses. Similarly, the possibility of multiple feeds to the tower could also be analyzed. This type of analysis is illustrated in Figure 5 below.

The pinch technique was also used to compare the benefits of a low pressure C2 heat pump that was integrated with the ethylene refrigeration system against those of a medium pressure standalone C2 heat pump and separate C2 refrigeration system. It was able to illustrate that the minimum work for the standalone system was 90% of the minimum work for the integrated system.

Application of the benchmarking techniques resulted in a cost-effective low energy design that greatly assisted the economies of scale of the mega-cracker without requiring any step-outs from proven equipment design and sizing.

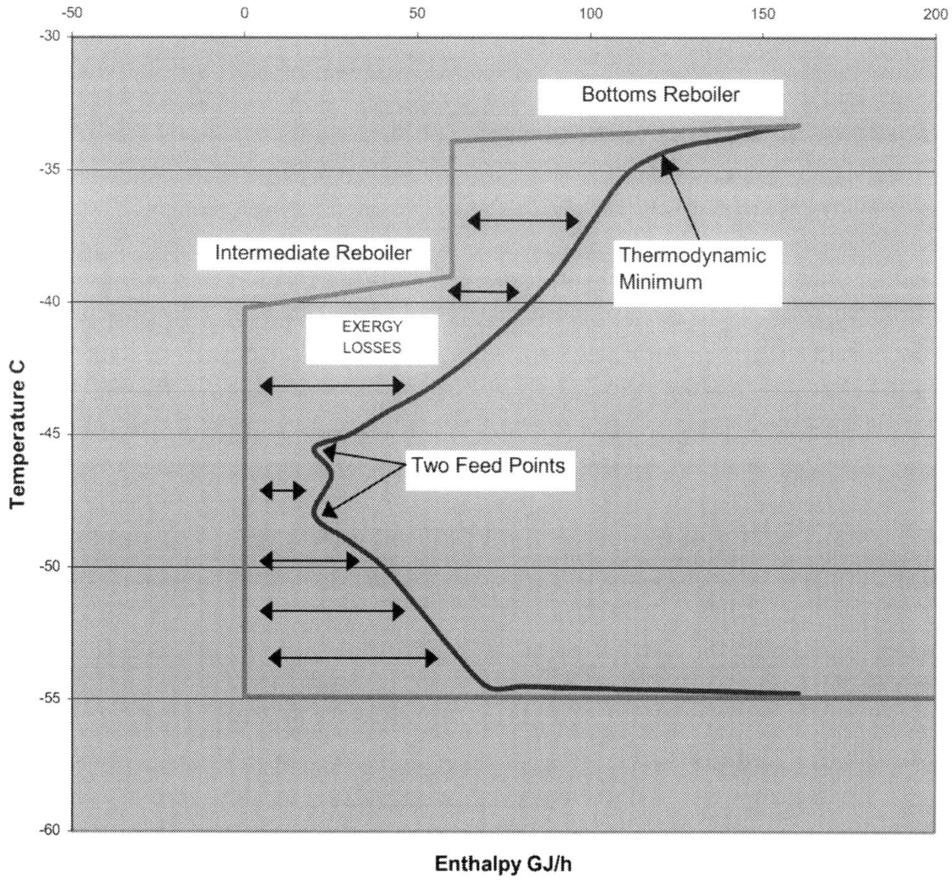

Figure 5: C2 Column Targeting

Energy Integration Impacts

One impact of a design that has a high degree of energy integration is that there are typically extensive process interactions. This can make the plant vulnerable to exporting even minor local process upsets across several process systems if not the whole plant. Heat balances are typically very tight and minor fluctuations can easily be amplified. A simple event such as a change in plant operating rate can lead to large disturbances if it isn't managed properly. Modern advanced process control systems can help to mitigate these issues by the use of energy balance management and feed-forward rate change applications. When these applications are not available the plant operators are left to manage some complex process interactions.

Dynamic modeling tools can be used to understand the timing of process interactions that can result from more integrated designs. These can be particularly effective if they are used to create process operator training simulators. The E3 project scope included the development of a dynamic training simulator and this certainly helped the operators to gain process understanding and to produce appropriate operating procedures to handle various abnormal events.

Process Equipment and Intensification

Brazed-Aluminum Platefin Heat Exchangers

Platefin heat exchange equipment has been used for many years within the ethylene industry. This type of equipment has a very high heat transfer surface to volume ratio, which lends itself to designing for the very small temperature differences that are desirable in order to minimize expensive refrigeration energy. Some of these heat exchanger applications would be very large if they were performed using shell and tube type equipment.

Perhaps the most traditional ethylene plant platefin exchanger application has been the use for cold gas heat exchange with incoming cracked-gas in the ethylene plant chilling train. These units typically involve multiple passes of multiple streams in several cores that are housed in an insulated cold-box. Some of the streams can include refrigerants that are circulated by thermosiphon. Pressure drop is a significant consideration with this equipment particularly as it can be susceptible to plugging.

While the platefin units are extremely compact, larger plants and shipping constraints can still cause size issues such as the need for parallel cold boxes. These might be avoided by carefully examining which units are driving the size of the cold box. One example that was observed during the E3 design was that the duty of one of the C2-refrigerated sections was determining the need for two parallel boxes. This heat load was placed in a separate core-in-kettle exchanger that was external to the cold box cores and this allowed a single box design to be maintained.

The core-in-kettle concept typically involves placing one or two platefin cores within a horizontal shell. Refrigerant or process liquid is vaporized on the shell side in order to cool or condense whatever streams flow through the other core passages. The core-in-kettle approach can be of great benefit for large plant designs. These units typically achieve duties that would require several very large shell and tube type kettles. The E3 design made use of platefins for large refrigerant loads in the chilling train as well as for the demethanizer and C2 fractionator reboiler services.

Figure 6: Core-in Kettle

Despite the great heat transfer benefits, there are several design and operating issues that are associated with the use of brazed aluminum heat exchange equipment. Pluggage is almost always a concern. There are two principal sources for this. The first is due to dirt and debris from the construction process and the second is due hydrate formation if moisture is present in a cold system. Construction debris can be removed by velocity blowing the piping and this is a generally accepted practice. One of the specific construction issues is the use of paper enclosures (or dams) to hold gas for inert gas welding of the piping. Unless this material is removed it will migrate to the inlet faces of any platefin cores and can result in significant pluggage. Physical removal can be achieved as fabrication proceeds but there will always be residual material at the closing welds. Rice paper is often used on the grounds that it can be dissolved in water during hydrotesting or subsequent line flushing. Some people have suggested avoiding the use of dams by inerting whole sections of piping. It is prudent to provide a strainer upstream of any platefin cores otherwise the cores themselves will act as filters for any solid material. The E3 installations included strainers upstream of the platefin cores (Figure 7 below).

Hydrate formation is always a risk and may arise from moisture that is present during initial commissioning or may occur due to operating events such as dryer breakthrough. It is advisable to apply some thought about the potential impacts of hydrate pluggages during the design phase and to develop some concepts for removing restrictions. Injection points for methanol or propanol should be identified and the possibility of various types of warm-up procedures should be considered.

Figure 7: Example of a Wedge-Wire Strainer – used upstream of platefin cores

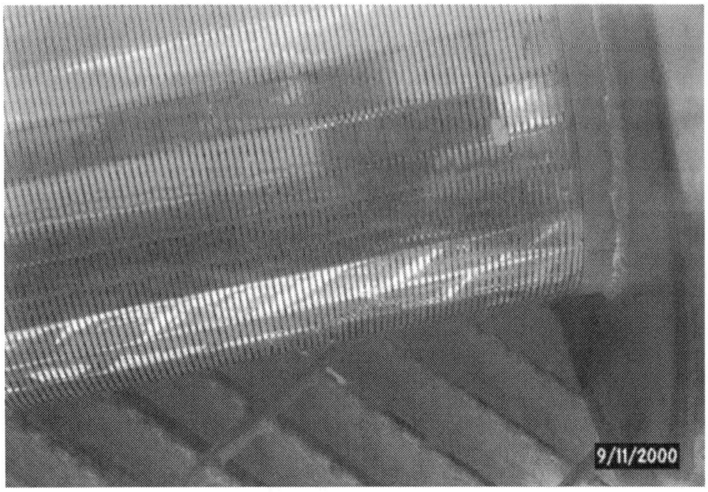

Enhanced Shell & Tube Heat Exchangers

Hi-flux heat exchangers have been used by the ethylene industry for more than 25 years. NOVA was a relatively early adopter of this technology in the early 1980's. The porous hi-flux coating provides a large number of heated sites for nucleate boiling to occur. This promotion of the nucleate boiling regime enhances the boiling heat transfer coefficient. The promotion can occur at relatively small temperature differences, which makes hi-flux particularly applicable to some of the large refrigeration duties in world-scale ethylene plants. Application of this enhancement allowed NOVA to service the C2 splitter condensing requirements of a 1980's world-scale ethylene plant with two kettle units where a conventional bare tube application would have required three.

Hi-flux technology can also incorporate a condensing heat transfer enhancement by increasing condensate adherence to the outside of the exchanger tubes. This is achieved by incorporating a fluted profile on the exterior diameter of the tube. NOVA has applied this technology to the outside of C2 splitter reboiler tubes in addition to using the nucleate boiling coating on the inside. This allowed the reboiler duty of the same 1980's plant to be accomplished in a single unit where two bare tube units had originally been planned.

Since the more recent introduction of core-in-kettle technology, hi-flux has perhaps lost some ground in some services, such as the C2 splitter condenser, where it had been predominant. For example, the condenser application that is described above could probably have been accomplished with a single core-in-kettle in unit. However, when there are concerns over contaminants, fouling and pluggage, hi-flux may still be the preferred choice. The E3 plant utilizes a single hi-flux heat exchanger unit to vaporize all of its incoming ethane feed from the supply pipeline. This was chosen because of concerns about pipeline contaminants such as compressor oil and water. It has proven to be a good choice because the control valves at this exchanger frequently show signs of pluggage. A downstream ethane superheat exchanger, which is a platefin, also shows signs of heat transfer fouling, whereas the hi-flux vaporizer continues to perform well. This is another case where the requirements for the mega-plant were met by a single enhanced unit, in this case a hi-flux, rather than the multiple standard bare-tube units that would otherwise have been required.

Compressors and Expanders

The efficiency of the refrigeration compression equipment has improved significantly from 70-82% in the 1970's and 1980's to 80-86% in the 1990's[1]. This has assisted in reducing overall refrigeration energy requirements and in achieving greater capacities with the same equipment. The adoption of dry-gas seals in refrigeration compressors has been a significant advantage in eliminating the fouling of heat exchange equipment and product contamination in direct heat pump designs.

Reliable performance of the turbo-expander equipment can be important to the refrigeration capacity and efficiency of mega-plant operation.
Turbo-expander equipment with oil-lubricated bearings has caused some operating difficulties. Migration of oil into the process has led to fouling of the platefin heat exchange equipment. Once this occurs it can be especially difficult to remove because of the extremely low temperatures (-140°C to -160°C) that are involved. The lubricating oil system provides additional complexity to the startup and operation of these units. Newer designs should consider the use of magnetic bearing technology as a means of eliminating the need for oil. NOVA first implemented this into a new turbo-expander as part of a plant revamp design in 1993. There were some initial teething troubles with the reliability of the shaft positional sensors for that particular application but the issues had been remedied by the time the E3 design was being developed. One consideration for the design of magnetic bearing expanders is the need for tight gas shutoff. Even small amounts of leakage past the isolation valves will cause the expander to continue turning because there is no friction when the shaft is in a levitated position.

Process Intensification

There are a number of process equipment intensifications that can be considered. These can include simple concepts such as stacking drums or incorporating them into tower shells. They can also include performing multiple functions within the same equipment envelope.

One example of equipment intensification is to make use of the high surface to volume ratio of platefin cores by incorporating them directly within distillation towers to provide the reboiler and condenser services. This not only effectively removes the footprint of this equipment from the plot but also reduces the amount of piping and valving that is required. The E3 design showed a significant capital cost saving when the C2 fractionator reboiler was incorporated into the sump of the tower. The initial design called for two external core-in-kettle units. The revised design eliminated the need for the kettles and all of the piping on the vaporization side of this service. It also opened up the plot around the base of the C2 fractionator.

A similar opportunity can be realized in the demethanizer (Figure 8). In the case of this tower the overhead condenser was integrated directly into the tower envelope by stacking a core-in-kettle arrangement on top of the tower. This type of arrangement has been practiced in the past at smaller duties with a vertical tube-type reflux condenser[2]. It enables the elimination of a separate reflux drum and two very low temperature pumps. The absence of this equipment and the reboiler opens up the plot area around the demethanizer.

Figure 8: Demethanizer Process Intensification

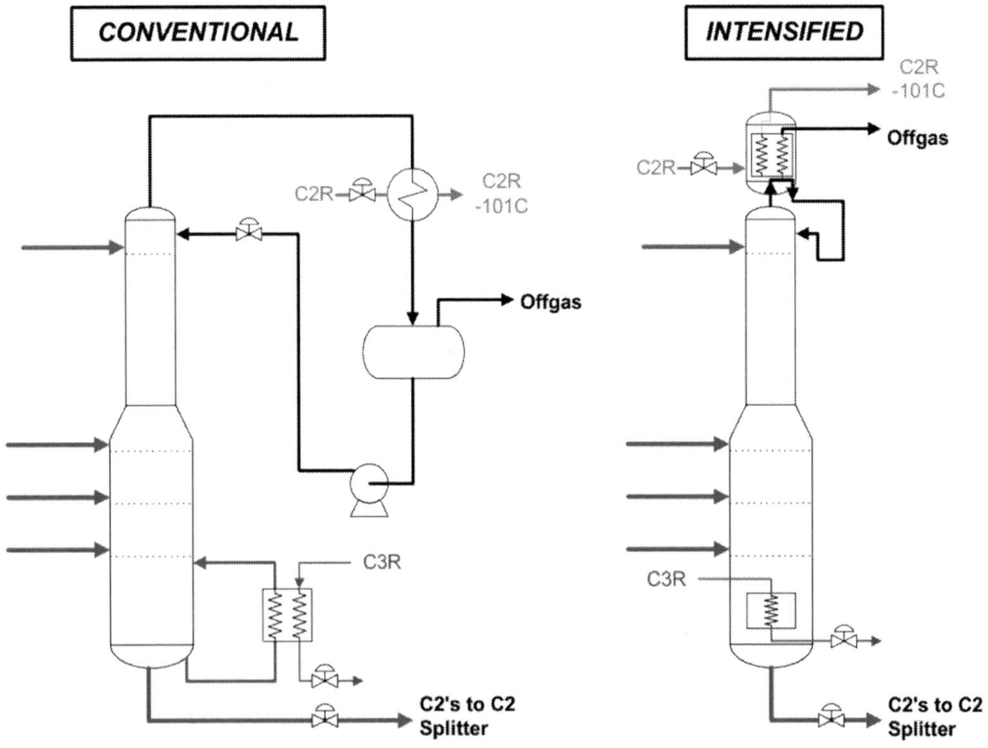

OPERATING EXPERIENCES

Equipment

As described previously, there are some specific issues with the design and operation of platefin heat exchanger equipment. NOVA has experienced problems related to some of these issues during both commissioning and early operations. There have also been occasional performance problems with the equipment during full commercial operation.

It is well understood that the cryogenic section of the plant needs to be dried adequately prior to chilling it down to operating temperatures. In the case of commissioning the E3 plant the dryout process was extensive and was planned in great detail. Nevertheless it became evident during commissioning that pockets of water had been left in piping and equipment. The first example of this was the pluggage of a platefin core exchanger in the C2 heat pump compressor circuit. Several unsuccessful attempts were made to clear this restriction by soaking the core with fairly large quantities of methanol. In the end it was decided that a warm-up process should be used to melt the restriction. This was achieved by running the compressor in a warm-gas recycle mode through the unrestricted passages of the core. The heat generated by operating in this mode was sufficient to thaw the restricted gas passages. Once the heat pump and C2 fractionator were brought to normal operation the next challenge was to decontaminate the methanol from the circuit. It took several hours before the methanol content of the product ethylene was reduced sufficiently that it could be sent to the product pipeline.

Another example of both moisture and debris causing restrictions to operations came during early operation of the plant. It was manifested as an inability to provide sufficient chilling from ethylene refrigeration. Troubleshooting indicated insufficient refrigerant thermosiphon circulations through platefin core passages of the main cold-box. Pluggage of the strainers upstream of these cores was suspected and indeed, a considerable amount of residual debris from welding dams was found when the strainers were opened. However, even with the strainers in clean condition it was not possible to establish satisfactory thermosiphon flowrates. In the end the system was opened at the refrigerant supply drums and the piping was inspected using a robot camera. Video from the camera showed appreciable amounts of snow coating the inside of the pipes and in some cases this was breaking off in relatively large chunks. In one case the camera was able to crawl right up to the inlet of the platefin and it was clearly evident that most of the flow passages were iced over. This was another platefin restriction that proved almost impossible to clear with methanol or propanol. It was decided to use another warm-up procedure to de-ice the core. This involved bringing in warm cracked-gas from the nearby E2 plant through a start-up inventory line. Once this gas was circulated through the cracked-gas passages of the cores the pluggages in the refrigerant passages melted and the water was removed.

Figure 9: Example of Welding-Dam Material Trapped by Strainer

Figure 10: Robot Camera View of C2 Refrigerant Platefin Inlets

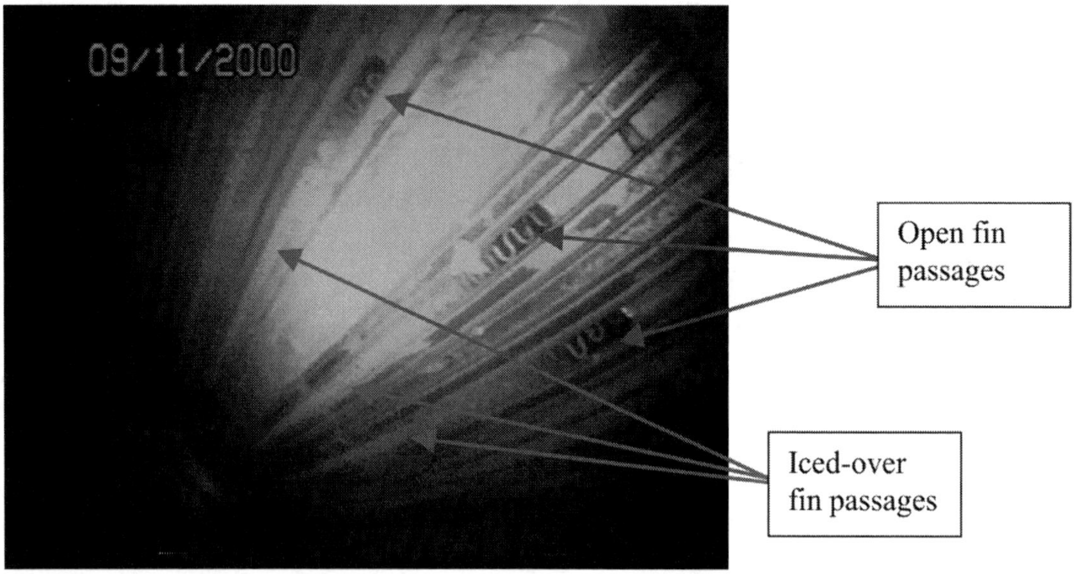

Further into the early operation of the plant another type of restriction became evident. There was a clear limit to how much cooling was being achieved on two core-in-kettle exchanger units. Initial troubleshooting around the valves and piping seemed to indicate that the restriction was within the units themselves. As the restriction was on the kettle side of these exchangers it didn't seem as likely that it would be due to a build-up of hydrates or construction debris. Eventually, it was noticed during a review of the fabrication drawings that there were some substantial impingement plates that ran the whole length of the kettle vapour space. These were intended to minimize any entrainment from the turbulent boiling liquid circulation as it exited the core and the vapour disengaged from the liquid upstream of the outlet nozzle. The impingement plate was manufactured and installed in sections with each plate being perforated to allow the vapor to exit toward the outlet nozzle. A misinterpretation of the fabrication drawings meant that the perforated area had been under-specified. The result was a substantial pressure drop across the impingement plates and as more load was put on the exchangers some of the plates buckled in order to allow vapour to escape to the outlet nozzle. This can clearly be seen in the following photographs.

Figure 11: Core-in-Kettle Impingement Plates

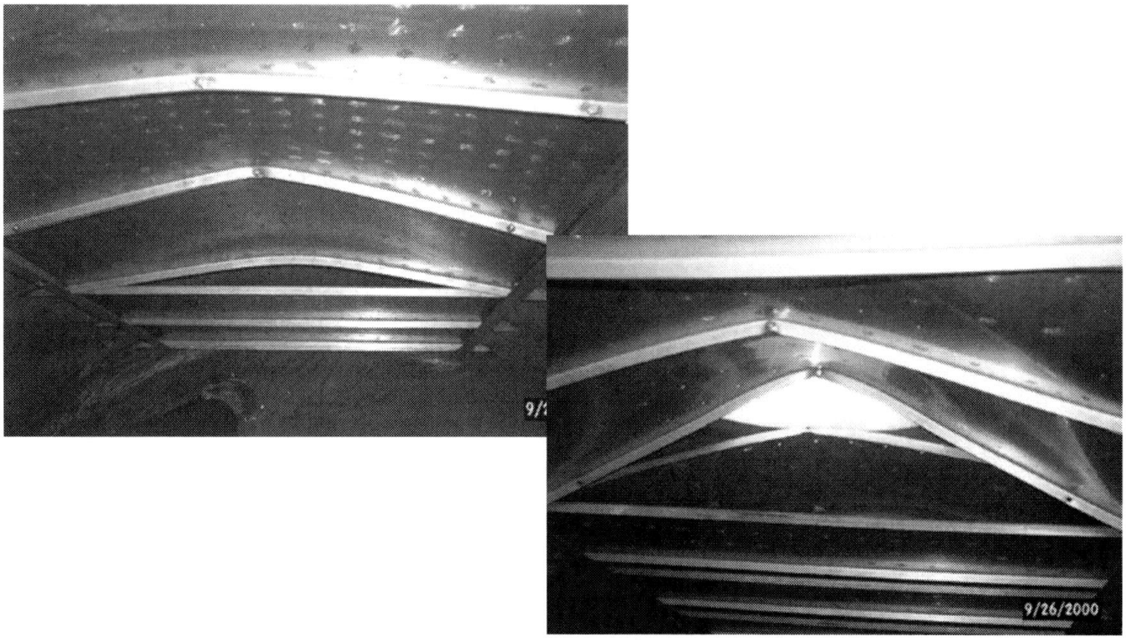

As described earlier, platefin exchangers are usually designed for relatively small temperature differences. Differences due to operating excursions or start-up and shutdown can sometimes be significant. The E3 plant experienced a failure of some passages on one platefin core application[3]. These failures were splits at bar-to-sheet brazed joints and they resulted in gas leaks to atmosphere. The failure occurred during some rapid pressure cycling where transient pressure spikes almost certainly exceeded the design of the equipment. However, one of the conclusions about the failure was that the failed joints had probably been previously weakened by cycles of temperature differences that were beyond the ALPEMA[4] recommended guidelines. Two key recommendations are that the temperature differential should not exceed 30 °C prior to flow being established and that the rate of temperature change should be limited to 2 °C per minute after flow is established. Both of these conditions had probably been exceeded during several plant startups prior to the failure incident. There are other similar units in the E3 plant where the temperature differential issue was recognized and the original piping design and operating procedures had been adapted in order to cope with it.

Performance of the turbo-expanders has been exceptionally good. They can readily be started by simply levitating the shafts magnetically and admitting gas to the process side. The older lubricated bearing expanders could be very difficult to start and the startup often resulted in lubricating oil entering the cold-box. One factor which can be an issue for this equipment is achieving adequate shut-off and isolation of the process gas. Even relatively small amounts of gas will cause the expanders to turn when the shaft is levitated because there is no frictional resistance.

Integration

Some significant negative process interactions were observed during the early operation of the E3 plant. One example of this was due to the very tight heat balance on the C2 heat pump feed system. This meant that the tower did not respond well to changes in plant rate. The resulting variations in feed enthalpy often resulted in cycles of heating-up and cooling-down of the tower over a period of several hours. Other examples were interactions between the recovery of refrigeration from the feed and product streams with operation of the refrigeration systems.

These interactions did lead to plant upsets and some instances of flaring of off-specification product. The implementation of some advanced process control schemes has helped to mitigate these issues and later operation proved much more stable.

Intensification

Operation of the reboilers within the demethanizer and C2 splitter towers has been satisfactory. Care has to be taken to ensure an adequate venting path for these units in order purge any non-condensable gases.

The integrated demethanizer condenser has also operated satisfactorily. One issue with this system is measurement of the reflux flow. The reflux pipe exits the condenser and runs outside before entering the tower. A flow measurement orifice was placed in this external section of piping. The reflux is a saturated liquid and there is relatively little head available to avoid flashing as the pressure drops across the flow orifice. A very wide range of flow measurements is observed particularly during startup of the system. Knowledge of this flow has not proven to be of any significant importance as performance is controlled by refrigerant flow to the condenser and online analysis of the tower overhead vent composition.

CONCLUSIONS

The E3 plant has been in operation for four years. It has demonstrated that a mega-plant ethylene capacity can be achieved with a low energy requirement and without the need for significantly larger equipment in the cold section of the plant. Design of the plant involved some thorough investigation of the process and refrigeration systems in terms of the process integration and intensification opportunities. The resulting reductions in equipment size and count were used to improve plot layout and this provided better construction and maintenance accessibility.

The approaches to large plant design can come with some operational risks to process stability and fouling. Although these are not necessarily new issues, the impacts can be greater for larger plants because of the quantity of material that may have to be flared when problems occur.

There is a continuing trend toward building larger ethylene plants. This will no doubt result in traditional equipment size boundaries being pushed further. The success of the E3 design suggests that a coupling of the increases in basic equipment capability with suitable integration techniques could lead to even larger plant capacities in the future. Ultimate limitations are likely to occur in areas of the plant outside of the cold section.

REFERENCES

1. Evolution of Ethylene Plant Compression Systems – Paper 85a
   S. Kapur; L.Schaider; H. Weyerman
   2000 AIChE Ethylene Producers Conference

2. Petrochemical Processes 1 Synthesis Gas Derivatives & Major Hydrocarbons
   A. Chauvel; G. Lefebvre
   Gulf Publishing 1989
   Page 151; Figure 2.19

3. Software Upgrade Triggers Events that Lead to Plant Shutdown – Paper 62d
   D. Gent
   2004 AIChE Ethylene Producers Conference

4. The Standards of the Brazed Aluminum Plate-Fin Heat Exchanger Manufacturers Association (ALPEMA) Second Edition 2000

SESSION 44

ROUNDTABLE DISCUSSION

ETHYLENE PLANT TECHNOLOGY – REFRIGERATION SYSTEMS

PRESENTED BY THE TECHNOLOGY SUBCOMMITTEE

The session consisted of a roundtable discussion covering many aspects of refrigeration technology. No papers were presented, however, the authors from Sessions 42 and 44, authorities in the fields of refrigeration equipment design, engineering and operation, served as panelists for the Roundtable.

The panel consisted of the following individuals:

- Thomas Pickett – Barnes and Click, Inc.
- William Hohlweg – Elliott Turbomachinery Co., Inc.
- Joseph Lillard – Mafi-Trench Corporation
- Qi Ma – ABB Lummus Global
- Julio Rios – Kellogg Brown and Root
- Dan Markussen – Chart Industries
- Jack Blanchard – CB&I
- David Gent – NOVA Chemicals Corporation

45a IEC 61511 and the Capital Project Process

Angela Summers, Mark Moderski

This year the process industry has reached an important milestone in process safety --- the acceptance of an internationally recognized standard for safety instrumented systems (SIS). This standard, IEC 61511, documents good engineering practice for the assessment, design, operation, maintenance, and management of SISs.

The foundation of the standard is established by several requirements in Part 1 Clauses 5 through 7, which cover the development of a management system aimed at ensuring that functional safety is achieved. The management system includes a quality assurance process for the entire SIS lifecycle, requiring the development of procedures, identification of resources, and acquisition of tools. For maximum benefit, the deliverables and quality control checks required by the standard should be integrated into the capital project process, addressing safety, environmental, plant productivity, and asset protection.

Industry has become inundated with a multitude of programs focusing on safety, quality, and cost performance. This paper introduces a protective management system, which builds upon the work process identified in IEC 61511. Typical capital project phases are integrated with the management system to yield one comprehensive program to efficiently manage process risk. Finally, the paper highlights areas where internal practices or guidelines should be developed to improve program performance and cost effectiveness.

45b Safety Instrumented Shutdown System Design Strategy for an Acetylene Hydrogenation Reactor System at Eastman Chemical Company

Kevin H McGuire, Mark Moderski

In 2004, Eastman Chemical Company installed a raw gas acetylene hydrogenation reactor system employing a Pd-based catalyst in one of their Longview, TX ethylene facilities. To prevent and protect against possible temperature excursions (runaway reactions), a safety instrumented shutdown system (SIS) that met ISA S-84 standards was designed and installed. This paper will provide an overview of the reactor system and the associated "shutdown triggers", the SIS design process including hazard reviews, instrumentation selection, and functional testing strategies, and lessons learned since start up of the new unit.

45c Dow St. Charles Use of SIS for Excess Pressure Management

Jim Verboon, Mark Moderski

Dow's St. Charles Olefins 1 unit was built in the early 1970's. It was originally built as part of the Union Carbide Corporation which merged with Dow in 2001.

Recently the plant has begun a project to make a number of improvements in the flare system and this project prompted an examination of excess pressure management capabilities. In several cases it was found that either pressure valve relief capacity was inadequate for worst case scenarios, or that the combined relief requirement during a plant wide loss of cooling would exceed flare capacity. As a solution to these deficiencies Dow will utilize SIS protection in lieu of additional safety valve capacity, or additional flare size capacity. SIS protection will be installed on five different distillation towers in order to achieve this protection.

This presentation will discuss the work process and requirements that Dow employes when using SIS for excess pressure management. The presentation will also discuss the specific solutions that will be installed in the Olefins 1 plant.

Global Ethylene Plant Performance Comparisons

Claire L. Cagnolatti
Manager of Petrochemicals Operations
Solomon Associates
Dallas, Texas 75240
claire.cagnolatti@SolomonOnline.com

Prepared for Presentation at the 2005 AIChE Spring National Meeting
17th Annual Ethylene Producers Conference
April 11-14, 2005
Atlanta, GA

Global Ethylene Plant Performance Comparisons

Major Trends in the Ethylene Industry

Solomon Associates (Solomon) is celebrating 25 years of competitive performance analyses (benchmarking). Solomon has conducted studies of fuels refineries since 1980 and olefin plants since 1987. Solomon Associates recently completed our tenth *Worldwide Olefin Plant Performance Analysis (Olefin Study)*, for operating year 2003. Currently, *Olefin Studies* are performed on a 2-year cycle, conducting the data analysis for the odd years' performance during the even years.

The *Worldwide Olefin Plant Performance Analysis for Operating Year 2003 (2003 Olefin Study)* was the largest and most comprehensive Olefin Study ever conducted. Total participation was 115 olefin plants, representing 70% of the ethylene-producing capacity in the world.[1] All major ethylene-producing regions of the world were represented in this study, including 68% of the Africa/Middle East ethylene capacity and 46% of Asian ethylene capacity.

Although actual numerical study results cannot be shared with the general public, there are some interesting study results that can be examined comparatively to show trends and peer group comparisons. This analysis will address economic trends, as well as other areas of general interest to the industry, such as personnel, energy, yields, maintenance/reliability, turnarounds, emissions, and capital expenditures. Of special interest is the impact of ever-increasing energy costs on the petrochemicals industry.

<u>Development of Trend Groups</u>

In reviewing study results over a period of years, we must eliminate the influence of changes in the study participation on those study-to-study results. To that end, we created a Trend Group of 50 olefin plants that have participated in each of the past 3 Olefin Studies (for operating years 1999, 2001, and 2003). Half of these Trend Group plants are from Europe, another 14 are from North America, and the rest are from other regions. Analyzing this Trend Group shows us measurable changes and trends in the industry using a consistent plant population.

Trend Group results from the *2003 Olefin Study* revealed that, as expected, the typical Trend Group plant in 2003 was larger than in 1999 by nearly 7%. Despite the passage of four years over the trend period, the average furnace technology actually advanced by three years. Ethylene capacity utilization for 2003 was up by 1.4 percentage points over 2001 levels, but was still 3.3 percentage points below the 1999 levels.

<u>Personnel Trends</u>

Productivity in the olefins industry continued to improve, with 2003 staffing levels in work hours per unit of output showing a reduction of just over 6% compared to 1999 levels. Some of this is because of increased capacity utilization as reported above, but the remainder is due to productivity improvements. As always, Solomon urges the **cautious use of personnel and productivity targets**, as it is very easy to incorrectly apply this data. Personnel requirements are plant-specific and require a detailed analysis of skills and experience levels to determine the number of work hours required to safely, reliably, and responsibly operate the highly technical equipment in an olefin plant.

---

[1] World Capacity excluding former Soviet Union.

Energy Trends

The Trend Group also showed better energy efficiency in 2003, with the Energy Efficiency Index (EEI) improving by 3 percentage points from 2001. This index measures the actual energy consumption level compared to a standard energy consumption calculated for each plant. Energy consumption per unit of production decreased, by just over 2%.

Rising energy costs continue to be a global concern for the industry, but North American olefin plants have suffered a particularly severe blow to their competitiveness as a direct result of energy cost increases since 2000. In our 1999 Olefin Study, North American olefin plants were competitive with their European and Asian counterparts with respect to energy costs per ton of production, as illustrated in Figure 1. However, energy consumption per ton of high-value chemicals (HVC) is higher in North America than in Europe, Asia or Latin America. As a percent of the 2003 Olefin Study average energy consumption, North American consumption was 111% of average, Europe was 95%, and Asia was 86%. Prior to 2000, this higher energy consumption was offset by a much lower energy cost per Btu, making North American plants' average energy cost exactly the same as Europe or Asia. Figure 1 shows that North American plants' net energy costs per ton of production have more than doubled since 1999, while those of Europe and Asia have increased by only 50%.

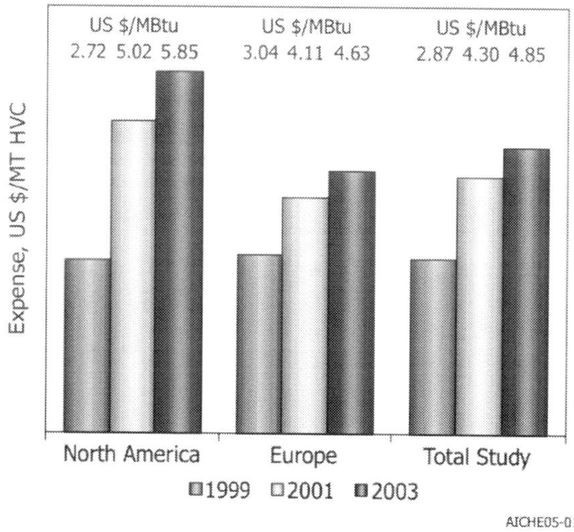

Figure 1. Energy Expense Trend

One could argue that the predominance of light feed plants in North America is the reason for their higher energy consumption. In the *2003 Olefin Study*, energy consumption analysis by feed type shows that the average ethane cracker in the study consumed 124% of the study average energy consumption, where the average naphtha cracker only consumed 95% of the study average. However, even after allowing for feed type differences, average energy consumption rates are significantly higher for the North American region, as shown in Figure 2.

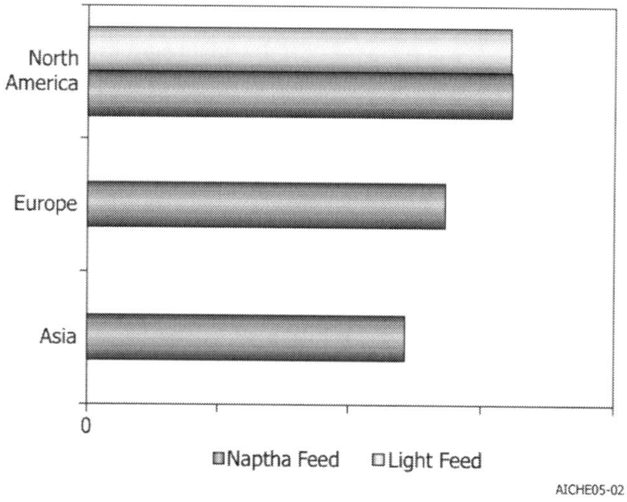

Figure 2. 2003 Energy Consumption per Unit of Production

Yield Trends

Yield performance is measured as the actual value of pyrolysis products as a percent of the benchmark yields as calculated by the PYPS[2] model. Yield performance improved by one percentage point from 1999 to 2003. This means that the actual yields as reported by participants in the 2003 Olefin Study are one percentage point closer to the benchmark yields than in 1999. This trend is certainly expected in light of the newer average furnace age (see page 2), but our studies have determined that yield performance is affected by many factors, including plant limits, degree of furnace control, product losses, and commercial issues surrounding co-products.

---

[2] Pyrolysis Yield Prediction System, an ABB Lummus Global product

Maintenance and Reliability Trends

Lost production due to plant-related causes increased by one percentage point of ethylene capacity from 1999 to 2003. Lost production due to non-operational causes—such as lack of demand, feedstock interruptions, alternate severity targets, or capital projects—more than doubled over that same period, with the largest increase occurring between 1999 and 2001.

Turnaround intervals continue to improve in the Trend Group, increasing by half a year between 1999 and 2003. This was accomplished while reducing the turnaround duration by about 2 days. Turnaround cost as a percent of replacement asset value dropped very slightly, by 0.1 percentage points over that same 4-year period, while the number of work hours attributed to the turnaround increased by nearly 9%. However, it should be noted that 2003 replacement values for plants outside the U.S. were affected by the weak U.S. dollar and were somewhat higher than in 1999 and 2001.

The number of plants that cited capital projects as the principle factor that set the timing for the most recent turnaround in 2003 was half of the number reported in 1999. The principle factor showing the highest increase over that same interval was the need for a regularly scheduled turnaround. The number of plants citing regulatory requirements or deterioration of plant performance as the main reason for their most recent turnaround remained the same from 1999 to 2003.

Plants in our Trend Group are reporting less capacity loss between turnarounds. Average ethylene plus propylene capacity loss dropped by 0.3 percentage points from 1999 to 2003.

NO_x Emissions Reduction

All three major regions represented in the Trend Group (North America, Europe, and Asia) showed a slight reduction in NO_x emissions in pyrolysis stack flue gas, reported as pounds of NO_x per million Btu fired. Europe showed a small decrease from 1999 to 2001, while North America and Asia showed steady reductions over the three studies. At least some of the North American plants' reductions were no doubt fostered by the Houston-Galveston Area NO_x Non-Attainment regulations.

Trends in Plant Facilities

From 1999 to 2003, the average megawatts of electricity generating capacity reported by plants that included cogeneration facilities as part of their olefin plant increased by one third. The average number of cogeneration units per plant, however, did not increase.

Capital investment is represented in our studies as a percent of replacement value. Total historical (past 5 years) capital investment reported by the Trend Group has steadily decreased from 1999 to 2003, and total capital spending over the next 5 years also showed a marked decrease. Figure 3 reveals that the drop is much more severe for North America.

For the total Trend Group, capital spent over the last 5 years for new plant capacity addition dropped by 84% from 1999 to 2003, and planned capital over the next 5 years for capacity expansion dropped 90% over the same interval. The representation of plants from Asia, Latin America, and Africa/Middle East in our Trend Group is too small to show trend data for those regions. However, in the

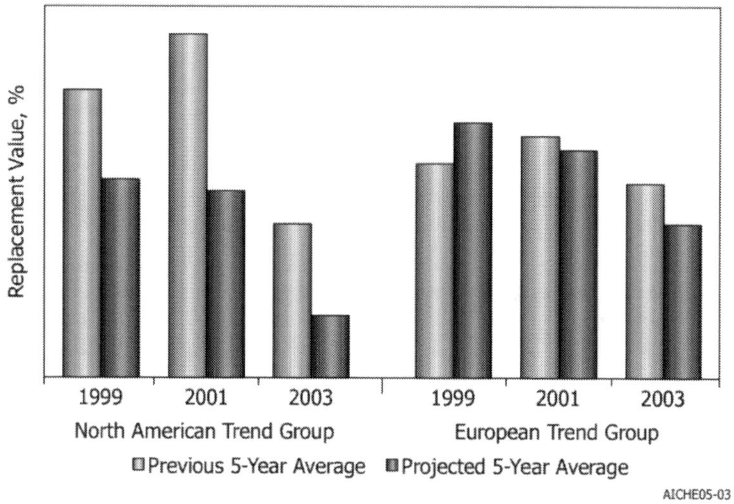

Figure 3. Historical and Future Capital Spending

2003 Olefin Study, a large percentage (78-96%) of the future capital spending reported for those regions is dedicated to capacity expansion, either as new plants or additions to existing plants. The highest figures for expansion capital as a percent of current plant replacement value were reported in Latin America, followed by Europe, and Africa/Middle East.

While historical capital spending for energy conservation increased 4.5 times and for regulatory and environmental compliance reasons increased 144% over this time period, planned future capital investments for these causes showed a decrease. Energy conservation spending for the next 5 years as reported in 2003 shows a 66% reduction compared to 1999, while predicted future regulatory and environmental capital spending over the same interval dropped 10%. One other category of capital spending showing a decrease over the 1999 to 2003 period is advanced process control and computers, which showed a decrease in both historical and predicted future spending.

Conclusions

Energy costs, energy consumption, and emissions levels are all related issues to the petrochemicals manufacturer, and all must be minimized to be economically competitive and to comply with regulatory requirements. Many European countries are already obligated by energy covenants to reduce energy usage, and the European Union is instituting emissions reduction targets for refineries and other industries that will be expanded to include the petrochemicals industry in the coming years. Kyoto Protocol targets call for energy use reductions, and specific regions of the United States are subject to emissions reductions in Ozone Non-Attainment Areas. With the high cost of energy and the large penalties for non-compliance with emissions regulations, energy use and its related issues will be high on the agenda for all participants in this industry.

Solomon Associates will continue to look for ways to assist our clients in their efforts toward performance improvement. Feedback from current and potential *Olefin Study* participants regarding issues to be included in our studies is always welcome.

Paper 47a

Modeling of Olefins Production

Prof. Gilbert F. Froment
Chemical Engineering Department
Texas A & M University

Prepared for Presentation at the 2005 Spring National Meeting in Atlanta, GA April 10-14
17th Annual Ethylene Producers Conference
Session 47 – Ethylene Plant Fundamentals –
Improvements in Modeling and Testing Reaction Systems and Furnace Performance

Modeling of Olefins Production

G.F.Froment
Chemical Engineering Department
Texas A & M University

The presentation deals with the modeling of olefins production processes along thermal as well as catalytic routes.
Predicting accurate and detailed process yields of the thermal route requires in the first place rate equations expressed in terms of the elementary steps of radical chemistry: initiation, radical isomerization and decomposition … and this leads to gigantic reaction networks that can only be generated by an efficient computer algorithm. The huge number of rate coefficients associated with the steps of the network is reduced to a relatively small set by means of thermodynamic constraints, the single event concept that accounts, within a given type of elementary step, for the effect of structure on the frequency factor and the Evans-Polanyi relationship that accounts for the structure effect on the activation energy.
With the rate equations inserted into the set of continuity-, energy - and pressure drop equations describing the process inside the cracking tubes, process aspects can be investigated provided a detailed heat transfer and flow model is available for the calculation of the temperature distribution in the furnace. With reliable kinetic equations for coke formation the run length and the effect of coking inhibitors can be predicted.
Catalytic processes on acid catalysts can be modeled along similar lines when expressed in terms of the elementary steps of carbocation chemistry. An example is given for MTO, the zeolite catalyzed methanol-to-olefins process. Multi-bed adiabatic and fluidized bed reactor configurations are investigated.

This paper was not available in time for publication. If you are interested in copy of this paper please contact: Prof. Gilbert F. Froment, Email: g.froment@che.tamu.edu

Paper 47b

Primary distribution products from the pyrolysis of cyclo-alkanes

Mario Dente, Tiziano Faravelli, Sauro Pierucci, Eliseo Ranzi

Professors of Chemical Engineering

CMIC, Politecnico di Milano, Italy

Simon Barendregt, Peter Valkenburg

Managers

Technip Benelux,
Zoetermeer,
The Netherlands

Prepared for Presentation at the 2005 Spring National Meeting in Atlanta, GA April 10-14
17th Annual Ethylene Producers Conference
Session 47 – Ethylene Plant Fundamentals –
Improvements in Modeling and Testing Reaction Systems and Furnace Performance

Primary distribution products from the pyrolysis of cyclo-alkanes

M. Dente, T. Faravelli, S. Pierucci, E. Ranzi
CMIC, Politecnico di Milano, Italy

S. Barendregt, P. Valkenburg
Technip Benelux, Zoetermeer,
The Netherlands

Abstract

Steam cracking of liquid feedstocks to produce light olefins is a process where the chemical complexity is dictated both by the characterization of the hydrocarbon mixture and by the complete definition of the kinetic mechanism. On the pyrolysis reactions of normal and branched alkanes, literature publication can be found. However, scarce attention the pyrolysis of cyclo-alkanes has received. Nevertheless, cyclo-alkanes, constituted both by alkyl cyclo-pentane and cyclo-hexane components, are always abundant in liquid feeds.
Aim of this paper is to present an example of a kinetic modeling in the area of characterization of primary distribution products from the pyrolysis of cyclo-alkanes. The complete results of the modeling effort for the steam pyrolysis process, here not reported, have been incorporated in the proprietary SPYRO program, largely used by ethylene companies. New reference kinetic parameters need to be singled out as well as simplifications to describe the pyrolysis of heavy liquid feedstocks need to be introduced. In fact, the complexity of the system becomes quite evident considering that the number of components in the primary decomposition path rapidly increases with the increasing of the molecular weight and becomes higher than 10^6 for carbon number larger than 16-17.

1.- Introduction

The modeling of steam cracking of liquid feedstocks to produce light olefins is a process where the chemical complexity is dictated both by the characterization of the hydrocarbon mixture and by the complete definition of the kinetic mechanism. On the pyrolysis reactions of normal and branched alkanes, literature publication can be found [1]. However, scarce attention the pyrolysis of cyclo-alkanes has received. Nevertheless, cyclo-alkanes, constituted both by alkyl cyclo-pentane and alkyl cyclo-hexane components, are always very abundant in liquid feeds. This paper discusses and present an example of the kinetic modeling addressed to the characterization of primary distribution products from the pyrolysis of cyclo-alkanes.
When treating cyclo-alkane components, the complexity of the overall reacting system greatly increases, due to the presence of new relevant reaction classes such as the ring opening and cyclo-addition reactions [2]. While the ring opening can take advantage of the reference kinetic values of the β-decomposition reactions of alkyl radicals, new reference kinetic parameters need to be singled out for the internal isomerization of cyclo-alkyl radicals as well as the cyclo-addition-reactions of alkenyl radicals. The result of the successive decomposition and cyclo addition reactions of these radicals gives rise to several different isomeric radicals. For instance, starting from cyclo-hexyl radical, different methyl-cyclopentyl as well as different normal and branched hexenyl radicals are obtained.

Furthermore, the pyrolysis of cyclo-alkanes is a chain radical process where also several molecular reactions attain a significant role. Namely, the molecular isomerizations between cyclo-alkanes and alkenes are typical examples of four and six center concerted reactions. The complexity of these reacting systems becomes higher due to the number of intermediate components in the primary decomposition paths rapidly increasing with the molecular weight. For this reason it was convenient to develop a specifically conceived computer code in order to evaluate the net result of these reaction mechanisms.
Finally, if the interest is the characterization of the ethylene yields in commercial crackers, the complexity of these reaction networks demands an experienced creative capability to simplify a detailed composition and the successive decomposition paths of the heavy liquid feedstocks (naphthas, kerosines and gasoils) with viable and reasonable computing efforts.

2.- Reaction mechanism and reference kinetic parameters.

Pyrolysis reactions proceed with a radical chain mechanism. The primary distribution products of the decomposition of cyclo-alkanes are the result of primary propagation reactions of radicals obtained via initiation and H-abstraction reactions. These subjects will be described in the next paragraphs and kinetic constants available in literature will be discussed. The amount of these parameters are only a limited portion of those required to completely describe the whole set of phenomena for the purpose of modeling the steam cracking process.

2.1 Initiation reactions of cyclo-alkanes

Heavy cyclo-alkanes are typically constituted by components with cyclo-pentane and cyclo-hexane rings with a partial degree of methylation and a more or less long alkyl side-chain. It is quite obvious to assume that the C-C cleavage in the side chain follows the same rules and applies the same reference kinetic parameters as the initiation reactions of normal and branched alkanes.
The formation of cyclo-alkyl radicals with the radical position on the ring is the result of a C-C bond cleavage in the alkylated position. For instance, the following model reaction:

$$CH_3\text{-cyclo-}C_6H_{11} \rightarrow CH_3 + \text{cyclo-}C_6H_{11}$$

Is similar to the initiation reaction of iso-butane and it is possible to refer to the same activation energy and to assume the following rate expression similar to the one suggested by Tsang [3]:

$$k=10^{16.7}\exp(-85000/RT) \quad [s^{-1}]$$

for isobutene. The reaction is quite faster than the one directly proposed by Brown and King [4] for the demethylation reaction of methyl-cyclo-hexane:

$$k=10^{16.15}\exp(-88000/RT) \ [s^{-1}]$$

Furthermore, new reactions involve the cleavage of a C-C bond inside the ring to form biradical intermediates which fast rearrange into molecular products: initiation, in this case, acts as an isomerization reaction to form alkenes.
From the above consideration, it derives that initiation of cyclo-alkanes is then turned into the initiation reactions of alkenes. In this situation the weakest C-C bonds in alkene

skeleton are those forming allyl type radicals. The following reactions are typical examples belonging to this class:

$1\text{-}C_4H_8 \rightarrow aC_3H_5 + CH_3$

$1\text{-}C_5H_{10} \rightarrow aC_3H_5 + C_2H_5$

$1\text{-}C_6H_{12} \rightarrow aC_3H_5 + nC_3H_7$

For alkenes heavier than pentene, the selected reference value is obtained by correcting the bond energy with the contribution given by the formation of a resonantly stabilized radical. This correction should be in line with the data proposed for two allyl radicals formation as in the case of 1,5 hexadiene initiation reaction:

$CH_2{=}CH\text{-}CH_2\text{-}CH_2\text{-}CH{=}CH_2 \rightarrow CH_2{=}CH\text{-}CH_2 + CH_2{=}CH\text{-}CH_2$

Dean [6] suggests $10^{15.2}$ exp (-57000/RT) which is also confirmed as $10^{14.8}$ exp (-57630/RT) by Roth et al. [7]
Initiation through the decomposition of different C-C bonds is significantly less important and is similar to the decomposition of alkane species:

1-hexene $\rightarrow CH_2{=}CH\text{-}CH_2\text{-}CH_2 + C_2H_5$

2.2- Propagation reactions of cyclo-alkyl and alkenyl radicals

H-Abstraction reactions follow the same rules and apply the same reference kinetic parameters as the same reactions of normal and branched alkanes.
The complete study of the primary propagation reactions of cyclo-alkyl-radicals requires the definition of a proper reaction path which involves intermediate and final products. As an example, fig. 1 refers this subject to a cyclo-hexyl radical (for drawing simplicity most of the dehydrogenation reactions are not reported).
These reaction paths follow elementary steps which apply according to very simple and predefined rules. They may be simply classified as: Isomerization and decomposition reactions.

Isomerization reactions include:

- Ring decomposition and the reverse cyclo-addition reactions
- 1-4, 1-5,1-6 H transfer reactions via 5, 6 and 7 member ring intermediates. These reactions are favored when forming allyl or similar resonantly stabilized radicals.

Decomposition reactions include:

- β decomposition reactions
- Dehydrogenation reactions

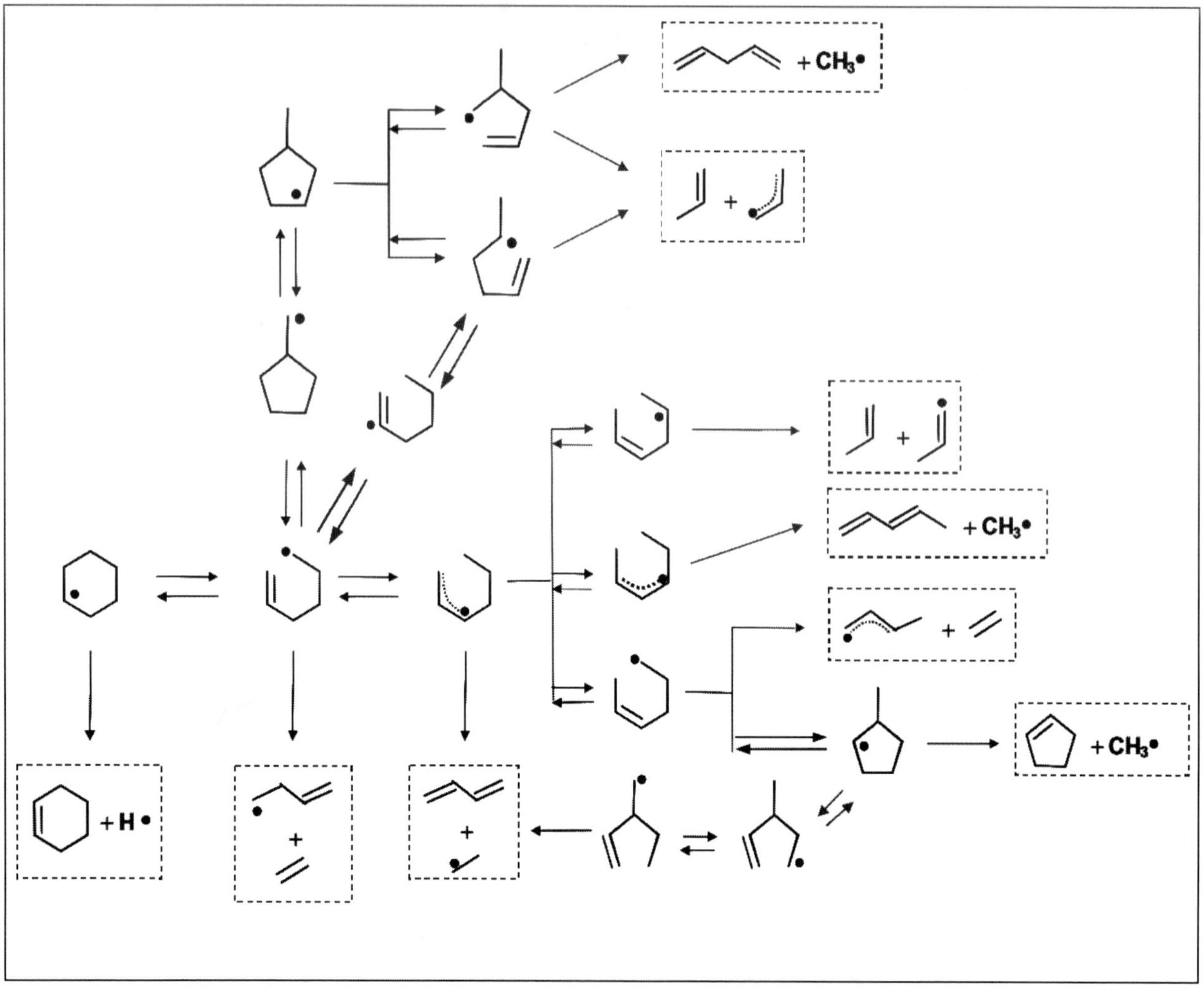

Fig. *1 Primary propagation path of cyclo-hexyl radical*

A simple inspection of the scheme reported in fig.1 demonstrates both that each reaction step falls into the list reported above and that the more the initial radical carbon number increases, the more is the number of possible reactions and greater is the pool of all the intermediate and final species.

Several kinetic parameters required by the description of propagation reactions can be directly taken from the analogous reactions of alkyl radicals and species. Dehydrogenation and some β-decomposition reactions of cyclo-alkyl and alkenyl radicals are simple examples. For instance, the kinetic parameters of the dehydrogenation reaction of cyclohexyl radical to form cyclohexene

are the same parameters used for the dehydrogenation reaction of 3-pentyl radical to form 2-pentene:

$$CH3\text{-}CH_2\text{-}CH\bullet\text{-}CH_2\text{-}CH_3 \rightarrow CH3\text{-}CH{=}CH\text{-}CH_2\text{-}CH_3 + H\bullet$$

Similarly, the kinetic parameters of the decomposition reaction of the 5-hexen-1-yl radical to form C_2H_4 and the 3-butenyl radical are the same parameters used for the β-decomposition reactions of primary alkyl radicals to form ethylene and smaller primary radicals. The presence of a double bond inside the radical does not affect the reaction itself, that is the double bond is far enough from the reaction zone.

The isomerization reaction to form the resonantly stabilized hexenyl radical:

$$\bullet CH_2\text{-}CH_2\text{-}CH_2\text{-}CH_2\text{-}CH{=}CH_2 \rightarrow CH_3\text{-}CH_2\text{-}CH_2\text{-}CH\bullet\text{-}CH{=}CH_2$$

applies the same estimation rules of alkyl radicals. The kinetic parameters should take into account the formation of a 5-membered ring intermediate, the H-abstraction reaction of two H atoms of allyl type, and the formation of the resonantly stabilized 1-hexen-3-yl radical. These facts explain why the reverse isomerization reaction needs a higher activation energy. As clearly depicted in Fig. 1, a new class of important reactions should be considered, that is: ring decomposition (e.g. cyclo-hexyl to form hexenyl radical) and the reverse cyclo-addition reaction.

The activation energies of ring decomposition, to form primary radicals, as proposed by Matheu et al. [8], are respectively 31500 and 28000 kcal/kmol for the 6 and 5 member ring rupture. The reverse cyclo-addition reactions (through the addition of a primary radical) require 8000 and 13500 kcal/kmol for the formation of a 5 and 6 membered ring respectively.

Successive isomerization and β-decomposition reactions of these radicals increase the complexity of the overall mechanism and allow to explain the primary formation of propene, butadiene, pentadiene, and cyclopentene from cyclo-hexane.

It is interesting to observe that propene, partially formed via methyl-cyclopentyl radicals, can also be obtained through the intermediate vinyl like radicals, rapidly transformed into more stable radicals. For instance, the following sequence of isomerization reactions can also take place and can explain the formation of propene, from cyclo-hexyl radicals:

$$\bullet CH_2\text{-}CH_2\text{-}CH_2\text{-}CH_2\text{-}CH{=}CH_2 \rightarrow [CH_3\text{-}CH_2\text{-}CH_2\text{-}CH_2\text{-}\bullet CH{=}CH]$$
$$\rightarrow CH_3\text{-}CH\bullet\text{-}CH_2\text{-}CH_2\text{-}CH{=}CH_2 \rightarrow CH_3\text{-}CH{=}CH_2 + \bullet CH_2\text{-}CH{=}CH_2$$

A sensitivity analysis shows that the formation of propene from this reaction path is of similar importance like the one that moves throughout the successive isomerization reactions of methyl-cyclo-pentyl radicals.

Figure 2 shows the selectivities of the different primary products, as a result of the isomerization and decomposition of the intermediate radicals. It appears evident the large importance of butadiene formation through the decomposition of 1-hexen-3yl radical and it is also relevant to observe that these selectivities are only a weak temperature dependent, in the analyzed temperature range. This result has been already well established since a long time for the large alkane radicals [9] and it has been also recently again proved for n-octyl radical [10].

3.- Primary distribution products from pyrolysis of cyclo-alkanes

Primary distribution products of hydrocarbons may be obtained by the use of a software dealing with the automatic generation of kinetic schemes. Ranzi et al (1983) described a program in such area named MAMA, which has been in the years deeply modified to cover larger radicals including olefins and naphthenic components. Appendix A provides details on the structure of the program.

3.1-Pyrolysis of cyclo-hexane and C_6H_{12} species

The overall mechanism of isomerization and decomposition of cyclo-hexyl radicals, already discussed and shown in fig.1, may be lumped and reduced to a single equivalent reaction, which provides the net result of all the intermediate propagation and decomposition steps:

R+ cyclo-hexane → RH + {cy-C_6H_{11}} → .34 (C_2H_5+C_4H_6) +.11 (cy-C_6H_{10}+H) + .02 (C_6H_{10}+H) +.23 (C_5H_8+CH_3) +.20 (C_2H_4+pC_4H_7)+.01(cy-C_5H_8+CH_3)+.01(C_2H_4+aC_4H_7) + .09 (C_3H_6+aC_3H_5)

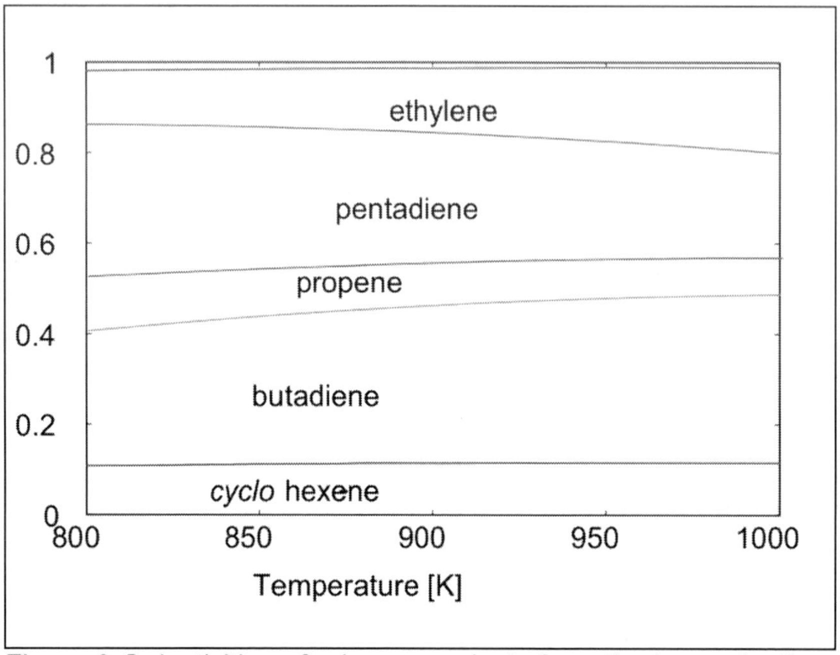

Figure 2 *Selectivities of primary products from the isomerization and decomposition of cyclo-hexyl radical.*

This is an example of a lumped or apparent reaction, where the intermediate radicals cy-C_6H_{11} are supposed instantaneously isomerized and decomposed into their end products.
In a similar way it is also possible to analyze the primary decomposition products from the decomposition of methyl-cyclo-pentane. Only a few reactions need to be added to the scheme of fig.1, because three methyl-cyclo-pentyl radicals are already present as formed from the successive isomerization reactions of cyclo-hexyl-radical. It is then sufficient to include only the tertiary radical with its isomerization and decomposition reactions to form methyl-allyl radical and ethylene:

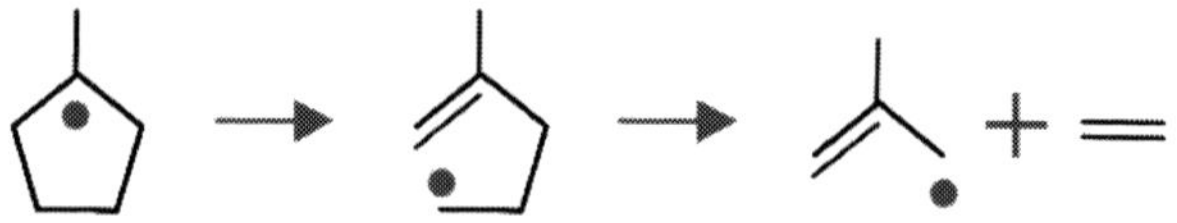

The net result of the H-abstraction, propagation and decomposition reactions of methyl-cyclo-pentane becomes:

R+ CH_3-cyclo-pentane → RH + Mix [CH_3-cyclo-pentyl]
Mix [CH_3-cyclo-pentyl]→ .34(C_3H_6+aC_3H_5)+ .25(C_2H_4+iC_4H_7)+.32(C_2H_4+aC_4H_7) + .03 (C_2H_5+C_4H_6) +.02 (cy-C_5H_8+CH_3) +.02(me-cy-C_5H_8+H)

Mix [CH_3-cyclo-pentyl] is the mixture of the four different alkyl radicals derived from the H abstraction reactions from methyl-cyclo-pentane:

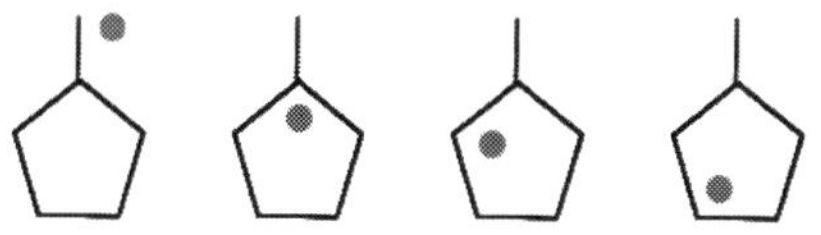

Also in this case the complex mechanism of successive isomerization and decomposition reactions is reduced to an equivalent or lumped reaction. The same considerations are applied when considering the H abstractions and successive decomposition reactions of the different alkenes with 6 carbon atoms.
Cyclization to form cyclohexene are of less important (an order of magnitude less) and consequently neglected to reduce the amount of components.

3.2 Pyrolysis of heavier cyclo-alkane species

Figure 3a and b show the distributions of small alkenes and radicals from the decomposition, at 1040 K, of the series of 1-methyl-4-alkyl-cyclo-hexanes versus the carbon number (from 1,4dimethyl-cyclohexane up to 1-methyl-4-heptyl-cyclohexane).

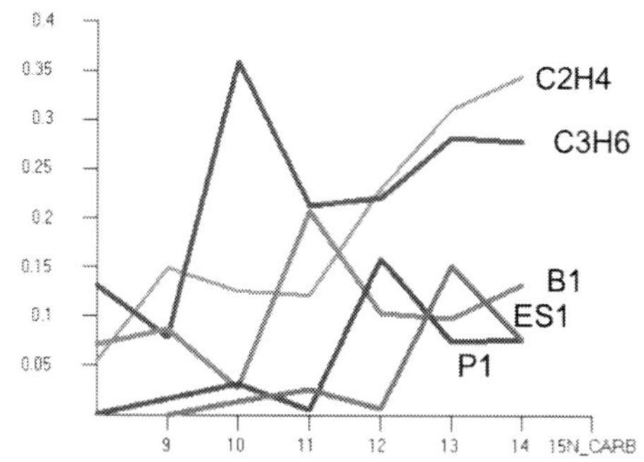

Figure 3 a
Primary yields of alkenes from the decomposition of 1-methyl-4-alkyl-cyclo-hexanes vs. carbon number

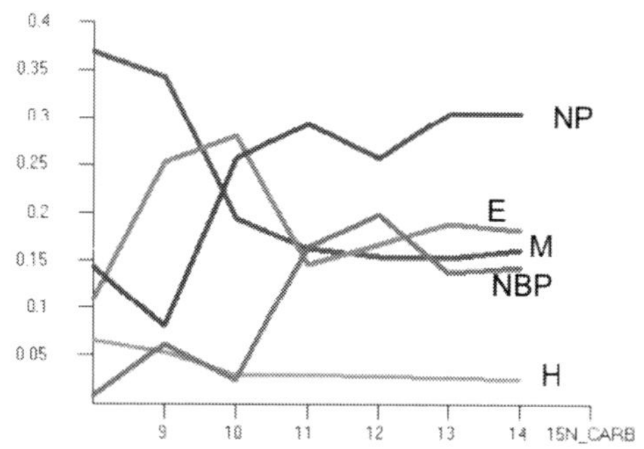

Figure 3b
Primary yields of small radicals from the decomposition of 1-methyl-4-alkyl-cyclo-hexanes vs. carbon number

The analysis of the primary decomposition and isomerization of unsaturated hydrocarbons is limited to 14 C atoms due to the large number of possible isomers and to the resulting dimension of the overall problem. Nevertheless, it is relevant to observe that small alkenes and radicals reach an asymptotic trend, for C atoms higher than 12-13. The complete and detailed description of heavier species soon becomes unmanageable, for this reason the stoichiometry of the primary reactions of heavier homologous species are simply obtained with a direct extrapolation.

4.- Lumped components of cyclo-alkane fractions

A further example of the complexity of the feed description is given in figure 4, which shows several cyclo-alkane components with 10 carbon atoms, 15 isomers with a cyclo-hexane ring and 17 with cyclo-pentane structure.
When studying and referring to the steam pyrolysis of these components, MAMA program can easily evaluate the primary propagation reactions as well as the net product distribution from the isomerization and decomposition mechanism of all the isomers. By using this procedure and maintaining this description level for the characterization of the steam cracking process of liquid feedstocks, the dimension of the overall kinetic mechanism fast becomes uncontrollable. This is a further evidence of the practical need to apply simplifications for the description of liquid and heavy hydrocarbon fractions.
As in the case of branched alkanes, also for cyclo-alkanes it is possible to observe a relative regularity of the virgin feeds, always due to the same diagenesis process. Cyclo-hexane rings are more abundant than the cyclo-pentane ones. It is possible and reasonable to assume a ratio 60 : 40 between the two classes of isomers. Moreover, the probability of methyl substitutions, both on the side chain and on the naphthenic carbon is about 25%. The experimental distributions of these components indicate that there is a prevailing presence of a single long alkyl chain and several methyl substitutions. For this reason, di-ethyl cyclohexanes, methyl-di-ethyl-cyclopentanes and ethyl-propyl-cyclopentanes are not reported in Figure 4.

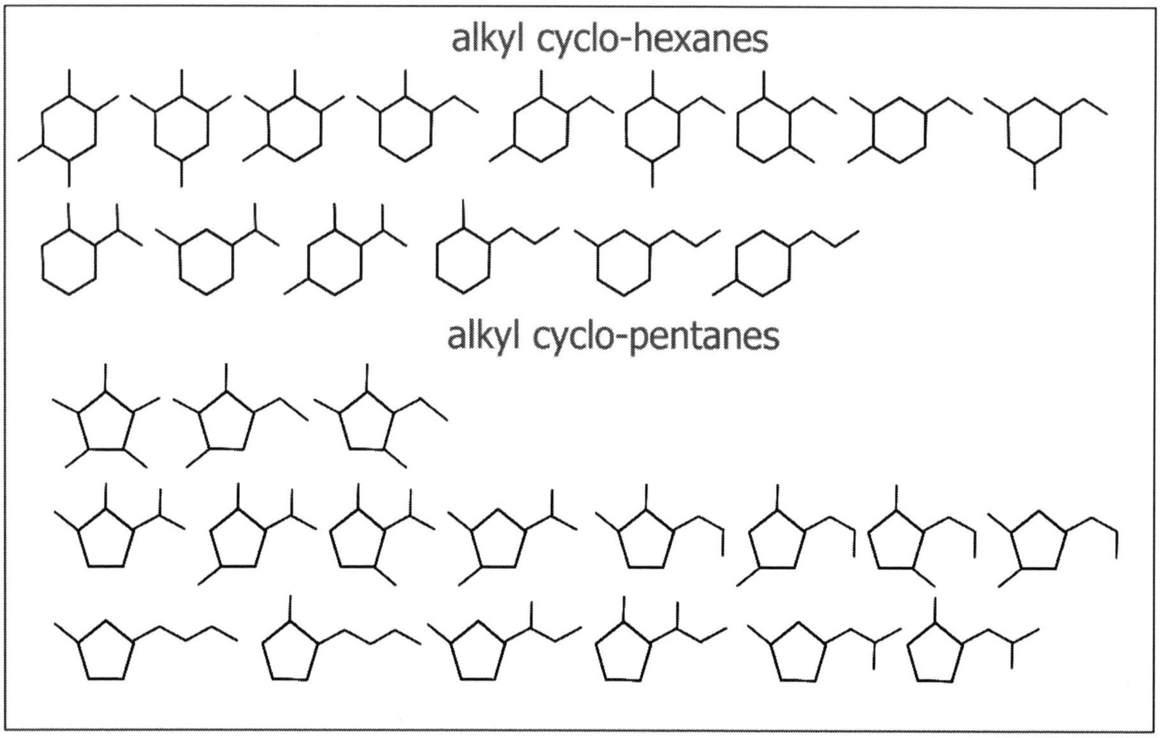

Figure 4 *Cyclo-alkane components with 10 carbon atoms*

On the basis of this and similar considerations, the fractions of the isomers of C10 cyclo-alkanes has been assumed as a single equivalent component with a default and fixed internal composition, as it will be better described in the next paragraph. Mama program analyzes the primary propagation reactions of all the real isomers of the mixture and it also groups and lumps all the stoichiometries and generates the equivalent stoichiometry of the lumped component.

In this case we refer to a lumped reaction with the meaning of a set of similar reactions, operating collectively, on similar molecular species, or {A} → {I} → {B} where {A}, {I}, and {B} are sets of similar reactants, intermediates, and products, respectively [11].

As an example, the lumped alkyl-cyclo-hexane component C10 $\{NAF_{Cy6}C_{10}\}$ standing for the following mixture:

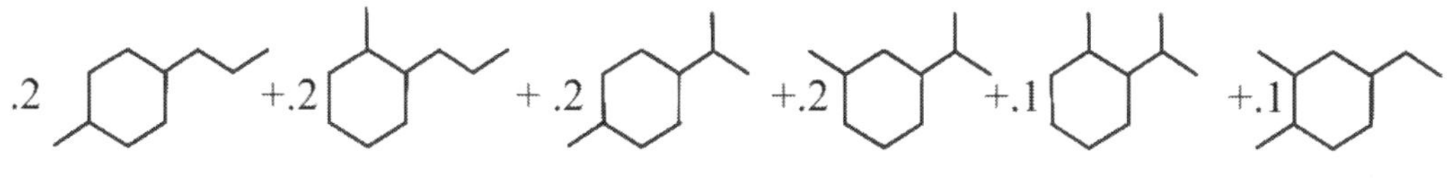

Processed by MAMA generator, gives a detailed product distribution after H-Abstraction which, for simplicity, is here partially reported up to the portion </=C4:

0.036	H	0.328	CH_3	0.185	C_2H_5	0.149	$1C_3H_7$
0.15	$2C_3H_7$	.092371	aC_3H_5	3.75E-04	vC_3H_5	8.63E-4	$1C_4H_9$
2.30E-02	$2C_4H_9$	1.20E-02	iC_4H_9	2.53E-02	aC_4H_7	1.50E-02	PC_4H_7
2.80E-02	iC_4H_7	9.95E-02	C_2H_4	2.45E-01	C_3H_6	3.31E-02	$1C_4H_8$
1.83E-02	$2C_4H_8$	7.74E-03	IC_4H_8	8.55E-02	C_4H_6	....	

In a similar way, when assuming the following internal composition for the lumped alkyl-cyclo-pentane C10 $\{NAF_{Cy5}C_{10}\}$ component:

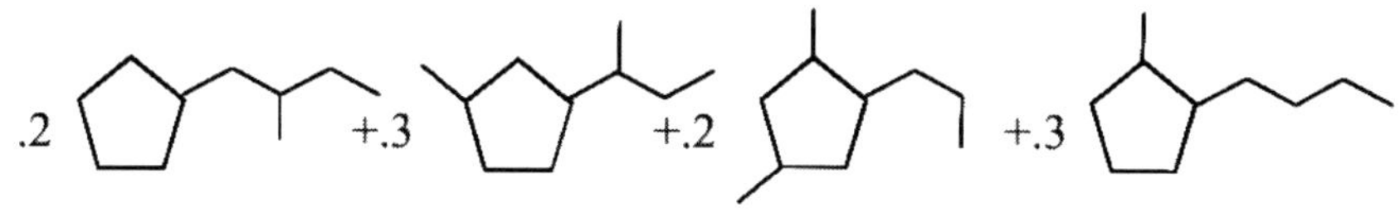

MAMA program gives the following primary products distribution (<= C4), as a weighted result of the H-abstraction reactions of the different isomers:

3.56E-02	H	0.159696	CH_3	0.149715	C_2H_5	2.36E-02	$1C_3H_7$
8.95E-02	$2C_3H_7$	0.138755	aC_3H_5	4.67E-04	vC_3H_5	1.64E-03	$1C_4H_9$
8.54E-02	$2C_4H_9$	1.46E-02	iC_4H_9	8.54E-02	aC_4H_7	4.62E-03	PC_4H_7
3.69E-02	$2C_4H_8$	4.24E-02	IC_4H_8	8.83E-02	C_4H_6		......

As already mentioned, the typical internal distribution in the cycl-alkane fractions of liquid feeds shows that alkyl-cyclo-hexanes prevail on the corresponding cyclo-pentanes. Thus, a single cyclo-alkane lumped components is obtained by simply mixing on 60/40 basis the previous $\{NAF_{Cy6}C_{10}\}$ and $\{NAF_{Cy5}C_{10}\}$ lumped components.

In this way, the single lumped component $\{NAFC_{10}\}$ represents the overall fraction of C10 cycloalkanes with the following internal composition:

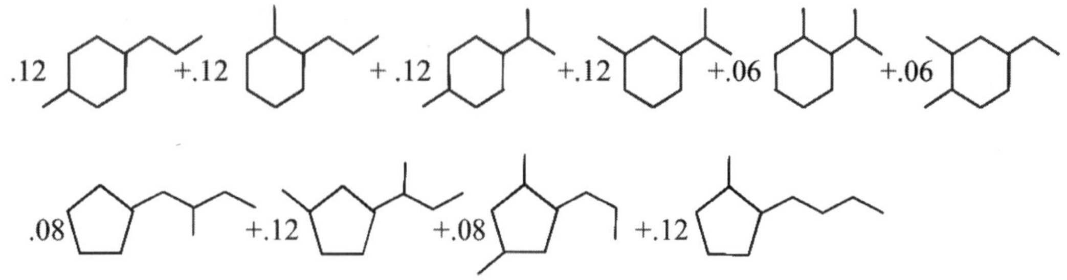

The primary products distribution is then obtained by combining the previous reactions of cyclo-C6 and cyclo-C5 lumped components:

5.- Conclusions

Comments useful for the modeling of pyrolysis reactions of cyclo-alkanes have been presented in this paper. Due to the complexity of the reaction network, the primary distribution products from the decomposition of cyclo-alkanes are obtained with the use of a kinetic generator: MAMA program, properly described in the Appendix.

Moreover, a proper attention has been devoted to the analysis of the different simplifications and lumping procedures unavoidably introduced in order to make manageable the description of the pyrolysis of heavy liquid feedstocks. In fact, the complexity of the detailed system becomes quite evident considering that the number of components in the primary decomposition path rapidly increases with the increasing of the molecular weight and becomes higher than 10^6 for carbon number larger than 16-17. A description and definition of lumped reaction and lumped species is also reported in the Appendix.

Appendix A.
MAMA Program: Automatic Generation of Primary Lumped Reactions

Lumped Reaction and Lumped Species

According with the glossary of the 'Workshop on combustion simulation databases for real transportation fuels' [11], it seems relevant to summarize the meaning of the lumping and lumped terms as used in the present work.

Lumped Reaction – A collection of elementary reactions expressed as an "equivalent" apparent single-step reaction. A lumped reaction may be a stepwise reaction, involving a consecutive set of elementary reactions, the simplest example being A → I → B, where I is a steady state intermediate. A lumped reaction may also be a set of related, but dissimilar reactions, operating as alternative or parallel pathways, or A → {In} → B, where {In} represents the set of parallel pathways. A lumped reaction may also be a set of similar reactions, operating collectively, on similar molecular species, or {A} → {I} → {B} where {A}, {I}, and {B} are sets of similar reactants, intermediates, and products, respectively.

Lumped Species – The sets of similar reactants {A}, intermediates {I} and products {B} in the lumped reaction {A} → {I} → {B} are a first example of lumped species. Thus, lumped species {A} can be defined as a weighted mixture of several isomers:

$$\{A\} = \sum x_j A_j$$

Lumped species reacts in a similar way and the internal distribution remains unchanged with the reactions.

The meaning of 'lumped species' is also extended in this work to similar species, for instance the cyclo alkane components with 10 Carbon atoms are a mixture of the two lumped components, alkyl- cyclo pentanes and cyclo hexanes:

$$\{NAFC_{10}\} = a\,\{NAF_{Cy6}C_{10}\} + .b\,\{NAF_{Cy5}C_{10}\}$$

Furthermore, for heavy species, it is possible and convenient to extend the concept also to homologous species with different number of carbon atoms. Thus, the lumped species $\{NAFC_{20}\}$ and $\{NAFC_{25}\}$ are used as reference species to describe also cyclo-alkane intermediate components with 21, 22, 23 and 24 carbon atoms. As a result, it is possible to consider $\{NAFC_{23}\}$ and its decomposition products with a simple linear splitting: 60% $\{NAFC_{25}\}$ and 40% $\{NAFC_{20}\}$.

This approach also applies to alkanes, alkenes, aromatics and polyaromatic species.

This extension is equivalent to what applied in the sectional methods for the description of soot formation, where individual species (section) represents a well defined mass range [12].

As already described in the paper, the complex reaction mechanism presented in figure 1, can be directly reduced and lumped to a single equivalent reaction where large intermediate radicals isomerize and decompose to form small radicals and unsaturated species. Therefore, the aim of this mechanism generator is to evaluate the set and weights of the end-products obtained by the decomposition of a large, heavy radical. Starting with an initial radical, the propagation path include isomerization and decomposition/dehydrogenation reactions with the formation of final products. The assumptions made with the generation of the primary propagation mechanism, correspondingly define the structure of the algebraic system which once solved, gives the distribution of the primary decomposition products. This subject will be treated as first in the next paragraph.

Due to the large number of components which can be involved in a propagation path, the concept of lumping of components as well as the classes of components which need to be lumped, will be then discussed.

The generation of a propagation path requires computing time which rapidly increases with the carbon number of the initial radical. However, the resulting generated pathway of a specific radical is independent from the formation mechanism of the radical itself and therefore the need to save the path in a database becomes evident. The description of the database structure will include details on the molecule representation, as well as on the way the reaction path is stored and recorded. Finally, before to summarize the whole functionality of the MAMA Generator, the algorithm for molecule homomorphism will be shortly described.

A.1-Evaluation of End-Product Composition

The automatic generation of the primary product distributions is based on the assumption that the propagation reactions of radicals are described with an autonomous system. I.e. H-abstraction, radical addition and termination reactions and any interactions of the intermediate radicals with the reacting mixture are disregarded. As a consequence, the end-products of this propagation mechanism are small radicals and unsaturated components (UC). The total production rate P_j of the jth product is simply expressed in the form:

$$P_j = \sum_{i \in R_j} k_{Di,j} Z_i + P_{j0} \qquad (1)$$

where P_{j0} is the initial production rate of the jth component j (normally =0 for end-products), $KD_{i,j}$ is the production rate in the ith reaction, Z_i is the concentration of the radical decomposed in the reaction i, and R_j is the set of decomposition reactions forming the jth product. For all the intermediate radical components, assuming quasi steady conditions, mass balance equations are written in the form [9]:

$$X_j\left(\sum_{i \in I_j} k_{Ij,i} + \sum_{i \in D_j} k_{Dj,i}\right) = \sum_{i \in I_j} k_{Ii,j} X_i + P_j \qquad (2)$$

where:

X_j = concentration of the jth component

I_j = set of isomerization reactions involving the jth component

D_j = set of reactions decomposing component j

$k_{Ij,i}$ = Isomerization rate of the jth component to form the ith component

$k_{Ii,j}$ = Isomerization rate of the ith component to form the jth component

$k_{DJ,i}$ = Decomposition rate of the jth component in the reaction i

P_j = total production rate as described in eq. 1, with $Z \in X$

The left side of eq. (2) represents the depletion of jth component due to its isomerization and decomposition. The right side is the formation contribution to jth component coming both from the isomerization of ith radical and from the decomposition of larger radicals.
At fixed temperature, the material balances turn into a system of linear equations in radical concentrations.

$$A\bar{X} - \bar{P}_0 = 0 \qquad (3)$$

where
A is the coefficient matrix (n,n),
X is the vector of radical concentrations (n),
P_0 is the vector of the initial production rates (n),
n is the number of the intermediate radical components.

It is worth to notice the following properties of this linear system:

- the size of the system rapidly increases with the carbon number of the initial radical, reaching values of several hundreds thousands when studying the primary propagation path of unsaturated components such as alkenes and cyclo-alkanes with more than 14-15 carbon atoms,
- the system is dominantly sparse being each equation (2) in few tens of unknowns,
- the overall system may be decomposed in subsystems of reduced size as shown in fig. A1.

Once the linear system is solved and the intermediate radical concentrations known, the end product distribution (eq.1) is directly obtained.
Fig. A1 reports in the upper portion the structure of the system (3), while in the lower one the set of equations (1) giving the total production of end-products. As mentioned, the two portions are solved disjointedly, in sequence, the upper portion followed by the lower one.
In the system (3), the unknowns are ordered according with decreasing carbon number C. It has to be underlined that a radical with Cn carbon atoms can only decompose generating radicals with Cn-2, Cn-3, ...carbon atoms. As a matter of facts, β-decomposition does not allow the formation of a radical with Cn-1 carbon atoms. This explains the presence of empty block in the structure of the system. Equations are ordered accordingly with decreasing carbon number too. The resulting structure of the system in fig. A1 suggests the most effective solution. For its nature, the block including the equations of the highest carbon number does not contain any production term (see eq. 1) except the initial values P_{j0}, i.e. the set of decomposition reactions R_j is empty. This block may be solved and the calculated values of the unknowns factorize the vertical band 'P'. 'P' contains the production terms of smaller radicals. The factorization renders as known all the production terms of the successive block with the highest C number. This block is solved and the results allow the factorization of the related vertical band. This procedure is then repeated till the end of the system. Finally, the production of the end-products are easily calculated being all the related unknowns available by the previous system solutions.

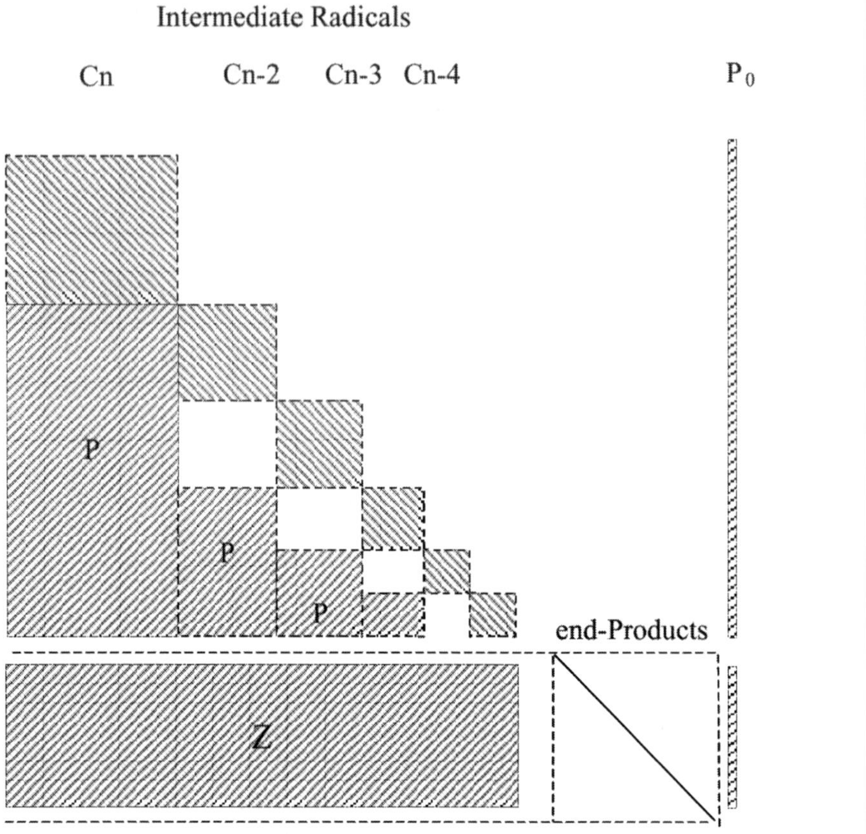

Fig. A1: *Structure of the system (3) and production rate of End-Products*

An important aspect derives by the assumption initially made: each radical evolves without interacting with the surrounding environment, in other words the decomposition reactions are considered irreversible, which results in the linearity of the algebraic system: the end-product composition of an initial radical is not dependent on the initial composition of any other intermediate radical.

In fact if $X_j(..,P_{j0},.)$ and $X_i(..,P_{i0},.)$ stand for the solution vectors of the above system (3) for a null vector P_0 except, respectively, for the initial production rates of component j and i, a trivial property of linear systems gives:

$$X_{j,i}\,(..,P_{i0},\,..P_{j0}..)= X_i(..,P_{i0},.) + X_j(..,P_{j0},.)$$

This property applied to the problem at hand turns into the statement that, given a mixtures of initial radicals at P_{10}, P_{20} ,... P_{R0} , their products distribution is equal to the sum of the distributions of each single radical. The advantage of this result is the possibility to consider product distributions of each radical as independent on the mixture is belonging to such as, for example, the initiation reactions which generated it. This fact anticipates an item relevant to the invariance of the propagation paths in front of the problem which they are included into. The propagation paths are built once for all and stored in a database and they are ready to be analyzed for all the classes of problems which need them.

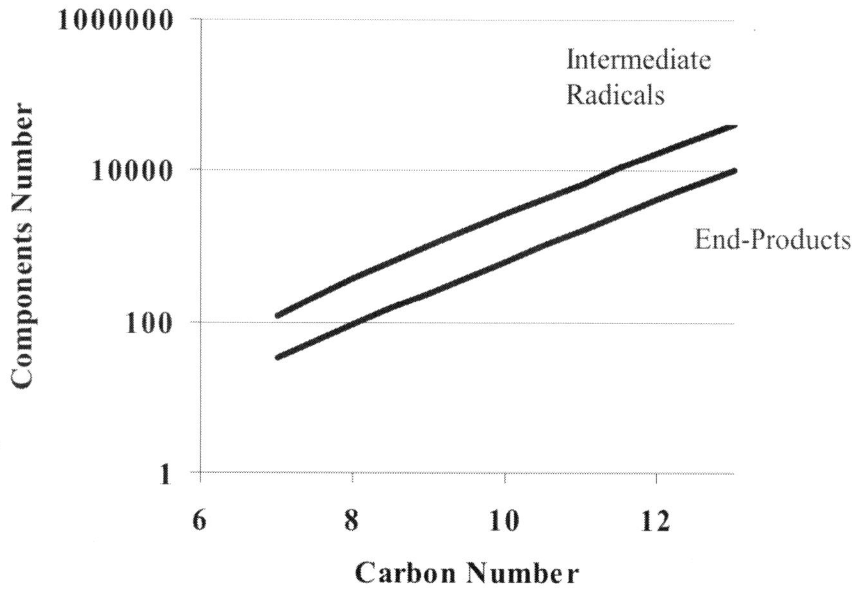

Fig. A2 : *Components number in the primary propagation reactions of normal-alkenes*

A.2-*Lumping of Components*

The evaluation of the end-product composition in a propagation chain, may require the solution of linear systems involving a large number of unknowns referring to the concentrations of intermediate radicals. Also the number of the end-products may be as large as almost the same magnitude.

Fig. A2 shows the number of components involved in the primary propagation reactions of normal-alkenes. These data clearly indicate that this number rapidly becomes very large increasing the carbon number of the initial alkene. This fact justifies more than the opportunity the necessity to turn into the lumping of components when dealing with detailed and mechanistic models describing the hydrocarbons pyrolysis of such heavy species.

However, the following conclusions may help to make more viable the solution of this complex problem:

- The amount of alkenes and unsaturated components in the end-products is large and it strongly suggests the introduction of lumping of components, as it will be described below.
- Being the end-products composition of an initial radical not dependent on the initial composition of any other intermediate radical, the simplification and lumping procedures will not refer to such components.
- The amount of small and stabilized radicals, in the end-products, is rather limited compared to the amount of the remaining components, so that the lumping might be avoided for this class or applied to a small portion of them.

If just a single radical is considered (see fig. A3), propagation reactions produce a smaller and/or stabilized radicals and a olefin and/or unsaturated component (UC). Each new radical is then processed by its primary propagation reactions (isomerization and decomposition) which turns into the production of new olefin or UC and a smaller radical. The produced olefins define the whole set of components involved in this chain, so that the lumping technique conveniently applies only to them.

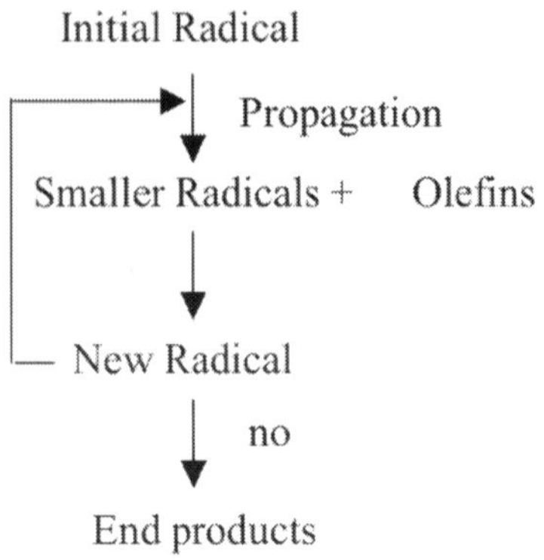

Fig. A3. *Schematic propagation reaction chain of radicals decomposition and formation of end-products.*

The grouping or lumping of components does not refer to unique rules, but it is only dependent on the successive uses of the kinetic scheme. As a consequence, the Automatic Generator does not contain any rule of lumping, however, the resulting final products may be conveniently lumped according with pre assigned rules.

A.3-*Database Structure*

As already anticipated early above, a key item of the automatic generator is a database storing the propagation paths of initial radicals. The choice of which radical has to be considered is dependent, as mentioned above, on the mathematical model (say steam-cracker simulator) which will include the results of the generator. Provided that all the possible isomers of hydrocarbons up to 6 carbon number are limited to few hundreds, let suppose that the database already stores the propagation paths of all the radicals included in this set. Now, the propagation path of a C7 hydrocarbon will produce new isomerization reactions and products with 7 carbon atoms, while the remaining reactions and products with a lower number of C atoms are already present in the existing database (so that there is no need to further describe them). In this way the database is enlarged to containing only the new portion created by this component, as it is sketched in fig.A4.

If the aim is to enlarge the database to include all the possible isomers with 7 carbon number, this may be reached with some effort, being the isomers still acceptably limited in their number. The aim becomes rather difficult and practically impossible with higher carbon number due to the rapid increase of the involved isomers set. A compromise is then required between comprehensiveness and size of the whole database.

In the scope of a steam cracker simulator, it has been realized, by experience, that the accumulation of propagation paths in the database by starting with mixture of radicals obtained by H-abstraction reactions of normal paraffins and normal olefins (with progressively increased carbon number), produces a whole information set which covers the description of a good percentage of components present in the kinetic scheme. The rational stands in the fact that isomerization reactions produce via successive ring formation and break, a set of isomers of alkenyl and cyclo-alkyl radicals which is close to the complete theoretical one. Dehydrogenation reactions, then, increase total the number of end products with a definite finger print of all the intermediate radicals.

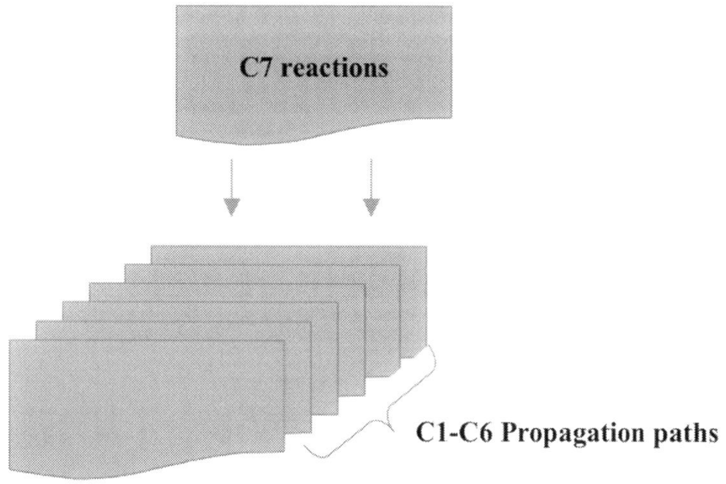

Fig. A4: *Sketch of the accumulation of new portions in the database.*

The computing time to create reaction paths follows a behavior similar to that already reported in figure A2: few seconds are required for radicals with carbon atoms lower than 5-6, while almost a 'week' is needed for olefins with carbon number larger than 16.
Due to the rapid increase of components and the consequent practical problem of storing all the data in an efficient way, it is worth to mention also the technique adopted for this purpose.
Being the propagation path a sequence of reactions steps, basic elements for the description of the whole path are those ones capable to describe a single reaction:

- description of the components molecular structure
- indication of the products and related reactions associated to the component.

As an example the decomposition reaction of 6-hepten-2-yl radical:

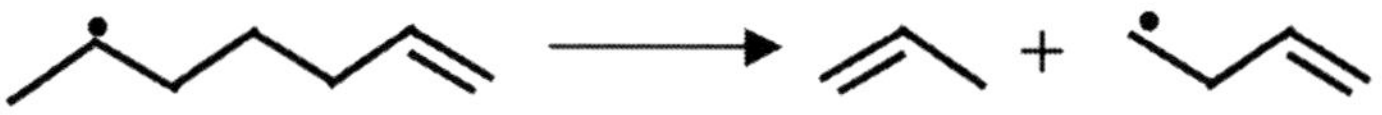

is characterized by the description of the molecular structures of the three components, the type of the reaction and the distinction between the reactant and the products.
Being the components normally involved in several different reactions, it is more convenient to collect all the component descriptions in a list, and to address their position in such a list. In the same manner, also the reactions may be grouped in a proper list.

Figure A5 provides an example of a possible presentation of the cited reaction as a sequence of addresses:

k(reactant), **i** (olefin product if any), **j** (radical) and **n** (reaction type).

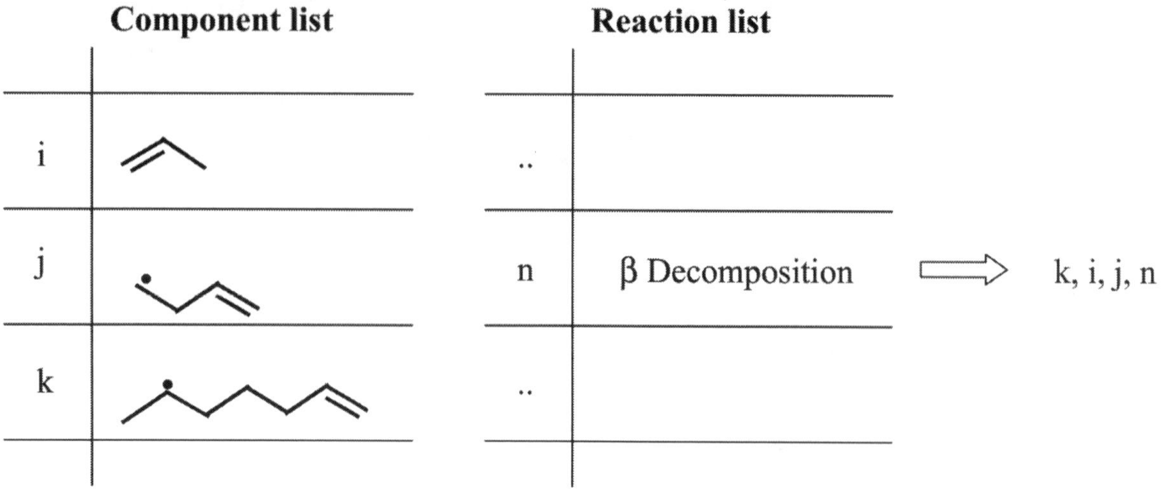

Fig. A5: schematic structure of elements apt to describe a reaction step

The molecule description is based on a sequence of integers: in fact dealing with hydrocarbons the molecule is univocally characterized by the carbon-carbon bonds and the type of the carbons. As an example, Tab. A1 reports the assumed correspondence between integer indexes and carbon types: they are used to represent a molecule, in a plane, as a sequence of integers.

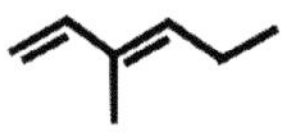 → 21 22 23 22 2 1
1

1 → $C-$	21 → $C=$	31 → $C\equiv$
2 → $-C-$	22 → $-C=$	32 → $-C\equiv$
3 → $-\underset{\vert}{C}-$	23 → $-\underset{\vert}{C}=$	42 → $=C=$
4 → $-\overset{\vert}{\underset{\vert}{C}}-$		

Table A1. Molecule description. Correspondence between integer numbers and carbon types

In a similar way, also cyclo-alkane with 5- or 6- member rings may be put into correspondence with integers which contain information about the ring members and their connections to the molecule chain.
It is worth to notice that in this type of molecule representation, it is not necessary to further specify the connections between C and H atoms [13].

A.4-*Molecule homomorphism algorithm*

Molecule homomorphism is a non ambiguous and effective statement whether two molecule descriptions refer to the same product or not. The importance or better the necessity of such a statement is obvious or, in a certain sense, it is a key statement on which is based the Automatic Generator. Inability to recognize identical structure may cause unnecessary calculations in a better situations, while most commonly they lead to infinite loops of calculations.

The molecule homomorphism is based on a 'branch and chain' algorithm. A branch is a linear portion (including rings) which is connected to the molecule chain at its ends. A molecule is then divided into branches (see fig.A6 a). Due to its linearity, it is assumed that the branch homomorphism, i.e. the statement whether two branches are identical, is a trivial problem. In facts two branches are checked for their length and position of possible unsaturated bonds. The molecule is represented as a graph whose nodes are the branches. The molecule homomorphism is then transformed into a graph homomorphism. Despite many algorithms present in literature, the one used by Steward [14] (for the definition of an admissible output set of an algebraic system of equations) may be slightly modified to solve the problem at hand. For low interconnected graphs, substitution matrixes are an efficient way to represent them (fig.A6 a). The homomorphism of two graphs analyses their representation with the matrixes S1 and S2 which start at the first row and column with a branch-edge, i.e. one of the two extreme points of a branch which is not connected to any other branch. While S1 description is maintained fixed, S2 is rearranged from top to bottom with the aim of creating a chain equal to S2 (fig.A6 b). The rearrangement involves an exhaustive analysis of all the possible chains obtained by connecting each branch of S2 with admissible branches, i.e. those which are equal to the corresponding one in S1.

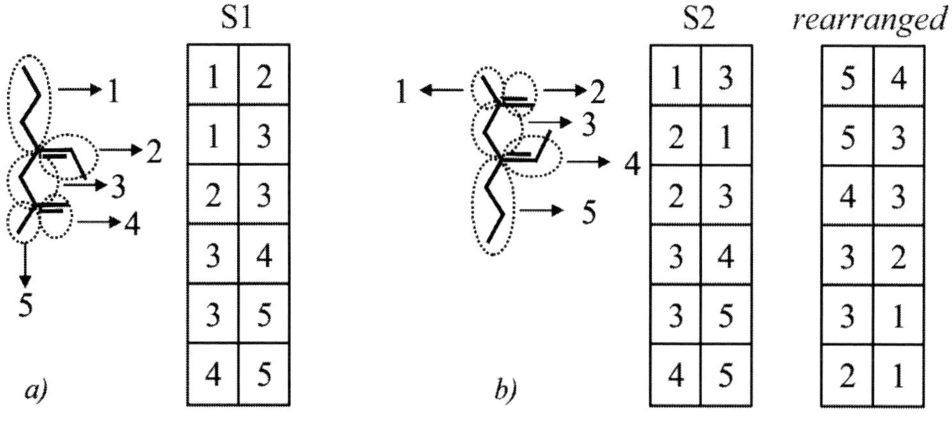

Figure 2. a) Molecule divided into branches and substitution matrix; b) Initial substitution matrix S2 of an equal molecule, and its content after rearrangement.

A.5-*Schematic functionality of the kinetic Generator*

The kinetic generator functionality is based on three elements having three different activities (see fig. A7).

- A pre-Processor which prepares the kinetic data and the initial radicals mixture according with specified user requirements (Initiation reactions, Addition ..etc..)
- The main Processor, which analyses each single radical of the mixture with the aim to calculate the end products composition. In the case of a radical not present in the database, the propagation path of the radical is created for the portion not present in the database, which is then updated with these new data.
- Finally a post-Processor lumps the end-products according with pre-assigned rules specified by the user.

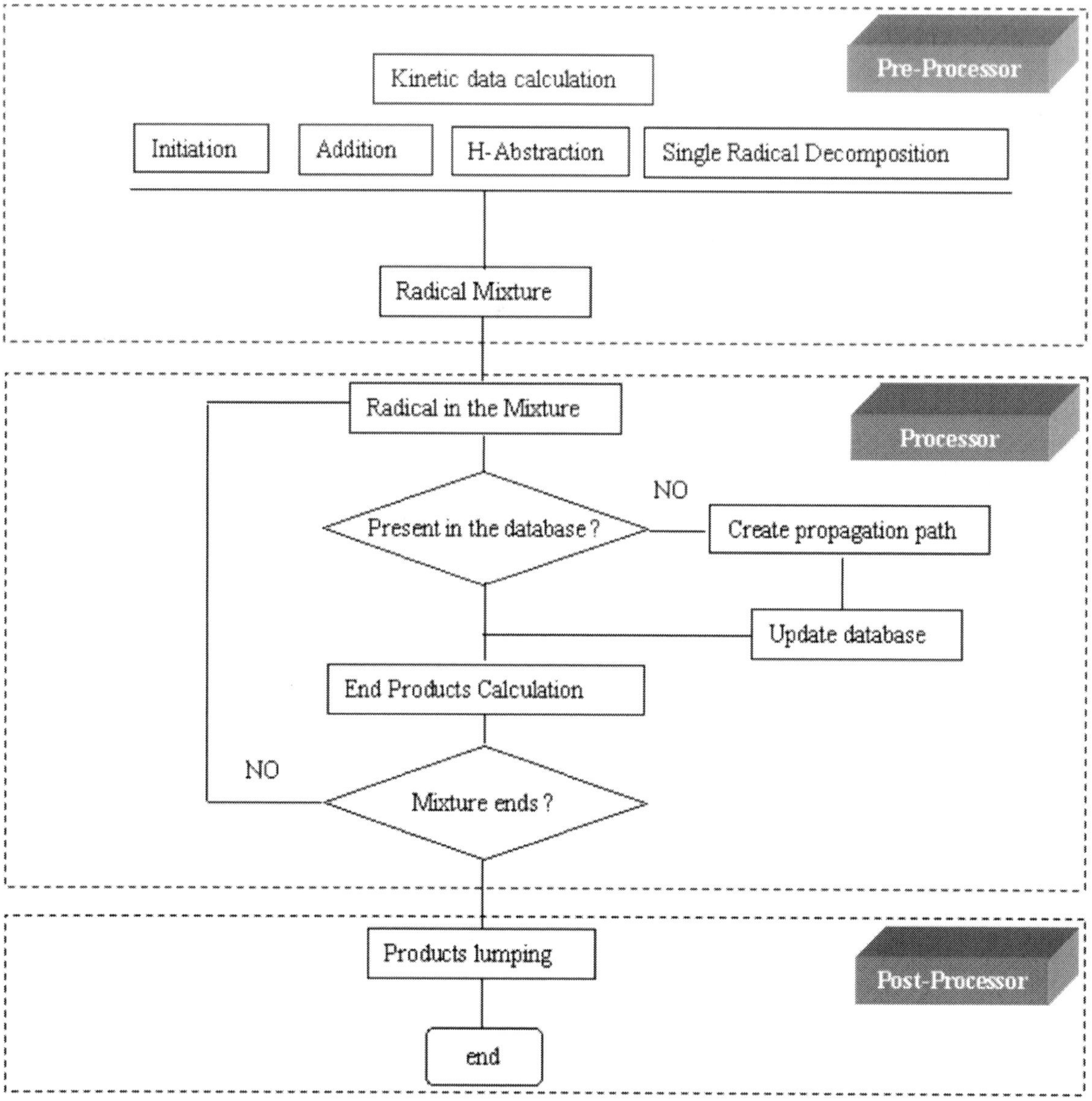

Fig. A7: *Schematic functionality of the kinetic Generator*

References

1. Ranzi, E., M. Dente, G. Biardi and S. Pierucci. *Initial product distributions from pyrolysis of normal and branched paraffins.* Ind. Eng. Chem. Fundam., 1983. **22**: p. 132.
2. Pierucci, S., E. Ranzi, and S. Barendregt, *A kinetic generator of hydrocarbon pyrolysis mechanisms.* ESCAPE 15, Barcellona, Spain, June 1-5, 2005.
3. Tsang, W., *Thermal Stability of Cyclohexane and 1-Hexene.* Int. J. Chem. Kinet., 1978. **10**: p. 1119.
4. Brown, T.C. and K.D. King, *Very low-pressure pyrolysis (VLPP) of methyl- and ethynyl- cyclopentanes and cyclohexanes.* Int. J. Chem. Kinet., 1989. **21**: p. 251.
5. Tsang, W., *Thermal Decomposition of Cyclopentane and Related Compounds.* Int. J. Chem. Kinet., 1978. **10**: p. 599.
6. Dean, A.M., *Predictions of pressure and temperature effects upon radical addition and recombination reactions.* J. Phys. Chem., 1985. **89**: p. 4600.
7. Roth, W.R., et al., *Die bildungsenthalpie des allyl- und methallyl-radikals.* Chem. Ber., 1991. **124**: p. 1453-1460.
8. Matheu, D.M., W.H. Green, and J.M. Grenda, *Capturing pressure-dependence in automated mechanism generation: Reactions through cycloalkyl intermediates.* International Journal of Chemical Kinetics, 2003. **35**: p. 95-119.
9. Dente, M. and E. Ranzi, eds. *Mathematical Modelling of Pyrolysis Reactions.* Pyrolysis: Theory and Industrial Practice, ed. L. F. Albright, B. L. Crines, and W. H. Corcoran. 1983, Academic Press: San Diego.
10. McGivervy, W.S., J.A. Manion, and W. Tsang, *Decomposition and isomerization of 1-octyl radicals.* WIP 30th International Symposium on Combustion. Chicago il. USA July, 2004: p. Paper 1F1-02.
11. Hudgens, J.W., Workshop on combustion simulation databases for real transportation fuels NIST Gaithersburg, MD, 2003.
12. Gelbard, F., Y. Tambour, and J.H. Seinfeld, J. Colloid Interface Sci, 1980. **76**(2): p. 541-556.
13. Wauters, S. and G.B. Marin, *Computer generation of a network of elementary steps for coke formation during the thermal cracking of hydrocarbons.* Chemical Engineering Journal, 2001). **82**: p. 267–279.
14. Steward, D.V., Soc. Ind. Applied Math. J., 1965. **2B**: p. 345.

Paper 47c

Estimation of the Selectivities of the First-order Reaction in Naphtha Pyrolysis Process Based on Fuzzy Matching Method

Xiaodan Gao, Bingzhen Chen, Xiaorong He
Department of Chemical Engineering
Tsinghua University, Beijing, China, 100084

Prepared for Presentation at the 2005 Spring National Meeting in Atlanta, GA April 10-14
17th Annual Ethylene Producers Conference
Session 47 – Ethylene Plant Fundamentals –
Improvements in Modeling and Testing Reaction Systems and Furnace Performance

Estimation of the Selectivities of the First-order Reaction in Naphtha Pyrolysis Process Based on Fuzzy Matching Method

Xiaodan Gao, Bingzhen Chen, Xiaorong He
Department of Chemical Engineering
Tsinghua University, Beijing, China, 100084

Abstract: A fuzzy matching based method for estimating the selectivities of the first-order reaction of the Kumar's model is proposed. First a database with a set of standard data of the selectivities should be established by calculating the historical samples with known feed characteristics, and then membership functions are defined for the expression of the similarity degree of the naphtha to be estimated to the standard ones. After the matching, the estimation can be carried out based on the matched standard examples possessing relatively large values of the membership function. In a case study on the naphtha pyrolysis, the yields calculated based on the estimated selectivities of the first-order reaction are very close to the experimental results, which demonstrate the effectiveness of the proposed method.
Keywords: Ethylene, Fuzzy matching, Naphtha, Pyrolysis

1. Introduction

Naphtha pyrolysis is one of the basic and important processes of chemical industry. For implementing the on-line or off-line optimization of the naphtha pyrolysis process a mathematic model of the process is needed. The kernel of this process optimization model is the reaction model. As known, Kumar's molecule reaction model[1] has been widely and satisfactorily used to predict the main yields of the product. But in this model the selectivities of the first-order reaction always changes along with the change in the composition of naphtha. So, before using this model, it is necessary to get the selectivities of the first-order reaction of naphtha pyrolysis through estimation or experiments. To do some experiments is very expensive and time-consuming. But so far there is not any efficient method to estimate the selectivities. To address this problem, the paper proposed a fuzzy matching based method for estimating the selectivities of the first-order reaction of the Kumar's model.

2. Establishment of the database

The aim of establishment of database is to memorize a set of standard data that obtained from practical production for the fuzzy matching. Each standard data includes some necessary characteristics of naphtha feed (PONA, ASTM distillation, density), operation condition of the pyrolysis process (steam/naphtha ratio, naphtha flow rate, temperature of coil inlet and outlet, pressure), and the relevant selectivities of the first-order reaction. Kumar's first-order reaction equation is given as follows:

$$C_X H_Y = z_1 H_2 + z_2 CH_4 + z_3 C_2H_6 + z_4 C_2H_4 + z_5 C_3H_8 + z_6 C_3H_6 + z_7 C_4H_{10} + z_8 C_4H_8$$
$$+ z_9 C_4H_6 + z_{10} C_4^+$$

The selectivities of the standard data are calculated from the characteristics of naphtha, operation condition of the process such, and the yield distribution of the main products provided by practical production.

3. Fuzzy matching

When the amount of the standard data is large enough, Fuzzy matching can be used to estimate the first-order reaction selectivities of the naphtha the yield distribution of which is unknown.

The aim of the fuzzy matching is to find out three standard data, and the characteristics of naphtha in the standard data found must be the most similar to the naphtha to be estimated. The degree of similarity can be described by the membership function. In the end, the first-order reaction selectivities of the naphtha unknown can be estimated by add the three set of selectivities found that multiplied with specific power.

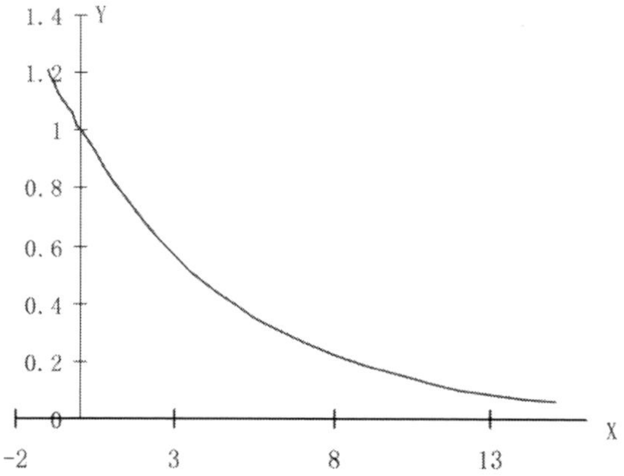

Fig.1 Graph of a membership function

Depend on industrial experience, five characteristics of naphtha have important influence on the first-order selectivities. They are listed as follows: (1) normal paraffins proportion (NP) (2) iso-paraffins proportion (IP) (3) average molecular weight (Ma) (4) characteristic factor (K) (5) relative density of 15℃ (d15). Each characteristic has a membership function, and the total membership function can be calculated by adding them together multiplied with special power.

Membership of each characteristic is:

$$P_i(s) = 2^{-\frac{|\Delta S^i|}{S^i_{0.5}}} \qquad i=1,\ 2,\ldots,5$$

In the equation, i represent the list number of each characteristic, ΔS^i is the characteristic difference of the standard data and naphtha unknown. $S^i_{0.5}$ is a parameter given from experience that can be adjusted based on the database, and its physical meaning is the value of ΔS^i when the value of membership function is 0.5.

The total membership function is $P(s) = \sum_{i=1}^{5} w_i (2^{-\frac{\Delta S^i}{S^i_{0.5}}})$. w_i is the power of each characteristic that can be given depending on the relationship degree of the characteristic and the first-order reaction selectivities, and $\sum_{i=1}^{5} w_i = 1$.

The total membership function $P(s)$ of each standard data is calculated and listed as: $p_1(s) \geq p_2(s) \geq \cdots \geq p_m(s)$, m is the amount number of the standard database, so $P_1(s), P_2(s), P_3(s)$ can be found to match. Their first-order reaction selectivities vectors are $\mathbf{z}(1), \mathbf{z}(2), \mathbf{z}(3)$.

At last, the first-order reaction selectivities of the naphtha unknown can be calculated as:

$$\mathbf{z} = \frac{P_1(s)}{P_1(s)+P_2(s)+P_3(s)}\mathbf{z}(1) + \frac{P_2(s)}{P_1(s)+P_2(s)+P_3(s)}\mathbf{z}(2) + \frac{P_3(s)}{P_1(s)+P_2(s)+P_3(s)}\mathbf{z}(3)$$

4. A industrial case

In a case study on naphtha pyrolysis, a standard database including fifteen samples is established and the fuzzy matching method is used to estimate the selectivities of the first-order reaction of some types of naphtha. The main product yield can be calculated by the method provided by Xuqiang[2-3].The result is listed in table 1.

Table 1 Comparison of realistic value and matching value calculated in a typical case

Characteristics and operation condition (From Beijing Chemical factory)			
Density(20℃)	0.6931	*N* (%)	24.5
IBP	30	*O* (%)	0
0.1	46	*A* (%)	5
0.5	92	Feed rate(kg/h)	12.5
0.9	136	Steam/naphtha	0.6
FBP	163	Outlet pressure	0.05
NP (%)	33.6	Inlet temperature	629
IP (%)	32.2	Outlet temperature	833

Comparison of calculation			
Yield distribution	Experienced	Calculated	Error%
C2H4	32.74	32.57	-0.52
CH4	13.16	12.96	-1.55
C3H8	0.43	0.43	-1.66
C3H6	15.31	15.42	0.77
C2H6	2.87	2.92	1.72
H2	0.78	0.76	-2.44
C4H8	2.7	2.67	-1.07
C4H6	3.84	3.79	-1.50
C4H10	0.32	0.30	-5.7

The yields calculated based on the estimated selectivities of the first-order reaction are very close to the experimental results, which demonstrate the effectiveness of the proposed method.

5. Conclusion

(1) The calculation result presents that the estimation method based on fuzzy matching can predict the main product distribution accurately.

(2) To ensure the precision of the matching, the characteristics of standard data should be similar to that of naphtha unknown, and the parameter of fuzzy matching should be adjusted at any moment depend on the difference of standard data and naphtha unknown.

Reference

[1]Kumar P, Kunzru D, Kinetics of Coke Deposition in Naphtha Pyrolysis, Can. J. Chem. Eng, 1985, 63 (8) : 598—603.

[2]Qiang Xu, Bingzhen Chen, Xiaorong He. [J] Computer and Applied Chemistry, 2001, 18 (3) , 223—228.
[3] Qiang Xu, Bingzhen Chen, Xiaorong He. [J] Computer and Applied Chemistry 2001, 18 (4) , 303—306.

Paper 47d

Accelerating innovation in olefins production by application of high throughput experimentation

Stephan Andreas Schunk, Alfred Haas, Jason King, Jens Klein

hte Aktiengesellschaft

Kurpfalzring 104

69123 Heidelberg

Prepared for Presentation at the 2005 Spring National Meeting in Atlanta, GA April 10-14
17th Annual Ethylene Producers Conference
Session 47 – Ethylene Plant Fundamentals –
Improvements in Modeling and Testing Reaction Systems and Furnace Performance

Accelerating innovation in olefins production by application of high throughput experimentation

Stephan Andreas Schunk, Alfred Haas, Jason King, Jens Klein

hte Aktiengesellschaft

Kurpfalzring 104

69123 Heidelberg

Tel.: 0049 6221 7497 111

e-mail: stephan.schunk@hte-company.de

http: www.hte-company.de

Introduction

Olefins production both from C1- and from higher carbon streams play a major role in the value chain of the petrochemical and chemical industry. Technologies for so-called high-throughput experimentation (HTE) to synthesize and test catalyst candidates span and reflect a whole range of technologies entailing software and hardware solutions that can aid in the development of new processes for olefin generation and conversion.

HTE technologies similar to those deployed to accelerate R&D in the pharmaceutical industry over recent decades, have found a place in non-pharma chemical R&D, addressing the demand for R&D groups to improve innovation productivity. Specifically this means expanding the scope of experimental search, while simultaneously reducing the unit cost and time per experiment. Key to such endeavors is the use of parallel arrangements of downscaled reactors, advanced analytics and a high degree of automation together with software environments that enable the scientist to scan large libraries of catalysts and reaction conditions and identify the most prospective combinations.

The development cycle in high throughput experimentation is naturally very focused and dedicated towards two goals: 1) the identifying and improving lead candidates within a given development program and 2) the reducing of the total number of potential lead candidates with each level of screening by application of certain performance thresholds.

The first goal is usually achieved by the "classical" HTE cycle which consists of the four steps of library **design**, the **make**-step in which libraries of materials are prepared, the **test**-step where the material's properties are evaluated and the **model**-step where the results that are obtained in the test phase are correlated with known properties of the materials and data collected during the make-step or data obtained from elsewhere (alternative sources of data can be data available by additional analysis of the materials candidates, data retrieved in the open literature or patent literature or data available from client databases). Figure 1 illustrates the self-perpetuating character of the circular approach, still it is not to be forgotten that human interference in this cycle, even if the workflow can be highly automated, is an important factor for success.

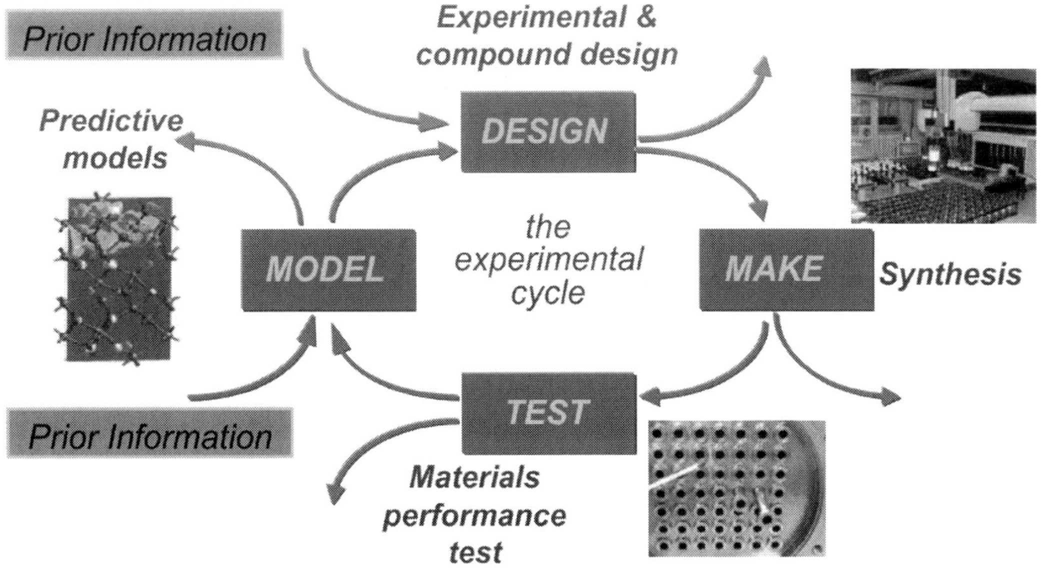

Figure 1: The "classical" HTE development cycle illustrating the workflow for the steps of design, make, test and model.

The second strategy that is used to narrow down the number of potential lead candidates mainly by discrimination in order to take a number of materials as low as possible to the next screening level, is called funneling strategy (see figure 2). In most screening programs in catalysis applications the number of potentially interesting materials at an initial point of the project will be a substantially larger number (typically on the order two or three magnitudes) than the one that is hopefully achieved at the target line of the project (reasonably one candidate material or a maximum of a handful of lead candidates is desired at the end of a testing campaign). Going along with the focusing of the number of materials that can provide a viable solution to the problem is usually the depth of information that is obtained following the screening program. Starting with Stage I, the so called exploratory screening Stage, several hundred materials can be tested within short time frames. The experimental conditions in this screening Stage can sometimes be off what would be considered reasonable process variables, the main focus rather being on a large number of candidate materials (i.e. a high diversity of the tested libraries) for evaluation. Stage II is especially different with regard to the process variables: here the focus is on proximity to the later industrial process in order to retrieve data that can aid late development stages also. Typically materials throughput is much lower in Stage II and data quality is usually close to conventional.

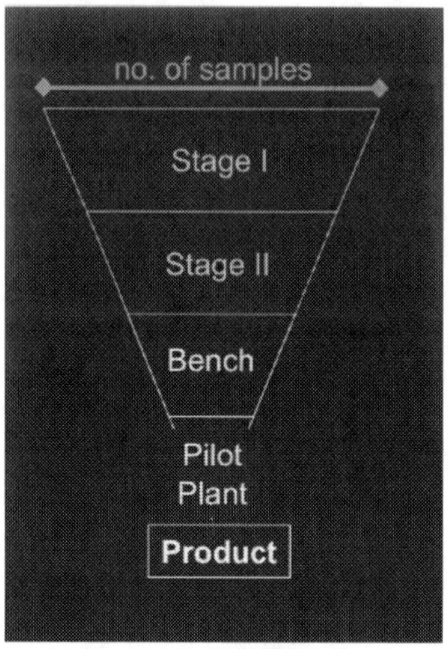

Figure 2: Illustration of the funneling strategy being applied to the different testing Stages.

Both approaches described above namely the HTE-cycle and the funneling strategy have for themselves a high value in HTE developments and are the main drivers with which a project can usually be significantly advanced . Still to a large extend both approaches rely on data modeling at different levels. The modeling steps can be a purely qualitative analysis of performance data, statistical analysis to link performance data with data obtained through preparative sequences or additional analysis data or even go to levels where structure activity relationships are established at the molecular level. Of course modeling on a molecular level, especially for heterogeneous catalyst systems has very challenging aspects to it, so in many cases more simple approaches that allow correlation and the retrieval of structure-activity relationships will be the preferred. Another important aspect of modeling is that it can not only be used to retrospectively evaluate the data obtained during catalyst screening but can also be efficiently employed to plan and direct the screening process. In this article we intend to discuss the pros and cons of a number of computational and screening approaches which have been employed to develop and find new catalysts for the generation of olefins from different feedstocks including recently obtained results as an illustration.

Experimental Planning and Data Handling using Modeling Approaches

Combinatorial Chemistry and HTE are powerful tools in the hands of a scientist, as they are a source for meaningful consistent records of data that would be hard to obtain via conventional methods within a reasonable timeframe. This blessing of fast data acquisition can turn into a curse if precautions are not taken to carefully plan the experiments ahead and the means of handling the data and analyzing them afterwards. The two essential elements that ensure a successful execution of ambitious projects on a rational and efficient basis are therefore tools that enable the scientist to carefully plan experiments and get the most out of the minimum number of experiments in combination with the possibility of fast and reliable automated data acquisition and retrieval from databases. Therefore experimental planning, modeling and data management are complementary skillsets for the pre- and post-experimental stages.

For the part of data-handling and databases we recommend the following literature for further studies [zitat födisch schunk sundermann finger saupe artikel[1]]. A number of programmable database systems are available on the market and can be customized to the actual project needs. The roots of these databases go back into the eighties where analytical laboratories started to digitize results by the aid of databases for the purpose of higher efficiency. The databases and visualization tools that are offered today by the market leading companies in HTE [2, 3] fulfill the requirements to master the challenges that the scientist is confronted with. It should be mentioned here that visualization tools are niche software products all by themselves with a high degree of specialization [4]. Generally these tools which are a great aid for the visualization of complex data packages interface nicely with most data formats extractable from databases.

In the stage of experimental design a number of different methods to carry out a sensible structuring of experiments are available. The choice of the different approaches will largely depend on the nature of the problem to be solved, the knowledge and the data already available, the data quality that can be obtained, the accuracy at which an answer is expected and the time available for the experimental campaign/the single experiment and evaluation of the data. In the following we will focus on the different approaches for experimental planning and modeling and discuss their pros and cons.

Fully randomized approaches

The number of approaches reflected in the literature that operate on a fully randomized basis is rather small [5, 6], still these approaches are an interesting way of dealing with the challenge of catalyst development via HTE technologies. Instead of a rationalized basis for the development of heterogeneous catalyst formulations an algorithm for the random production of mixed metal oxides via impregnation of a combustible exo-template was developed. The only input from the experimentation side was the boundary condition that the oxides that would possibly be obtained via the synthetic pathway had to be non volatile. The screening campaigns were focused on two different reactions: CO oxidation and ethylene production from ethane by oxidative dehydrogenation. A large number of catalyst formulations were synthesized via the efficient synthetic route and could be tested through the fast photoacoustic analysis technique. It has to be stressed here that the randomized approach was only taken for the initial library setup, the library optimization was done on a rational basis by the experimenter. Although this approach may appear to be exotic with regard to highly knowledge driven research approaches, it is a valuable methodology when it comes to the question of non-prejudiced search for new element combinations for target reactions.

Descriptor driven approaches

Descriptors are vectors or scalars that describe certain properties of a molecular or solid state compound (the descriptors for molecular compounds are also described as SMILES or broad SMILES [7] or as QSAR-approach [8] a similar approach was developed by Klanner et al. for the description of solid state materials [9, 10]). In their work the approach was employed for the oxidation of propylene not being focused on one single target product but more generally looking at all possible potential target products of the oxidation reaction with the goal of establishing general descriptor approaches for propylene oxidation. Apart from the fact that the descriptors can be used as a correlation and database for general properties they can be used for experimental planning as the display of the molecular or solid entity as vectorial or scalar units enables the scientist to rationalize the potentially different behavior of molecular units and solid state compounds with regard to a certain chemical test. These descriptors can therefore be employed very efficiently for library planning as they give the scientist the possibility to oversee the quality of a library of molecular or solid state catalysts with regard to its quality regarding diversity or clustering of catalyst species. It should be mentioned here that it is of course evident that the same or similar descriptors can be applied for molecular catalysts and molecular compounds that are potential candidates for a screening versus their bio-activity. The descriptors needed for solid state compounds differentiate substantially from the latter and go far beyond them with regard to the numerical description of physico-chemical properties. One has to keep in mind that the descriptor approach is novel for solid state compounds and it will take more time to get to the state that the descriptor approach for molecular compounds has already reached. Figure 3 shows a display of descriptors that can be utilized together with molecular and solid state compounds.

What is the most useful way to take use of the descriptor approach? The descriptor approach is extremely useful when it is applied together with algorithms seeking to cover the experimental space equidistantly with points (an excellent example for optimal coverage is the optimal coverage algorithm which also be discussed below [11]). Through the vectorial and scalar display of the properties of the catalyst candidates the experimenter can get an overview of the diversity of the library and the coverage of the multidimensional space being spanned by the properties of the catalyst candidates. It is clear that this approach is useful especially if a large number of samples is intended to be screened. The drawback of this method is connected to the generation of the descriptors that need to be developed usually need a pre-experimental evaluation stage.

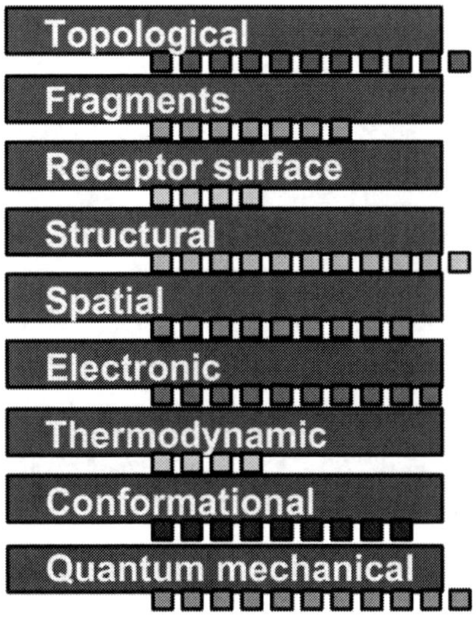

Figure 3: List of descriptors that can be employed for describing molecular and solid state compounds.

Approaches based on "classical" statistical designs

The application of classical statistical designs has a long tradition in the evaluation of properties of inorganic solid state compounds also in catalysis. Generally when statistical approaches like factorial designs, Box&Hunter, Hunter, Plackatt-Burmann, Box-Behnken, Taguchi or even NK model designs are employed it is very important to be aware of the nature of the variables and the expected results/answers from the experiment (please see classic textbooks for elemental statistics and basic design of experiments for example [12]). Not only that the scientist has to be aware what the control and response variables are but during the experimental planning also has to figure whether the design is suitable for the questions that he wants to answer with the experimental approach. In addition the chemical know-how is essential for the correct application of a statistical design. Only when the experimenter has an overview over all of the variables and can also exclude variations that do not make sense from a chemical point of view and add additional points to the design where it is obvious that the density of points is not sufficient, the success of the design strategy is ensured. Figure 4 shows a graphic representation of common factorial designs.

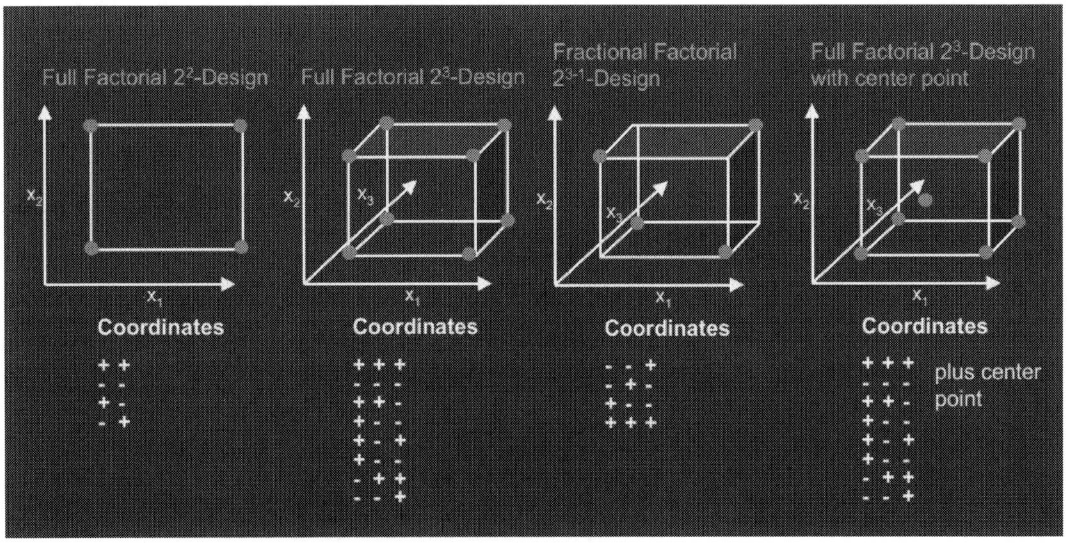

Figure 4: Graphic representation of factorial designs.

Figure 5 illustrates with three examples why the chemical know-how plays such an important role for the planning of an experiment. The highest risks are associated to an inadequate choice of the density of experimental points. The only way to overcome this problem is by carefully consulting chemical know-how in order to carefully plan the experiment and also carefully evaluate the data obtained during the experiment. It is though obvious that not all of the behavior of a newly explored system can be foreseen and a risk of missing singularities or sharp global optima in the behavior will always remain.

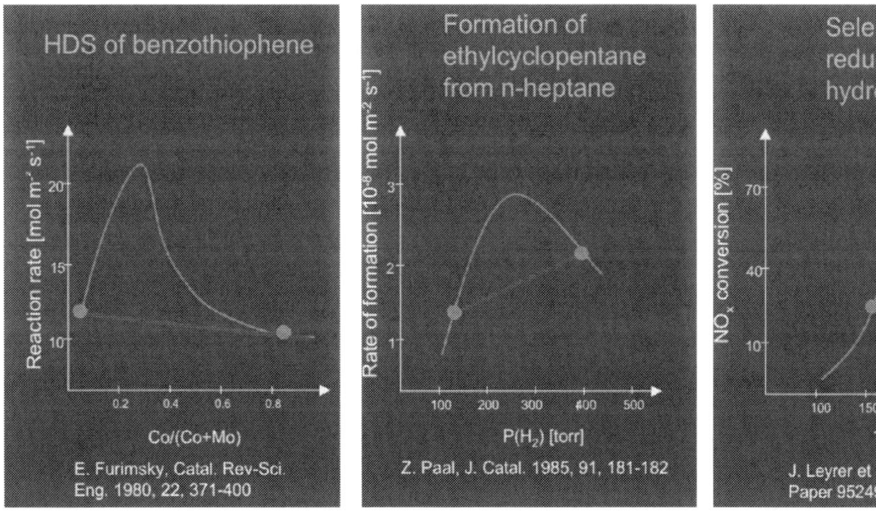

Figure 5: Three illustrative examples why the density of points is essential for the experimental planning stage. It can in all three cases be seen that the wrong choice of experimental points can lead to missing the opportunity of finding the global optimum.

The above described processes for classical statistical designs play an important role in complex chemical systems where it may be difficult to establish models on a molecular basis. A good example for such a catalytic process is the C1-based methanol to olefins (MTO) technology, which was originally established by Mobil [13]. The MTG process leads to a product which contains 75 - 80% of light olefins in mixture with water, paraffins, aromatics and dimethylether (DME). Methanol is initially converted to DME and water over the micro-porous catalyst and this reaction is followed by oligomerization reactions and hydrogen transfer, with the final product of the process being gasoline-range aromatics (see figure 6). The choice of catalyst and process conditions (temperature, space velocity and partial pressure of the feed methanol) will mainly govern the products obtained, naturally with the focus being on light olefins. Although MTO is a well studied process, the crucial C-C coupling step starting from methanol as a C1-Hydrocarbon is not well understood at present. Shape-selectivity plays an important role and the medium pore ZSM-5 materials show superior yields for larger olefins and low coke formation. Small pore zeolites like erionite and chabazite will usually produce more light olefins like ethylene and more coke. Especially remarkable is the performance of SAPO-materials which are superior for high yields of ethylene and propylene. It is clear that the nature of the zeolitic material employed plays an essential role, but approaches to molecular modeling were so far only partially able to reveal the reasons for the catalytic behavior. One of the major reasons for this is potentially the fact that a) all of the zeolite materials employed have to be formulated into a matrix that has usually also an impact on the catalytic behavior of the material and b) it is common to dealuminate the zeolites before exposing them to process conditions. Both of these preparative factors make the catalysts less accessible to description on a molecular level. These reasons make the application of statistical approaches very sensible in the context of refinery related applications. An additional factor for the performance is the coke formation and time on stream behavior. hte Aktiengesellschaft having been active in the field of petroleum chemicals has developed a proprietary testing platform for low to high pressure refinery related reactions which allows full monitoring time on stream behavior and independent regeneration of the catalyst test candidates [14].

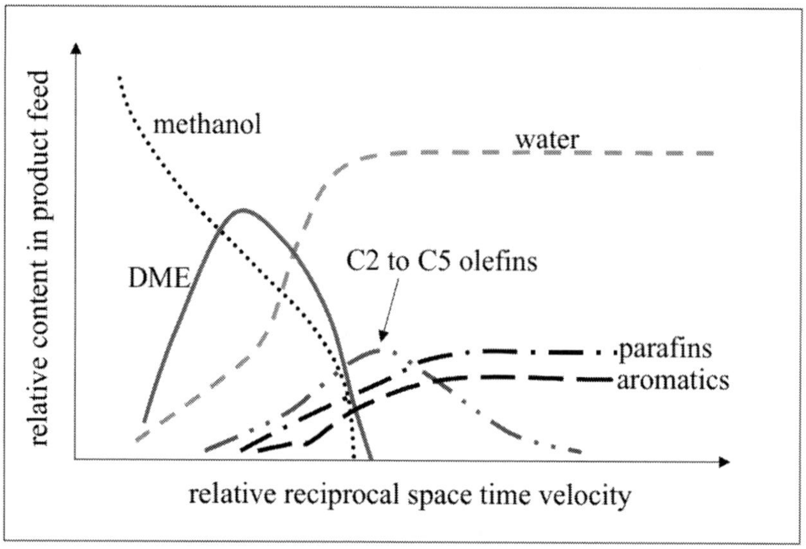

Figure 6: Product distribution of the MTO reaction as a function of the reciprocal space hourly velocity for an HZSM-5 catalyst [15].

Still different in nature but also concerned with the optimal distribution of points in multidimensional space are optimal coverage algorithms [11]. Bergh et al [16, 17] have employed these algorithms for library generation and graphical evaluation for ethane dehydrogenation to ethylene and ethane oxidation to acetic acid. The results are conclusive as a clear correlation between the compositional data and the performance data of the libraries can be observed. Still one has to keep in mind that this approach relies very much on a compositional chemical space which may not directly translatable and/or connected to a structural space; this may lead to a non optimal coverage of points over the structural coordinates. In many cases this can be acceptable still it has to be kept in mind that the optimal coverage of points on the compositional coordinate does not necessarily lead directly to a structure-property relationship.

Other techniques for finding global optima

The search for optima within a given experimental space can also be explored by different methodologies than those that we have discussed before. We want to highlight two of them in this context namely genetic algorithms and neural networks.

The principle of genetic algorithms is largely connected to the idea of evolutionary improvement of the properties of a given specimen within a defined space of variables. A genetic algorithm is, in other words, an algorithm which permutes and converts variables in search of a global maximum or minimum of a multidimensional parameter space. Within that context genetic algorithm have to be seen under the same precautions that apply for statistical designs as listed above. Also here the parameterization and the chemical know-how play a large role in the set up an adequate parameter space and the choice of the right variables. The rules according to which the algorithm operates are the selection of "winners" from a given set of samples and their potential "improvement" via "mutation" and "cross-over". The so called selection of "winners" is an algorithm which tests whether the system has reached a global maximum or minimum and suggests new experiments to confirm and reach the predicted minimum or maximum. Both "mutation" and "cross-over" are algorithms by which new experiments are created choosing from a given set of experimental parameters. Figure 7 illustrates the approach graphically.

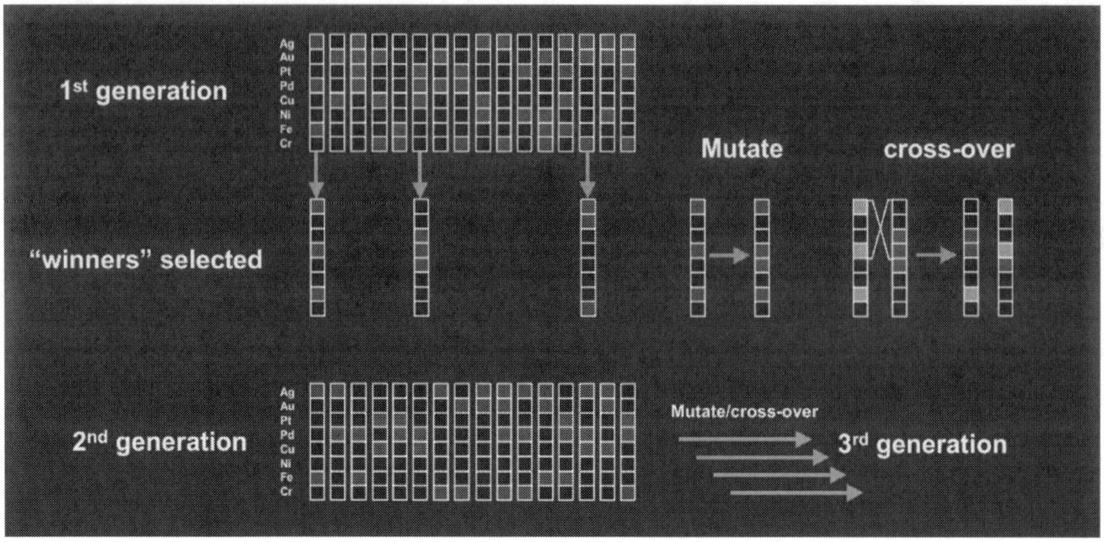

Figure 7: Graphical illustration of the evolutionary approach for optimization.

Genetic algorithms have been successfully tested for applications in heterogeneous catalysis, especially for the oxidative dehydrogenation of propane, [18] and although the results of this example and the general approach look very promising one should keep in mind that generally several thousand experimental data points are required to reach a global maximum or minimum. The original challenges that were the target applications of this algorithm were improvements in production chains or other logistical tasks where a single multi hour run delivers generally several thousand data points. Though if sufficient data points can be generated in the catalytic testing genetic algorithms are an interesting approach for optimization. Within the work of Wolf et al [18] a base library of redox metal oxides with metal binding energies from -400 to -200 kJ/mol were chosen. The "survivors" of the first test (or "evolution") were then combined with a basic metal oxide, an acidic metal oxide and a metal oxide on which O_2 can dissociate to generate binary and ternary compositions. Within five "generations" of catalyst modifications propylene yield could be improved from 2% to 9%. The outcome as the optimum of the application of the genetic algorithm for the dehydrogenation of propane to propylene were new catalytic formulations based on Mg-V-Ga-Mo mixed oxides.

Even more sophisticated in their mathematical functionality are neural networks: these are mathematical models which target at the correlation between input and output data by application an underlying system of neurons and connections (as depicted in figure 8). The neurons and connections are also named "hidden layers", generally the experimenter has no influence on the structure of these layers. Neural networks were designed to imitate the functionality of biological neural networks, so similarly to these systems they also have to undergo a learning or training process prior to predicting optimum compositions for obtaining global optima.

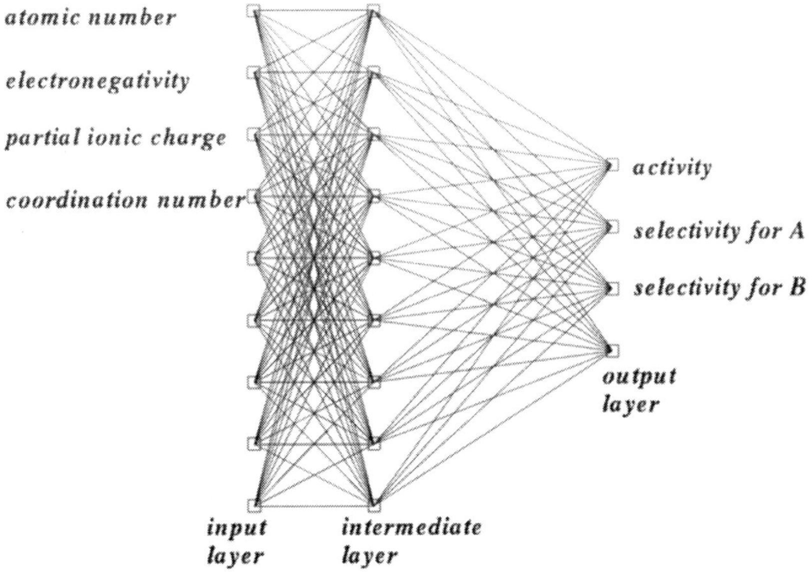

Figure 8: Principal construction of a neural network. The input and output data are examples of potential data sources of interest.

Again neural networks are an interesting approach for system optimization, still one has to take into account that a) the training phase requires a certain amount of time and experience (both over- and under trained networks will tend to give false readouts) and b) generally the data correlation will be logical but non-deterministic, meaning without insight of the underlying physico-chemical correlation. From the point of view of the author further development work is required to make neural networks into a powerful tool for HTE.

Summary and Outlook

Modeling techniques as well as experimental design techniques are valuable tools that can aid in progressing projects executed with the aid of HTE technology. The choice of the methodology which is employed to generate virtual libraries and model and evaluate test results is crucial and has to be adapted appropriately to the catalytic challenge. Further progress is to be expected on the forefront of this research topic and the future will show which design criteria and modeling techniques will become the workhorses in HTE.

1 R. Födisch, K. E. Finger, S. A. Schunk, A. Sundermann, M. Saupe: " Requirements and solution approaches for software architectures supporting High Throughput Experimentation", accepted for QSAR & Combinatorial Science

2 www.hte-company.de

3 www.symyx.com

4 www.spotfire.com

5 T. Johann, A. Brenner, M. Schwickardi, O. Busch, F. Marlow, S. Schunk, F. Schüth Angewandte Chemie, 2002, 41, 2966

6 T. Johann, A. Brenner, M. Schwickardi, O. Busch, F. Marlow, S. Schunk, F. Schüth Catalysis Today 2003, 81, 449

7 J. Hinze and U. Welz, "Broad Smiles", In J. Gasteiger, (editor), Software Development in Chemistry 10, page 59. Gesellschaft Deutscher Chemiker (GDCh), Frankfurt/Main, 1996

8 C. Hansch, A. Leo, D. Hoekman, Exploring QSAR - Hydrophobic, Electronic, and Steric Constants. American Chemical Society, Washington, D.C, 1995

9 C. Klanner, D. Farrusseng, L. Baumes, C. Mirodatos, F. Schüth, QSAR & Combinatorial Science 2003, 22, 729

10 C. Klanner, D. Farrusseng, L. Baumes, M. Lengliz, C. Mirodatos, F. Schüth, Angew. Chem. in press

11 D. Bem, R. D. Gillespie, E. J. Erlandson, L. A. Harmon, S. G. Schlosser, A. Y. Vayada, Combinatorial Experimental Design Using the Optimal-Coverage Algorithm in Experimental Design for Combinatorial and High Throughput Materials Development, Editor J. N. Cawse, Wiley-Interscience, Hoboken, 2003, 89

12 Statistical Design and Analysis of Experiments: With Applications to Engineering and Science, 2nd Edition, Robert L. Mason, Richard F. Gunst, James L. Hess, Wiley-Interscience, Hoboken, 2003

13 P. M. M. Blauwolf, J. W. Gosselink, E. P. Kieffer, S. T. Sie, W. H. J. Stork in Catalysis and Zeolites, Fundamentals and Applications, J. Weitkamp and L. Puppe (Editors), Springer 1999, 437

14 S. A. Schunk, D. Demuth, A. Cross, O. Gerlach, A. Haas, J. Klein, J. M. Newsam, A. Sundermann, W. Stichert, W. Strehlau, U. Vietze, T. Zech in High Throughput Screening in Heterogeneous Catalysis, A. Hagemeyer, P. Strassr, A. F. Volpe (Editors), Wiley-VCH 2004, 19

15 C. D. Chang, Catalysis Reviews Science and Engineering 1983, 25, 1

16 S. Bergh, P. Cong, B. Ehnebuske, S. Guan, A. Hagemeyer, H. Lin, Y. Liu, C. G. Lugmair, H. W. Turner, A. F. Volpe Jr., W. H. Weinberg, L. Woo, J. Zysk Topics in Catalysis 2002, 23, 65

17 S. Bergh, S. Guan, A. Hagemeyer, C. Lugmair, H. Turner, A. F. Volpe Jr., W. H. Weinberg, G. Mott Applied Catalysis 2003, 254, 67

18 D. Wolf, O. Buyevskaya, M. Baerns, Applied Catalysis 2000, 200, 63

Paper 47e

Towards Comprehensive CFD Modeling of Lean Premixed Ultra-Low NO_x Burners in Process Heaters

Q. Tang,

B. Adams, M. Bockelie, M. Cremer, M. Denison, C. Montgomery, A. Sarofim

Reaction Engineering International

D. J. Brown

Shaw Stone & Webster, Inc.,

Prepared for Presentation at the 2005 Spring National Meeting in Atlanta, GA April 10-14
17th Annual Ethylene Producers Conference
Session 47 – Ethylene Plant Fundamentals –
Improvements in Modeling and Testing Reaction Systems and Furnace Performance

Towards Comprehensive CFD Modeling of Lean Premixed Ultra-Low NO_x Burners in Process Heaters

Q. Tang, B. Adams, M. Bockelie, M. Cremer, M. Denison, C. Montgomery, A. Sarofim
Reaction Engineering International
http://www.reaction-eng.com

D. J. Brown
Shaw Stone & Webster, Inc.
http://www.shawgrp.com/StoneWebster/index.cfm

Abstract

Air quality emissions regulations, NO_x in particular, are impacting the chemical process industry and forcing adoption of expensive selective catalytic reduction (SCR) units or retrofit of less expensive but relatively unproven ultra-low NO_x burners. CFD modeling provides a potentially cost-effective method for evaluating NO_x emissions and furnace performance for new burner technologies, thus minimizing furnace start-up times and unforeseen performance impacts. However, existing CFD models are incapable of accurately predicting NO_x emissions due to the complex geometries, turbulent mixing, lean premixed combustion chemistry, heat transfer and low NO_x levels in the new generation of burners.

This paper reviews a new computational tool developed by REI specifically designed to model lean premixed combustion associated with low NO_x burners in process heaters/furnaces. The new tool is based on a Hybrid Scalar Transport Probability Density Function Solver and includes adaptive mesh refinement to capture near-burner mixing, reduced chemical kinetic mechanisms with *in-situ* adaptive tabulation to compute finite rate chemistry, a combination of conventional Eulerian turbulence modeling with Monte-Carlo PDF methods to model turbulent reactions and mixing, and a matrix-free Newton-Krylov method to reduce solver computational time and improve robustness. Predicted flame characteristics and emission levels for low NO_x burners in a test furnace will be discussed and compared with measured and observed data.

1. Introduction

Air quality emissions regulations resulting from the National Ambient Air Quality Standards (NAAQS) are requiring increasingly stringent ozone emissions from numerous furnaces in the chemical process industry. NO_x emissions have been identified as a major contributor to ground level ozone, and have been targeted for significant reduction. These reductions are impacting a large number of industrial boilers and heaters in non-attainment areas, approximately 1900 in the Houston-Galveston area alone, for example, and are forcing furnace operators to adopt expensive selective catalytic reduction (SCR) technology (an SCR retrofit can exceed $3M for a large ethylene cracking furnace) or retrofit less expensive but relatively unproven ultra-low NO_x burners. These ultra-low NO_x burners can provide a relatively inexpensive solution to NO_x reduction requirements, but reductions must be achieved without compromising the heat release profiles or flame 'quality' of the burners: it is not uncommon for ultra-low NO_x burners to have wider and longer flames than conventional burners or even, in the worst case, for flames to impinge directly on the heat absorbing process tubes. CFD modeling provides a potentially cost-effective method for evaluating NO_x emissions and furnace performance for new burner technologies, thus minimizing furnace start-ups times and unforeseen performance impacts. However, existing CFD models have shown poor accuracy in predicting furnace NO_x emissions due to the combination of complex geometries, turbulent mixing, lean premixed combustion, and low NO_x levels in the new generation of low NOx burners.

Existing CFD software has been used to study more general process furnace characteristics such as furnace exit temperatures, heat transfer to process tubes, and flame shapes, as well as specific combustion components such as single burners (Brown, *et. al* 2003, Brown, *et. al* 1997). However, current commercially available CFD tools are generally limited in one or more of the following capabilities necessary for accurate low NO_x process heater modeling:

- sufficient computational points to resolve the detailed burner geometry, multiple fuel mixing zones, flame interactions between burners, and full furnace geometry;
- combustion sub-models to predict fuel-lean, premixed, turbulent combustion;
- radiative heat transfer sub-models to describe gas-wall-tube heat exchange;
- flow and chemistry sub-models to account for turbulence-chemistry interactions; and
- finite-rate chemical kinetics sub-models to account for NOx reactions.

Developing software that provides all of these capabilities in a single modeling tool has proven challenging due to limitations in computer capacity and descriptions of detailed turbulence and chemistry.

This paper describes a new computational tool developed by Reaction Engineering International (REI) that is specifically designed to model lean premixed combustion associated with ultra-low NO_x burners in process furnaces. The new tool is based on a Hybrid Scalar Transport Probability Density Function Solver and includes:

- adaptive mesh refinement to capture near-burner mixing;
- reduced chemical kinetic mechanisms with *in-situ* adaptive tabulation to compute finite rate chemistry;
- a combination of conventional Eulerian turbulence modeling with Monte-Carlo PDF methods to model turbulent reactions and mixing; and
- a matrix-free Newton-Krylov method to reduce solver computational time and improve robustness.

The remainder of the paper is organized as follows: A brief review of relevant turbulent combustion models is provided first followed by details of the included sub-models and their implementation into the new tool. This is followed by sections that provide simulation results for a well characterized, partially premixed methane flame and an industrial scale furnace firing a low NO_x burner, respectively. Last, conclusions and planned future work are summarized.

2. Modeling Turbulent Combustion

Most practical combustion systems involve turbulent flames, which are characterized by various complex and rate-controlling processes associated with turbulent transport, mixing, chemical kinetics, and convective and radiative heat transfer, among others. The success of any methodology used in modeling turbulent combustion depends a great deal on how well it can model these processes, which often strongly interact with each other at various disparate timescales and length scales.

In combustion processes which are characterized by fast chemistry, use of the flamelet model (Peter, 1984; Pitsch and Steiner, 2000) or conditional moment closure (CMC) (Bilger, 1993) based on a conserved scalar is known to be quite accurate in making qualitative as well as quantitative predictions. However, these methods may not be viable in slow chemistry regimes such as the formation and destruction of NO_x. The flamelet model requires the specification of a scalar dissipation rate and assumes the shape of the PDF at the sub-grid level. The first-order CMC model ignores any fluctuations about the conditional mean (Raman *et al.*, 2003).

Probability density function (PDF) methods have been demonstrated to be a successful modeling approach for turbulent combustion (Pope, 1985, Lindstedt *et al.*, 2000, Tang *et al*, 2000, Xu and Pope, 2000). The approach offers an important advantage over other methods in that chemical reactions appear in closed form in the PDF equations. As a consequence, realistic combustion chemistry can be incorporated without the need for closure approximations pertaining to the reactions. Therefore, PDF methods are able to accurately describe finite-rate chemistry effects and turbulent-chemistry interactions in turbulent flames.

There are basically two categories of PDF methods: the joint scalar PDF (or composition PDF) method and the joint velocity-turbulent frequency-composition PDF method. Both methods have been used for modeling turbulent combustion. The methods are general, apply equally well to premixed, partially premixed, and non-premixed combustors. In this study, the joint scalar PDF method is selected as the foundation of the advanced modeling tools for modeling practical lean premixed combustion problems due to its simpler implementation and ease of handling the complex 3-D geometry needed to model industrial combustion equipments. In this approach, the modeled transport equation for the joint scalar PDF equation is solved, and a coupled or uncoupled RANS solver is applied to solve modeled momentum equations. In addition to the unclosed terms in the traditional RANS equations, the molecular flux terms in the PDF equations also need to be modeled. It is straightforward to combine this method with well-established RANS solvers. In addition, the statistical error due to randomness is reduced because the convection is not stochastic. The implementation of well-established physics models for other processes such as radiative heat transfer, and the treatment of complex boundary conditions are straightforward, which is important for comprehensive modeling of industrial combustion systems.

New approaches such as large eddy simulation (LES) and direct numerical simulation (DNS) have been applied to combustion modeling (e.g., Pitsch and Steiner, 2000, Veynante and Vervisch, 2002) in recent years. These methods have the potential to resolve turbulent combustion phenomena in unprecedented detail. However, these methods have not reached the state of maturity required for implementation in a model intended for practical combustors with complex geometry and chemistry. For a more thorough review of turbulent combustion models and their use in modeling turbulent combustion processes, see the textbook by Peters (2000) or the recent review paper by Hilbert *et al.* (2004).

3. Description of New Modeling Tool

This section describes the new modeling tool. Presented are:

1) An overview on the structure of the tool and the overall solution strategy;
2) The adaptive mesh refinement (AMR) turbulent flow solver;
3) The approach for problem initialization;
4) Sub-models and solution scheme used within the PDF solver. Described are the sub-models used to treat turbulence diffusion, molecular mixing, and heat transfer processes, respectively;
5) A brief explanation on the reduced chemical mechanism;
6) The *in-situ* adaptive tabulation (ISAT) algorithm that incorporates realistic combustion chemistry into PDF calculations is discussed.

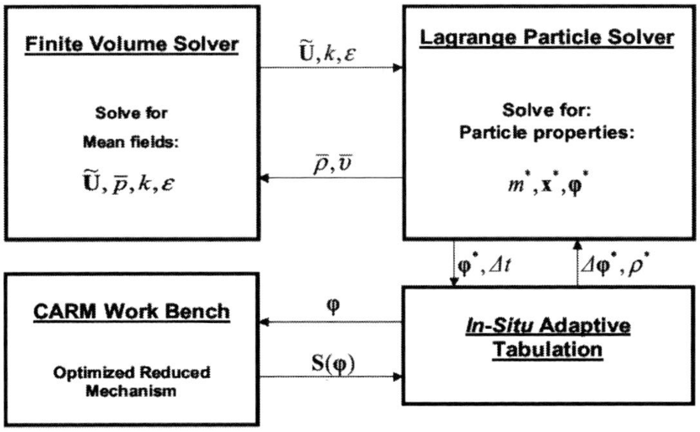

Figure 3.1 Modules in new CFD modeling tool.

3.1. Overview

The modeling tool solves the joint scalar PDF transport equations for statistically stationary turbulent reactive flows. In the PDF approach, the high dimensional joint probability density function is represented by a large ensemble of notional 'fluid' particles. The particle properties evolve according to particle sub-models such that the evolution of the statistics of the particle ensemble corresponds to the modeled PDF evolution.

The new modeling tool consists of four major components, namely an Eulerian finite-volume flow solver, a Lagrangian particle solver, an *in-situ* adaptive tabulation (ISAT) (Pope, 1997) module for reaction source term evaluation, and a chemical mechanism generator (CARM work bench) (Montgomery et al., 2002). The relationships among different components are illustrated in Figure 3.1.

During the simulation, the conservation equations for mean mass, momentum, pressure, and turbulent kinetic energy and dissipation rate are solved by the finite volume flow solver while the Lagrangian particle solver is employed to solve the joint composition PDF transport equation. A reduced chemical mechanism (i.e., a FORTRAN subroutine) is generated by CARM offline, and is linked to the other modules to form the executable. The highly efficient *in-situ* adaptive tabulation module is called by the particle solver to evaluate the reduced mechanism when solving the joint PDF equations.

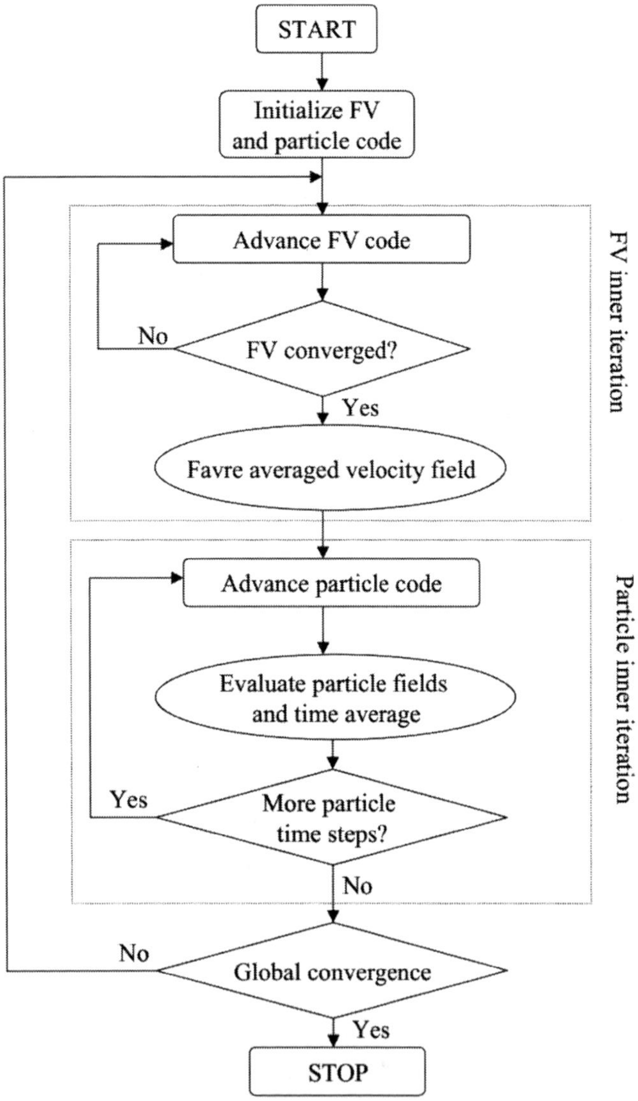

Figure 3.2 Solution scheme - hybrid FV/particle method.

Figure 3.2 shows the solution scheme adopted in the modeling tool. The FV and particle methods are periodically used in the hybrid algorithm to solve their respective equations. The hybrid FV/particle algorithm can be designed to run in a tightly or loosely coupled manner. In a loosely coupled algorithm, the outer iteration is completed by running the FV code for a specified number of iterations (steps) or until convergence and the particle code is run for a specified number of time steps. Note that tight coupling, in which both FV and particle codes are run for a single step to complete an outer iteration, is in fact a special case of loose coupling.

The overall solution sequence can be summarized as follows.

1) The Eulerian finite volume solver solves for mean velocity, pressure, turbulent kinetic energy and turbulent dissipation.

2) Given these mean fields, the Lagrangian particle solver solves for the joint composition mass density function.
3) Effects of thermo-chemistry are solved in the Lagrangian algorithm.
4) Mean thermo-chemical properties that influence the flow field, like density and molecular viscosity fields, are then coupled back to the finite-volume method that solves for the new turbulent flow field. Subsequent sub-model calculations are performed until a stable solution is reached.

3.2. Unstructured Grid Flow Solver with Adaptive Mesh Refinement

An AMR flow solver developed by REI is used for the 3-D finite volume solver. The solver is an adaptive grid code that uses an orthogonal grid with local mesh refinement. A structured base grid is input and grid refinements are performed. Cells that are tagged for refinement are divided into eight smaller cells. Cells are tagged for refinement either manually via user input, automatically to resolve geometry, or based on gradients in field variables in the solution. A typical AMR grid is shown in Figure 3.3.

The flow model of the AMR code uses a SIMPLEM algorithm (Acharya and Moukalled, 1989). This algorithm is similar to SIMPLER (Patankar, 1980) except that the mass-satisfying flow field defined on the cell faces is calculated from the pressure equation directly in lieu of a pressure correction equation (Acharya and Moukalled, 1989). The momentum equations are solved in a collocated manner so that the velocity field is stored at the cell centers and velocities from the discretized equations, without the pressure gradient source term, are interpolated to the cell faces via a momentum interpolation. The pressure field, also stored at cell centers, is solved from the continuity equation discretized over the cells. The resulting pressure gradient sources are then added to the interpolated face velocities resulting in a mass satisfying flow field at the faces. The discretized partially differential equations are solved using an algebraic multi-grid (AMG) strategy.

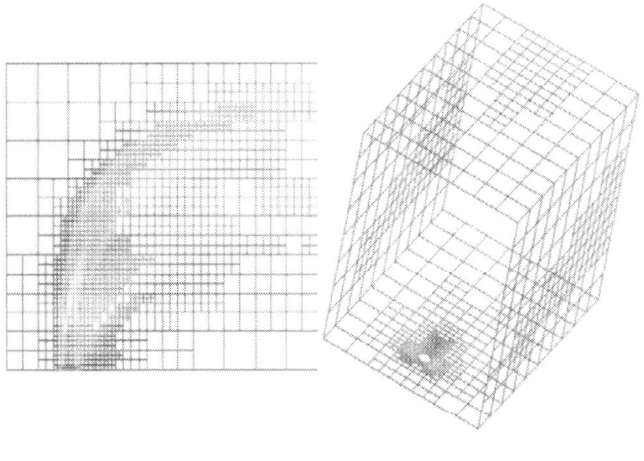

Figure 3.3 AMR solution for an injector

3.3. Problem Initialization

In the particle-based Monte Carlo method, the fluid is represented by a large number of "particles" having properties that evolve according to the model stochastic differential equations. For a typical 3-D turbulent combustion simulation, millions of particles are tracked in the simulation to achieve a quality solution. Even with the latest numerical techniques and computational power, the calculation can be very time consuming. Hence, a good problem initialization strategy using a simple, yet sufficiently accurate, turbulent combustion model is crucial for the success of the PDF method in practice.

The AMR code includes an assumed PDF turbulent combustion model based on mixture fraction and its variance. For non-premixed, diffusion-limited combustion the approach works well. However, its extension and application to premixed combustion with multi-step chemistry is awkward because of the assumption of statistical independence for each additional reaction variable and the significant increase in the required computational effort. To avoid this difficulty, an eddy dissipation concept (EDC) model has been implemented to provide an initial estimate of the solution for pre-mixed combustion.

The implemented EDC model is an extension to the EDC model of Gran and Magnussen (1996), and can account for multi-step chemistry. The EDC model is based on the assumption that chemical reactions occur in the regions where the dissipation of turbulence energy takes place. In flows from moderate to intense turbulence, these regions are concentrated in isolated regions, occupying only a small fraction of the flow. These regions consist of fine structures whose characteristic dimensions are of the order of the Kolmogorov length scale in one or two dimensions. Based on this assumption, the mean reaction rate is expressed as a function of the volume fraction of the fine structure regions, the lifetime of the fine structure, and the species mean mass fraction. In the implementation of the modified EDC model, a pseudo time splitting scheme is applied in the AMR code to solve the mean species and enthalpy transport equations, where a convection-diffusion step and a reaction step, which solves the EDC mean reaction rates, are performed to update the scalar fields. Note that the ISAT algorithm is applied in the reaction step, which speeds up the calculation significantly.

The work of implementing another solution scheme based on a Newton-Krylov solution strategy for the EDC model is in-progress. Unlike the traditional "Picard" iteration scheme used in the time splitting scheme, the Newton-Krylov based solver solves the species transport equation simultaneously. Previous study has shown that the Newton-Krylov solver can cut the CPU time by a factor of two compared to the Picard iteration (Wang, *et al.*, 2002)

3.4. Sub-models for Lean Premixed Combustion

3.4.1 Problem Formation

In the transport PDF approach, species concentration and specific enthalpy are treated as random variables, and the transport of their joint PDF rather than their finite moments is considered. Once the PDF is determined, the mean of any quantity can be evaluated exactly from the PDF, if it is a function of the species concentration and/or specific enthalpy.

The composition vector is defined as $\boldsymbol{\psi} \equiv (\psi_1, \psi_2, \ldots, \psi_{ns}, \psi_h)^T$, where $\psi_1, \psi_2, \cdots, \psi_{ns}$ are species mass fractions and n_S is the number of species, and ψ_h is the specific enthalpy of the mixture. The sub-model calculates the local one-point, one-time, joint composition PDF f_φ. The transport equation for f_φ can be derived from the general equations for conservation of species mass fraction and energy for low-Mach-number flow. Following the procedure outlined by Pope (1985), the transport equation for f_φ can be expressed as

$$\begin{aligned} &\frac{\partial}{\partial t}\left[\rho(\boldsymbol{\psi}) f_\varphi\right]+\frac{\partial}{\partial x_i}\left[\rho(\boldsymbol{\psi}) \bar{U}_i f_\varphi\right]+\frac{\partial}{\partial \psi_\alpha}\left[\rho(\boldsymbol{\psi}) S_\alpha(\boldsymbol{\psi}) f_\varphi\right] \\ &\quad=\frac{\partial}{\partial \psi_\alpha}\left[\left\langle\left.\frac{\partial \mathbf{J}_i^\alpha}{\partial x_i}\right| \boldsymbol{\psi}\right\rangle f_\varphi\right]-\frac{\partial}{\partial x_i}\left[\rho(\boldsymbol{\psi})\left\langle u_i'' \mid \boldsymbol{\psi}\right\rangle f_\varphi\right]+\frac{\partial}{\partial \psi_h}\left[\left\langle\nabla \cdot \mathbf{q} \mid \boldsymbol{\psi}\right\rangle f_\varphi\right] \end{aligned} \tag{1}$$

where ρ, $\bar{U}$, u'', S, $\mathbf{J}$ and $\mathbf{q}$ denote density, mean velocity, turbulent fluctuation velocity, chemical reaction source term, diffusion flux and radiation heat flux, respectively. The notation of $\langle A \mid B \rangle$ is the expectation of the conditional probability of event A, given that event B occurs. All terms on the left-hand-side of Equation (1) appear in closed forms. The first term on the right-hand-side represents transport in composition space due to molecular mixing. The second term represents transport in physical space due to turbulent velocity fluctuations. The third term represents transport in composition space due to radiation heat loss. These three terms need to be modeled. The approach for modeling these terms is provided below.

3.4.2 Turbulent Diffusion Model

The transport term due to turbulent velocity fluctuations is modeled using the gradient-diffusion model. Detail about this model can be found in Pope (1985).

3.4.3 Mixing Model

Two mixing models are implemented to model the transport of probability density function in composition space by molecular fluxes, namely the interaction-by-exchange-with-the-mean (IEM) model (Dopazo, 1975) and the Euclidean Minimum Spanning Tree (EMST) model (Subramaniam, 1997, Ren *et al.*, 2002).

IEM: The IEM model has the effect of moving the value of the instantaneous composition towards the mean composition at a controlled rate and thus reduces the distribution in the composition values. The model is simple and easy to implement, but could be problematic when there are strong interactions between turbulent mixing and finite-rate chemical reaction (Subramaniam, 1997).

EMST: The EMST model is a complicated particle-interaction model designed to overcome shortcomings of simpler models. A full description of the EMST model can be found in Subramaniam (1997) and Ren *et al.* (2003). The EMST model is a localized mixing model, and was successfully used in several joint PDF calculations of Piloted-Jet flames and Bluff-body flames to predict local extinction and re-ignition (Tang *et al.*, 2000, Xu and Pope, 2000, Liu, 2004).

The mixing frequency required by the IEM and the EMST mixing models is obtained from the $\kappa-\varepsilon$ turbulent model (Jones and Launder, 1973) contained in the finite volume flow solver.

3.4.4 Wall Heat Transfer Model

Radiation heat transfer and near-wall convection/conduction heat transfer must be accurately represented in modeling wall-bounded combustion flows that are typical in industrial combustion equipment. Hence, models to handle the heat transfer between the combustion gas and wall have been implemented into the hybrid PDF solver.

The heat transfer between gas phase and wall due to convection and conduction is described by a linear model (Fox, 2004) of the form

$$\frac{dT^{(n)}}{dt} = C_w\left(T_w - T^{(n)}\right) \tag{2}$$

where $T^{(n)}$, T_w are particle temperature and wall temperature, respectively. C_w is a coefficient chosen to yield the desired heat transfer rate, which will vary from cell to cell and change with time as the flow evolves. It should be noted that the maximum heat transfer will occur when $C_w = \infty$ (i.e., $T^{(n)} = T_w$), and that $C_w = 0$ corresponds to a zero-flux wall.

3.4.5 Radiation Heat Transfer Model

Radiation is typically the most significant mode of heat transfer in practical combustion equipment. Accurately simulating radiation heat transfer to specific regions in a system requires a model that can account for both absorbing-emitting radiation processes and complex system geometries, including arbitrary structures such as convective tube passes. REI's model utilizes the discrete-ordinates method (DOM) that has proven to be a viable choice for modeling radiation in combustion systems, both in terms of computational efficiency and accuracy (Adams and Smith, 1995). The model has been implemented into the AMR flow solver and is modified to couple with the Monte Carlo PDF solver.

The discrete-ordinates method solves the integral-differential radiation heat transfer equation (RTE) in a number of discrete angular directions spanning the total solid angle of 4π steradians. The discrete-ordinates representation of the mean radiation heat transfer equation can be written in Cartesian coordinates as

$$\mu_m \frac{\partial \langle I_m \rangle}{\partial x} + \eta_m \frac{\partial \langle I_m \rangle}{\partial y} + \xi_m \frac{\partial \langle I_m \rangle}{\partial z} = -\langle k_{ag} \rangle \langle I_m \rangle + \langle e_{bg} \rangle \tag{3}$$

where μ_m, η_m, and ξ_m are the direction cosines for the discrete angular direction Ω_m, $\langle I_m \rangle$ is the mean radiation intensity in the direction Ω_m, $\langle k_{ag} \rangle$ is the mean gas absorption coefficients, and $\langle e_{bg} \rangle$ is the mean gas blackbody radiation intensity, which can be calculated as

$$e_{bg} = \frac{\sigma}{\pi} k_{ag} T_g^{\,4} \tag{4}$$

where $\sigma = 5.669 \times 10^{-8} W/m^2K^4$ is the Stefan-Boltzmann constant and T_g is the gas temperature. The radiation heat loss rate per unit volume that forms the radiation source term in the energy transport equation can be expressed as

$$Q_r = \nabla \cdot \mathbf{q} = 4\pi e_{bg} - k_{ag} \sum_{m=1}^{M} w_m I_m \tag{5}$$

where w_m is the quadrature weight for the m-th direction, and M is the total number of discrete directions. In traditional finite volume CFD codes, Equation (3) is solved together with the energy transport equation in an iterative manner where $\langle I_m \rangle$ and $\langle e_{bg} \rangle$ are updated periodically to achieve a converged temperature field.

To model the radiation term in the joint composition PDF equation, we substitute Equation (5) into the third term in the right-hand side of equation (1) and split it into two terms

$$\frac{\partial}{\partial \psi_h}\left[\left\langle \left(4\sigma k_{ag} T_g^{\,4} - k_{ag} \sum_{m=1}^{M} w_m I_m\right) \middle| \boldsymbol{\psi} \right\rangle f_{\boldsymbol{\varphi}}\right] = \frac{\partial}{\partial \psi_h}\left[4\sigma \langle k_{ag} T_g^{\,4} \rangle f_{\boldsymbol{\varphi}}\right] - \frac{\partial}{\partial \psi_h}\left[\left\langle \left(k_{ag} \sum_{m=1}^{M} w_m I_m\right) \middle| \boldsymbol{\psi} \right\rangle f_{\boldsymbol{\varphi}}\right]$$

(6)

The first term on the right-hand side of equation (6) represents radiative emission and appears in closed form because the conditional mean can be calculated from local composition exactly. The second term on the right-hand side of equation (6) needs to be modeled because I_m is not a part of the composition space, and the conditional mean cannot be calculated by local composition. The model adopted here is the optically thin eddy approximation (Kabashnikov and Myasnikova, 1985) that suggests that the local incident radiation intensity I_m is only weakly correlated to the local radiative properties. This leads to

$$\frac{\partial}{\partial \psi_h}\left[\left\langle\left(k_{ag}\sum_{m=1}^{M} w_m I_m\right)\middle|\psi\right\rangle f_\varphi\right] \approx \frac{\partial}{\partial \psi_h}\left[\left(k_{ag}\sum_{m=1}^{M} w_m \langle I_m\rangle\right) f_\varphi\right] \tag{7}$$

Equation (7) can be coupled with equation (3) to solve for the change in particle enthalpy due to radiation and hence the mean temperature field. $\langle I_m\rangle$ in equation (7) is solved from equation (3) using a finite difference scheme, while $\langle e_{bg}\rangle$ in equation (3) is obtained from the ensemble mean of particle radiation intensity determined by particle composition via equation (4).

3.4.5 Solution Strategy

The joint composition PDF is solved using a modified Euler scheme (Chao and Pope, 2002) to sequentially calculate the change in particle properties due to the simultaneous processes of mixing, etc.

1) The particles are first convected for half a time step.
2) The changes in particle properties due to mixing, reaction, and radiation are then calculated for the full time step using the Strang's operator splitting scheme.
3) Finally, the particles are convected to their final positions for the time step.

A local time-stepping strategy (Muradoglu and Pope, 2001) is adopted in the modified Euler scheme where the time steps to advance particles are determined by local residence and mixing times. A distributed computing algorithm based on domain partition of particles (Cao, *et al.*, 2005) is used in the Monte-Carlo particle solver using Message-Passing-Interface (MPI).

In the FV solver, convergence is measured by residuals and a FV solution is considered to be converged when the magnitude of residuals is smaller than a specified tolerance value. Here the residual is defined as the mean of absolute residuals of the continuity, momentum, turbulent kinetic energy, and turbulent kinetic energy dissipation equations averaged over the entire computation domain.

The global convergence of the particle algorithm is monitored by the changes in the particle mean fields. These changes are typically measured at important physical locations and serve the same purpose as the residuals in the FV solver. In addition, overall mass and energy balances are also used to help determine convergence.

3.5. Reduced Mechanism Based on Quasi-Steady State Assumption

The reduced mechanism is constructed using the quasi-steady-state assumption (QSSA) introduced by Bodenstein (1906), where given a detailed mechanism with n_S chemical species, a small number of n_C species are retained to represent the dynamics of the detailed system, while the other $n_{SS} = n_S - n_C$ species that are associated with fast processes are assumed to be in steady state with their net chemical production rates being set to zero. Then, in the turbulent combustion calculation, the relevant equations are solved for the n_C unsteady-state species instead of for the n_S species. Because the computational cost increases at least linearly with the number of species represented, substantial gains can be achieved for $n_C << n_S$. The technique has been used extensively to hydrogen and methane combustion. Many examples can be found in the volumes edited by Smooke (1991), and Peters and Rogg (1993).

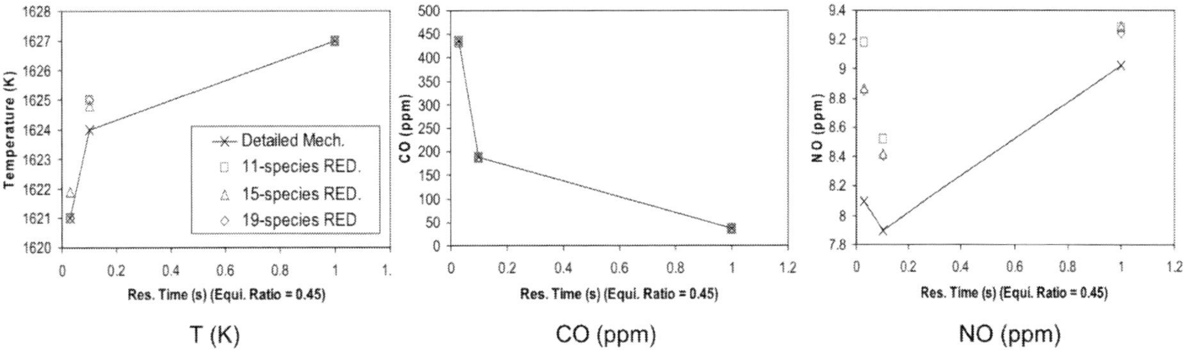

Figure 3.4 Comparisons between detailed and three reduced mechanisms using Perfectly Stirred Reactor (PSR) solutions under lean premixed combustion condition. (Equivalence ratio is 0.45)

CARM (Chen, 1997) is a computer code that automates this process. It uses a set of input test problems, typically PSR and PFR solutions, to evaluate the error in the SS assumption for each species in the detailed mechanism under conditions of interest. REI has developed a fully integrated computational problem solving environment (PSE) based around CARM for creating, testing and optimizing reduced chemical kinetic mechanisms for combustion simulation and implementing these mechanisms into commercial and research CFD codes (Montgomery, *et al.*, 2002a; Montgomery *et al.*, 2003; Montgomery *et al.*, 2004). The CARM-produced reduced mechanisms represent the original detailed mechanisms faithfully over a broad thermo-chemical parameter space. REI has been quite successful in applying reduced mechanisms for simulations of combustion and pollutant formation (Cremer, *et al.*, 2000; Cremer, *et al.*, 2001; Montgomery, *et al.*, 2001; Montgomery, *et al.*, 2002b; Montgomery *et al.*, 2004).

Three reduced mechanisms with different numbers of species have been developed in this project for modeling lean premixed gas combustion. Figure 3.4 compares the calculation results of a Perfectly Stirred Reactor (PSR) using a 64 species detailed mechanism (Glarborg *et al.*, 1998; Djurisic, 2004) and the reduced mechanisms derived from the detailed mechanism. It can be seen that all three reduced mechanisms lead to excellent predictions in flame temperatures and CO concentrations. The NO predictions are good, with the largest error being about 15% when using the 11-species reduced mechanism. Another important observation (not shown in the figures) is that the independency of NO concentrations to residence times is well captured at equivalence ratio 0.45, which demonstrates the potential of using these mechanisms to model lean premixed gas burners which commonly operate at about this equivalence ratio in order to achieve single digit NO_x emissions. These reduced mechanisms will be used in modeling the very low-NO_x levels in the ultra low NO_x burners for process heater.

3.6. *In-Situ* Adaptive Tabulation

ISAT is a storage/retrieval algorithm for the efficient implementation of combustion chemistry. In a typical turbulent combustion calculation, the combustion chemistry has to be computed at least 10^9 times, e.g., on each of 1,000 iterations (or time steps) for each of 1,000,000 cells (or particles). The idea of storage/retrieval is to avoid performing these expensive chemistry calculations 10^9 times by storing the information from many fewer calculations, and then (when possible with sufficient accuracy) to retrieve the information 10^9 times. A detailed description of the ISAT algorithm can be found in Pope (1997), and Fox (2003). Some of the important features of ISAT are outlined in the following.

- The tabulation table is built up in situ (as the turbulent combustion calculation is performed) rather than as a pre-processing stage. In this way, only the region of the composition space that is accessed in the calculation is tabulated.
- Interpolation in the table is performed using a linear approximation.
- The interpolation error ε is controlled relative to a specified tolerance εtol. If a table entry is found such that ε is less than εtol, the linear approximation ("retrieval") is used. Otherwise, a "grow" of the ellipsoid of accuracy is performed, or a new table entry is generated ("add").

Successful applications of ISAT to realistic turbulent combustion modeling have been reported recently, including: particle-based PDF calculations of methane air jet flames (James *et al.*, 1999, Tang *et al.*, 2000, Xu and Pope, 2000, Raman *et al.*, 2004) using reduced mechanism; and finite-volume-based three-dimensional transient CFD for HCCI engines with a detailed n-heptane mechanism (Embouazza *et al.*, 2002). REI has successfully implemented the ISAT algorithm into VULCAN, a NASA CFD code, to simulate high-speed hydrocarbon combustion using reduced mechanism (Zhao *et al.*, 2003), and into beta versions of REI's proprietary multiphase, turbulent reacting CFD code, GLACIER, used for modeling coal combustion problems. Significant speed-ups ranging from 10 to more than 1000 compared to direct solving the chemical systems have been reported in these calculations with modest storage requirements and tabulation errors being controlled under certain limits.

4. Model Evaluation – Laboratory Flame

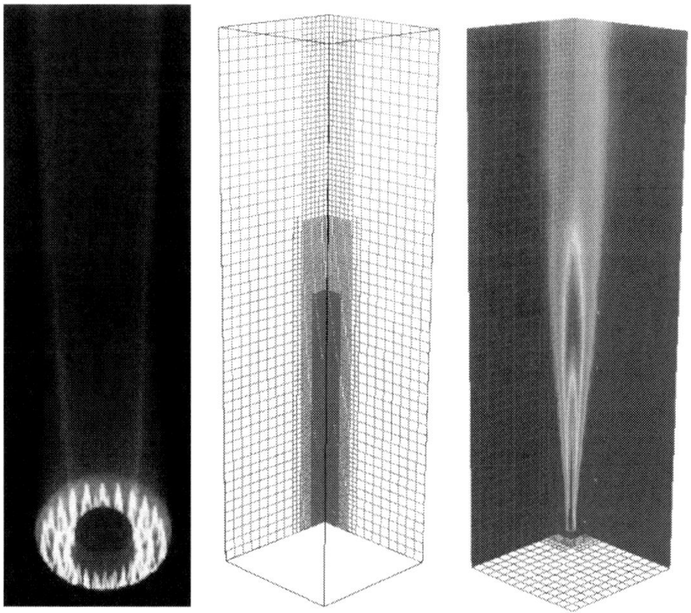

(a) (b) (c)

Figure 4.1 (a) Sandia piloted-jet flame (Barlow and Frank, 1998); (b) Calculation domain and mesh refinement;
(c) Calculated temperature contours.

The partially premixed Sandia D flame (Barlow and Frank, 1998) shown in Figure 4.1(a) is selected as the target flame in this study for model evaluation. The burner has a fuel nozzle of diameter D = 7.2mm and a premixed pilot that extends to a diameter of 18.2 mm. The jet fuel is 25% CH_4 and 75% air by volume and the pilot burns a lean pre-mixture of C_2H_2, H_2, air, CO_2, and N_2 with the same nominal enthalpy and equilibrium composition as methane/air at an equivalence ratio of 0.77. The bulk velocities of the fuel jet are 49.6m/s with a jet Reynolds number of 22,400. The pilot jet velocity is 11.9m/s and the co-flow velocity is 0.9m/s. Strong effects of finite-rate chemistry are observed in the flame, where local extinction is visible at the axial location of 15-jet diameter.

Extensive experimental measure-ments have been conducted and the data is well documented at: (http://www.ca.sandia.gov/TNF).

Because the flame is essentially an axisymmetric flame and the new modeling tool is designed for three-dimensional simulations, one quarter of the flame is calculated with two vertical symmetric planes extending from the inlet plane to the outlet plane. Figure 4.1 (b) shows the calculation domain. The structured base grid is $14 \times 14 \times 52$. Local mesh refinement can be seen in Figure 4.1 (b) to resolve the geometries of fuel jet and pilot jet nozzles, and the refinements are extended into the flow to get better resolution in flow field and turbulent mixing. Three levels of mesh refinement are used, resulting in a grid with about 110K cells.

A CARM generated 16-species 12-step reduced mechanism for methane combustion (http://www.ca.sandia.gov/TNF) was used in the simulation and is listed in table 4.1. 40 particles per cell were used in the Monte Carlo PDF calculation, resulting in about 4 million total particles. The ISAT error tolerance was set to be 1.0×10^{-4} with 1000 Mb storage space.

Table 4.1 Reduced mechanism for methane combustion.

Species	H_2 H O_2 OH H_2O CH_3 CH_4 CO CO_2 CH_2O C_2H_2 C_2H_4 C_2H_6 NO HCN N_2
Reaction Step	(R1) $H + 0.50\ O_2 = OH$ (R2) $H_2 + 0.50\ O_2 = H + OH$ (R3) $0.50\ O_2 + 0.50\ C_2H_2 = H + CO$ (R4) $0.50\ O_2 + CH_3 + C_2H_4 = CH_4 + CH_2O + 0.50\ C_2H_2$ (R5) $0.50\ O_2 + 2CH_3 = H_2 + CH_4 + CO$ (R6) $0.50\ O_2 + 2CH_3 = CH_4 + CH_2O$ (R7) $0.50\ O_2 + CH_3 = H + CH_2O$ (R8) $0.50\ O_2 + CO = CO_2$ (R9) $0.50\ O_2 + C_2H_6 = CH_4 + CH_2O$ (R10) $H + OH = H_2O$ (R11) $H + CH_4 + NO + HCN = 0.50\ O_2 + 2CH_3 + N_2$ (R12) $H + 0.50\ O_2 + CH_4 + HCN = 2CH_3 + NO$

Figure 4.1 (c) shows the calculated temperature contours in the two symmetric planes and the inlet plane. The calculated velocity field and mixture fraction are compared with the measurements. Excellent agreement is achieved as evidenced in Figure 4.2, where the centerline velocity profiles and mixture fraction profiles (symbols for measurements and lines for calculation results) are shown in the two plots respectively.

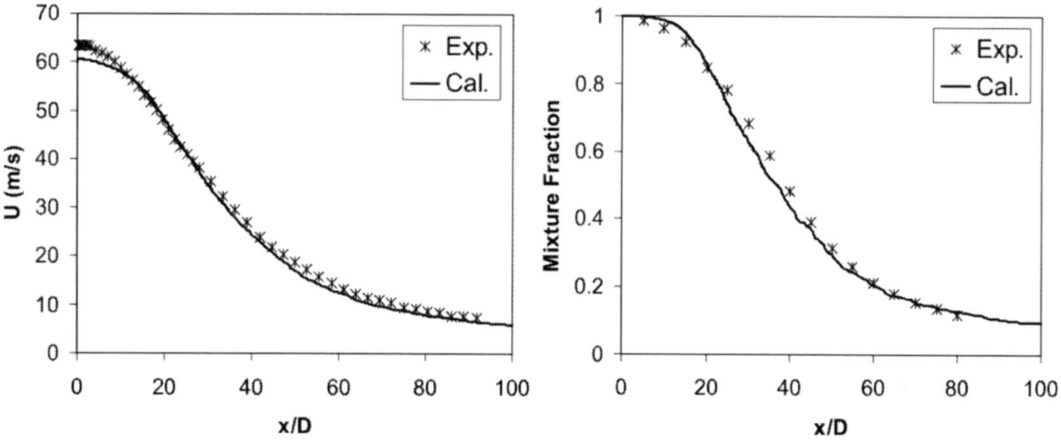

Figure 4.2 Centerline profiles of axial velocity and mixture fraction

Comparisons of the predicted and measured mean and variance of temperature and NO mass fraction profiles at different axial locations along one of the symmetric planes are presented in the following.

Figure 4.3 shows comparisons between the measured and predicted profiles of the Favre mean temperature and its variance at different axial locations ($x/D = 7.5, 15, 30, 45$). These comparisons show that the predicted temperatures were well represented. The temperature profiles up to *x/D* = 15, where most local extinction caused by strong interactions between turbulence and reaction took place, were in excellent agreement with the measurements. Slightly over-predictions of mean

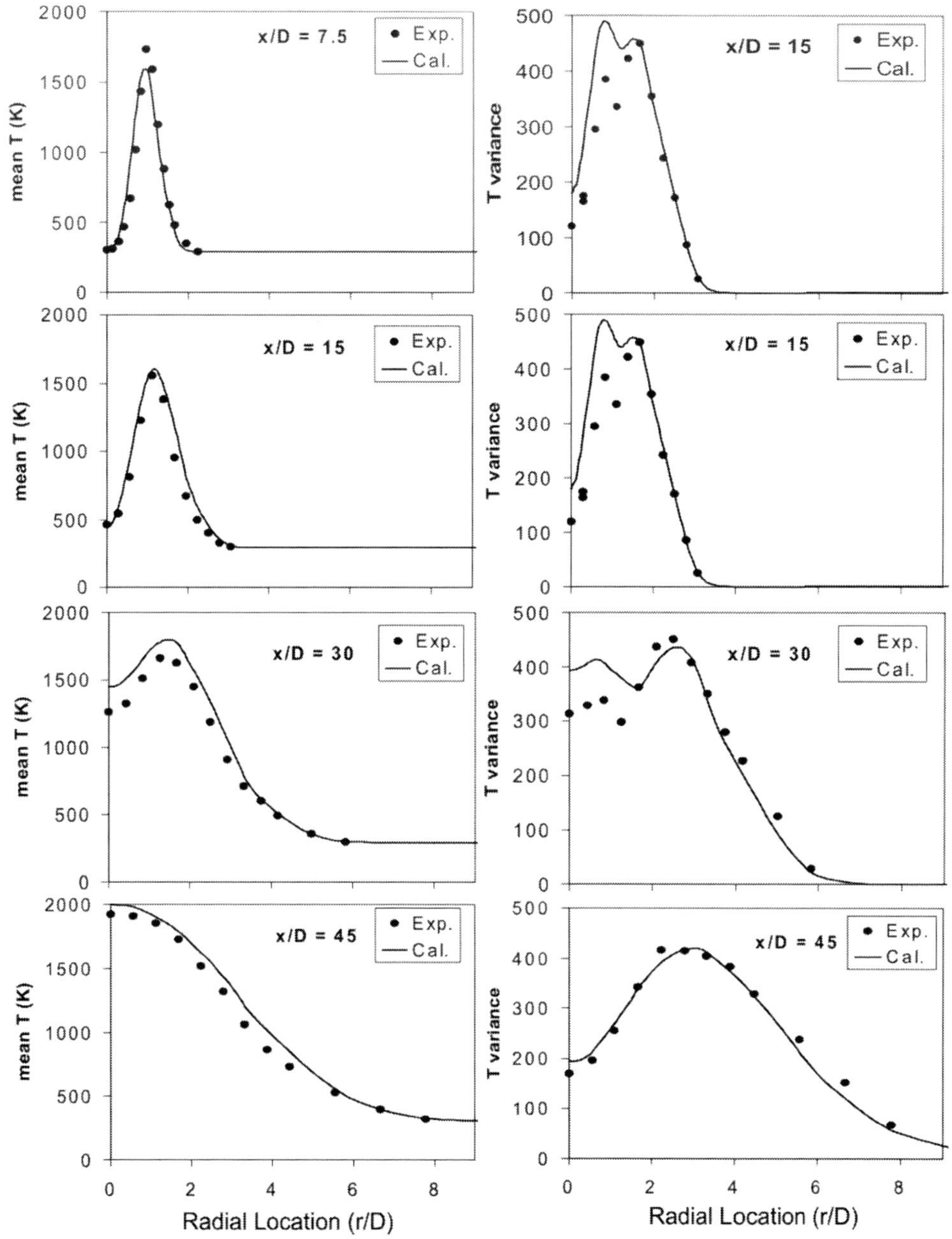

Figure 4.3 Measurements and prediction of mean temperature (left) and

temperature can be found at *x/D* = 30 and 45. Insufficient grid resolution and/or inadequacies of the $\kappa - \varepsilon$ turbulence model might be the reason that caused the over-prediction of downstream temperature. Nevertheless, the overall agreement between the calculation results and data in temperature is very good.

The transport PDF model has the capability of calculating all one-point, one-time statistics. Shown in the right column of Figure 4.3 are the temperature variance profiles that describe the turbulent mixing process. The saddle shape of the temperature variance profiles, which is caused by the shear layers located between the fuel jet and the pilot jet, and the pilot jet and the co-flow, is well captured by the PDF simulation.

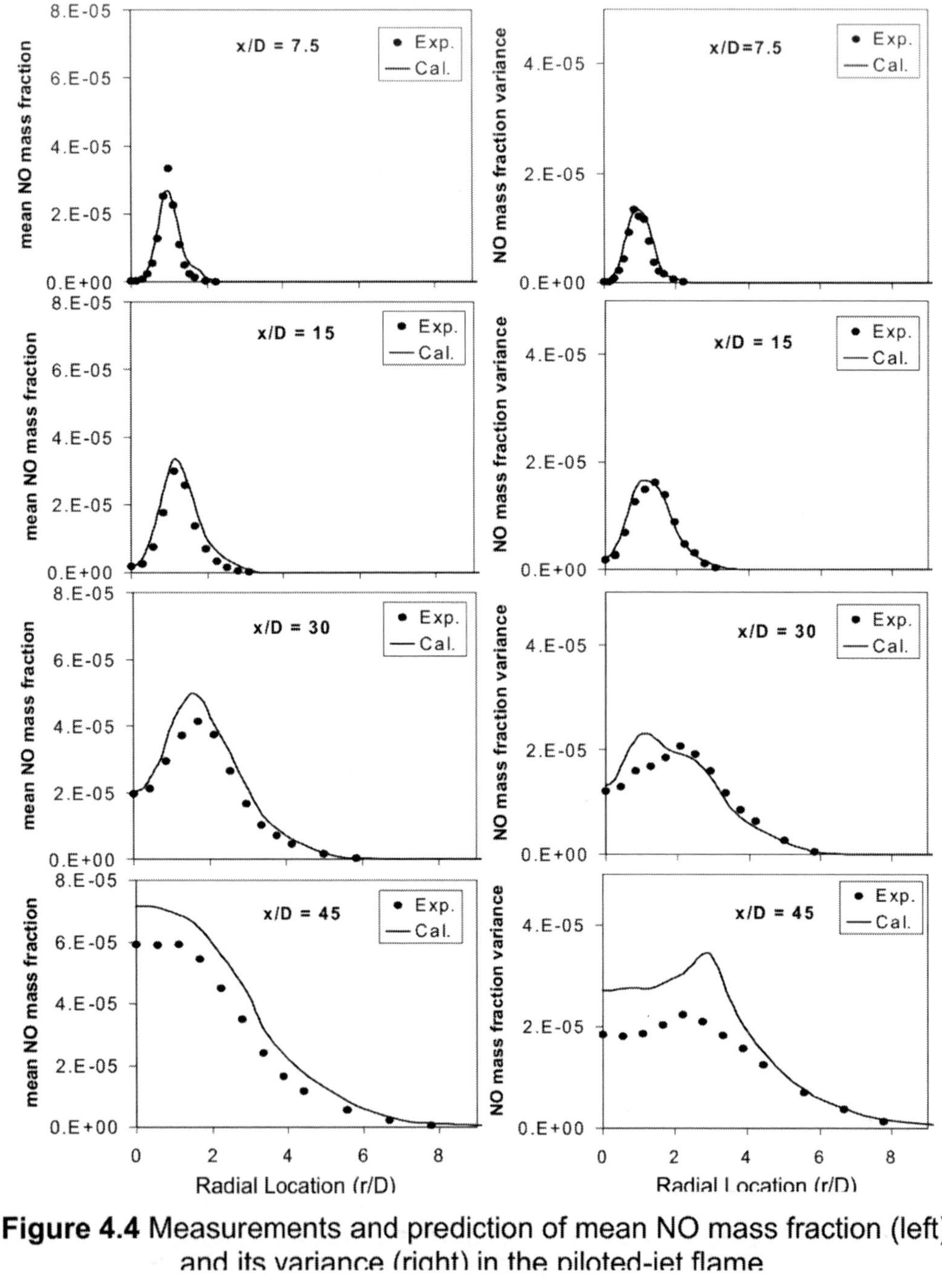

Figure 4.4 Measurements and prediction of mean NO mass fraction (left) and its variance (right) in the piloted-jet flame

Figure 4.4 shows radial profiles of the Favre mean and variance of NO mass fractions at various axial locations on one of the symmetric planes. It can be seen that the agreement between the calculations and the experimental data is in general very good. The peak value of mean NO mass fractions were slightly over-predicted at *x/D* = 7.5, and under-predicted at *x/D* = 30 and 45, but otherwise the calculated levels of the mean mass fraction and its variance are in very good agreement with the data.

The ISAT algorithm has played a key role in making the three-dimensional calculation of the D flame possible. Figure 4.5 shows the performance of the ISAT algorithm where the number of operations (i.e., retrieval, grow and add; see section 3.6 for detail) is drawn against the number of calls to ISAT by the PDF solver during the calculation. It can be seen that, later in the calculation, the table was filled up (i.e., the "add" line became flat) and direct integration was performed if necessary. However, the majority of the operations were still "retrieval". In other words, later in the calculation, given a chemical mechanism evaluation request to ISAT the probability of performing "retrieval" was more than 0.995.

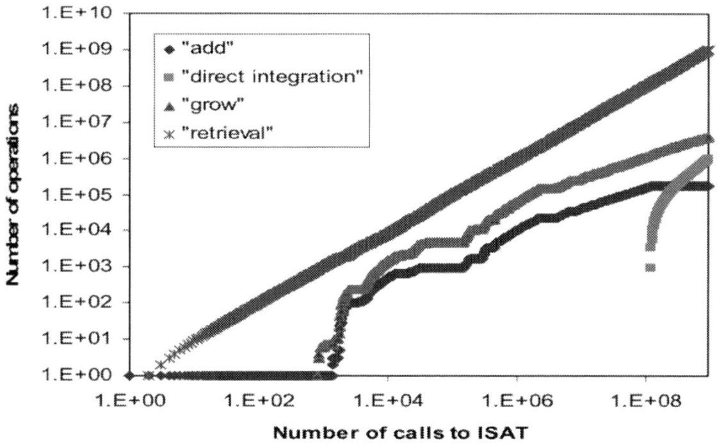

Figure 4.5 ISAT build-up history.

In the ISAT algorithm, the "retrieval" operation (i.e., linear interpolation) is about 3 orders of magnitude faster than the "add" and the "grow" operations because the later ones involve direct integrations of the stiff ordinary differential equation (ODE) system that describes the chemical kinetics. The overall ISAT speed-up factor for this calculation was about 200. Roughly two weeks were required to achieve a converged solution for the case running in serial mode on a 64-bit Opteron PC; without ISAT, this simulation would have required years to complete.

The three dimensional calculation of the Sandia D flame demonstrates the capability of the new CFD tool in terms of accuracy, efficiency and robustness for modeling practical turbulent combustion problems that have strong finite-rate chemistry effects.

5. Model Application – Single Burner Test Furnace

The new modeling tool has been used to model a single burner test furnace (SBTF) based on an actual Performance Demonstration undertaken at the John Zink Company's International Research and Development Test Center, in Tulsa, OK. The purpose of the SBTF is to provide a test bed for low NO_x gas burners commonly used in process heaters. The SBTF provides a means to study the impact of different firing conditions on burner performance. In addition, the SBTF results can be used to assist in modeling the burner conditions in full sized process heater furnaces containing a large number of burners.

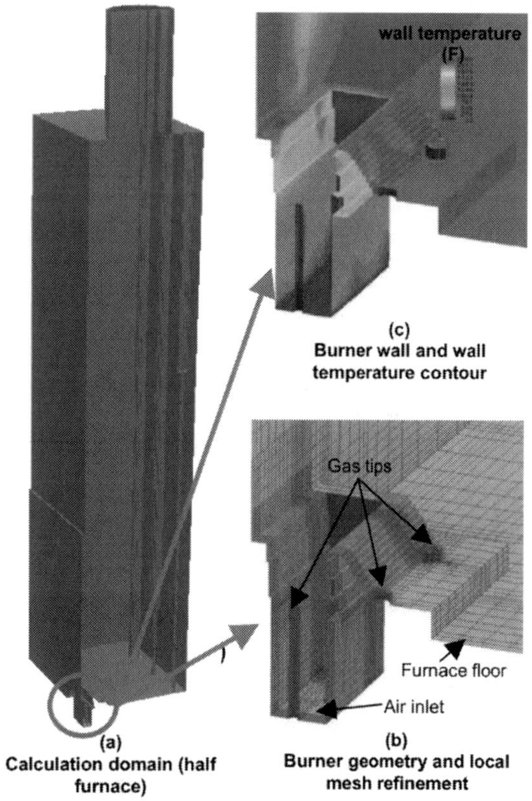

Figure 5.1 SBTF calculation domain and mesh refinement.

Due to the symmetric location of the burner in the furnace, only half of the furnace was modeled (see Figure 5.1 (a)). Some details of the low NO_x burner are shown in Figure 5.1 (b) and (c), where the burner wall surface is colored by calculated wall temperature. The SBTF consists of a rectangular furnace with one up-fired burner located on the furnace floor along one wall (Figure 5.1 (a)). The burner is a natural draft (single fuel) gas burner. The burner consists of eight gas tips: 3 so-called primary tips are mounted inside the throat of the burner tile, close to the centerline; 3 so-called staged gas tips are positioned outside the side of the tile and fire up the outside of the angled front of the tile; and 2 additional primary gas tips are positioned outside the tile at each end firing mainly horizontally through an aperture in the tile into the air stream flowing up through the throat of the tile. Each gas tip consists of several minute gas jets, each oriented so as to enhance fuel-air mixing and to re-circulate flue gas as a mechanism for NO_x reduction. The furnace also contains five partially insulated, water-cooled tubes located along the furnace wall opposite the burner (three of them are shown in Figure 5.1 (a)). The furnace has a circular exit located approximately at the center of the furnace (the blue area shown in Figure 5.1 (a)).

The dimensions of the calculation domain are approximately 2.0 m in width, 1.75 m in depth, and 10.0 m in height. The base grid is $34 \times 36 \times 77$. To resolve the smallest fuel jet nozzle (that has a diameter of about 1.5 mm) nine levels of local mesh refinements are applied; hence, the inlet flow area of each jet contains at least 10×10 cells. Mesh refinement is also used to resolve the process heater tube geometries. To better represent the intense turbulent mixing of the primary fuel jets and air flow, and the staged fuel jets and the flue gas, three levels of manual mesh refinements are applied to the cells inside the burner well and above the staged gas tips. The total number of cells after mesh refinement is about 250 K.

The simulated case was a normal firing rate of 7.5 MMBtu/hr (2.2 MW). In the test, the burner fired a mixture of 40mol% hydrogen and 60 mol% methane. The fuel mass flow rate is about 0.04 kg/sec and the air mass flow rate is about 0.8 kg/sec. The excess oxygen is about 2.1 mol% (dry basis). Both the fuel flow and the air flow are at a temperature of 283.15 K.

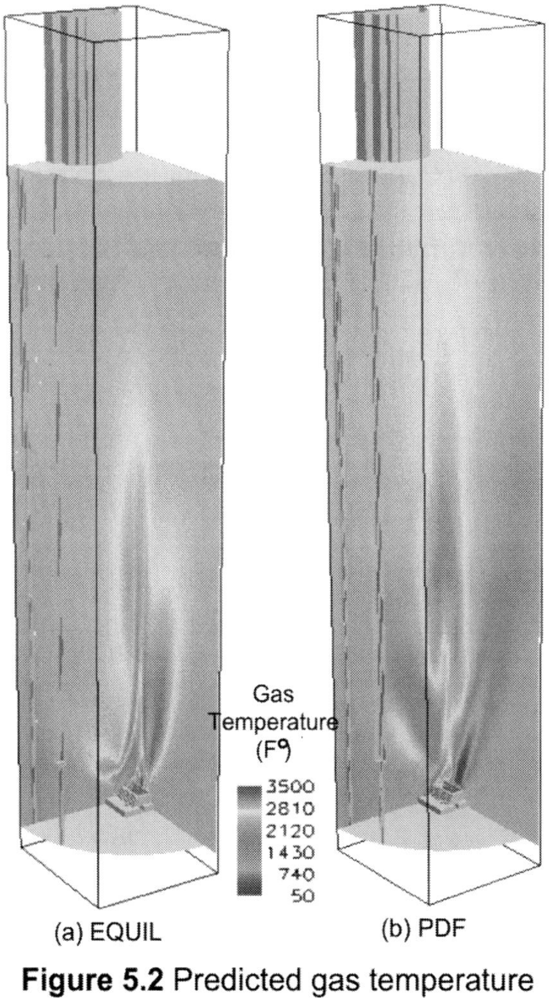

Figure 5.2 Predicted gas temperature contours in the SBTF.

The burner achieved low NO_x emission level (about 50 ppvm at furnace outlet) using the staged combustion strategy. The overall equivalent ratio of the burner was about 1.0, but the actual fuel/air mixing process was quite complicated inside and around the burner, and the combustion process involved different modes: non-premixed, partially premixed and premixed. For the simulated case, the fuel was almost exactly equally split (50:50) between the staged tips and the primary tips. The fuel from the three primary gas tips located inside the burner well first mixed with the air stream to form a lean pre-mixture and then ignited. The staged fuel was first diluted by the re-circulating flue gas, and would not burn until it met the lean combustion products exiting the burner well.

Two simulations were performed. The first one is denoted as "EQUIL" where the assumed PDF model implemented in the original AMR code was used. The second one is denoted as "PDF" where the transport composition PDF model implemented in the new modeling tool was used. A nine species reduced mechanism and the same ISAT parameters as those adopted in the piloted-jet flame simulation were used in the calculations. The PDF simulation used 25 particles per cell. Calculation results are presented below, including the predicted temperature, NO and CO concentration. Comparisons with the measurement data are also presented.

Figure 5.2 shows the predicted gas temperature contours on the symmetric plane and a plane that is parallel to the fired wall and the distance is 0.03 m. The furnace and burner walls are represented in gray color in the figure. It can be seen that the EQUIL and the PDF calculations gave completely different flame structures. This is not surprising considering that the assumed PDF model is in effect a "mixed is burnt" model. The EQUIL calculation indicates two separate flames (in the calculation domain) corresponding to the two primary gas tips firing upwards (because of symmetry considerations the computational results shown are actually for one-and-a-half tips). The PDF simulation predicts only a single flame, which originates from the center primary gas tip. The overall flame length predicted by the PDF simulation was much longer than that predicted by the EQUIL. The peak temperature in the furnace calculated by the PDF was about 200 K lower than that calculated by the EQUIL. It should be pointed out that, just above the staged tips close to the symmetric plane, both calculations predicted a high temperature region that points towards the cooling tubes representing of the process tubes in real furnaces. The high temperature region points to the tubes predicted by the PDF is much longer and has a slightly lower temperature level compared those predicted by the EQUIL calculation. A close look at the calculated flow field around the burner indicates that this separated high temperature region is caused by the cyclone-type circulation zone created by the staged fuel jets outside the burner wall, which engulfs the oxygen rich primary fuel-air mixture from the burner well exit to mix with the hot staged fuel-flue gas mixture, and leads to the ignition of part of the staged fuel earlier than expected.

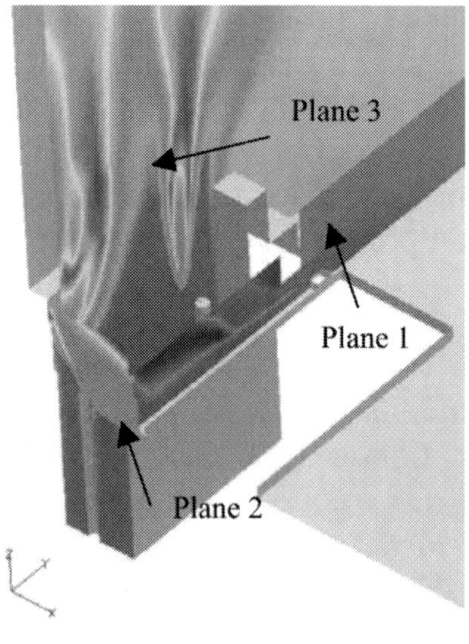

(a) EQUIL

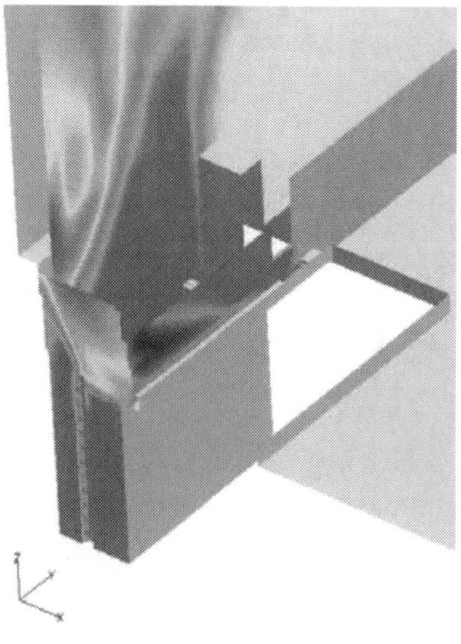
(b) PDF

Figure 5.3 Predicted gas temperature contours around the burner

Figure 5.3 illustrates the temperature contours on three planes inside the burner (only half of the burner is shown). Plane 1 is parallel to the fired wall and crosses the primary gas tip firing horizontally into the air flow; Plane 2 is the symmetric plane; and plane 3 is parallel to the fired wall. The fuel/air mixture generated by the gas tip firing horizontally is the first to ignite due to the slow fuel jet velocities, and hence long residence times (see Plane 1 in Figure 5.3). The hot combustion products are carried by the jet momentum towards the burner center, and ignite the fuel/air mixture formed by the center primary gas tip firing upwards (see Plane 2 and 3 in Figure 5.3). This mixing and reaction processes were captured by both calculations. The mixing between the cold staged gas streams with the hot combustion products circulated within the furnace was also predicted by both calculations.

The major difference between the two calculations is that the fuel/air mixture created by the side primary gas tip (firing upwards) ignites inside the burner well in the EQUIL prediction (see Plane 3 in Figure 5.3), whereas in the PDF prediction the pre-mixture remained cold well above the burner. This is not surprising because the EQUIL calculation used an assumed PDF approach that is based on a fast chemistry assumption (i.e., "mixed is burnt"). PDF prediction is a more reliable description of the actual flame structure generated by the low NO_x burner because this model can treat finite-rate chemistry and turbulence-chemical reaction interactions more accurately. This is also evidenced by the predictions of the NO level in the furnace.

There are limited experimental data available, including the mass weighted O_2, NO, and CO concentrations (dry basis) at the outlet. Temperatures were measured at two points inside the furnace: the furnace gas temperature at approximately 18 feet above the burner was measured with a suction pyrometer mounted horizontally through the south wall of the heater (denoted by T1); a second suction pyrometer was mounted through the floor to measure the gas temperature in the center of the furnace about 1 foot above the floor (denoted by T2).

Both calculations predicted the O_2 level at outlet correctly, which is 2.1% by volume. Comparisons between the other measured and predicted quantities are shown in Table 5.1. CO and NO concentrations are corrected to 2.1% O_2 on dry basis. The following observations can be made:

- There is good agreement between the measured and predicted gas temperatures for both simulations. The gas temperatures predicted by the PDF simulation have slightly better agreement with the measurements than do the gas temperatures predicted by the EQUIL simulation.
- The predicted outlet NO concentration by the PDF model is in much better agreement with the measured value than the value predicted by the EQUIL model. The PDF simulation under-predicted the outlet NO by only about 3 ppvm (the measured value is 53 ppvm). The EQUIL simulation over-predicts the outlet NO by a factor of more than ten.
- Both calculations over-predicted the CO concentration at the outlet by about 9 ppm while the data indicates almost no CO at the outlet.

Table 5.1 Comparisons between measured and predicted quantities in SBTF.

	T1 (K)	T2 (K)	Outlet NO (ppvm)	Outlet CO (ppvm)
Measurements	1430	1373	53	1
EQUIL	1499	1264	534	10
PDF	1451	1395	50	9

6. Conclusions and Future Work

In this paper an advanced CFD modeling tool for simulating ultra-low NO_x burners in process furnaces was presented. The tool combines state-of-the-art techniques for comprehensive modeling of practical turbulent combustion problems, including: a Monte Carlo joint scalar PDF method for modeling turbulence-chemistry interactions; a reduced chemical kinetic mechanism for describing finite-rate chemical kinetics; an *in-situ* adaptive tabulation algorithm for incorporating realistic combustion chemistry accurately and efficiently; and an adaptive mesh refinement finite volume flow solver for treating complex geometries of practical combustion equipments. Other models such as a radiation heat transfer model based on a discrete ordinate method were also developed to be used in the Monte Carlo PDF simulation.

The new modeling tool was demonstrated on the Sandia D Flame, a well characterized, partially premixed methane-air flame. Extensive comparisons between the experimental data and the predictions were made and good agreement was achieved including pollutant emission level such as NO. The results demonstrated the capability of the tool to simulate complex turbulent combustion problems with strong finite-rate chemistry effects. This test problem also highlighted the benefit of using an ISAT algorithm to incorporate a reduced mechanism into the joint PDF calculation; compared to the CPU time required for direct evaluation of the reduced mechanism, the ISAT method provided a speedup factor of about 200. A novel distributed computing scheme based on particle domain decomposition was implemented in the Monte-Carlo PDF solver, which also greatly improved the overall efficiency of the code. Hence, the new modeling tool is capable of simulating practical combustion problems at a reasonable computational cost.

Comprehensive simulations of an industrial scale single burner test furnace firing a low NO_x gas burner were presented. By design, this test furnace is intended to simulate the performance of the low NO_x burner in an actual process heater furnace. The adaptive mesh refinement strategy used in the flow solver made it possible to resolve the fine detail of the burner as well as the furnace, or spanning more than four orders of magnitude in length scale, while using a tractable number of grid cells. Simulation results were presented for using the joint PDF model and a traditional assumed PDF model. Comparison of the results demonstrated the ability of the new modeling tool to accurately predict the finite-rate chemistry effects and NO emission accurately. Altogether, the new tool represents a significant improvement in modeling ability for lean pre-mixed combustion systems that are intended to have low NO_x emissions.

Several improvements to the new modeling tool are in-progress. The new capabilities are intended to improve the accuracy and robustness and decrease the computational cost of using the new modeling tool. Items in progress include a distributed computing capability, using domain decomposition, to reduce simulation turn around time of the flow solver, and a matrix-free Newton Krylov flow solver to speed up the problem initialization process. Future plans also call for applying the new CFD tool to simulate lean premixed combustion for ultra-low NO_x burners and to simulate a process furnace containing multiple burners.

Acknowledgements

Funding for this work was provided by the National Science Foundation under SBIR Phase II grant no. DMI-0216590 (NSF Program Manager: Dr. Errol Arkilic) and by Stone and Webster, Inc., a Shaw Group Company. In addition, we would like to acknowledge the assistance of Dr. Roberto Ruiz, Dr. Dick Waibel, and Dr. Christopher Jian of the John Zink Company, LLC, Koch Industries, for providing data on the single burner test furnace. CFD graphics in the paper are produced by Fieldview of Intelligent Light.

References

Acharya, S. and Moukalled, F.H., *Numerical Heat Transfer*, Part B, 15:131-152, 1989.

Adams, B.R., and Smith, P.J., *Combust. Sci. and Tech.*, 109:121-140, 1995.

Bilger, R.W., *Physics of Fluids*, 5(2):436-444, 1993.

Brown, D., Ma, J., and Adams, B., *Paper 68C, AIChE Spring National Meeting*, New Orleans, LA, 2003.

Brown, D. Cremer, M., Smith P., and Waibel R., *paper 32D, AIChE Spring National Meeting*, Houston, TX, 1997.

Cao, R., Pope, S.B., and Masri, A.R., *Combust. Flame*, to be published, 2005.

Cremer, M.A., Montgomery, C.J., Wang, D.H., Heap, M. P., Chen, J.-Y., *Proc. Combust. Inst.* 28:2427-2434, 2000.

Cremer, M.A., Wang, D.H., Montgomery, C.J., Adams, B. R., *2001 Joint AFRC/JFRC/IEA International Combustion Symposium*, Kauai, Hawaii, September 9-12, 2001.

Dopazo, C., *Phys. Fluids A*, 18:389, 1975.

Embouazza, M., Haworth, M.C., and Darabiha, N., *SAE paper No. 2002-0102773*, 2002.

Fox, R.O., *Computational Models for Turbulent Reacting Flows*, Cambridge University Press, Cambridge, 2003.

Gran, I.R., and Magnussen, B.F., *Combust. Sci. Tech.*, 119:171-190, 1996.
Hilbert, R. Tap, F., El-Rabii, H., and Thevenin, D., *Prog. Energy Combust. Sci.*, 30:61, 2004.
James, S., Anand, M.S., Razdan, M.K., and Pope, S.B., *The 44th ASME Gas Turbine and AeroEngine Technical Congress*, Indianapolis, IN, USA, 1999.
Jones, W.P. and Launder, B.E., *Int. J. Heat Mass Transfer*, 16:1119-1130, 1973.
Kabashnikov, V.P., and Myasnikova, G.I., Heat Transfer-Sov Res., 17(6):116-125, 1985.
Lindstedt, R.P., Louloudi, S.A., and Vaos, E.M., *Proc. Combust. Inst.*, 28:149-156, 2000.
Montgomery, C. J., Swensen, D. A., Harding, T. V., Cremer, M. A., and Bockelie, M. J., A *Advances in Engineering Software*, 33:59-70, 2002a.
Montgomery, C. J., Cremer, M. A., Chen, J.-Y., Westbrook, C. K., and Maurice, L. Q., *Journal of Propulsion and Power*, 18:192-198, 2002b.
Montgomery, C., Swensen, D., Shino, D., Yang, C., Brunson, S., Parkinson, A., and Chen, J.-Y., *2003 SIAM Conference on Computational Science and Engineering*, San Diego, CA, Feb. 10-13, 2003.
Montgomery, C.J., Tang, Q., Yang, C., Chen, J.-Y., Goldin, G.M., Kong, S.-C., and Reitz, R.D., and Parkinson, A.R., *NSF Design, Service, and Manufacturing Grantees and Research Conference*, Dallas, TX, Jan. 5-8, 2004.
Muradoglu, M., and Pope, S.B., *AIAA Journal*, 40:1755-1763, 2001.
Patankar, S.V., *Numerical Heat Transfer and Fluid Flow*, Hemisphere Publishing Company, New York, 1980.
Peters, N., *Prog. Energy Combust. Sci.,* 10:319-339, 1984.
Peters, N., *Turbulent Combustion*, Cambridge University Press, Cambridge, 2000.
Pitsch, H. and Steiner, H., *Proc. Comb. Inst.*, 28:41-49, 2000.
Pope, S.B., *Prog. Energy Combust. Sci.*, 11:119-192, 1985.
Pope, S.B., *Combust. Theo. Modelling*, 1:41-63, 1997.
Raman, V., Pitsch, H., and Fox, R.O., *Annual Research Briefs*, Center for Turbulence Research, 2003.
Raman, V., Fox, R.O., and Harvey, A.D., *Combust. Flame*, 136:327, 2004.
Ren, Z., Subramaniam, S., and Pope, S.B., http://mae.cornell.edu/~laniu/emst, 2002
Subramaniam, S., *PhD thesis*, Cornell University, 1997.
Subramaniam, S., and Pope, S.B., *Combust. Flame*, 115:487, 1998.
Tang, Q., Xu, J., and Pope, S.B., *Proc. Combust Inst.*, 28:133-139, 2000.
TNF website, http://www.ca.sandia.gov/TNF/abstract.html
Veynante, D. and Vervisch, L., *Prog. Energy Combust. Sci.*, 28(3):193-266, 2002.
Wang, D.H., Bockelie, M.J., Cremer, M.A., and Chen, J.-Y., *PVP-VOL. 448-1, Computational Technologies for Fluid/Thermal/Structural/Chemical Systems with Industrial Applications*, 1:113-120, 2002.
Xu, J. and Pope, S.B., *Combust. Flame*, 123:281, 2000.
Zhao, W., Montgomery, C.J., Cremer, M.A., Adams, B.R., Eklund, D.R., and Chen, J-Y., *2003 ASME International Mechanical Engineering Congress and Exposition*, Washington, D.C., Nov. 16-21, 2003.

Paper 47f

Carburization Resistance of High-Cr, High-Ni Weld Overlayed Furnace Tubes for Ethylene Pyrolysis

T. Maeda
Daido Steel Co. Ltd. Nagoya, Japan

F.X. Terwijn
Daido Steel Co. Ltd.

Prepared for Presentation at the 2005 Spring National Meeting in Atlanta, GA April 10-14
17th Annual Ethylene Producers Conference
Session 47 – Ethylene Plant Fundamentals –
Improvements in Modeling and Testing Reaction Systems and Furnace Performance

Carburization Resistance of High-Cr, High-Ni Weld Overlayed Furnace Tubes for Ethylene Pyrolysis

T. Maeda
Daido Steel Co. Ltd. Nagoya, Japan

F.X. Terwijn
Daido Steel Co. Ltd.

ABSTRACT

The carburization behavior of furnace tubes with high-Cr, high-Ni alloy weld overlay has been evaluated. A thick layer of 45 % Cr, 50 % Ni, 1 % Mo alloy was welded on the inner surface of conventional HP-mod. type cast tubes using the Plasma Powder Welding (PPW) technique. The evaluation was conducted both in laboratory under accelerated conditions and in the field tests. The results show that if the PPW layer was applied, no carburization was observed in the base material under the conditions after 4 years operation where severe carburization occurred in conventional tubes already after 1 year. Using the Larson-Miller parameter and the equilibrium phase diagrams of the PPW and HP-mod. alloy compositions, the carburization tube life are predicted to be more than three times longer than that of conventional tubes. This significant tube life improvement can be explained by the very slow carburization of the PPW layer due to the presence of a continuous chromium oxide surface layer which is a barrier for carbon diffusion from the surface into the PPW layer in combination with the effect of the PPW / base material interface which is a barrier to carbon diffusion into the base material. The continuous, self healing and anti-catalytic chromium oxide surface layer is also expected to suppress the catalytic coke formation which results in longer run length.

Key Words: ethylene, pyrolysis, furnace tubes, carburization, remaining life, catalytic coke

INTRODUCTION

Ethylene pyrolysis furnace tubes are typically constructed of Fe-Ni-Cr cast alloys, such as HK40, HP mod. or HP micro alloys. However, to meet the increasing demand for better furnace tube alloys to handle higher temperatures in ethylene pyrolysis, a wide range of modified alloys have been produced. These modifications have involved additions of niobium, tungsten, molybdenum and silicon. Some also involved increases in nickel and / or chromium. It has been found that compositional variations can also improve carburization resistance. Carburization is a major controlling factor for the performance and lifetime of ethylene pyrolysis furnace tubes.

An alloy is considered carburized if the carbon content after exposure in a process environment exceeds the original carbon content of that alloy. Carburization results in the formation of carbide precipitates which give rise to internal stresses by volume increase and an associated degradation of mechanical properties such as toughness and creep ductility.

From a thermodynamic point of view an alloy can be carburized if the carbon activity of the process environment is higher than the carbon activity at the alloy surface. The carbon activity of the process environment is a defined thermodynamic quantity. For the simple decomposition reaction of methane, the carbon activity is defined as follows:

$$CH_4 <=> C + 2H_2. \quad (1)$$

$$a_c = K\, p_{CH_4} / p^2_{H_2} \quad (2)$$

The constant "K" in equation (2) is the equilibrium constant for reaction (1). Carbon activity can also be defined on the basis of the Boudouard reaction of the water/gas shift reaction, but these reactions are less relevant for ethylene pyrolysis conditions. Data on the carbon activity of commercial alloys are very limited. For carbon steel the carbon activity can for example be estimated by assuming that it is in equilibrium with Fe_3C (cementite). The generally higher carbon activity in the process atmosphere is the driving force for carbon diffusion into the tube material.

Carbon cannot diffuse through oxide layers because carbon has no solubility in technically relevant metal oxides. To avoid carbon diffusion through the surface of an alloy, a tight, continuous and stable oxide layer is needed. This type of layer can only be formed on alloys with a high chromium content and with stabilizing alloying element additions. The layer must also have self healing properties in case of damage caused by creep, erosion or deformation.

Carbon diffusion into the bulk material can be reduced by the forming of a stable carbide network along the grain boundaries in combination with a small grain size. The requirement for minimizing grain size in centri-cast materials to reduce carbon diffusion is counter to the general requirement for optimizing high temperature strength and creep properties where course grain sizes are preferred.

The PPW overlay welded ethylene pyrolysis furnace tubes have chromium content high enough to form, with other alloying elements, a stable continuous, tenacious chromium oxide layer. The 2 mm thick PPW layer acts as a buffer to ensure a long term supply of chromium for the formation of the protective scale. Furthermore, the unique PPW overlay process results in a layer with a very small grain size to minimize carbon diffusion. The mechanical properties are maintained by the conventional centri-cast base material.

PLANT EXPOSURE TESTS

PPW overlay welded tubes have now been installed and operated at various plants. The plant feed stocks consisted of ethane-naphtha mixtures or naphtha. The longest exposure time is now 4 years.

At the process side, the PPW material is typically covered with a chromium oxide scale, which contains virtually no iron oxides. Typically, HK or HP type materials show a scale, which is a mixture of chromium oxide and iron oxides.

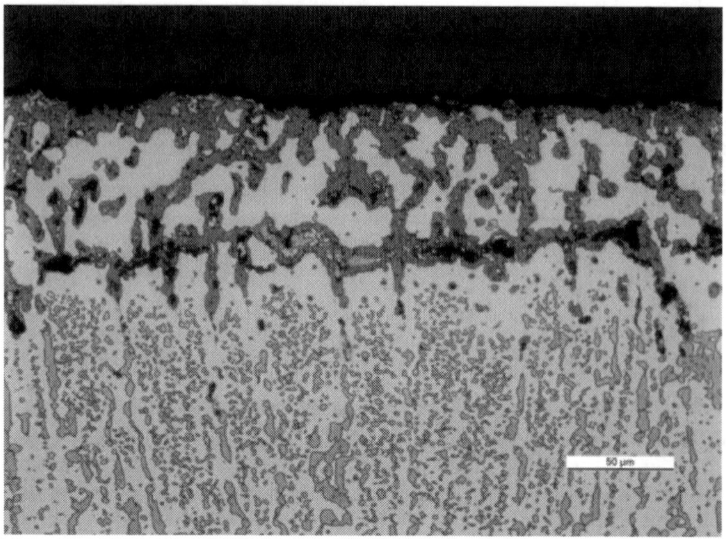

FIGURE 1 - Metallographic cross section of PPW layer after two years of operation

Figure 1 shows a cross section through the PPW overlay welded material at the process side. The top layer (about 50 µm) predominantly consists of metal and an interconnected network of chromium oxides. Below this top layer the material shows the presence of chromium carbides to the interface with the base material (HPmod.) as can be seen in Figure 2.

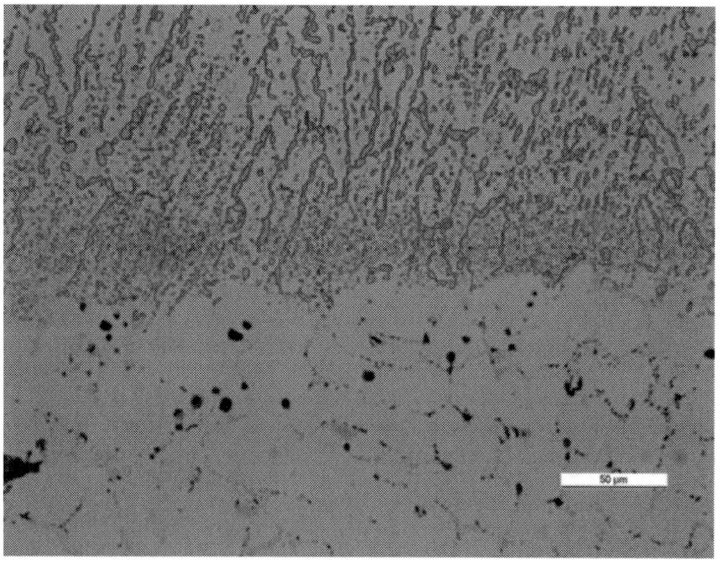

FIGURE 2 - Transition of PPW overlay weld material to HPmod. base material after two year operation.

The average carbon content in the PPW layer is about 0.9%. The carbon content in the base material has remained at the initial level of around 0.4%. PPW tubes which had been in operation for 4 years, did not show a significant increase with regard to the carbon content in the PPW layer or the base material with respect to the PPW tubes exposed for two years.

The carbon distribution over the cross section is also reflected by the hardness profile. For the PPW overlay welded layer the Vickers hardness is in the range of 350 – 250Hv. At the transition to the base material this hardness rapidly drops to values in the range of 150 – 175Hv.

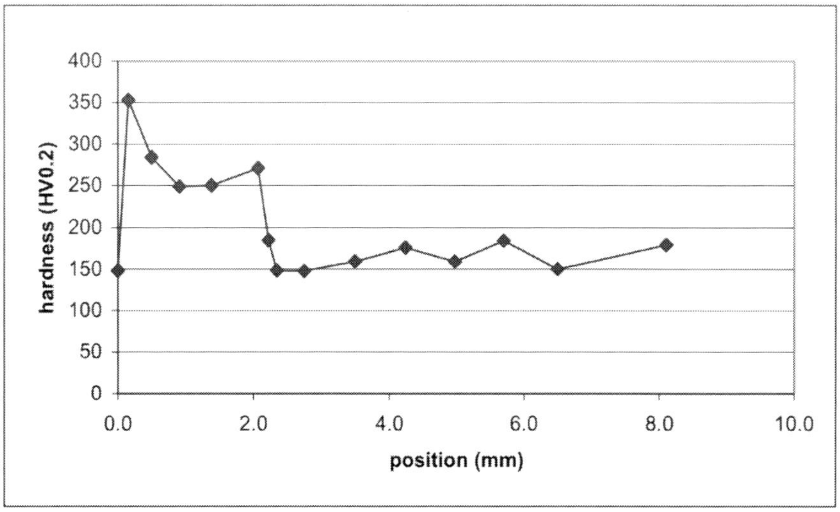

FIGURE 3 - Hardness profile over cross section. The transition from the PPW layer to the base material is at 2 mm.

The elemental distribution over the cross section was also determined. The results are shown in Figure 4.

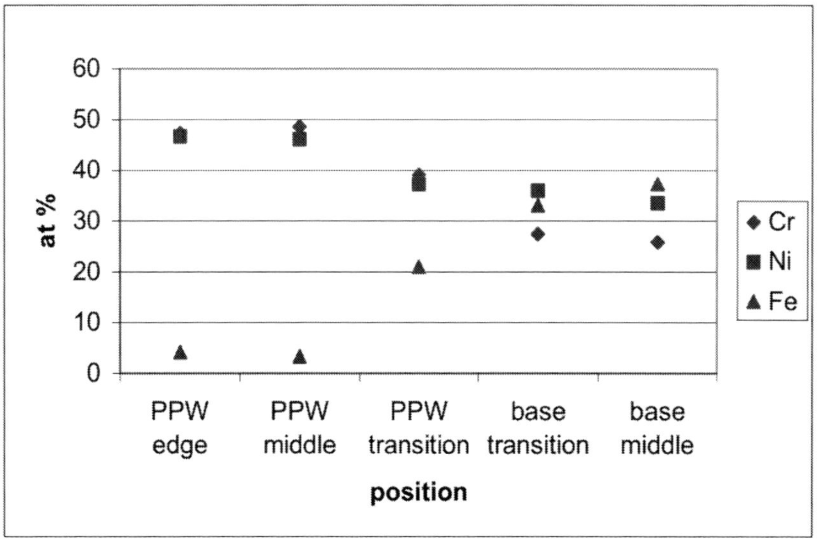

FIGURE 4 - Element distribution across the tube section after two years of operation

There is virtually no diffusion of Cr from the PPW layer into the base material or diffusion of iron from the base material into the PPW layer. At the transition the composition changes due to the mixing of the two materials during the overlay welding.

The phenomena as described here are also observed for furnace tubes which have been in service for about 1 year as well as for tubes which have been in service for 4 years, irrespective of the type of feed stock.

LABORATORY TESTING

A series of laboratory carburization tests were conducted using samples of HPmod. base material and HPmod. material with a PPW layer applied. These tests were carried out using a pack carburization facility. These data were used to supplement and validate the field trial data.

The results are plotted in Figures 5 and 6 in the form of carbon content versus the Larson-Miller parameter for HPmod. and PPW cases respectively. In both cases the carbon content, derived from the respective phase equilibrium diagrams, at which graphite first begins to form within the microstructure is indicated.

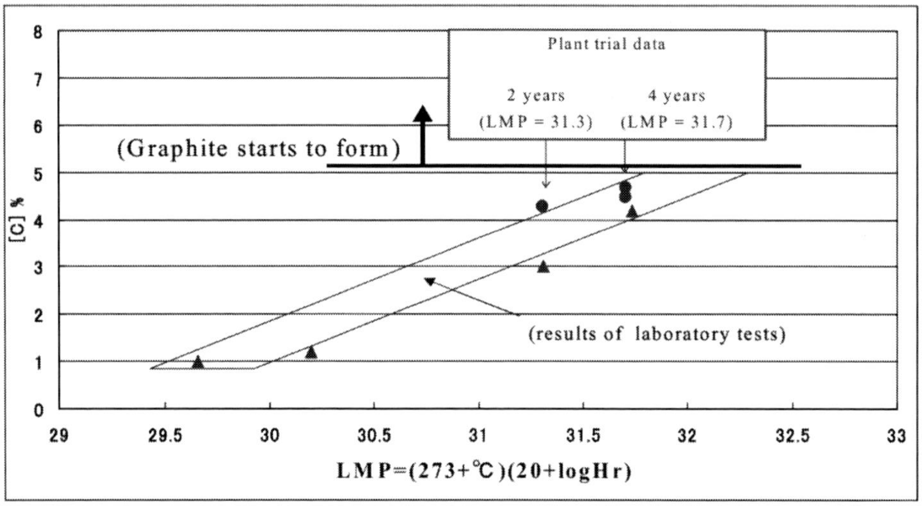

FIGURE 5 - Results of the field tests of HPmod. alloy in comparison to the laboratory test data

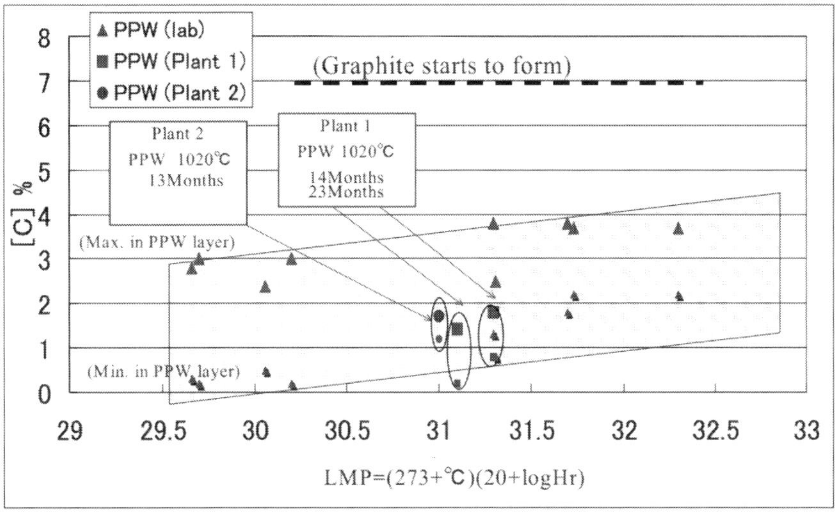

FIGURE 6 - Results of the field tests of PPW in comparison to the laboratory test data

The samples from the field trial HPmod. tubes exposed for 2 and 4 years were fully carburized with a carbon content of around 4.5 % on the internal surface and 1.0 to 2.5 % on the outer surface.

The PPW field tested tube showed up to 1.8 % carbon on the inside diameter of the PPW layer and up to 1.6 % carbon on the interface between the base material and the PPW layer after 48 months of exposure. The HPmod. base material showed no carburization at all after 48 months of operation (Fig.7). In addition, the PPW furnace tubes, unlike the conventional HPmod. tubes showed no distortion.

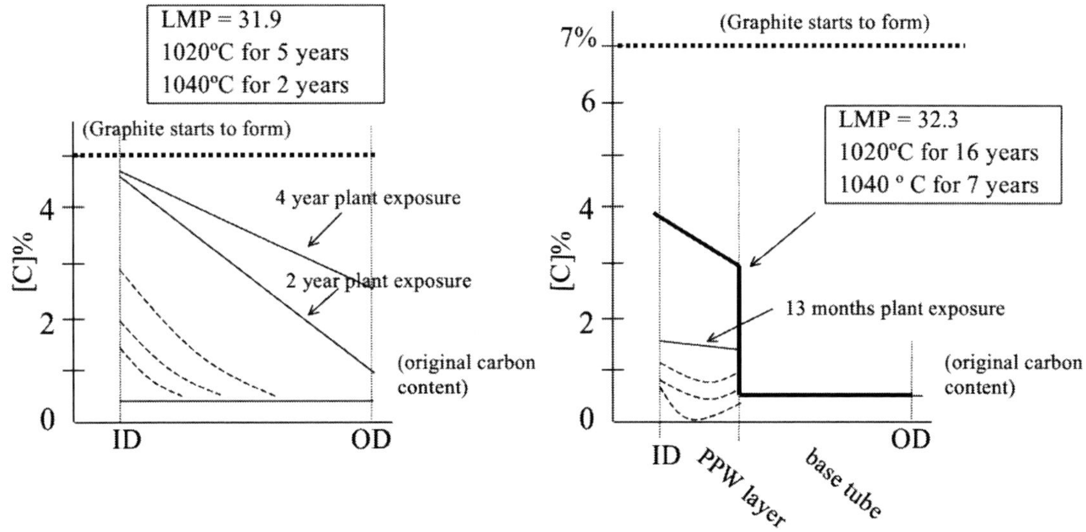

FIGURE 7-Comparison of carbon diffusion in HPmod. And PPW tubes and the implication for tube life

REPRESENTING CARBURIZATION PERFORMANCE

The carburization characteristics of the PPW coated tubes that have been determined metallographically have been used to establish an empirical predictive model that describes carburization life of a PPW coated ethylene tube.

The approach uses a Larson Miller type relationship linking carburization (carbon content), temperature and time. Analogous to the use of Larson Miller techniques for creep life assessment its application to carburization allows an estimate of tube life to be established based on known operating parameters.

The phase equilibrium diagrams for conventional HPmod. and PPW material have been established and can be used to determine the life limiting carburization level for each material. This is defined as the carbon content at which free graphite forms within the microstructure. For HPmod. material the value is 5% carbon, whilst for PPW it is 7% carbon.

The carburization distribution for the HPmod. and PPW tubes included in this study is presented in Figure 7. It can be seen that for similar exposure conditions carburization in the HP tube progresses to a higher carbon content and to a greater depth than in the PPW tube. Specifically, the PPW material acts as a "carbon sink," with, after 2 years operation, no rise in carbon content within the base tube material. Reference to the phase equilibrium thermodynamics indicates that diffusion of carbon from the PPW material to the base material will only start to occur when the carbon content in the PPW material exceeds 3%. Consequently, the Larson Miller approach can be used to estimate the time required at a particular service temperature before graphite first begins to form.

If a conventional HPmod. tube is considered initially, for the limiting Larson Miller parameter of 31.9, reference to Figure 7 indicates that graphite will begin to form at the surface of the tube after five years at a mean metal temperature of 1020C or after two years at a mean metal temperature of 1040C.

If the PPW tube case is now considered with the Larson Miller value of 32.5 corresponding to the 3% carbon content at the PPW base material interface at which carburization of the base tube material begins it can be seen from figure 7 that at a mean metal temperature of 1020C carburization of the base material will commence after 16 years and at 1040C after seven years. Hence, the total tube life is estimated to be the sum of the time required to reach a carbon content of 3% at the PPW/base material interface plus the carburization life of the conventional tube material. Hence, on this basis the predicted tube lives are shown in Table 1.

TABLE 1 Comparison of predicted tube lives of conventional HPmod. Alloy tubes and PPW'ed overlay tubes

Tube type	Temp (C)	
	1020	1040
	Life (yrs)	Life (yrs)
Conventional HPmod.	5.0	2.0
PPW'ed	21.0	9.0

No tubes in service have yet to reach these lives, however, the laboratory work and plant experience (4 years service) to date has provided a provisional validation of the theoretical basis on which these predictions are founded.

In addition to the realization of extended service lives the improvement in carburization resistance can alternatively be used to allow operation at higher tube temperatures whilst maintaining the tube life expectations associated with conventional materials. Hence, PPW technology can be used in a flexible mode to optimize productivity or tube life depending on prevailing economic criteria.

CONCLUSIONS

The laboratory and plant trials have demonstrated the following

- The laboratory and plant trials of PPW'ed overlay ethylene pyrolysis tubes has demonstrated that the layer effectively protects the base material against carburization
- The carbon pick-up rate in the PPW layer decreases and appears to stabilize with increasing service exposure.
- The Larson Miller approach has been used to quantify the carburization behavior of the PPW layer. This approach, validated by experimental data indicates that an increased metal skin temperature for the tubes can be tolerated without sacrificing current tube life expectations, or, alternatively, increased tube life may be realized.
- The anti-carburization properties of the PPW overlay welded ethylene furnace tubes results in more reliable furnace operation due to longer lifetime without unexpected tube repair / replacements and prevent tube damages during unexpected shutdowns due to brittle behavior of the tubes.
- The anti-carburization properties of the PPW'ed overlay prevent shadowing / distortion of the tubes caused by deformation related to non-equal, non-circumferential carburization in conventional tubes, which results in unexpected tube replacements.

Paper 47g

CENTRALLOY® 60 HT

Alloy Development for High Temperature Operation and Reduced Coking

Dr. Dietlinde Jakobi
Schmidt + Clemens Group
Kaiserau, Germany

Dr. Benno Ganser
Schmidt + Clemens Group
Kaiserau, Germany

Key words: heat resistant alloys, ethylene pyrolysis furnace, carburization, oxidation, creep, aluminumoxide scale, centrifugal cast tubes

Prepared for Presentation at the 2005 Spring National Meeting in Atlanta, GA April 10-14
17th Annual Ethylene Producers Conference
Session 47 – Ethylene Plant Fundamentals –
Improvements in Modeling and Testing Reaction Systems and Furnace Performance

CENTRALLOY® 60 HT

Alloy Development for High Temperature Operation and Reduced Coking

ABSTRACT

In 2003 S+C launched a new High Performance Alloy named "Centralloy® 60HT" for commercial application. Compared to our standard Centralloy® products, the resulting properties of Centralloy® 60HT are:

Thermal and mechanical resistance up to 1200°C tube wall temperature, due to an unique balance of strengthening elements which acts as the "backbone" in order to achieve application temperatures beyond 1150°C.

The special metallurgy of Centralloy® 60 HT *reduces the formation of catalytic coke significantly*. The Aluminum content in the alloy forms a tight Aluminum oxide scale and thus provides a barrier between the active Nickel and Iron catalyst of the bare tube and the hydrocarbon. This reduces catalytic coke formation and lengthens the time a furnace can operate between decokes.

Aluminiumoxide scales are known to be thermodynamically, the most stable protective scales of metallic materials. The slow growth rate of the scale ensures the materials survival in long-term service applications up to temperatures well above 1150°C.

The resistance of conventional materials against carburization is provided by the Ni-content in the alloys matrix and the presence of protective oxide layers (Chromium oxide layer and a Silicon oxide subscale). Therefore, modern materials have a high Nickel content (well above 40%), Chromium contents of a minimum of 25-30% and contain 1,5-2,5%Si.

In carburizing atmospheres at temperatures above 1100°C, the protective chromia scale fails by conversion to carbides, allowing ingress of carbon (carburization).

Centralloy® 60 HT *excludes carbon ingress almost completely from the alloy* by forming a stable (up to > 1200°C) Aluminumoxide scale. Additionally, the high Nickel content reduces the diffusion rate of carbon into the alloy matrix.

Despite the highest degree of "metallurgically active" alloying, the material can be machined, bent (by induction bending) and welded like any other Centralloy® product.

A wide range of field tests, started in 2003 in the Petrochemical Industry, show very promising results and the first full furnace applications have been installed in steam crackers during 2004.

1. Introduction

Developments in steam cracking plant process engineering have led to increasingly severe operating conditions. The upper limit of design temperature for the steam cracking process is determined by the metallurgy of the furnace tubes. Therefore, highly-alloyed centrifugal cast alloys are required with adequate creep strength and structural stability up to very high operating temperatures, but must also possess good oxidation, carburization and coking resistance, good machinability and excellent weldability (even after years of service).

2. Development of Heat-Resisting Alloys

Pyrolysis coils in ethylene cracking furnaces are exposed to very severe conditions, e.g. start/stop- and decoke cycles, oxidizing and nitriding flue gases at the outside and carburizing atmospheres at the tube inside surface. The maximum service temperature of alloys presently used in ethylene cracking furnaces is 1150°C and the aim is to target even higher operating temperatures. The principle alloys for ethylene cracking furnaces are shown in ***Table 1***.

Table 1: Typical Alloys for Pyrolysis Furnace Tubes

S+C Grade Centralloy	Common Name	C (%)	Cr (%)	Ni (%)	Nb (%)	W (%)	other
G 4848	HK 40	0.4	25	20	-	-	-
G 4868	-	0.5	30	30	-	-	-
G 4868 Micro	-	0.5	30	30	-	-	microalloy additions
G 4852	HP-Nb	0.4	25	35	1	-	-
G 4852 Micro	HP-Micro	0.4	25	35	1	-	microalloy additions
G 4879	28/48	0,4	28	48	-	5	-
ET 45 Micro	35/45 Micro	0.4	35	45	1	-	microalloy additions

The required ***creep strength*** for these alloys can be obtained by using solid solution formers and carbide formers.

The primary function of solid *solution forming elements* is to impart strength to the matrix by primarily increasing the resistance to the movement of dislocations, this can be achieved by elements with high melting points and high hardening coefficients, e.g. Tungsten, Molybdenum, and Titanium.

The presence of a network of uniformly distributed discrete *carbide particles* along the grain boundaries of an alloy is desirable as it prevents grain boundary sliding and migration. But the optimum effect on properties depends upon the structure and morphology of carbides. Moderate and high Chromium contents (≥ 18 %) lead to the formation of M_7C_3 carbides during solidification. These M_7C_3 carbides transform during even a short time exposure at service temperature to "primary" and also fine "secondary" $M_{23}C_6$ carbides.

The network of primary carbides which is surrounding the austenite grains remains in place throughout service, only changing in form (coarsening) with increased time and temperature. They are, therefore, creep inhibiting barriers. The secondary carbides act to inhibit dislocation movement. With increasing service time and temperature , the secondary $M_{23}C_6$ carbides both re-dissolve and agglomerate thus losing their "barrier" effect.

The cubic MC carbides (stabilized e.g. by niobium and titanium) are the most stable type of carbides. The presence of strong carbide forming elements results in the formation of MC carbides during solidification. In some alloys, a little MC carbide may also precipitate in the solid state where M is Titanium, Niobium, Molybdenum, Tungsten or Chromium.

The good ***oxidation, carburization and coking resistance*** of the "standard" tube materials for cracking furnace tubes (Ni/Cr + Si) relies on the formation of protective scales composed of an outer spinel layer (Mn,Cr-oxide), an inner Chromium oxide layer (Cr_2O_3) and a Silicon oxide subscale.
Up to a certain temperature limit, the Chromium oxides protect the material reliably against high temperature corrosion attack, but if certain limits are exceeded, the protective Chromium oxide layer is transformed to non-protective Chromium carbides (internal surface of tubes), strong nitriding occurs or even Chromium oxide evaporation (from outer tube surface):

Above 1100°C strong *nitriding* respectively internal nitride formation occurs from the outer wall of the radiant tube (flue gas side). Nitrogen is transferred through the Chromium oxide scale and the precipitation of nitrides occurs. The precipitation in most cases starts with the conversion of carbides into carbo-nitrides $M_2(C,N)$, $M(C,N)$ and $M_6(C,N)$ which grow by uptake of Chromium and Nitrogen. These Nitrogen containing very brittle phases can be detected (dependent on the tube material composition) up to a distance of about 2 mm from the outer tube surface.

Also, lack of oxygen in the flue gas plays a role. Under reducing conditions (often caused by flame impingement of badly adjusted burners) this can result in severe loss of wall thickness by alternate oxidation and nitriding. The nitrides then cause spallation of the oxides. As a result a thick layer of oxides (up to 10-20 cm thick!) can be found on the furnace floor ("oxide shedding").

Chromia evaporation may become a problem for Chromium oxide forming alloys at high surface temperatures >1050°C and high oxygen pressures (from the outer tube surface), as in the flue gases of pyrolysis furnaces. Strong Chromium depletion in the alloy subsurface regions can be the consequence and also the formation of a wide carbide free zone which advances into the alloy. In this area there are no carbide precipitates available for strengthening the alloy, and the "sound wall thickness" of the alloy is reduced.
This formation of Chromium (and carbide) depleted zones certainly causes a reduction in creep resistance, tube elongation is the consequence. In addition, the evaporated chromia species cause catalyst poisoning in the SCR (Selective Catalytic Reduction) units which are used recently in the EU and in the US for NO_x removal from the cracker off gas.

Carburization is a carbon ingress and internal carbide formation which occurs at high temperatures (T>800°C) in carbonaceous atmospheres (internal tube/ coil surface). The carburization rate is related to the carbon activity of the gas (occurs at carbon activities $a_C \leq 1$) and progresses exponentially in relation to temperature. Carbon pickup increases the metal volume resulting in internally induced stresses. Carburization progresses from the inner tube surface, causing compressive stresses in the inner wall (carburized tube material) and tensile stresses at the outside (non- carburized tube material). Intergranular cracking results, starting from the middle of the tube wall. Carburization should be negligibly slow at temperatures below 1000°C for the usual materials with 25% Cr, since they form protective oxide scales which are nearly impermeable for carbon diffusion from the atmosphere (carbon solubility in these oxides is extremely low).

However, at temperatures above 1050°C and high carbon activities $a_c \geq 1$ the Chromium oxides are converted to non-protective carbide phases, and these surface carbides do not provide protection against internal carburization.

Brittle fracture is one of the main mechanisms which leads to ethylene cracking tube failures, it occurs most frequently when the tubes are already a few years old (carburized, nitrided) *and* when the furnace is in the second half of its operation run.

During normal operation a coke layer is deposited at the inner surface of the tube. During a temperature drop the ethylene cracking tube shrinks onto this "coke tube". Because the thermal expansion coefficient of the metal is much higher than the coefficient of the coke, the metal-tube shrinks onto the "coke-tube". Due to the high compressive strength of coke, the metal cannot crush the coke. This causes high tensile stresses in the tube metal.
The temperature drop between normal operation and decoking is limited to about 100-200°C, but during a furnace trip such a temperature drop can be 500-1000°C. The strain during such a furnace trip (0.75-1.5%) equals the rupture ductility of aged, carburized and nitrided material between RT and ~600°C.

Since materials tend to crack when their rupture ductility is reached, so of course will ethylene furnace tubes crack. Because aged, nitrided and carburized furnace tube materials are brittle at these temperatures, they crack brittle by splitting of these carbides and subsequently the tubes.

Coking Resistance

Catalytic and amorphous coke formation occurs at the internal surface of the ethylene cracking tubes. The deposition of coke filaments starting from catalytically active Iron and Nickel particles (the catalytic coke formation) can be suppressed and the adherence of the amorphous coke can be reduced by protective and dense, strongly adhering oxide layers. The most modern Chromium oxide forming material is the alloy ET45 Micro (45Ni/35Cr Micro) – in this alloy the oxide scale adherence has been successfully improved by rare earth element additions.

In particular the catalytic coke formation substantially increases with the tube metal temperatures (of the tubes), this being a function of cracking severity and coil outlet temperature (COT). Therefore, gas crackers are typically more affected by catalytic coking – due to the generally higher COT – than Naphtha crackers.

Again, the amount of catalytic coke formation is significantly increased, if the tube metal temperatures exceed 1050°C, as the Chromium oxide is converted under high carbon activity into non-protective scales.

3. Centralloy® 60 HT

The new alloy Centralloy® 60 HT is a cast nickel base alloy containing about 60 weight% Nickel, 25 to 30 weight% Chromium, 2 to 5 weight% Aluminum plus additions of Niobium, Titanium and other minor alloying elements.

Aluminum additions lead to the generation of - compared to Chromium oxide - a thermodynamically more stable Aluminum oxide layer which is formed at relatively low oxygen partial pressures and is stable over a wide range of temperatures.
The dense, strongly adherent Aluminum oxide layer formed at the surface of the alloy Centralloy® 60 HT offers therefore excellent protection against oxidation and carburization at temperatures beyond 1150°C and reduces significantly the coking rates in carbon rich atmospheres.

In addition to this, the alloy has excellent structural stability and very good high temperature stress rupture strength.

High Temperature Corrosion Resistance

Cyclic Oxidation Behavior
The cyclic oxidation behavior of the alloy Centralloy® 60 HT has been investigated for different temperatures in comparison to "standard" alloys for ethylene cracking furnace tubes.

The standard materials suffered rapid weight loss (spallation of the oxide layers) at 1150°C and also at 1000°C and the alloy Centralloy® 60 HT showed superior resistance to cyclic oxidation at both temperatures – due to aluminum oxide formation at or near the alloy surface.

The results of the examinations performed at 1150°C with the alloy Centralloy® 60 HT and the standard alloys G 4852 Micro (HP Micro) and ET 45 Micro (35/45 Micro) are shown in ***Picture 1***.

The standard alloys based on Chromium/ Silicon oxide protection suffered from high weight loss after a few cycles at 1150°C. The weight of the samples from the alloy Centralloy® 60 HT showed a slight weight increase after the first cycles (parabolic growth rate of the oxide layer) and nearly no further change in weight over 250 cycles – this means that no spallation of the protective oxide layer occurs when the alloy 60 HT is exposed to cyclic conditions at 1150°C.

Carburization Resistance
Isothermal carburization tests have been performed in Methane/ Hydrogen atmospheres at different temperatures for different exposure times. ***Picture 2*** shows the weight change for the different "standard" alloys in comparison to the Centralloy® 60 HT.

The reducing test conditions do not allow the formation of protective oxide layers, and due to this reason, the weight change measured in these carburization tests corresponds to the diffusion rate of carbon into the alloys matrix – and this is directly linked to the matrix Nickel content.

Therefore, the highest amount of carbon pick-up occurred in the alloy G 4852 Micro (with 35% Ni), alloy ET 45 Micro showed less carbon uptake (due to 10% higher Ni content) and the alloy Centralloy® 60 HT (20% higher Ni content) showed at all temperatures/ exposure times a very low weight increase and therefore an excellent resistance to carbon diffusion into the alloys matrix.

Coking Resistance
The coking tests performed on the alloy Centralloy® 60 HT showed that the coking phenomena on an Aluminum oxide forming Nickel base alloy is very complex. The coking resistance is strongly dependant on the alloys Aluminum content. By increasing the Aluminum content by around double, the coking rate was reduced by a factor of four. This is, again, attributed to the formation of a dense, defect free aluminum oxide layer.

Creep Rupture Strength
Extensive creep rupture testing performed on the alloy Centralloy® 60 HT demonstrated the significant strengthening which has been achieved by additions of carbide forming elements and solid solution strengtheners. The Larson Miller curves for the standard alloys and the new alloy Centralloy® 60 HT is shown in ***Picture 3***.

Manufacturing Characteristics
Despite the high Aluminum content, the alloy can be machined, bent (by induction bending) and it can be welded like typical standard alloys currently in use for ethylene cracking furnaces. Filler materials are commercially available.

Field Trials
A wide range of field tests with the alloy Centralloy® 60 HT which started during February 2003 show very promising results. The first full furnace applications installed in steam crackers started in 2004 (Linde Pyrocrack 1.1, and Lummus SRT V). As shown in ***Picture 4 and 5***, the effect of Centralloy® 60 HT on the coking rate was significant.

4. Conclusions

Due to the formation of a dense, protective and stable (up to very high temperatures) Aluminum oxide layer combined with the carefully designed alloys matrix, the new High Performance Alloy Centralloy® 60 HT has the following advantages (compared to the S+C standard Centralloy® products):

- The diffusion of Nitrogen from the outer tube surface (trough the oxide layer) into the alloys matrix (the Nitriding) is reduced, therefore also the amount of brittle phases precipitated in the subsurface areas. This means that the aged alloy is less susceptible to brittle failures.

- The good carburization resistance of the alloy Centralloy® 60 HT which is resulting from the barrier effect of the oxide layer (reduced carbon diffusion trough the oxide layer) combined with the low diffusion rate of carbon in the alloys matrix. A lower volume of carbides precipitated in the alloys matrix (less carburization) means that the aged alloy has a ductility comparable to the "as cast" alloy and it is therefore less susceptible to brittle failures during furnace trips.

- Less Chromium oxide evaporation occurs and due to this, less creep problems caused by a Chromium depleted area ("carbide free zone") with low creep strength and less problems due to creep elongation at high operation temperatures (well above 1100°C). In addition to this, poisoning of the SCR units does not occur.

- Significant reduction of catalytic coke formation, as the tight Aluminum oxide scale provides a barrier between active Nickel and Iron catalyst on the tube surface and the hydrocarbon, therefore the run-length of the ethylene cracking furnaces is increased.

In addition to this, the special metallurgy of the alloy Centralloy® 60 HT results in excellent creep and creep rupture strength, the material can be machined, bent (induction bending) and welded like any other Centralloy® product.

The wide range of field tests in the Petrochemical Industry, together with full furnace applications, continues to show very promising results.

References

Jakobi, D.; Gommans, R.; *"Typical pyrolysis coil failures and how to combat them"* AIChE 2002 Spring National Meeting New Orleans, Paper no. **98e**, 2002.

Bagnoli, D. L.; Krupowicz, J.J.; *"Experiences with Ethylene Heater Tube Carburization"*, Corrosion 92, Paper No. 307, (1992).

Grabke, H. J. *"Carburization - A high Temperature Corrosion Phenomenon"*, MTI Publications No. 52, St. Louis, Elsevier Science, Amsterdam, New York 1998, ISBN 1-57698-015-4.

Grabke, H. J.; *"Corrosion by carbonaceous gases, carburization and metal dusting, and methods of prevention"*, Materials at high temperatures **17** (4) 483-487 (2000).

Guttmann, V.; Bürgel, R.; *"The Creep Behavior of HK-40 and Alloy 800H in a Carburizing Environment"*, Corrosion and Mechanical Stress at High Temperatures, Applied Science Publishers, Ltd. London, 1980.

Guttmann, V.; Beck, K. E.; Bürgel, R.; *"A Comparison of the Creep Behavior of Various Austenitic Steels under Carburizing Conditions"*, Mat.-wiss. U. Werkstofftech. **19**, 104.111 (1988)

Hendrix, D. E.; *"Comparative Performance of Six Cast Tube Alloys in an Ethylene Pyrolysis Test"*, Corrosion 98, Paper No. 430, Houston/TX USA, NACE International, 1998.

Kirchheiner, R.; Jimenez, J. L.; *"Correlation of Oxidation, Carburization and Metal Dusting; Controlling Corrosion by Corrosion"*, Corrosion 2001, Paper No. 01374, NACE Conference, Houston/TX, USA, NACE International, 2001.

Petkovic-Luton, R.; Ramanarayanan, T.; *"Mixed oxidant attack of high temperature alloys in carbon- and oxygen containing environments"*, Oxidation of Metals, **34** (1990) p.381.

Ramanarayanan, T. A.; Petrovic, R. A.; J. D. Mumford, A. Ozekcin, "*Carburization of high chromium alloys*" Materials and Corrosion **49** (1998).

Grabke,H.J.; Jakobi, D.; *"High temperature corrosion of cracking tubes"* ", Materials and Corrosion **53**, 494-499 (2002)

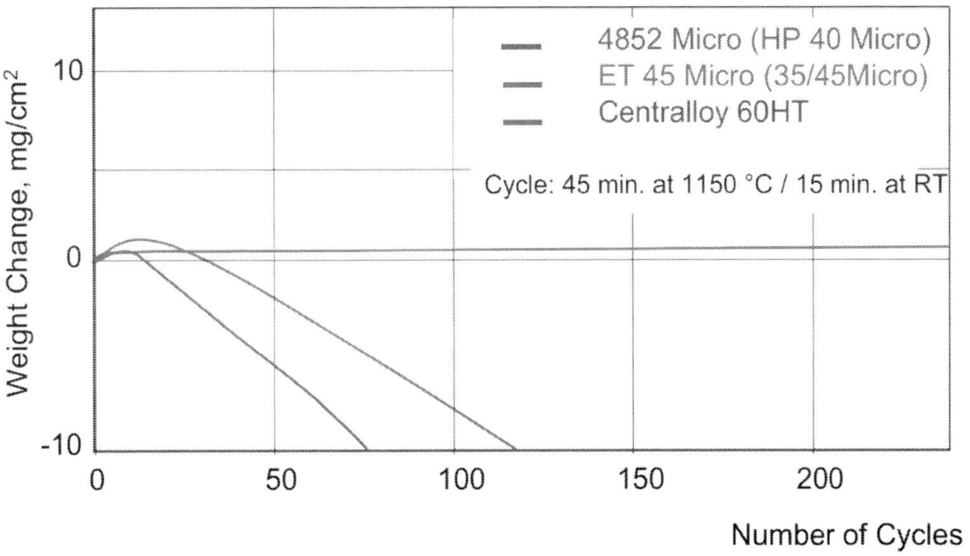

Picture 1: Cyclic Oxidation Resistance at 1150°C in Air.

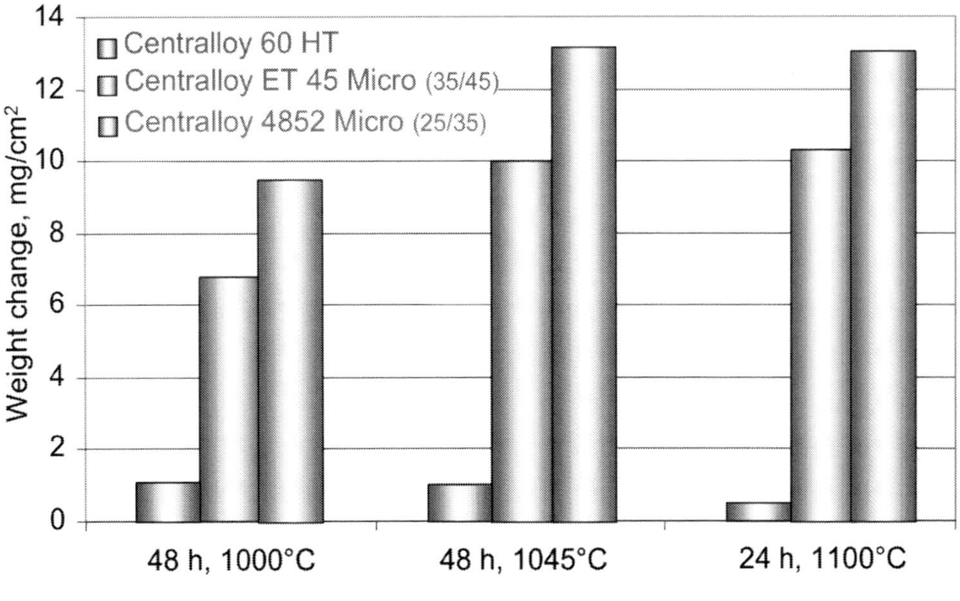

Picture 2: Isothermal Carburization Resistance

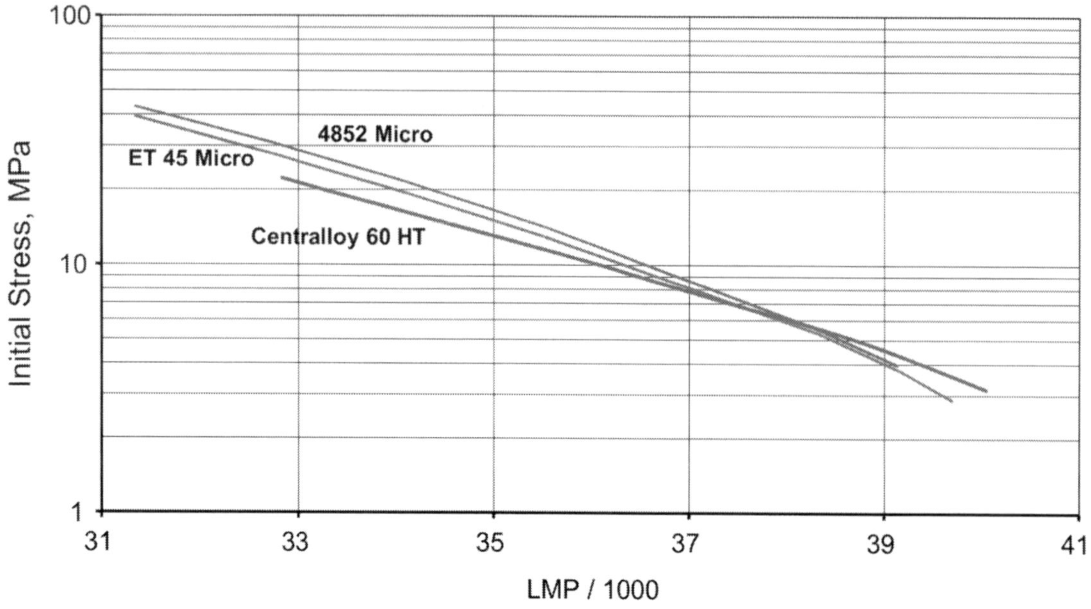

Picture 3: Parametric Stress Rupture Strength

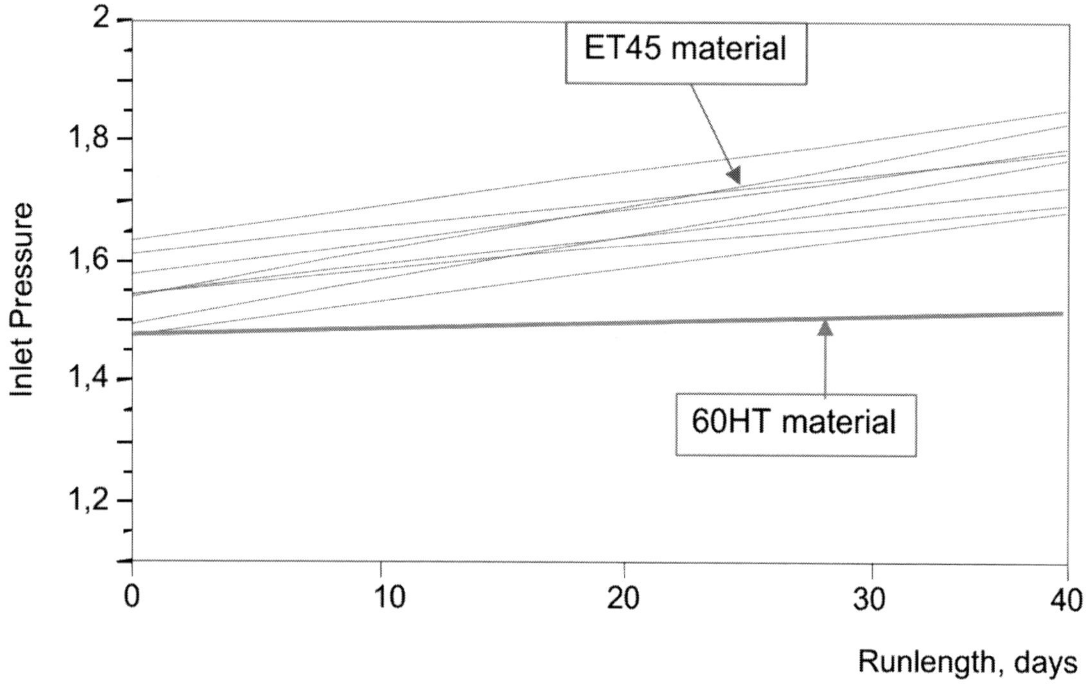

Picture 4: Pressure increase during field trials over 20 run-lengths.

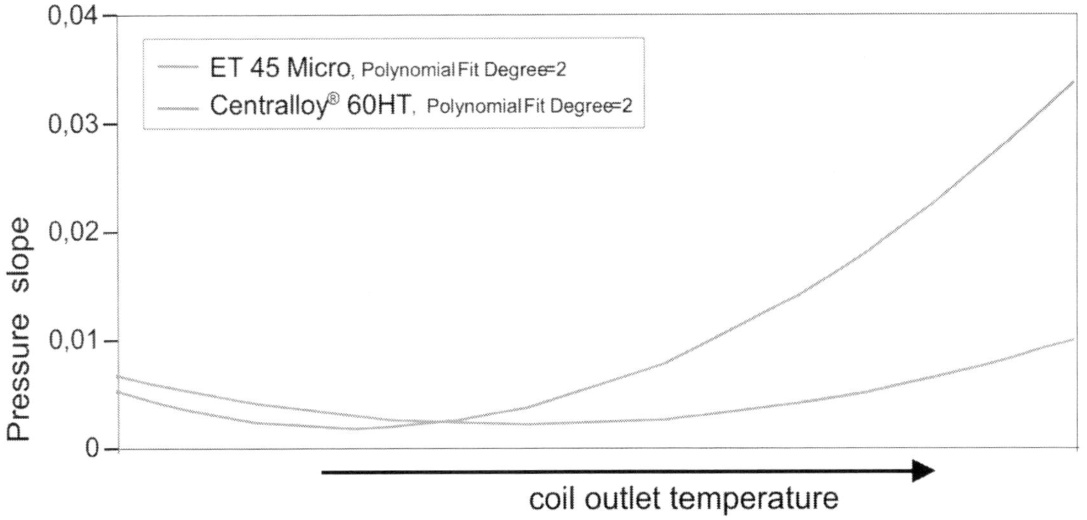

Picture 5: Results of furnace trials: Centralloy® 60 HT coking rate vs. severity

Leak Detection And Repair (LDAR) Maintenance Program

Gerard M. Lassere

Shell Chemical Company
15536 River Road
Norco, La 70079

gerard.lassere@shell.com

Prepared for presentation at AIChE Spring National Meeting/Ethylene Producers Conference
April 10, 2005 – April 14, 2005
Session TA015

Abstract Outline

Leak Detection And Repair (LDAR) Maintenance Program

Leak Detection and Repair (LDAR) is a program implemented to comply with environmental regulations for reducing the fugitive emissions of targeted chemicals into the environment. The purpose of this paper is to describe how an extensive leak detection and repair program is maintained and utilized to prevent fugitive emission and limit releases to the atmosphere.

Process components subject to LDAR are periodically monitored to detect leaks, which are required to be repaired within a predefined time period. Most of the fugitive emissions, volatile organic compounds (VOC), hazardous air pollutants (HAP), are invisible to the human eye.

This paper will outline the steps taken to detect, repair and maintain components that have been identified as fugitive emissions leakers. The following will be discussed:

⇒ **Overview -** a summary of the LDAR program

⇒ **EPA Method 21** – guidelines to monitor and detect leakers, as well as the type of analyzers needed to detect leakers.

⇒ **Type of Components** – equipment that's subject to the LDAR program.

⇒ **Repair Methods** – requirements to repair leakers in a predefined time period.

⇒ **Extraordinary Repairs** – online repair of leakers.

⇒ **Emission Reduction Techniques** – embracing best practices and new technology developments.

Gerard M. Lassere
Shell Chemical Company

Overview

Everyday thousands of substances are released in the air. The results are often pleasant and soothing. Sometimes, they are not so pleasant. And sometimes, they can be quite a nuisance.

The chemical manufacturing processes, just like any manufacturing process, consist of miles of piping and numerous components which convey petrochemical feedstocks, water, steam, and intermediate and final products. There is potential for small, almost invisible leaks from piping connection points and other moving parts and equipment that make up the processes.

As good neighbors and concerned citizens, Norco maintains an extensive leak detection and repair program to minimize and prevent emissions and to limit releases to the atmosphere. As a leading industry in Louisiana, Shell Norco continuously strives to meet and exceed all state and federal requirements while protecting our community. Our environmental performance is critical to sustainable development. Sustainable development is Shell's pledge to use resources in such a way that we do not diminish economic, social, and environmental options for future generations. Leak detection and repair **(LDAR)** is our program for compliance with environmental regulations to reduce emissions of targeted chemicals into the environment.

LDAR programs are generally comprised of four processes :

- Identify components to be included in the program
- Conduct routine monitoring of identified components
- Repair any leaking components
- Report monitoring results

Fugitive volatile organic compounds (VOC) and hazardous air pollutants (HAP) are regulated under (3) different types of standards. They are as follows:

1. Performance standards
2. Equipment standards
3. Work practice standards

The LDAR program falls under the work practice standards. Process components are monitored for leaks at periodic intervals. Component identification is a fundamental element of effective leak detection and repair. If any components are identified as "leaking," then appropriate repair measures are then implemented within the required time interval. The purpose of LDAR is to reduce fugitive VOC's and HAP's emissions in the facility. Fugitive emissions sources of VOC's and HAP's are those sources whose emissions are not confined to a stack or a vent pipe.

Simply defined, our Leak Detection and Repair Program consist of two phases:

1. Monitoring potential fugitive emissions sources within our process units to detect leaks and
2. Repairing or replacing a leaking component within a predefined time period.

Type of Components

At Norco, more than 60,000 monitorable components fall under one of five federal regulations that target the control of fugitive emissions and these 60,000 monitorable components maybe monitored up to 4 times per year with a leak detection instrument. The environmental regulations that define Leak Detection and Repair programs have very specific standards that must be applied to a monitoring and repair program. These regulatory standards define the:

- Chemical streams that are required to be monitored,
- Types of components to be monitored,
- Parts per million concentration indicating a leak,
- Frequency of monitoring,
- Method of monitoring,
- Specific actions if a leak is detected,
- Length of time to repair, and
- Recordkeeping and reporting.

The types of process equipment components within the plant, which are subject to the LDAR program, are as follows:

- Pumps in light liquid service
- Valves in gas and light liquid service
- Pressure relief valves in gas service
- Open-ended lines
- Compressors
- Sampling connections
- Connectors in gas and light liquid service
- Pumps and valves in heavy liquid service, pressure relief devices in light liquid and heavy liquid service, and flanges and other connectors.

EPA Method 21

EPA and some state environmental agencies have established a specific method by which LDAR personnel monitor and detect leaks. These agencies also have established the level of the screening component's VOC concentration which determines a leak. If the measured VOC concentration at the component is above the level defining a leak, the component must be repaired or replaced to reduce the concentration to below the acceptable level.

Equipment is screened with a portable analyzer to determine actual concentrations of VOC fugitive emissions. Screening procedures should involve:

1. A survey of the component at any interface where leakage could occur, and

2. If a meter reading increases, the leak area should slowly be surveyed to determine the specific source of the leak. The analyzer probe should be held right at the interface to the maximum leak area at least 2 times the instrument response time. If the maximum instrument response exceeds the applicable leak definition, then the components are considered to be leaking.

Method 21 describes specific screening procedures for each major type of process equipment.

- Valves – the most common source of valve leakage is at the seal between the valve stem and the housing. Sample at the interface where the stem exits the packing gland

- Flanges and other Connectors – Sample around the circumference of the flange-gasket interface or other threaded connections.

- Pumps and Compressors – Sample around the circumference of the pump or compressors shaft / seal interface. In addition to the shaft / seal interface, sample all other joints on the pump or compressor housing where leakage could occur.

- Pressure Relief Devices – Sample at approximately the center of the area where the pressure relief device exhausts to the atmosphere.

Method 21 does not specify any particular type or make of portable analyzer, but does state performance specifications for any analyzer that is to be used for leak testing. At Norco the current analyzer being used is the Cosmos XP-316. The Cosmos utilizes a catalytic oxidation sensor that has a measurement range of 0 – 10,000 ppm.

Repair Method

If a leak is detected and repairs are necessary, the 1st attempt at repair must be made in 5 days. Final repairs must be completed and documented within 15 days of initial detection. After repair of a component, follow-up monitoring is conducted to verify that the component has been repaired.

If the leaking component cannot be removed from service for immediate repair, by regulation, it is placed on delay of repair status until the component can be safely isolated or the process unit is completely down for unit turnaround.

Repair methods include but are not limited to:

- Tightening of packing bolts
- Tightening of valve bonnets bolts
- Tightening of flange / connector bolts
- Component replacement
- Replace of seals on pumps / compressors
- Injection of lubricants into lubricated packing

Information reporting is a significant element of the LDAR program. All components must be identified and all historical data kept. All leaks and repairs must be recorded and reported to the agencies twice a year.

Extraordinary Repairs

Repairing valves under pressure or online leak repair is simply that. It is the process of sealing pressurized leaks while under continuous pressure. Valves are usually the single largest source of fugitive emissions. Valve packing leaks can be repacked online very quickly and economically without shutdown or having to isolate the valve itself. Emissions from a single piece of equipment are usually small. However, cumulative emissions can be very large. The drill and tap method of repair on valves under operating process can be used as an alternative method of repair in lieu of putting valves on the delay of repair list. This procedure is accomplished by drilling into the valve packing gland and injecting a sealant in the body of the packing gland while the valve is online. The sealant used to seal these leaks is very similar to the original packing material supplied from the valve manufacturer and will never harden or ruin the valve. In most cases, the valve will not have to be replaced during an outage. Proper inspection and safety procedures must be followed before this procedure is allowed.

Emission Reduction Techniques

Leak detection and repair is at the forefront of Norco's environmental concerns. It is based on the premise that there is the potential for leaks, and on our commitment to prevent and minimize those leaks effectively. We are continuously striving to improve our program. There are two ways to approach fugitive emission leak detection and repair programs:

1. Reactive – find it and fix it approach, or
2. Proactive - take an aggressive approach and make things better.

Some improvements include the use of:

- Live Loading of control and block valves using Belleville washers
- Use of low emissions packing and pump seals
- Use of Bellows Seal valves
- Use of a lower leak limit than required

- More frequent monitoring – monthly vs quarterly
- Evaluating pump seals – “Bad Actors” - equipping pumps with dual mechanical seals with barrier fluid and seal pots
- Having project engineers minimize the use of valves and flanges on new projects.

Compressor Dry Gas Seals Reduce Emissions

By Joe Delrahim
John Crane Inc. 6400 Oakton St
Morton Grove, IL. USA, 60053
1-847 967 3735
jdelrahim@johncrane.com

Prepared for Presentation at:
2005 AIChE Spring National Meeting, Atlanta, Georgia,: April 10-14
17th Annual Ethylene Producers Conference
Session TA015–Ethylene Plant Maintenance–Maintenance of Emission Control Devices

Abstract:

As a result of increasing worldwide concerns about the affect of the level of toxic gas emission on everyday life, today's global and local environmental regulations are intended to reduce the level of plants emission. Under these regulations, the plants have responsibility for lowering and following the fugitive emission limit standards within their respective area. One factor in reducing emission to the atmosphere, as related to a variety of components, but particularly compressor applications, is the selection of optimal sealing technology. This paper describes the use of an innovative sealing technology that has already proven to greatly reduce the fugitive emission from compressors as compared to conventional sealing systems.

Compressor Sealing Concepts:

Although the main function of a compressor seal is to confine the process within a compressor casing, the seal can also serve to protect the environment against process emission. Different sealing concepts have been developed to accomplish this task and a brief description of the various devises follows:

Labyrinth Seals:

A labyrinth (fig 1) is the simplest method to seal a rotating shaft. The labyrinth clearance is closely controlled to limit the process leakage or emission, though leakage can be vented if the process is non-hazardous gas. The leakage flow rate across the labyrinth, which may measure as high as 100 scfm, is a function of clearance, size and sealing pressure.

Figure 1. Labyrinth Seal

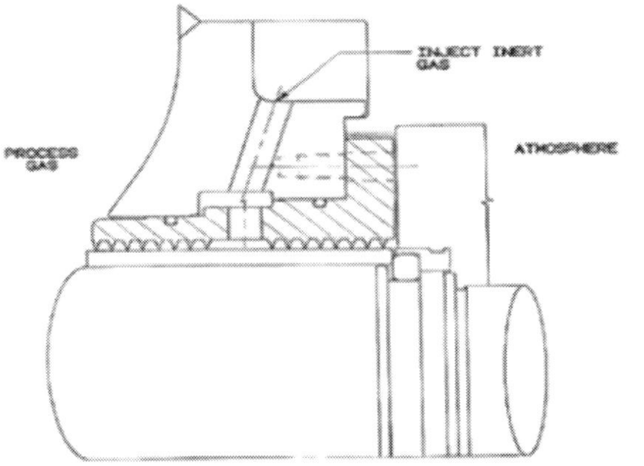

Carbon Ring Seals:

Carbon ring seals leakage is typically 50 percent less that that of a labyrinth seal because of the smaller clearances between the rings and the rotating shaft. These types of seals are primarily used on low temperature/pressure applications.

Figure 2. Carbon Bushing Seal

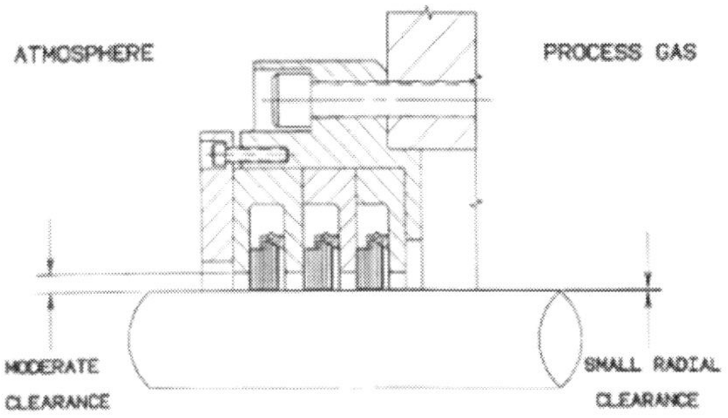

Bushing Seals:

Bushing seals are always used in conjunction with buffered oil and the oil pressure must be maintained above the sealing pressure. The amount of oil leakage depends on shaft size and sealing pressure, though oil leakage may be as high as 10 to 20 gallons per day. The oil leakage will increase as the clearance between the bushing and shaft increases due to the radial contact. Process oil contamination and degassing the contaminated oil is a major concern when this devise is specified.

Figure 3. Bushing Seal

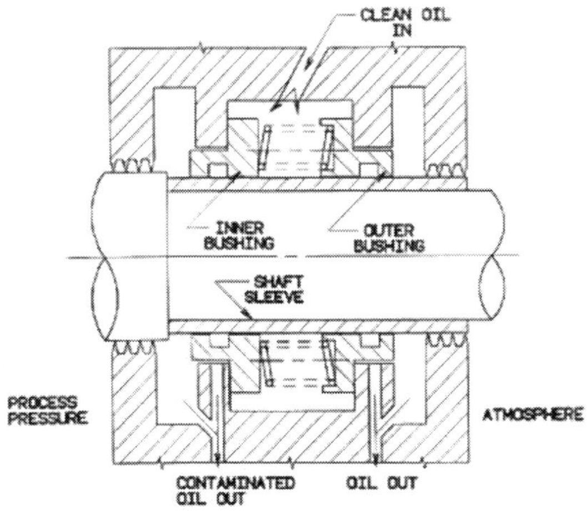

Pump Bushing Seals:

Buffered liquid such as oil or water is always used for a pump bushing seal. The buffer pressure must be higher than the process sealing pressure and the leakage flow rate is a function of the operating conditions and shaft size. Static leakage can be high until the shaft starts turning and the leakage flow rate to the trap may be as high as 1 to 5 gallons per day. Leakage is then degasified and returned to a reservoir.

Figure 4. Pump Bushing Seal

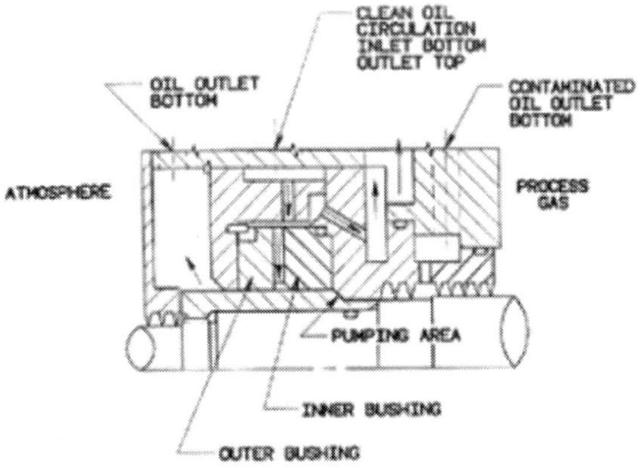

Circumferential Seals:

A circumferential sealing devise may contain a series of one or multiple segmented carbon rings held together with a spring. The seal may operate directly with process fluid and leakage is typically a function of pressure, speed and quantity of carbon segments in a series arrangement. Normal leakage is 1 scfm and each segment is typically capable of sealing 100 psig. This type of seal may operate directly in a sealing process or buffer gas.

Figure 5. Circumferential Seal

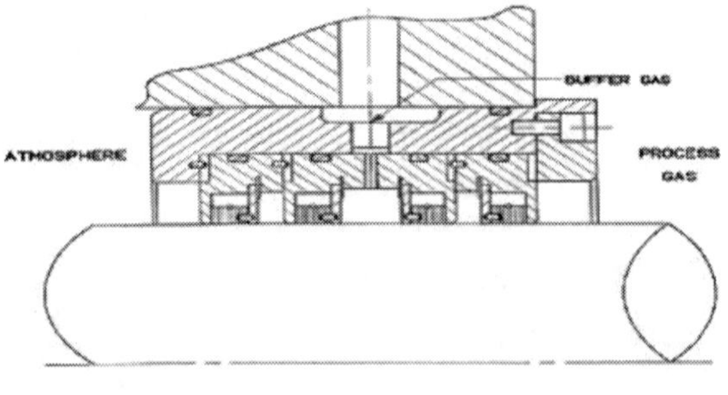

Contacting Face Seals

This devise requires oil buffer at pressure above the sealing process pressure. Oil leakage passes the seal face and is drained to a reservoir and separated from the gas. Speed and pressure are limited to the seal size, though oil leakage may be as high as 8 gallons or more per day. The seal oil support system reliability is vital to the operation of this devise. Process fluid contaminated from oil during the operation or massive seal failure has been a major reliability concern. Degassing or disposing of oil mixed with process gas has forced many plants to seek alternative solutions to this sealing devise.

Figure 6. Contacting Face Seal

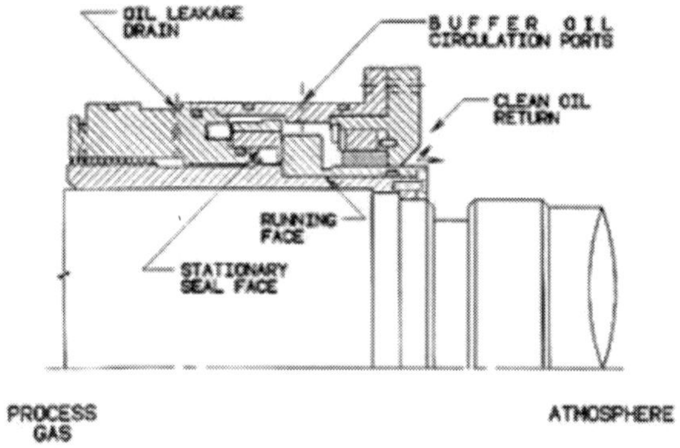

Non-contacting spiral groove gas seal technology:

Sealing technology continues to evolve in response to demands from regulatory agencies to reduce emissions. This global concern about fugitive emissions is forcing virtually all industries to use any and all available seal and system technologies to meet current and expected emission regulations. One such solution is the non-contacting spiral groove gas seal. This sealing concept was introduced to the compressor industry in the early 1980s in response to the growing demand for increased operating design as well as to comply with regulations focused on Volatile Organic Compounds (VOC). As a result, discarding or degassing contaminated oil became expensive and process material, such as ethylene fluid contaminated with buffer oil, was no longer acceptable.

The basic premise for the non-contacting dry running gas seal is based on face separation resulting from hydrodynamic lift. The seal face separation is the only way to eliminate the effects of friction due to contact. And the most efficient method found to separate the seal faces is by spiral grooving in the rotating face against the stationary faces.

Figure 7. Graph

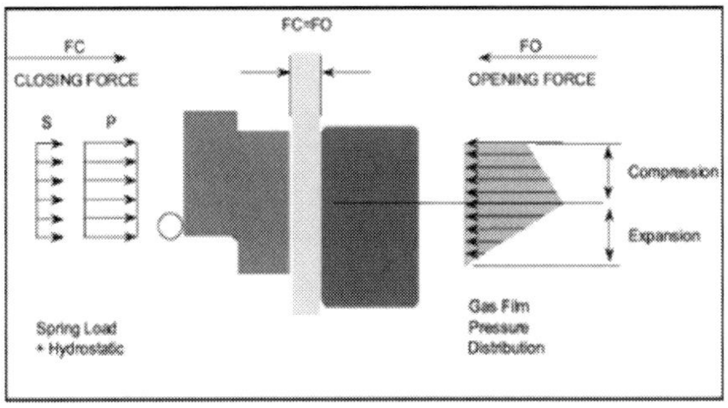

The pressure distribution at the seal face is illustrated in fig 7, and the sealing dam at the inside of the seal face allows the gas to compress and provide a necessary force to separate the seal faces (fig 8). This face separation results in slight leakage across the faces. This leakage flow rate is a function of sealing pressure and speed (fig 8).

Figure 8. Pressure distribution

Tandem Arrangement Gas Seals:

These are commonly used on Ethylene and Propylene applications, as shown in fig 9. The first seal operates as a primary seal while the second seal functions as a back-up devise. Process gas from the discharge line of the unit is piped to the control system supply line and this control system then regulates and filters the buffer flow before it is injected to the primary seal. Leakage across the primary seal is piped to the flare. The pressure and/or leakage flow rate are monitored and recorded to ensure that the seals function properly. To comply with emission requirements, an intermediate labyrinth is installed between the primary and secondary seals. A controlled nitrogen flow at a rate of one acfm and two to three psi above flare pressure is injected between the two seals. As a result, emission across the secondary seals contains only inert nitrogen gas moisture. This sealing arrangement meets and often exceeds the fugitive emission regulation issued by the EPA and the California South Coat Air Quality Regulation Management District directive limiting fugitive VOC emission to 1,000 PPM in 1991

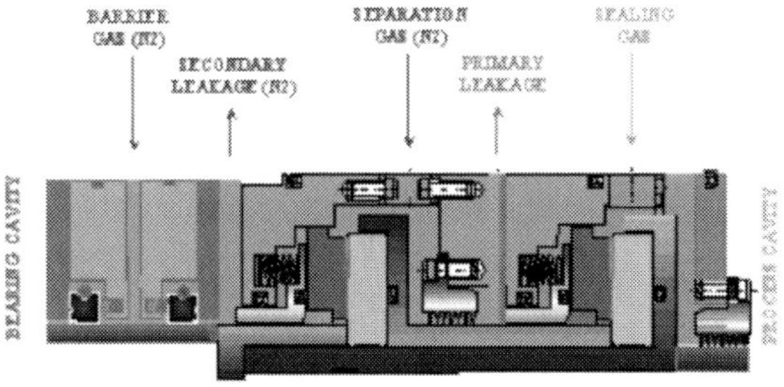

Figure 9. Tandem Seal

Double Seal Arrangements:

This seal consist of two seals facing each other (Fig 10). Nitrogen or inert buffer gas at 20 to 30 psi above process gas pressure is injected between the two seals.

The double seal is normally used to provide zero process emission where small amounts of Nitrogen can be tolerated in the process.

Tandem seal arrangement is normally specified for refrigerant applications such as propylene and ethylene to avoid process being contaminated from buffer nitrogen leakage.

Figure 10. Double Seal

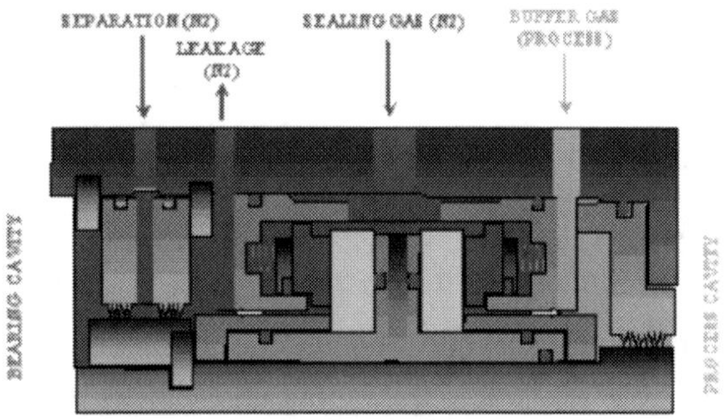

Summary:

Global and local environmental agencies continue to investigate the best available control technology to set a guideline for future fugitive emission standards. The development of non-contacting spiral grove technology has been regarded as the most outstanding achievement in the science of sealing technology. Further, with an understanding of a seal's environmental and triobological issues, users experience a significant increase in rotating equipment Mean Time Between Failures (MTBF) without compromising the concerns for process emission and environmental regulation.

TA015C

EPC 2005
Misapplication of Dry Running Seals in Low Temperature Process Service

Louis J. Trahan

E. I. DuPont
Sabine River Works
FM 1006
Orange, Texas 77631

lou.j.trahan@usa.dupont.com

Prepared for presentation to the AIChE Spring National Meeting- Ethylene Producers Conference April 2005

"Unpublished"

EPC 2005

Misapplication of Dry Running Seals in Low Temperature Process Service

Louis J. Trahan
Sr. Mechanical Engineer
E. I. DuPont

Lee W. Allison
Maintenance Area Superintendent
E. I. Dupont

David Lowe
Senior Sales Engineer IV
Flowserve

EPC 2005
Misapplication of Dry Running Seals in Low Temperature Process Service

Abstract

Repeated seal failures and high maintenance costs led to a Root Cause Failure Analysis which showed that the seal and seal support system design was not applicable for the ethylene product pump service.

While this technology is applicable in some cases, it is not intended for every service. Specifically, pumps that experience "large" axial shaft movements may encounter problems with lightly loaded mechanical seals. This paper will present Dupont's experience in its Sabine River Works ethylene plant.

EPC 2005

DuPont Sabine River Works
System Details

- ***3 Pacific Multistage Pumps***
  - ***Model 3S-A***
    - 9 stage horizontal barrel
    - 500 hp electric drive, 3600 rpm
    - pumps 180,000 pph or 500 gpm of ethylene
    - ***Kingsbury thrust bearings***
    - Two pumps are twins (403/S) and third pump (403B) is similar
    - 2 of 3 required to meet production
    - Use water to melt ice at gland

EPC 2005
DuPont Sabine River Works
GA403B

EPC 2005
DuPont Sabine River Works
GA403/S

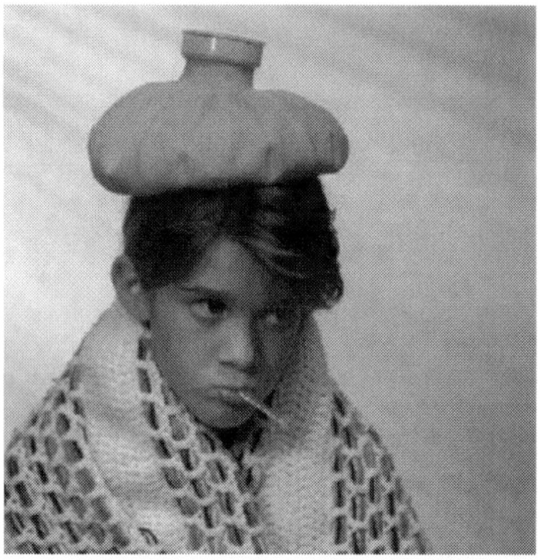

EPC 2005
Statement of Problem

- *Hydrates were problem*
- *Dry running seal design chosen*
- *Installed during major turnaround in '99*
- *Shortly after and until 2003 multiple seal failures*

EPC 2005
403/S/B Product Pumps
Maintenance & Repair Cost

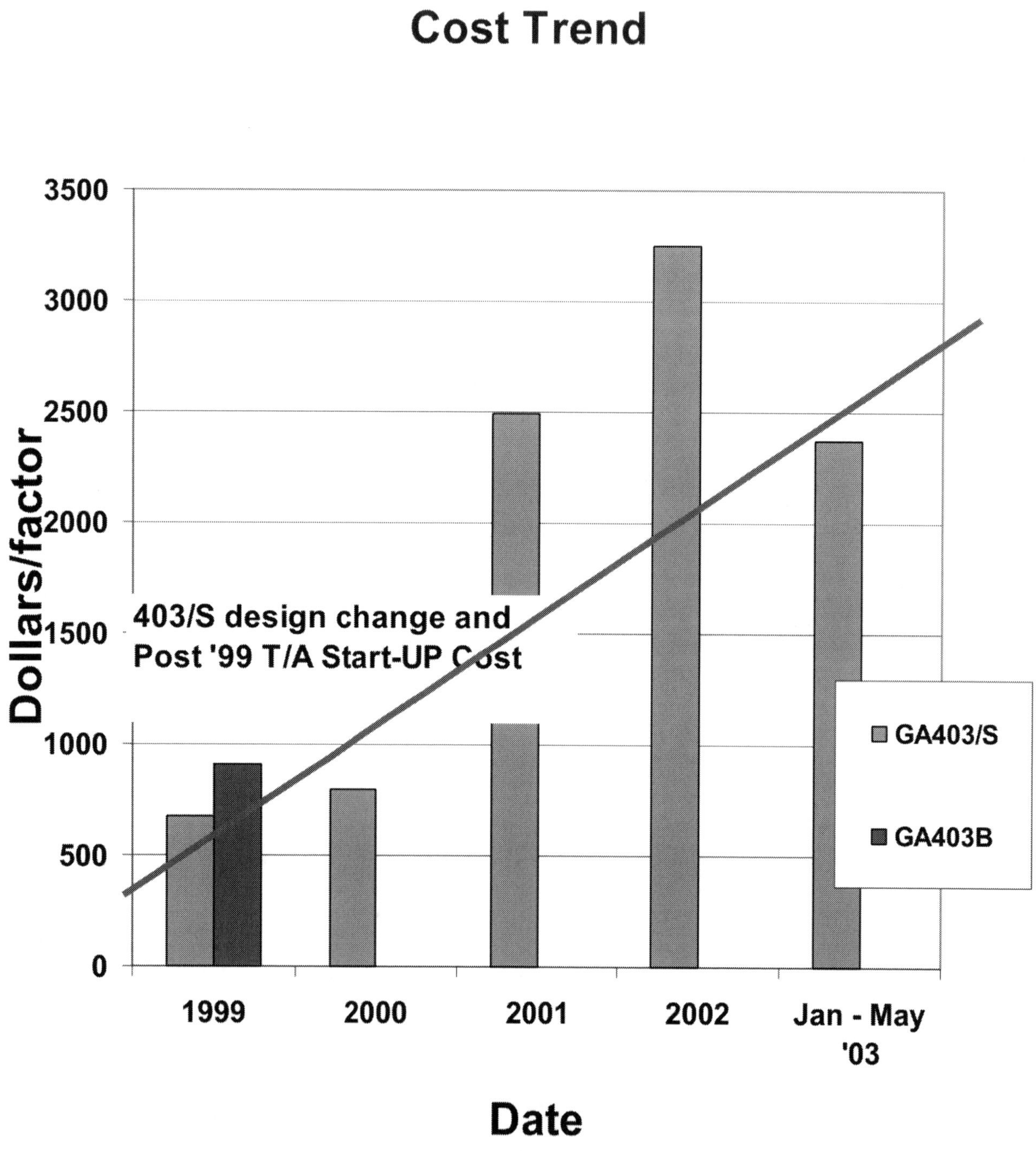

EPC 2005
Failure Analysis

- ***Why Low Reliability?***

  •Choices?

  1 - Simple solution

  •Make 403 & 403S just like 403B

  •Possible unknown causes

  •Miss something

  •Systems are not identical

  2 - Conduct a Root Cause Failure Analysis (RCFA)

  •Chose to do an RCFA

EPC 2005
Failure Analysis

RCFA Results:

- ***Physical Causes***
  - Seal can't stand large axial loads
  - Seal has an atmospheric cavity
  - Water on gland
- ***Human Causes***
  - People like to make adjustments
    - More or less water at gland

EPC 2005
Failure Analysis

RCFA Results: (cont.)

- ***System Causes***

  - Axial movement allows for water intrusion

  - Water used on pump @ below freezing –bad match

  - No other successful users found w/Kingsbury bearings

EPC 2005
Recommendations

RCFA Recommendations:

•Change seals to tandem w/barrier fluid

•Add throttle bushing to seal design

•Design a simple tool to aid seal installation

•Modify Maintenance Procedures to reflect changes

EPC 2005
New Seal Design

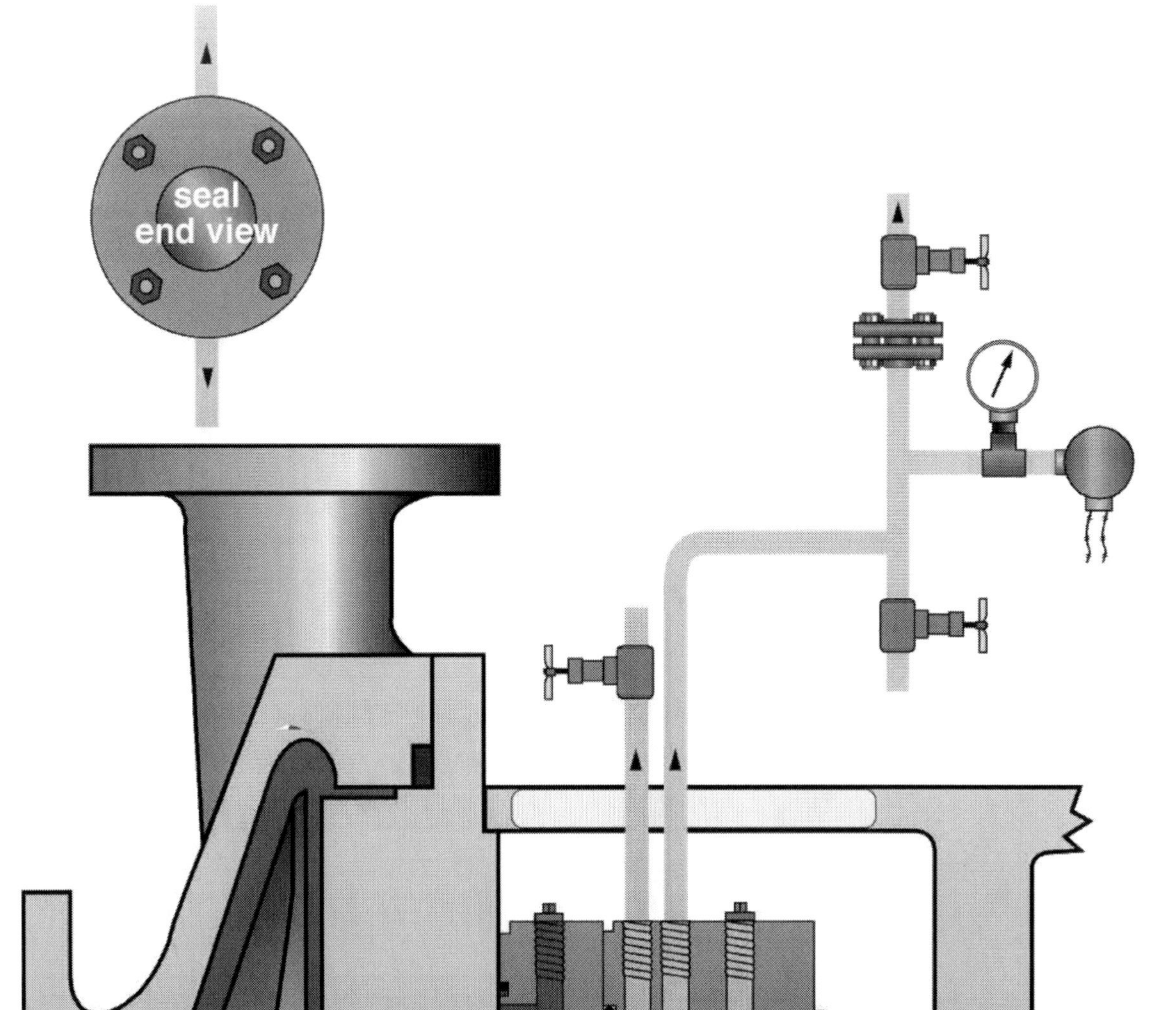

EPC 2005
Product Pumps
Maintenance & Repair Cost
Cost Trend

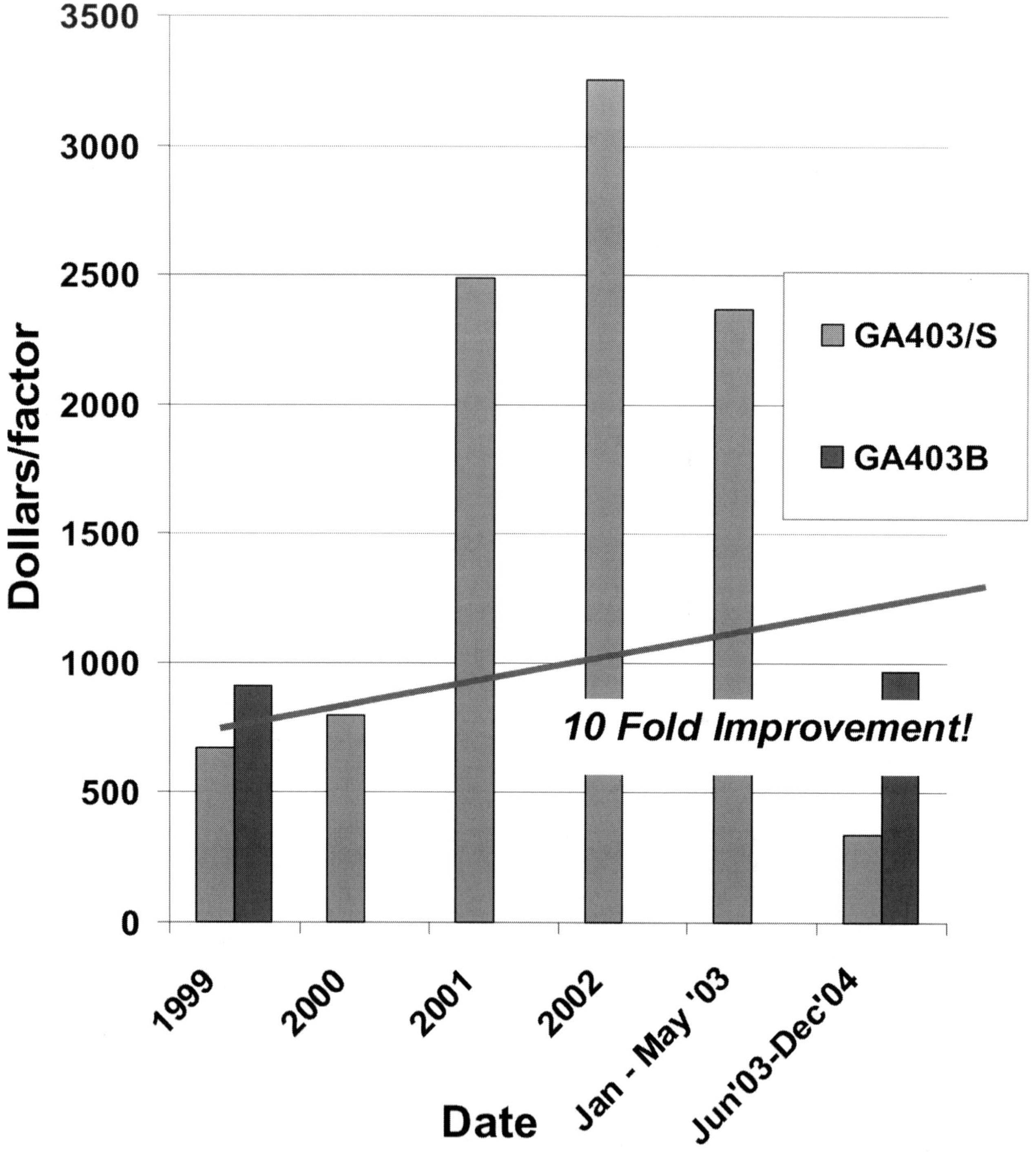

EPC 2005
Misapplication of Dry Running Seals in Low Temperature Process Service
Conclusions

WHAT ARE CEMs AND WHY DO WE NEED THEM

James E. Ferrell
BASF Corporation
Port Arthur, Texas, U.S.A.

Prepared for Presentation at:
2005 AIChE Spring National Meeting, Atlanta, GA,: April 10 -14
17th Annual Ethylene Producers' Conference
Session TA015 - Ethylene Plant Maintenance - Maintenance of Emission Control Devices

UNPUBLISHED

WHAT ARE CEMs AND WHY DO WE NEED THEM

Abstract

Today's business environment has become well rooted in computer technology and enhancements to the point that meetings and training sessions involving persons from all parts of the country are now being conducted via the internet. This same technological trend has made its way to the heart of manufacturing as the specifications for monitoring emissions have tightened. It has become increasingly difficult to detect impurities at concentrations in the part per billion range and to turn-around the results of the assay within 10 to 15 minutes. This type of rigorous analysis has to be completed using continuous emission monitors (CEMs). This paper presents the use of continuous emission monitors in the Ethylene industry, the challenges associated with their use, and the advantages and disadvantages of maintaining these highly sophisticated machines.

WHAT ARE CEMs AND WHY DO WE NEED THEM

Introduction

Looking back through thirty-five years of instrumentation and analytical work in the chemical industry, no one piece of instrumentation has presented more of a challenge in terms of installation and application in the olefins industry than the on-line analyzer.

In retrospect at the one component of hardware that has made analyzer work worthwhile and taken a small amount of the high anxiety from the daily stress level has been the microprocessor. Most people would agree that when the microprocessor was introduced, it made all things instrumentation wise possible and a new world of analytical equipment came on the scene. The analyzer went from mechanical operations to electronic time control functions and all of us breathed a little easier in our maintenance efforts. Components were monitored and separated with greater speed and accuracy. Setup time and maintenance became much less stressful and a new breed of instrument personnel referred to as "analyzer technicians" were pushed to the forefront of the Olefins maintenance crews. That may sound somewhat arrogant, but when compliance is based on concentrations in parts per billion, one usually wants the best technicians available informing them that the analysis is correct. Never before have specifications and emission limits been so tight.

Continuous Emissions Monitors

Continuous emission monitors, or CEMs as they are most commonly referenced, are defined as the total equipment required for the determination of a gas concentration or emission rate. CEMs are used to monitor limits of components emitted from stacks as emissions to the atmosphere. The technology for these systems has become very advanced over the last ten years starting with the sampling systems, detection systems, safety systems, microprocessors and software package that compiles and stores the data.

The use of continuous emissions monitors is critical to today's industry leaders. Monitoring stacks with an analyzer is not new, but monitoring stacks for environmental compliance has become increasingly widespread and is a requirement in most permits and federal and state rules and regulations. The CEM systems have as much of an effect on the health of the plant as does product purity monitoring.

Application of CEMs in Determining Emissions

As the requirements for determining emissions have tightened, those entities supplying analyzers have been pushed to manufacture an analyzer that can monitor components in the very low parts per million and billion range; thereby, making the analyzers much more sensitive and thus more difficult to maintain. Also, the pure gas suppliers (for carrier gasses) and standard bottle manufacturers have been called upon to step up their quality of service and supplies. Carrier gases must be ultra pure and standards precise in their concentration of the components of interest. The days of the grandfathered stacks are quickly coming to an end and monitoring for compounds that are potentially harmful to the environment are and should be at the front of all plant concerns. CEMs have been installed in accordance with Title 40 of the Code of Federal Regulations (CFR) Part 60 Subpart Db (for boilers) and Subpart GG (for Cogeneration facilities). Also, specific permit requirements instruct the Port Arthur site to place monitors on the furnaces, boilers, and the thermal oxidizer. CEMs are used to monitor carbon monoxide (CO), oxides of nitrogen (NO_x), and sulfur dioxide (SO_2) via a total sulfur analyzer. (In order to reduce NO_x emissions from some of the sources at this facility, selective catalytic reductions (SCR) systems were installed.) One NO_x CEM, analyzing two streams, before and after the stack and is used to determine ammonia (NH_3) slip in combination with the amount of ammonia injected into the SCR. The ability to rapidly switch from one sample stream to another while maintaining the integrity of the sample stream in question was a must for several of the applications at the Port Arthur site. Currently, one CEM services two sources and allows one to alternate between streams when analyzing NO_x, CO and oxygen (O_2).

Maintaining CEMs

Every effort to maintain and prevent downtime of the CEM is taken; collection of data and sample integrity are the usual areas of focus. For this reason, a good environmental software package is key to collecting and compiling data for the federal and state governments environmental records. A good software package should be able to collect data directly from the CEM, gather data from external process systems, perform calculations, generate reports and store data with a sufficient amount of memory. It is a good idea to execute the software from a main frame given the amount of data collected and stored.

Most analyzers used to monitor stacks from major suppliers are good. The biggest problems most often observed are with the sample systems. A good sample system will prevent the CEM from experiencing downtime due to "no sample detection" when the sample temperature is not accurate, the sample line is plugged, when moisture interferes with the sample, or if the calibration failed. The sample system used most often involves a heat traced sample line made of Teflon, in a two tube bundle, mounted to the stack and extended into the stack, in-situ. One tube is connected directly to the probe and the sample is drawn to the analyzer via a poppet valve sample pump. The sample is passed through a membrane filter and a sample cooler set at 4°C to remove the moisture before it is analyzed. To assist in verifying the sample system, the entire sample system is flooded with a protocol standard gas, which is greater than the amount normally drawn from the stacks. This process uses a large amount of standard and zero gas and is expensive, but is the best way to prove that the sample system is working properly.

Advantages and Disadvantages of CEMs

A key advantage of using CEMs is that one measures the stack gases directly. The actual concentration of the compounds of interest in the stack at any given moment in time can be determined with an on-line analyzer. This direct evaluation of the gas stream of interest allows for prompt response to events that could lead to non-compliance situations. Secondly, CEMs allow for continuous monitoring of emission limitations versus the one time evaluation obtained with a stack test. And lastly, CEMs have increased the accountability and creditability of the analysis and performance of the equipment/emissions control devices from which emission limitations are applicable. Numbers are crunched and used by the process systems in fuel consumption and overall equipment control.

However as with any system, there are also advantages and disadvantages. One major disadvantage of operating a CEM is making sure that unit downtime is entered into the software. Manpower is also a constraint when maintaining these systems. Many hours are spent trying to recover the number of hours of downtime of an analyzer due to equipment malfunctions, loss of carrier gases and/or process gasses, as well as problems with the sampling system that may have skewed the data. While the uniqueness and complexity of the CEMs are a virtue in an Ethylene Plant, these same attributes make these systems manpower intensive. The cumulative hours employed by environmental and analyzer technicians tending to CEMs can be as much as 8 to 12 hours per day, five to seven days a week. They require extensive maintenance and preventive maintenance programs are instituted as daily activities. The cost of the systems are significant plus the State requirement to conduct cylinder gas audits (CGAs), relative accuracy test audits (RATAs), and calibration drifts add to the maintenance and up-keep of these analyzers. Lastly, a disadvantage observed in the Port Arthur plant is the climate. The area is very warm and humid and these weather conditions affect the condition of the sampling stream, i.e. trying to maintain a sample temperature of 4°C, which add to the mechanical and operational problems encountered with the CEMs. The sample cooler installed on most systems has a refrigeration system that helps maintain the desired temperature of the sample stream given the Port Arthur climate.

While there may appear to be a number of disadvantages, the ability to determine compliance with federal and state rules and regulations based upon real-time data out weigh the disadvantages listed above.

Conclusion

The effort, that has been put forth to design/develop applications for CEMs that would meet the requirements for monitoring emission limitations imposed by federal and state regulations, as well as installation, implementation, and maintaining a reliable CEM program at a modern Ethylene facility has been tremendous. The rewards of this system; however, have significantly improved the response time and action of plant personnel in responding to process conditions that could potentially lead to non-compliance events. For example, responding to NO_x warnings, by adjusting the percent O_2 and/or damper controls on heaters or adjusting NH_3 flow when using a SCR. CEMs used on heaters and furnaces can also assist Operations in determining fouled burner tips. This is seen by a significant increase in CO. The data obtained from the CEMs has also been used to determine emission fees. One could reduce emission fees by 40% using actual data as obtained from the CEMs versus the maximum emission limits obtained with a one-time stack sample or the permitted rates. Instituting a continuous emission monitoring program at the Port Arthur Ethylene Cracker and associated facilities has come with great benefits.

Acknowlegements

I greatly appreciate all the efforts of those I have worked with over the last five years to implement and maintain the analyzer systems here at the Port Arthur site. A variety of technical people have worked countless hours to achieve our present status.

My hat is off to the BASF technicians listed here:

Mike McQuerry, David Forrest, Glen Sterling, Stephen Ballard, Rodney Youngblood, Bert Miller, Gerald Gasson, Debra Bucannan and many other maintenance technicians that have made the systems work.

Special Recognition must be given to the Environmental Department for not only their technical assistance in Environmental matters but also their eager assistance in whatever capacity they could help.

Terrance Young, Aaron Hebert and Frederick Scott with IEA, Inc.; Houston, TX

My special thanks goes to Jannetta Bowden Ned who heads the environmental department for the Port Arthur site, whose expertise on environmental matters and assistance is invaluable in environmental issues faced by the analyzer technicians. I greatly appreciate her insight.

SUCCESSFUL SEALING OF HEAT EXCHANGERS DUE TO THE IMPLEMENTATION OF NEW TECHNOLOGY IN A GASKET SYSTEM

Julie L. Simonton
Applications Engineer
Lamons Gasket Company
7300 Airport Blvd., Houston, TX 77061
Phone: (713) 547-9561
Fax: (713) 374-7311
julies@lamonsgasket.com

David W. Reeves
Reliability Analyst
Chevron El Segundo Refinery
El Segundo, California
Phone: (510) 242-2241
Fax: (510) 242-5353
dree@chevrontexaco.com

Prepared for Presentation at 2005 Ethylene Producers Conference, April 14, 2005, Maintenance Emission Control Devices

UNPUBLISHED

ABSTRACT

Research into the historical use and subsequent failure of double-jacketed type gaskets in heat exchangers has yielded the characteristics necessary for leak elimination and the realization that maintaining a seal cannot be achieved by gasket specification alone. An evaluation of the heat exchanger, stud load, tightening method, gasket specification, proper installation procedures, process conditions, as well as stud selection must each be carefully considered to consistently create an effective seal.

This paper highlights field test data from a refinery on the differential thermal expansion of flanges and how the gasketed connection is affected. Laboratory test data, specifically Ra.S.T. data, which mimics the radial shearing effects on a gasket in a heat exchanger, as well as verifying the affects on the gaskets in the field, is also presented. The details of the new technology in a gasket system for heat exchangers, perfected in conjunction with Chevron and Lamons Gasket Company, will be presented along with its successful implementation at a major petrochemical refinery.

INTRODUCTION

The key factors contributing to leaky heat exchangers are gasket load loss and the inability of the gasket to handle differential expansion, or movement between the flanges. Actual refinery test data and laboratory data have verified the magnitude of the differential expansion present in a typical heat exchanger and compared the performance of several different gasket types when subjected to these typical operating conditions.

The results have identified several factors: gasket selection, stud selection, stud loads, and tightening methods, that must be considered to successfully create and maintain a seal. The findings indicate that sealing is dependent upon more than just gasket selection; the implementation of a gasket system is vital to the success of a connection.

GASKET SELECTION

In addition to meeting the obvious chemical and temperature requirements of the specific application, the gasket chosen requires the ductility necessary to accommodate the differential radial expansion between the flanges of the heat exchanger. Differential expansion occurs as flanges of different mass heat up and cool down at different rates due to temperature variations, causing radial shearing forces across the gasket. Two flanged connections were examined at the Chevron facility in El Segundo, California, to determine the actual level of differential thermal expansion as seen in the field. **Figure 1** shows a connection with minimal differential expansion were both flanges experience movement consistently at approximately the same rate. **Figure 2** shows the movement typical in a heat exchanger tube sheet.

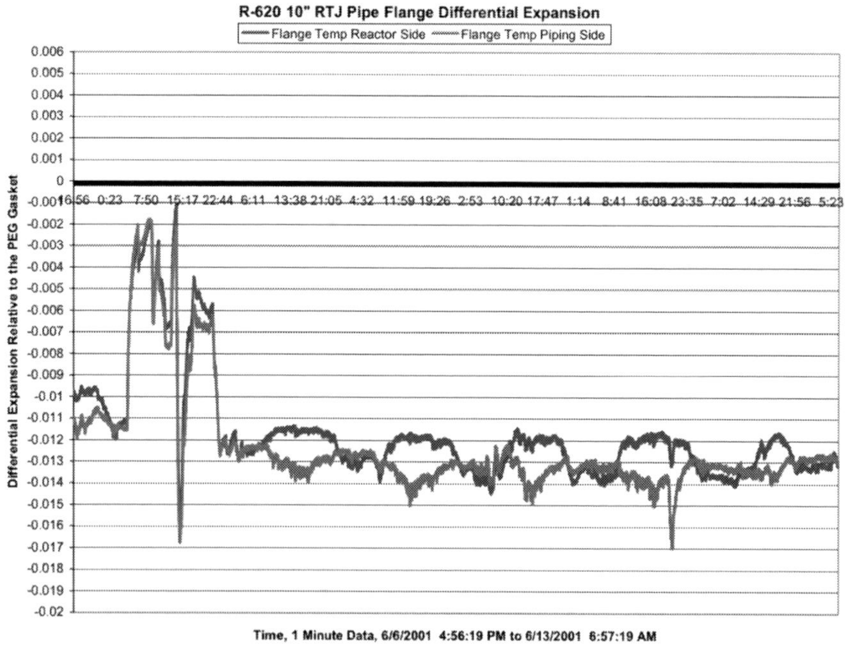

Figure 1. Actual measured movement between mating flanges with respect to the gasket in R-620, a 10" RTJ pipe flange at Chevron El Segundo.

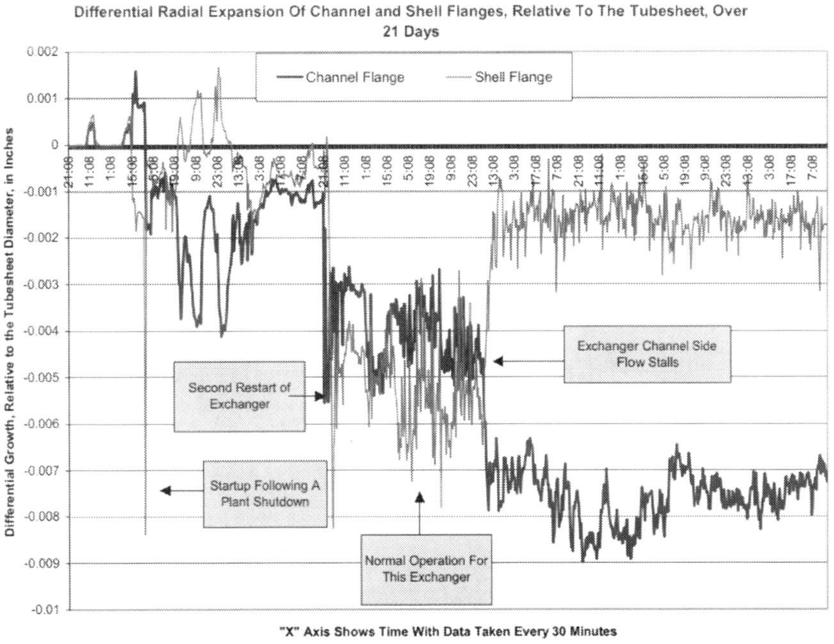

Figure 2. Actual measured movement of both the shell and channel flanges with respect to the tube sheet in the E-510 Re-boiler in Chevron El Segundo, CA.

The differential movement between the shell and channel flanges of the heat exchanger emphasizes the necessity of the specified gasket to accommodate this movement. Consider the following scenario: A heat exchanger generates 14,000 cycles a year of differential movement between flanges of between 0.002 and 0.010 inches, the total distance traveled in one year is 2 feet 4 inches. Most double-jacketed gaskets are manufactured of metal 0.16" thick and lack the flexibility necessary to overcome these radial shearing forces caused by the differential radial expansion typically seen in heat exchanger applications with thermal cycling.

The movement and radial shearing effects contribute to three (3) modes of failure, (1) gasket placed in shear, (2) gasket surface destroyed, and (3) redistribution of filler, as described in Brown and Reeves Failure of Heat Exchanger Gaskets Due to Differential Radial Expansion of the Mating Flanges.

Dr. Warren Brown designed testing equipment to mimic the radial shearing effects exhibited by heat exchangers in the field to be recreated in a controlled laboratory environment. The Radial Shear Tightness Test, or Ra.S.T. Test, uses thermal cycles to experimentally generate the differential radial expansion and shearing effects across the gasket. Several different gasket types were submitted anonymously submitted by gasket manufacturers for test. The gasket types tested included the double-jacketed (DJ) type, corrugated metal with graphite facing (CG), spiral wound type gasket with graphite filler (SW), and grooved metal with graphite facing (PG). Specific test parameters can be found in Brown *The Suitably Of Various Gasket Types For Heat Exchanger Service.*

Gasket relaxation and leakage were measured with the corrugated metal graphite covered gasket, specifically the Lamons style CMG-EX, exhibiting a low level of relaxation and no leakage detected, depicted in **Figure 3** and **Figure 4**.

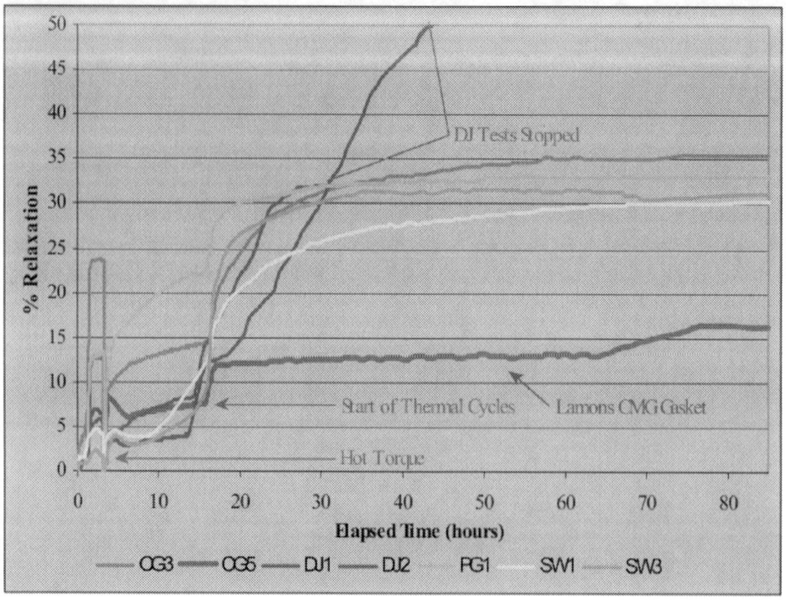

Figure 3. Gasket Relaxation as measured in the Ra.S.T. Test. The Lamons CMG-EX, labeled as CG5, exhibited a minimal amount of leakage.

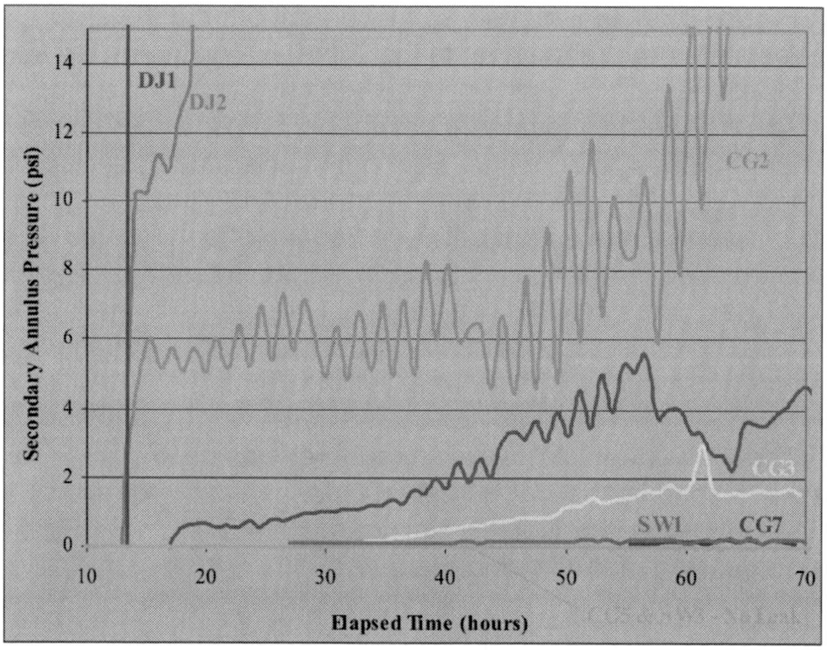

Figure 4. Leakage as measured in the RaST Test. The Lamons CMG-EX, labeled as CG5, exhibited no measured leakage.

With low relaxation levels and no measured leakage in the RAST test, the CMG-EX is typically the best choice in gasket selection for heat exchanger applications.

BOLT LOAD AND GASKET RELAXATION

The amount of bolt load applied to each bolt has a direct effect on the amount of gasket stress achieved and maintained. Without adequate bolt load, the stress on the gasket would not be enough to overcome the operating conditions and the connection would inevitably fail. For this reason, maintaining bolt load is paramount.

It is a fact that all gaskets will relax upon heat up. Gasket relaxation is a key factor in maintaining bolt load. For every 0.001" a gasket crushes or relaxes, 8,000 PSI of bolt stress is lost. Stud loads must be set high enough to compensate this relaxation.

Field test data in **Figure 5** shows the Lamons CMG-EX will relax approximately 30%. By setting higher stud loads, a "reserve stud load" will remain after the initial gasket relaxation, preparing for "long term" gasket relaxation.

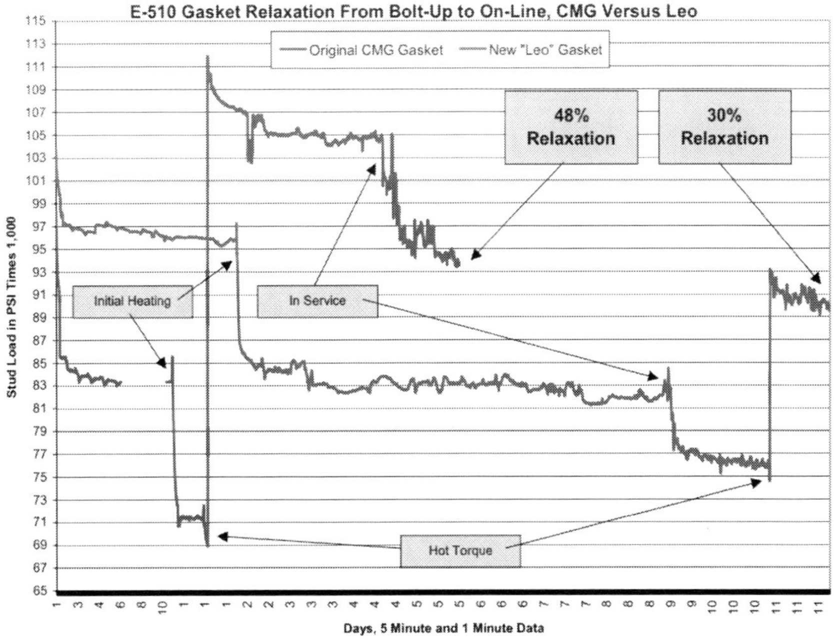

Figure 5. The amount of calculated gasket relaxation of the Lamons original CMG compared with the Lamons CMG-EX in the E-510 Re-boiler at Chevron El Segundo, CA.

Consider the following scenario: A corrugated metal gasket was specified for use in an application where the engineers calculated the minimum bolt load necessary to maintain a seal was 30,000 PSI. Targeting a bolt load of 50,000 PSI and assuming an accuracy of ± 30% applied torque using a torque wrench, the initial bolt stress would range from 35,000 PSI to 65,000 PSI. After 35% gasket relaxation, the stud load decreases to a range of 22,750 PSI to 42,250 PSI. The lower end of the range is less than the minimum bolt load necessary to maintain a seal and has a high potential for leaks. Increasing the targeted bolt load to 78,000 PSI, assuming the same accuracy as before, the initial bolt stress would range from 101,400 PSI to 54,600 PSI. After 35% gasket relaxation, the bolt load decreases to a range of 35,490 PSI to 65,910 PSI, which exceeds the minimum bolt load necessary to maintain a seal.

Bolt loads are not only affected by gasket relaxation short term, but also long term gasket relaxation. Actually, it may take as many as twenty-two (22) months for bolt load values to stabilize as seen in **Figure 6**:

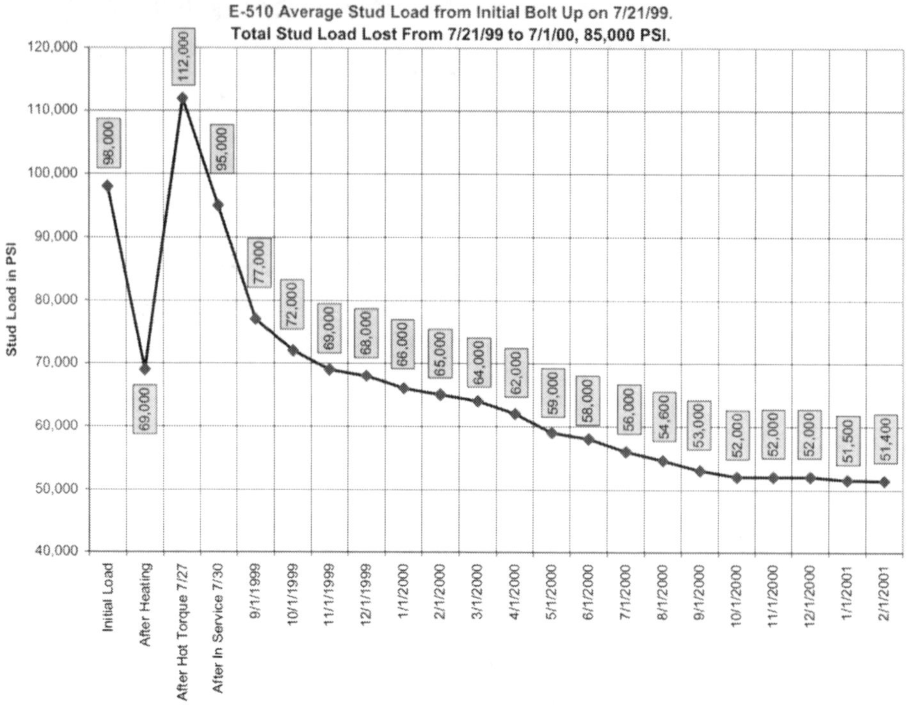

Figure 6. Actual measured average stud load from 7/27/1999 thru 2/1/2001 on the E-510 Re-boiler at Chevron El Segundo, CA.

Anticipating long-term gasket relaxation, setting higher initial stud loads and specifying a "hot" torque, the stud load loss can be compensated for and a seal maintained.

The bolt loads should be calculated to achieve a gasket stress of 13,000 –30,000 PSI over the entire area of the gasket.

THE IMPORTANCE OF AN ANTI-SEIZE LUBRICANT

The use of adequate bolt lubricant cannot be overstated! Lubrication decreases the friction factor associated with the stud thereby increasing the accuracy of initial stud loads. Lubrication also prevents galling and aids in disassembly.

A few studies have focused on the friction factors associated with different types of anti-seize, Warren Brown. Of the four major types of anti-seize, nickel based, copper basked, molybdenum (moly) disulfide based, and ceramic based, the moly-based lubricant seems to work well.

NEW BOLTS vs. USED BOLTS

The use of new bolts will provide a more consistent vessel in which to apply torque. Bolts, especially those exposed to environmental elements do not exhibit uniform corrosion, yielding, or contaminant build-up. For this reason, the reliability of used bolts with an applied torque is extremely low.

Consider the following sixty-four (64) bolt floating head connections. **Figure 7** shows the use of Used Fine Thread Studs, B7 grade, targeting a bolt stress of approximately 32,000 PSI bolt stress. The amount of stress measure in each bolt after the final torque is sporadic and inconsistent. **Figure 8** shows the use of NEW Coarse Thread Bolts targeting the same 32,000 PSI bolt stress. In this case, the stress measured in each bolt after the final torque is more consistent.

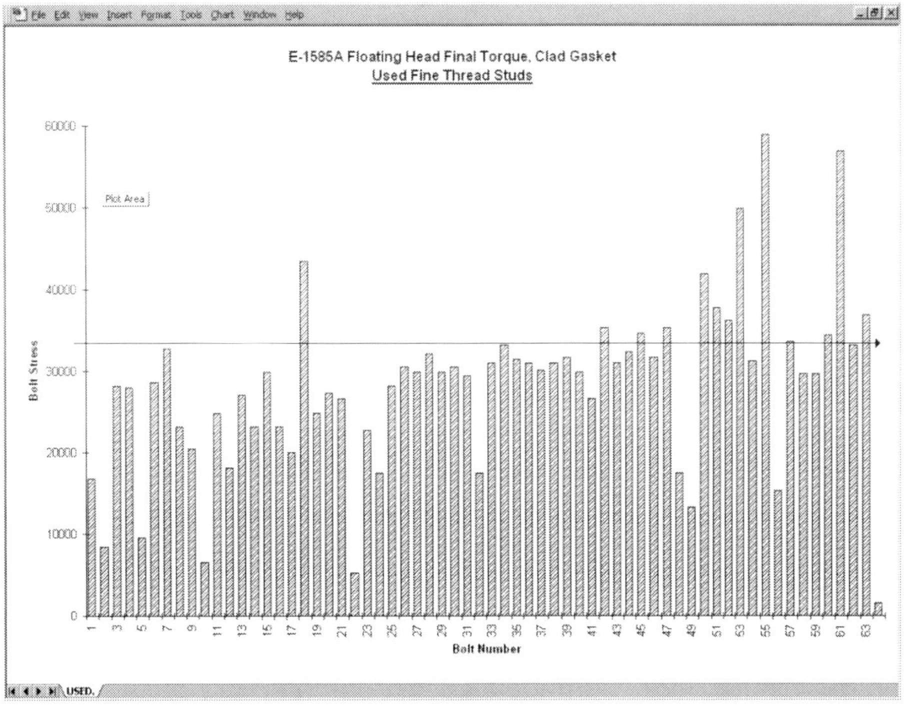

Figure 7. Actual measured bolt stress in each of the 64 used fine thread bolts in E-1585A floating head connection in Chevron El Segundo, CA.

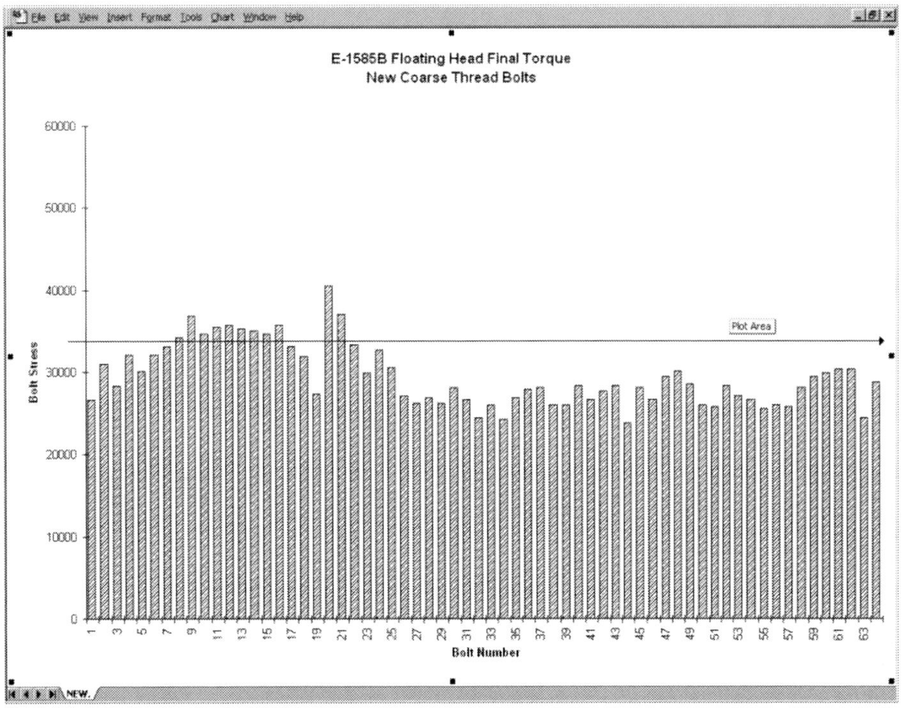

Figure 8. Actual measured bolt stress in each of the 64 new coarse thread bolts in E-1565B floating head connection in Chevron El Segundo, CA.

BOLTING COMBINATION

The bolting combination of NEW studs, new nuts, moly based anti-seize, and new ASTM F 436 hardened washers, creates the ideal situation to accurately achieve the targeted bolt torque value. The use of hardened washers creates a flat surface for the nut to rotate which decreases the friction between the nut and the flange.

HOT TORQUE

During initial heat up, a corrugated metal gasket will exhibit a significant amount of relaxation, causing a decrease in bolt load. To recover the load lost during this initial stage, it is recommended that after the final torque has been applied to the bolts, a "hot torque" be performed between 250°F and 400 °F. The success of the corrugated metal gasket in heat exchanger applications has been when a hot torque has been performed.

IMPLEMENTATION AT A MAJOR PETROCHEMICAL REFINERY: CHEVRON EL SEGUNDO

In 1999, armed with the knowledge of the effects of radial expansion and the failure mechanisms of double-jacketed type gaskets in heat exchangers, along with the gasket selection criteria as explored earlier, the corrugated metal type gasket was specified for the majority of the 1,200+ heat exchangers located on site of a major petrochemical refinery. Bolt load values were calculated to compensate for the expected gasket relaxation and to maintain a gasket stress of 13,000 PSI to 30,000 PSI. To ensure an accurate load application, new studs, nuts, hardened washers, and a Moly based lubricant were specified for each bolted connection as well as a mandate for torque wrench usage and a hot torque.

With the implementation of this new gasket system, this major petro-chemical refinery is running virtually leak free, meeting not only the Federal emission regulations, but also those mandated the state of California.

CONCLUSIONS

The failure of double-jacketed type gaskets in heat exchanger applications has led to evaluation of the heat exchanger both in the field and in the laboratory. The findings indicate the success of a heat exchanger is contingent upon not only the gasket, but the bolts, bolt loads, and torquing procedures associated with installation procedures. Collectively, consideration of each factor has yielded a gasket system that has been implemented at a major petrochemical refinery with leak free results.

REFERENCES

Brown, W., Reeves, D., 2001, "Failure of Heat Exchanger Gaskets due to Differential Radial Expansion of the Mating Flanges", Proceedings of the ASME PVP 2001, ASME, Atlanta, USA, 416, pp. 119-122

Brown, W. 2002, "The Suitability of Various Gasket Types for Heat Exchanger Service", Proceedings of the ASME PVP 2002, ASME, Vancouver, Canada, 433, pp. 45-51

Brown, W. 2004, "Efficient Assembly Of Pressure Vessel Bolted Joints"", Proceedings of the ASME PVP 2004, ASME, San Diego, USA

Paper TA015F

ADVANCES IN LOW FUGITIVE EMISSIONS FOR VALVES

SUBMITTED BY:

CHARLES POGUE, VICE PRESIDENT US SALES
VELAN VALVE CORPORATION, HOUSTON, TEXAS

MIREK HUBACEK, SENIOR MANAGER QT VALVES
VELAN, INC., MONTREAL, QC

PRESENTER: CHARLES POGUE

Prepared for Presentation at:
2005 AIChE Spring National Meeting, Atlanta, Georgia,: April 10-14
17th Annual Ethylene Producers Conference
Session TA015 – Ethylene Plant Maintenance
Maintenance of Emission Control Devices

UNPUBLISHED

ABSTRACT

- Low Fugitive Emission requirements for valves are driven by national standards or regulations limiting leakage of harmful substances outside the pressure retaining envelope of the valve.

- Various end users of valves have varying ideas of how to ensure a valve meet these standards. This results in a fragmentation of packing design requirements resulting in unnecessary higher cost and inconsistencies in the industry.

- Velan Valve has developed a way to consistently manufacture standard valves with the confidence to guaranty a valve off the distributor shelf for less than 50 ppm.

- Complications could arise in the future due to new "standards" that are in the developing stages by organizations like ISO and API. These new standards coupled with standards imposed by the likes of VDI (the old TA Luft) in Germany and others may result in not only different FE limits but also different testing methods. This will negatively affect a valve manufacturers' lead time and overall cost that ultimately will affect end users.

INTRODUCTION

- Today there are in existence numerous valve manufacturers that install different packing manufacturers, have varying stem and stuffing box finishes, and varying degrees of manufacturing capabilities. These variations are customer driven, industry driven and/or performance driven. Are these variations necessary?

- What drives the performance of a valve in regards to low fugitive emission and helping the end user install valves that enable them to meet FE limits? How can a valve manufacturer consistently produce valves with low FE limits?
- For valve manufacturers, the result of the fragmentation of conflicting standards regarding FE limits or testing methods is time consuming and costly. This isn't necessary but how can we address the situation?

In the USA, the Clean Air Act was the driving force for the limit setting standards. The Clean Air Act is a federal law with the aim to improve the air quality. The CAA requires the U.S. Environmental Protection Agency (EPA) to set standards and regulate emissions of toxic air pollutants.

Currently the limit for those pollutants is set at 500 ppm though some states, like California, require 100 ppm. The EPA developed leak detection approach is "Method 21". This is a local leak detection method called "sniffing". There is a picture on the following page.

This method uses a leak detector with a probe and a pump taking air samples adjacent to the expected leakage area (packing or gasket in valves) including the eventual component leakage. The test media is usually He or CH_4. The method only locates and classifies leaks but it is not to be used as a direct measure of mass emission.

VALVES

The best available technology for low fugitive emissions in valves is in bellows seal valves. You will get zero ppm as this technology prevents media from even reaching the packing. Packing is still installed because in the unlikely event of a bellows failure, there is a back-up seal. If the valve has a welded bonnet then there is no leak path for concern.

In the state of California for new valve installations it is required to install bellows seal valves or the "best available technology". This is however very expensive to do on an everyday basis and not economically feasible for all applications. Plants need to be able to use "standard" valves and still meet federal FE limits.

The possible leak paths in a valve are at the packing area or the bonnet gasket area. In order to achieve the toughest FE goals (100 ppm) valve manufacturers must concentrate on these two areas of the valve. The packing area is the most inconsistent by valve manufacturer and therefore the most controversial.

STUFFING BOX

There are many packing manufacturers available and many of them perform very well. Like anything else there are different levels of quality in each that carry corresponding prices. Valve manufacturers must find the optimum quality for a price that allows them to remain competitive. Packing is only one piece of the puzzle in obtaining low fugitive emissions, contrary to some misconceptions in the industry.

It is equally important that a valve manufacture maintains strict tolerances in machining the i.d. of the stuffing box and the stem. The stem moves within the packing so there are critical details to maintain consistency.

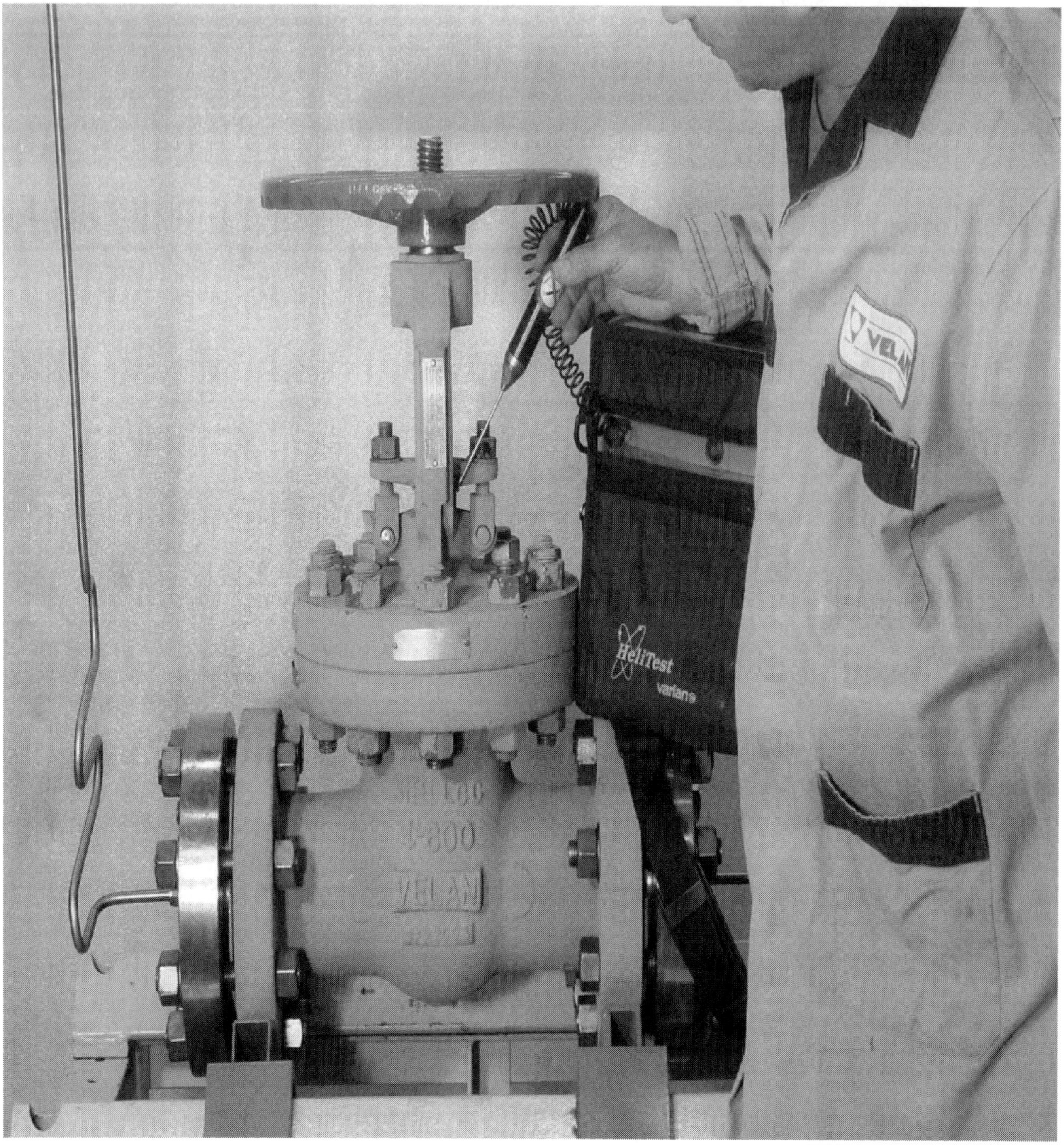

Local leak measurement – Sniffing method

Through years of extensive cycle testing, Velan Valve has found the following information vital to exceed our own manufacturing criteria of less than 50 ppm off-the-shelf:

- The optimum stem finish for our chosen packing arrangement is 8 rms consistently.
- It is important to closely monitor the straightness of the stem. The straighter the stem the less wear on the packing.
- It is equally important to closely monitor the cylindricity of the stem. The round stem must be perfectly round to reduce premature packing wear.
- Velan has found that the stem threads should be burnished and not machine cut. This eliminates excess metal shavings and rough spots that could damage the packing even at the installation stage.
- The stuffing box itself needs to be slightly rougher that the stem finish to help to keep the packing in place but not so rough as to damage the packing itself. The optimum finish for the stuffing box we have found to be 64 rms.

VALVE PACKING

As previously mentioned, there are several excellent packing manufacturers and packing types. There is no one, great miracle packing that will cover the sins of inconsistent valve stem detail. It is not just the packing that will get you where you need to be for FE limits. You need the whole package. When end users try to dictate a packing manufacturer and type it creates industry problems.

Valve manufacturers simply cannot afford to test every packing brand and packing type available. It is not economically feasible, for one thing, and the optimum stem finish for one packing type may not be the optimum for another. Also, as we at Velan individually pre-compress each packing ring prior to installation, not all packing types can handle the pre-compression. I will get into packing detail later.

VELAN LOW FE HISTORY

A.K. Velan's original goal in 1993 was to offer our valves with a guaranty of less than 100ppm off the distributor shelf. Our current guaranty is for less than 50 ppm off the distributor shelf.

We found early on through stringent cycle testing (at least 100 cycles per size and in some cases over 5000 cycles) and using the same testing devices used by the EPA, that our valves could reach our less than 100 ppm goal consistently. **It is important to note** that during extensive cycling tests it was found that after leak paths developed, leakage could be reduced or eliminated by retightening the gland bolts. Just like your car, if you want it to perform as it did new you must perform the proper maintenance. The more a valve is cycled, the greater the chance of creating a leak path.

The only thing we found that needed changing at that time was the bonnet gasket on the oval bonnet design on 150# class gate valves (all other pressure classes have round bonnets and a spiral wound stainless steel/graphite gasket did and still performs well as it is totally encapsulated between the valve body and bonnet).

The oval bonnet design in 150# class gate valves made it difficult to consistently keep the FE limits low. Through a series of more cycle testing valves with other gaskets we found that by using a corrugated steel gasket with graphite filler channels, we could achieve zero ppm.

As we strived to improve the performance and reduce further the FE guaranty we discovered a way to reduce the emissions and prolong the life of the packing. By compressing packing after installation, there is not an even compression (either the top or the bottom rings are more compressed) and the life of the packing is cut short. We found by individually pre-compressing each packing ring prior to installation the compression was more uniform throughout the packing chamber and did not put more stress on any one packing ring.

This revelation allowed us to offer a guaranty of less than 50 ppm for valves with carbon graphite packing after going through more extensive cycle testing. Our optimum packing arrangement is as follows:

- Five packing ring design with graphite inner rings and carbon graphite outer rings
- Inner rings (pre-compressed individually) are John Crane 237-I
- Outer rings (pre-compressed individually) are Garlock 1303
  - **NOTE:** If it is determined that temperatures will reach 650 degrees F or higher then we use Garlock 1303-dry.

We have successfully used other brands of packing but as a standard this has performed extremely well. It is interesting to note that we install Garlock 9000 EVSP packing in our valves as standard in the U.K. This is a good packing that has had good test results (as positive as our standard packing). As our distributor(s) in the U.K. stock for the refineries that all want Garlock 9000 EVSP, we can standardized in our Portugese manufacturing plant with this packing. As we don't have to make design modifications (i.e. stem finish) and after extensive testing, it makes sense in this case to manufacture these valves with the packing requested. If each refinery had their own "brand" preference, you could see the complications and cost increase possibilities that could occur (not to mention the effect on availability).

FE STANDARDS

While there are limits established in the U.S. and regulated by the EPA, there are no regulatory requirements on how we as valve manufacturers obtain positive results. There is a method for testing (Method 21).

There are standards organizations that are developing standards for fugitive emission limits that will not only have various limits but will have various testing methods. The fragmentation of these limits and testing methods will not only negatively increase cost for valve manufacturers (and ultimately the end user) but will negatively affect valve availability to the industry.

For example, VDI in Germany (the old TA Luft) have very strict emissions standards (it correlates to approximately less than 1 ppm). Special valve design is required (added cost). But more concerning are standards that could affect several countries.

ISO is developing a standard for fugitive emissions and now API is developing a standard. The limits may not be the constant and there are two methods of testing out there.

LOCAL TEST METHOD
This method locates and classifies the leak and not a direct measure of mass emission. Method "21" is an example.

TOTAL LEAK RATE MEASUREMENT "GLOBAL METHOD"
This is a measurement of the entire valve packing leakage.

- The valve is pressurized with the test gas, typically He, and the packing chamber is encapsulated with an enclosure and evacuated.
- Any leak through the packing is detected by the leak detector.
- This technique provides the highest sensitivity to a lead especially in case the leak has similar value as the background.
- The downsides are that it will not locate the leak site (stem or packing chamber) and it is expensive.

There is a picture of what this test would look like with an encapsulation around the packing chamber at the end. As you can imagine, the cost to custom build these encapsulations for various sizes of valves alone is substantial.

ISO is looking at the Global method of testing. API is looking at the Local method. However, API is also going to introduce one fundamental difference from all previous standards, namely it will introduce Corrosion and Load Transfer Tests.

There is a real need for simple and "Type" and "Production" FE test standards. Based on Velan's experience of over 12 years in FE testing there should be:

1. One FE standard for On-off valves and one for control valves.
2. For On-off valves 100 cycles is sufficient for "Type" tests.
3. For Production testing (when deemed necessary) the ISO/D 15848-2 is the best available plan that we have seen. Initial cost is still high.

Total leak measurement – Global method

CONCLUSION

For low FE in valves, manufacturers must have precise and consistent tolerances.

- The stem
  - Finish (a smooth 8 rms for Velan and our packing choice)
  - Perfectly round stem
  - Extremely straight stem
  - Threads are less likely to damage packing in installation if they are burnished instead of "machine cut"
- The packing chamber (64 rms)

The packing in the valve is NOT the only driving force for successful low emission results. The best results we have found are as follows:

- Flexible braided graphite jacketed with Inconel filament reinforced outer rings (Velan uses Garlock 1303).
- Die-formed flexible graphite inner rings (Velan uses John Crane 237-I)
- Precise pre-compression of the packing rings individually prior to installation enabled Velan to reduce our FE guaranty from less than 100 ppm to less than 50 ppm.

There are several FE standards in existence in the World and all vary to such a degree that there is no way to test to one protocol and qualify to another. The financial impact on valve manufacturers is enormous. The money and time spent on this impossible endeavor would be better spent in R&D, product improvements and keeping up to the highest quality standards that our customers require today.

48g Dual Seal Systems for Controlling Emissions from Machinery

Frank Huntington, Larry Bayer

The majority of fluid handling machines used in the chemcial processing industry have some form of mechanical sealing system to control leakage where a shaft penetrates the pressure containment. Thses sealing systems range from fairly simple single seals with simple support systems to very complex multiple seals that often have complex support systems. The complexity of the sealing system depends on several factors, including allowable leakage of process fluid to the environment, costs of ownership of sealing system and maintainability.

This paper focuses on sealing systems that have two mechanical seals in a seal assembly, often called "Double" or "Tandem" seals. A brief history will be provided and a review of current commercial technology will include the strengths and weaknesses of seal system options. This review will emphasize the emission control capabilities of the options and the maintainability issues that affect the emissions and cost of ownership of these systems.

AUTHOR INDEX

AUTHOR INDEX

AUTHOR INDEX

American Institute of Chemical Engineers
3 Park Avenue
New York, New York 10016-5991

ISBN: 978-1-60423-460-2